Research Methodology and Scientific Writing

C. George Thomas

Research Methodology and Scientific Writing

Second Edition

Ane Books Pvt. Ltd.

Springer

C. George Thomas
Kerala Agricultural University
Thrissur, Kerala, India

ISBN 978-3-030-64864-0 ISBN 978-3-030-64865-7 (eBook)
https://doi.org/10.1007/978-3-030-64865-7

Jointly published with ANE Books India
The print edition is not for sale in South Asia (India, Pakistan, Sri Lanka, Bangladesh, Nepal and Bhutan) and Africa. Customers from South Asia and Africa can please order the print book from: ANE Books India Pvt. Ltd.
ISBN of the Co-Publisher's edition: 978-9-388-26448-8

This Springer imprint is published by the registered company Springer Nature Switzerland AG
The registered company address is: Gewerbestrasse 11, 6330 Cham, Switzerland

Preface to the Second Edition

I am very happy that the first edition of the book *Research Methodology and Scientific Writing* was a hit among both researchers and research students. The first edition was published in 2015, by ANE Books, and a revised edition became a necessity within this short period because of the accumulation of huge quantum of new information in the subjects covered since 2015. Before attempting the revision, I have taken into consideration responses from the publishers, colleagues, and students.

This thoroughly revised and expanded edition is with several new features. The book comprises of 24 chapters, and each chapter is further divided into sections of specific units. Considering the demand of the readers, four new chapters have been added. A chapter exclusively on 'units and numbers' has been added considering the necessity of adopting only SI units in scientific measurements. I expect that a good number of students who are currently pursuing higher studies may join as lecturers or assistant professors. A chapter on 'Guidelines for Successful Lecturing' has been added for those who would choose a teaching career. A chapter has been set apart for 'Note making, Note-taking, and Assignments', which can boost the learning outcomes of students. The chapter 'Plagiarism: Prevention and Cure' has been added to warn researchers the perils of plagiarism and to take corrective steps.

Other chapters have been restructured with addition of new materials. By splitting the chapter on 'Approaches to Research', an additional chapter 'Major Research Methods' has been added with more inputs on the section, participatory learning and action. The chapter 'Information sources' has been split into two, 'Publications and the Library' and 'Academic Databases'. The chapter on 'Academic Databases; has been rewritten by deleting some non-functional databases and adding several useful ones with the latest trends and weblinks. Similarly, the chapter on 'Scientific Writing: Improve Your Writing Skills' has been split into three with additional information. The chapter on 'Preparation of Thesis and Research Papers' has been split into two, devoting a chapter entirely for thesis writing. The chapter on 'References: How to Cite and List Correctly' has been restructured by adding new trends in referencing with several new examples.

The revised edition has the same goals as the previous edition. With the restructured and additional contents, I am sure you will find the book more informative and useful.

Thrissur, India C. George Thomas

Preface to the First Edition

Thesis or dissertation is a mandatory requirement for the award of postgraduate degrees in most disciplines of science and technology. Because successful completion of research work within the stipulated time is necessary for the award of degrees, pressure on students as well as research guides would be very high. However, majority of students enrolling for a research degree have little knowledge or experience of research and may not be clear about the exact topic they wish to investigate. As research guides are often occupied with other work, they may not have enough time to instruct their wards on various aspects of practical research. Often, the students have to learn every step by themselves. Having recognized this issue, many universities in India have introduced compulsory coursework on 'research methodology' for their postgraduate students.

Recently, the University Grants Commission (UGC) has recommended compulsory course works of one semester for Ph.D. and specified that one course must be on 'research methodology'. Similarly, the Indian Council of Agricultural Research (ICAR) has suggested certain common compulsory courses for postgraduate education in agriculture and related disciplines in the country, which include 'Technical Writing and Communication Skills' and 'Library and Information Services'

In Kerala Agricultural University, a compulsory course on research methodology with a common syllabus has been in existence for a long time, and I have been associating with it for the last 16 years. My search for reference books to teach this course revealed certain interesting facts. Most of the books on 'research methodology' available in the library are written for social sciences! Probably, the major reason is the peculiar research methods, especially of the qualitative type, social scientists follow. Although there is no dearth of books on experimental design and statistics, books on how to plan, write, and speak on research are miserably low. This compelling situation prompted me to write a book covering all the practical aspects of research.

The book, Research Methodology and Scientific Writing, is primarily meant as a handbook for beginners in research—research students as well as young scientists. It can also be used as a textbook wherever a course on 'research methodology' or 'scientific writing and presentation' (or on similar subjects) is offered. The book deals mostly with interdisciplinary fields such as finding research problems, writing research proposals, obtaining funds for research, selecting research designs, searching literature and review, collection of data and analysis, preparation of thesis, writing research papers for journals, citation and listing of references, preparation of visual material, oral and poster presentation in conferences, and ethical issues in research. Besides introducing library and its various features in a lucid style, the latest on the use of information technology in retrieving and managing information through various means have also been provided. Another major highlight is the inclusion of tips to improve writing skills of research students. A brief introduction to historical and philosophical aspects of research including methods and tools of science has also been given.

The driving force behind this book is the numerous postgraduate students I have taught during the last several years. I acknowledge the help and cooperation from several of my colleagues and friends. I remember with gratitude Dr. E. Ahmed, Professor of Agricultural Extension (presently, Director, Centre for E-Learning), who asked me to join the research methodology team at the College of Horticulture, Vellanikkara. I am thankful to Dr. Meera V. Menon, Associate Professor, and my colleague in the Department of Agronomy for going through the manuscript and offering several useful suggestions. My thanks are also due to Dr. C. T. Abraham, Professor and Head, Department of Agronomy, for his encouragement. I also acknowledge with thanks the permission given by the Kerala Agricultural University to print and publish the book.

I appreciate the cooperation and sincere efforts of Mr. Sunil Saxena, Mr. Jai Raj Kapoor, and their team at Ane Books, New Delhi, for bringing out the book in time.

I earnestly request the readers to feel free to bring to my attention discrepancies and omissions, if any, and suggestions for improvement.

Thrissur, India C. George Thomas

Contents

About the Author

Dr. C. George Thomas is Retired Professor and Head in the Department of Agronomy, Kerala Agricultural University, Thrissur. He held the post of Associate Dean, College of Horticulture as well. He has more than 36 years of experience in teaching, research, and guiding students. Dr. Thomas obtained M.Sc. (Ag.) from Kerala Agricultural University and Ph.D. from Indian Agricultural Research Institute, New Delhi. A renowned science writer, Dr. Thomas has authored 16 books, 90 research papers, and over 250 other articles. He is a recipient of the Karshaka Bharathi award of Govt. of Kerala for his contributions towards farm journalism.

Chapter 1
Research: The Search for Knowledge

We know very little, and yet it is astonishing that we know so much, and still more astonishing that so little knowledge can give us so much power.
Bertrand Russell (1872–1970)

Human beings have been curious from the very beginning. Primitive humans observed matters concerning the universe, and changed their way of living, settlement, food habits, social institutions, and many others over time. In fact, these were the results of 'research' done by our ancestors on the processes happening in nature, which helped them to learn many lessons and draw several conclusions. Our ancestors used observation as the primary means to understand various phenomena. They invented many devices and tools by trial and error. Discoveries or inventions by accident were galore. Nevertheless, the pursuit of humans to unravel the mysteries of the universe had to face many challenges.

Earlier efforts of our ancestors to explain the operation of the universe paved way for primitive religious concepts; and for many phenomena, they attributed the cause to supernatural powers. This gave birth to religions and priests. At least for some part of history, religions became the authority on everything, even on science! Yet, there were many 'researchers' among them who could observe cause–effect relationships for various phenomena and processes. They also discovered that under certain conditions, events could be predicted with reasonable accuracy. The saddest part of the story is that often these findings and explanations were simply rejected, if they seemed to conflict with the prevailing religious dogmas. Nobody was allowed to question religious authority, and those who dared faced the consequences—sometimes even death! In course of time, however, humans could break the religious hold on matters concerning science. This emboldened them to have a quantum leap in science and technology, and they succeeded in offering accurate explanations for innumerable phenomena. In fact, the accumulated knowledge over the centuries was the result of 'research' done by our ancestors.

This process of research in search of more and more scientific knowledge is still going on in all the scientific disciplines.

C. G. Thomas, *Research Methodology and Scientific Writing*,
https://doi.org/10.1007/978-3-030-64865-7_1

1.1 Acquiring Knowledge

What is known is knowledge. From the day we are born, we begin to acquire and refine knowledge in many ways. Acquiring knowledge and sharing it with others are widely recognized as the basis for improving one's power, especially reputation and influence in the society. Every addition to scientific knowledge is an accession to human powers. Francis Bacon (1561–1626), considered as the 'father of scientific method', recognized this power of knowledge and said, 'Knowledge is power', implying that with knowledge one's capability to succeed in life would surely increase. Bertrand Russell (1872–1970), a renowned philosopher of the twentieth century, recognized the power of knowledge; he said: 'We know very little, and yet it is astonishing that we know so much, and still more astonishing that so little knowledge can give us so much power'.

Knowledge can refer to both theoretical and practical understanding of a subject, which includes facts, descriptions, information, and skills acquired through experience or learnt through books or other means. In other words, knowledge can be *implicit knowledge* as with practical skills or *explicit knowledge* as with the theoretical understanding of a subject. *Information* is knowledge communicated through any media such as sensible statements, opinions, facts, concepts, or ideas. Information becomes knowledge only when it is conceived and understood.

The knowledge we acquire may be a priori (a Latin term, meaning 'prior to') or a posteriori (posterior to). Note that the knowledge known independent of experience is a priori. It is non-empirical or arrived at beforehand without experience. In contrast, a posteriori knowledge is knowledge known only by experience. It is empirical or gained only after we have firm experiences.

We consider a priori knowledge as true, because *deductive reasoning* is used to arrive at that conclusion using valid arguments. For example, the knowledge that 7 plus 5 is equal to 12 is known without direct experience. Anybody with some arithmetical understanding will be able to tell the answer. In other words, we acquire a priori knowledge not through experience but by reason alone. Most of the equations in mathematics are examples of a priori knowledge, as they are self-revealing. Similarly, the statement, 'all dogs are mammals' is an a priori truth. By knowing the definitions of the word 'dogs' and 'mammals', one can use reason to establish that the statement, 'all dogs are mammals' is true without the need to examine all dogs for mammalian characters.

From *inductive reasoning* based on empirical evidences, we gather a posteriori knowledge. For example, a statement such as: 'this rose flower is fragrant' cannot be considered true through reason alone. You cannot be sure whether the rose flower in question actually possesses fragrance through reason; for that, you must have direct experience—you have to smell the flowers. Note that unlike a priori knowledge, this type of a posteriori statements can be faulty; the rose in the statement may not have fragrance! Although certain scientific disciplines such as physics treat all knowledge as empirical or a posteriori, some disciplines such as mathematics use logic and reasoning. Rationalists suppose that knowledge is primarily attained by a priori processes or is inherent, but for empiricists knowledge is a posteriori. Most scientific disciplines now utilize both a priori and a posteriori knowledge.

The discussion on knowledge leads us to recognize that acquiring knowledge is through acquaintance and through the description of characteristics of materials or phenomena. We learn many things through perception and sensation. However, most of our knowledge is by description. Knowledge may also take the form of beliefs and judgements. Some beliefs may be supported by evidences, but many are simply beliefs! The beliefs that are supported by evidences are 'justified beliefs'. According to the great philosopher Plato, knowledge is 'justified true belief'.

Kerlinger (1986) quoting Charles Sanders Peirce (1839–1914), a renowned American philosopher, described four methods of acquiring knowledge or fixing beliefs.

Method of tenacity: Probably because of our upbringing and socialization pattern, we believe and accept certain things to be true. These are being taught or thrust upon us from early childhood. Among the public, the more frequent the repetition of the belief, the more the enhancement of its validity. It is difficult to change such beliefs even in the face of conflicting evidences, and sometimes, fresh ideas may evolve from false beliefs or superstitions! For example, most religious beliefs, spirituality, omens, superstitions, and astrology existing in the society are through the method of tenacity or beliefs. People adhere to such beliefs, and it is very difficult to change them. The root cause of many superstitions in the society is because of this human habit of clinging to such dogma.

Method of authority: In olden days, nobody was allowed to question the authority. The authority might be kings, priests, or other leaders. People often take for granted the information passed on to them based on authority. Even now, the authority has some sway over the people. For common people, a statement must be true, if it is in the religious books. Most people also accept as true without questioning what the kings (now, political heads!), judges, priests, oracles, celebrities, leaders, or teachers say. The authority is considered infallible.

Although the 'method of authority' has several problems, it is considered to be superior to the 'method of tenacity'. Acquiring knowledge through authority is common among people because of some reasons. Most people are conditioned from birth by their parents with suggestions to listen, trust, and obey authorities. In most cases, authoritarian method is the fastest and most efficient method of transmitting knowledge. However, authoritarian knowledge should be corroborated by evidences and reasoning to consider it reliable. As far as science is considered, if we accept the authority of politicians or public figures like poets, artists, or other celebrities, that would be the end of science! The norm must be to accept only 'experts' in their own fields as authorities in science.

Method of intuition: The method of intuition or a priori method of acquiring knowledge is considered to be superior to the two already mentioned. As already discussed, it is called a priori method as reasoning is done from what is 'prior' or 'before'. The propositions accepted by the method of intuition are self-evident. However, in intuitive propositions, reason is considered as the criterion of truth rather than experience. It attempts to reason from cause to effect or from observed fact to another fact or principle not observed. However, if two reputed individuals reach different conclusions based on intuition, it would be difficult to decide for sure whose judgement is correct.

Method of science: The method of science or scientific method is a practical methodology of acquiring knowledge by framing specific questions and systematically finding answers. Among the four methods of acquiring knowledge, scientific method is the most reliable. In this method, the questions formulated and answers predicted are scrutinised based on observation, measurement, verification, and evaluation. The method of science is based on empirical and measurable evidences rather than beliefs or arguments. This unique characteristic distinguishes science from other methods of faith, authority, or intuition. Scientific method has also the rare characteristic of self-correction, which no other method has; in science, theories and laws are revised based on new evidences (see Sect. 1.8).

1.2 Science and Technology

Can you define science exactly? In fact, the word ‘science’ is from *Scientia,* a Latin term, meaning knowledge; and it was originally used to mean simply knowledge. In Sanskrit, the word for science is *Vijnana*, which is associated with the processes of discernment and understanding. However, science as we understand today is much more than simply knowledge.

A major distinguishing feature of scientific knowledge in comparison to other kinds of knowledge is that it is not static. Scientific knowledge is constantly revised or expanded through the processes of observation and experimentation. Therefore, one can say that science includes a ‘content’ part and a ‘process’ part of acquiring or improving this content. Scientific knowledge hitherto known to us developed as the content of science. This ‘content’ is what we usually read and learn from science textbooks comprising of descriptions, facts, principles, theories, laws, and their relationships (Sect. 1.7). The ‘content’ of science develops through the ‘process’ of science involving repeated observation and experimentation done in a methodical manner. This ‘process’ is nothing but *research* using scientific method (Sect. 1.8). The stipulation of methodical approach to acquire knowledge in science is important because this distinguishes science from other knowledge systems prevalent in the society.

The above discussion leads us to define science as follows: Science is a body of knowledge attained through repeated observation and experimentation done methodically, which is always amenable to correction, modification, or improvement upon getting better evidences.

Science is closely related to *philosophy*. For a considerable period in the history of science, both ‘science’ and ‘philosophy’ were considered the same. Some distinctions between philosophy and science, however, began to surface during the late modern period. By eighteenth century, the use of the term ‘science’ in the sense of any systematic field of study began to get acceptance, distancing itself from 'philosophy' (For more historical details, see Sect. 1.5; and for a discussion on science and philosophy, see Sect. 2.1).

Do not confuse 'science' with 'common sense' although both have some similarities. Thomas Henry Huxley (1825–95) gave a definition for science based on its similarity with common science. According to him, 'Science is nothing but trained and organized common sense differing from the latter only as a veteran may differ from a raw recruit: and its methods differ from those of common sense only as far as the guardsman's cut and thrust differ from the manne'. If you consider both as knowledge, then, common sense is knowledge attained from information around us, but scientific knowledge is through observation and experimentation on a particular subject.

Broad Classification of Science Disciplines

It is common to divide scientific disciplines into natural sciences, social sciences, and humanities.

Natural sciences: The disciplines categorized as natural sciences deal with the study of the physical world and its phenomena through empirical methods. Natural sciences are concerned with the processes and phenomena that would occur in the absence of humans, and include the disciplines such as biology, chemistry, physics, astronomy, and geology.

Social sciences: Social science disciplines study the processes and phenomena arising from the activity of humans, for example: sociology, economics, and anthropology. Social scientists also make use of the scientific method to study these processes.

Humanities: Humanities deal with the study of culture arising out of human activities and achievements, for example: literature, art, languages, and theatre. In humanities, generally a critical or analytical approach is employed to study human culture.

Although social sciences make use of the scientific method, in academic circles, these are often considered as 'arts' along with humanities; and many people object to consider these disciplines as sciences, because, according to them, empirical methods are seldom possible to apply in such disciplines.

Applied Science, Technology, and Engineering

Natural science disciplines are usually described as *basic sciences* and *applied sciences*. The disciplines that come under basic sciences, also called pure sciences, develop knowledge to predict, explain, and understand various phenomena occurring in the natural world. On the other hand, applied science disciplines apply knowledge from basic sciences to solve practical issues and develop applications. For example, agricultural entomology and medical entomology are applied fields of entomology contributing towards agricultural or medical knowledge for pest control applications.

Applied sciences, *technology*, and *engineering* are offshoots of science. Applied sciences utilize existing scientific knowledge and associated applications for specific purposes. However, technology is development or improvement of a device or appliance using already known principles and methods. Pure sciences help us to understand the world better through knowledge creation, but applied sciences and technology are for controlling the world and for making it a better place for human beings.

Some distinctions between *technology* and *engineering* are also needed. Technology is the field that applies science to solve problems with a commercial end in mind. The word is from *Technologia*, a Greek word, meaning systematic treatment of an art. The term 'technology' is generally used to explain new inventions and devices using recently derived scientific principles and processes. Although synonymously used in many circles, 'engineering' is a much broader term than 'technology'. It is used to denote an activity that applies human qualities such as imagination, critical evaluation, and intellectual thinking to existing human knowledge bases to create or utilize technology safely and efficiently. In addition to traditional branches such as civil engineering and mechanical engineering, now we have branches like bioengineering and genetic engineering, which may revolutionize life on earth in the future.

1.3 What is Research?

The literal meaning of 'research' is meticulous search. However, in popular usage, research is a systematic search for answering a particular question, solving a problem, or gathering information, especially for a project, literary work, cinema, or a television series. In academic fields, the term research is used to denote activities such as defining, redefining, and solving problems; observing facts and their interpretation; formulation of hypotheses and their testing through experiments; revision of existing theories and laws; and practical application of information already generated. Before the emergence of modern science, experimentation and scientific method were unheard terms, and the 'research' was mainly through *logical reasoning*. It is, therefore, quite natural that some of the basic distinctions in logic have carried over into contemporary research. Consequently, the *inductive* and *deductive* methods of reasoning (see Sect. 1.6 and Chap. 14) became part of modern science and research. *Logic*, *reason*, and *evidences* are essential parts of modern research.

Research in natural sciences and social sciences employ different methodologies and approaches based on the nature of disciplines. Objective measurements are far easier in natural sciences but comparatively difficult in social sciences. For example, in disciplines such as physics and chemistry, it is relatively easy to keep the conditions under control during experiments, and the results obtained by any researcher are testable and repeatable. However, as social sciences deal with phenomena arising from the activity of humans, it calls for a different approach to research. Understanding and studying human behaviour is a complex process, and therefore, the context of social science research keeps changing. This problem is apparent in most applied sciences such as agriculture and medicine too, which are also the creation of humans.

In most scientific disciplines, research is used as a tool to understand and solve problems for the benefit of people. Discovery of antibiotics for the treatment of diseases affecting humans or the discovery of fertilizers to take care of depleting soil fertility are examples. Similar to the above simple examples, several other examples of discoveries and inventions that made the life of humankind easier and comfortable can be quoted.

Research is also used as a management tool. Most managers use appropriate research techniques for routine as well as strategic long-term management. Research is used as a tool in defence for chalking out better strategies. In everyday life, even non-scientists such as journalists, judges, lawyers, police, scriptwriters, and artists make use of research to solve and settle issues related to their profession.

Motives for Doing Research

In earlier days, our ancestors dabbled in research because of necessity and the quest to unravel the unknown. At present, many people take up research as a career. Yet, some others pursue research as a hobby and become amateur scientists because of some specific interests. One or more of the following may be the motives in doing research.

Excitement of discovery: Many want to become scientists or researchers enthused by the chance of adding something new to the existing knowledge base. The excitement of discovery or invention is a strong motive to take up research. In fact, most of the breakthroughs happened in science were due to this quality of a few enthusiasts.

Pursuit of prestige: Scientists, who pursue a career in research, usually commands respect in society. The theories, principles, and other findings they make increase the status of scientists and may bring fame and glory to the researchers.

In depth knowledge on a subject: The person who undertakes research on a subject gets an opportunity to study that subject in depth.

Service to society: The society derives many benefits out of research. By using appropriate research techniques, the causes and remedies for many problems currently affecting the society such as food security and climate change can be found out.

Need for publications: For a successful career in an academic or research institution, publications are necessary especially to get promotions. In fact, many institutions rate their scientists based on the number and quality of publications made out of research.

Obtaining a higher degree: Students in agriculture or fields such as medicine and engineering need to submit a thesis or dissertation to get Master's degree. However, for a doctorate degree in any discipline, research and consequent report in the form of a thesis is mandatory.

Better management: Most managers use appropriate research techniques for routine as well as strategic long-term management.

A tool in defence: Research is used as a tool in defence for chalking out better strategies.

Research in everyday life: In everyday life, even no-scientists such as journalists, lawyers, police officers, scriptwriters, and novelists make use of research to solve and settle issues related to their profession.

Benefits of Conducting Research

The benefits of research are many to the society as well as the person who conducts research. Whatever be the motives of the researchers, the society enjoys the following benefits:

Expanding frontiers of knowledge: The 'body of knowledge' of various scientific disciplines expands through research.

New inventions and discoveries: This is the main objective of research in applied sciences and technologies.

Solving problems affecting the society: By using appropriate research techniques, the causes of many problems currently affecting the society can be found out and remedies suggested.

Increasing efficiency and reducing costs: Efficient machines and better problem solving increase the efficiency of work and reduce the cost of production.

Research strives to make life easier: Examples include better transport and communication facilities that made the world a better place to live in.

Luxury and comfort: Examples are many. Research for better houses, house materials, transportation, comfortable clothes, and many others provide luxury and comfort for a better living.

Infotainment: Information plus entertainment is infotainment. There are umpteen opportunities for entertainment now because of spurt in technologies. Computers, Internet, and smart phones made information retrieval and its dissemination fast and cheap.

Economic growth: Research is also needed for speeding up business, efficient factories, new products, new marketing strategies, and so on to ensure economic growth.

Characteristics of Good Research

The body of knowledge comprising various disciplines grows through research. Technological advancements in any field are through research efforts of numerous people. Good research is characterized by certain attributes.

Research is based on the work of others: Research is an activity based on the work of others. This does not mean that you are copying the work of others, but look to the work that has already been done to provide a basis for what and how you might conduct your work.

Research is a blend of logic and imagination: Research is guided by the rules of logical reasoning, and the logical processes of induction and deduction. Imagination and thought are used for making hypotheses and theories.

Research tries to identify and avoid bias: Evidences can be biased. Bias can occur during the planning of experiment, its implementation, data collection, interpretation, and reporting. Sometimes, nationality, gender, ethnicity, age, and political views of the researchers may influence them to go for biased evidences or interpretations. Possible sources of bias and how bias is likely to influence evidences and interpretations must be understood and precautions must be taken.

Repeatability: Repeatability is an important characteristic of good research. If we repeat the research, we should get the same results.

Research must be generalisable to other settings: Research is universal in nature. The findings and results obtained in one setting must be suited to other settings also.

Research is systematic: Scientific research is systematic and structured with specified steps in a sequence. Although it does not rule out creative thinking, it rejects mere speculation and intuition in arriving at conclusions.

Research generates new questions: An enquiry into a new phenomenon generates new questions, which must also be answered.

Research is an apolitical activity: Research is not authoritarian and should not have any 'politics' behind it. It should be undertaken for the betterment of society and not for selfish or destructive purposes.

1.4 Qualities of a Researcher

Researchers may have some inherent qualities. They should strive to attain qualities, which they may not have. The following are some qualities, which all the researchers must try to develop. Among these, the first two—the scientific attitude and research attitude—are the prime qualities every researcher should possess.

Scientific attitude: Attitudes are predispositions to react positively or negatively to some object. Scientific attitude is the attitude of a person to promote the use of scientific approaches and rationales to react to any object. Researchers should not succumb to superstitions, unfounded information, word of mouth, or pseudoscience.

Research aptitude: Not all those coming to the field of research may have a research aptitude. Researchers should be inquisitive and curious about things and events around them. They should have genuine interest in the subject and a mindset to unravel the unknown. In other words, they should have a 'research mind'.

Persistence: To become a successful scientist, one must be persistent and continue ahead of research through publications, especially peer-reviewed research papers. Persistence means the traits such as perseverance, patience, tenacity, thoroughness, and determination to achieve something. A researcher should be patient enough to wait for the results. Sometimes, they may not get the desired results within the stipulated time. Certain investigations are time consuming, involve drudgery, and require the help of many people. Undue hurry will not yield anything.

Self-motivation: Self-motivation is an important quality for scientists. Self-motivated scientists produce more results than others do.

Courage to ask questions: Researchers should ask questions about things, which they cannot agree or do not understand. They should ask 'what', 'how', 'when', and 'why'; and find answers by critically observing, experimenting, consulting, discussing, and reasoning.

Skepticism and receptivity: A critical attitude is essential for all scientists. Do not simply accept the data and interpretations. Evaluate them with an open mind. Strike a balance between skepticism and receptivity. You must always entertain new concepts and ideas rather than confronting them with a negative response or criticism.

Objectivity: The conclusion drawn by a researcher should be objective, and that it should be based on findings revealed through actual data. The researcher should strive to remain impartial to the outcome of the research, and ensure that prejudiced or emotional leanings of a researcher should not in anyway influence the conclusion.

Industriousness: Research is not an easy job. Those who are lazy and longing for comfort and luxury will not become a successful researcher, and the society will not gain anything from them. Researchers should aspire to make new discoveries and inventions by sustained and dedicated work. There is no retirement for good researchers.

Honesty and truthfulness: Researchers should be honest in their approaches. They should follow the established ethics, and should not commit any fraud or misconduct. They will record honestly the observations and experimental results, and try to avoid all types of plagiarism.

Open-mindedness: A researcher should be impartial. They should not have any pre-conceived notions or biases towards researchable issues. They will only be guided by facts, reasons, and logic.

Above-average intelligence: Scientists need not be super intelligent. It is often noted that intellectuals without the other needed qualities of a scientist seldom achieve anything in science. However, a scientist with above-average intelligence can succeed by hard work and performing better in other qualities required for scientists.

Knowledge: Researchers should be proficient in their subject. They should be willing to collect all the relevant literature connected with the researchable topic, read them critically, analyze them, and learn them.

Imagination: Imagination is essential for insight and for problem solving. Most scientists are surprisingly imaginative; however, when imagination is combined with both determination and a vision, the result can be wonderful!

Self-confidence: Self-confidence encourages motivation to tackle challenges and foster optimism.

Search for perfection: The researcher will repeat experiments carefully and systematically, if required, and will not manipulate results under any circumstances.

Team spirit: Certain problems cannot be solved by an individual scientist because of the vastness of the problem. Sometimes, the problem may be multidisciplinary, and a team comprising of experts from different disciplines is required. The researcher should be able to work in a team where personal ego or dislikes will not have a place.

1.5 History of Research

The age of earth is estimated to be about 4.6 billion years, and it is believed that life in the form of bacteria originated some 3.5 billion years ago. Human like beings (*Homo erectus*), the immediate ancestors of humans (*Homo sapiens*), first appeared on earth approximately 1.5 million years ago. It is also postulated that the first form of *Homo sapiens* evolved from these ancestors some 0.35 million years ago. Neanderthals and Cro-Magnons were early *Homo sapiens*. They transformed eventually to the present day modern humans (*Homo sapiens sapiens*) by about 40,000 years ago. Primitive humans lived on hunting, fishing, and by gathering food from nature. They could not continue this lifestyle for long because of population pressure and its effect on carrying capacity of their dwelling place. Humans with their experiences in hunting and gathering learned many characteristics of wild plants and animals. In the meantime, they also discovered and learnt the use of fire. As the pressure of population on their existing resources increased and their means of sustenance started decreasing quantitatively and qualitatively, they had to look for some other means to produce food. This search for alternate food production culminated in the discovery of agriculture.

The origin of agriculture was with the first domestication of plants and animals. The food gatherers and hunters became farmers by this development, which happened somewhere around 8000 BC by the end of the Mesolithic Age or the Middle Stone Age. The socioeconomic revolution initiated by this new development became famous in anthropological and archaeological literature by **Gordon Childe**'s phrase, '*Neolithic revolution*'. According to **Mark Cohen**, an archaeologist specialized in ancient cultures of coastal Peru, agriculture developed strictly as a response to population pressure (Cohen 1977). This gave credence to the argument that agriculture was born out of necessity. It is estimated that world human population was about 5 million only when agriculture began to take shape. Presently, after 10,000 years, it is more than 7800 million or 7.8 billion! (World population crossed 7.7 billion in Nov. 2018).

Before the arrival of agriculture, humans were probably contented with the satisfaction of their basic requirements of food and shelter only. The quest for secondary needs including clothing developed gradually, which called for more sophisticated technologies. In the primitive way of life, when humans depended solely on hunting, fishing, and gathering, they had limited time to devote to cultural activities, because the greater part of the day was spent in search of food to prevent starvation. Therefore, it is often argued that most of the characteristics of modern development coincided only with permanent settlement.

Humans have been inquisitive and learned many things based on observation and by trial and error. Our ancestors tried to explain various natural phenomena, paving the way for primitive religious concepts and religions. Some persons were designated as priests who explained every phenomenon as God's creation under authority. Priests began to claim special channels of communication with the gods leading to the establishment of a system of religious authority passed on from one generation to another. In course of time, a rigid dogma of nature's processes developed. Any deviation from religious teachings was not allowed, and those who dared to break the authority away from traditions faced dangerous consequences. This retarded the search for truth for centuries. The oppression and curbs from religious authorities on freethinking and scientific pursuits were most rampant in Europe during the 'Middle Ages' (fifth to fifteenth century), and this period is called the 'Dark Ages' of Europe. It is distressing to note that humans could wait up to about sixteenth century to break the religious hold on matters concerning the universe and offer accurate explanations for various phenomena!

The Birth of Science

In ancient Europe, '*philosophy*' was the term used to refer all types of knowledge including science. Before the eighteenth century, the present-day scientists were referred to as 'natural philosophers'; that is, philosophers who study nature, and the more common term for science was 'natural philosophy'. During the eighteenth century, the word '***science***' began to catch up in the sense of any systematic field of study. In the nineteenth century, people began to differentiate science from other forms of knowledge. The word 'science' became more popular after the establishment of the British Association for the Advancement of Science in 1831.

Philosophy and science, as we know them today, originated with the early Greek civilization. The period from 600 BC to AD 200 saw many eminent philosopher scientists; and for science, this period was a golden period. Around 300 BC to 400 BC, many Greek philosophers proposed natural explanations for phenomena that had previously been explained only by reference to supernatural myths or tradition. People in ancient times observed and believed *apparent truths*. The Greek philosopher–scientists proposed remarkably precise explanations for several natural phenomena. Along with these, they also proposed many mistaken explanations. This in no way, however, prevented them from continuing their pursuits to learn about nature. The Greeks believed that knowledge could be derived from *absolute truths* or *axioms*.

Great thinkers of Greek golden period, especially **Socrates**, **Plato**, **Aristotle**, **Hippocrates**, **Democritus**, **Aristarchus**, **Pythagoras**, **Euclid**, **Archimedes**, **Ptolemy**, and **Galen** made important contributions. An overview of important contributions of the most prominent Greek thinkers is mentioned below.

Socrates (469 BC–399 BC): Socrates was a great Greek scholar who taught people to question everything before accepting, and he maintained that self-knowledge is more important than knowledge of the universe. However, the establishment was not tolerant to Socrates, and he was brought to trial and sentenced to death by poisoning on charges that he was an atheist and corrupting youth.

Hippocrates (460 BC–370 BC): Probably, Hippocrates is the most outstanding physician in the history of medical science, who is known as the 'father of medicine' in recognition to his long-lasting contributions to the field. He is credited to developing medical science as a rational method. He is also associated with medical ethics, by coining what is now known as *Hippocratic Oath*, still being administered to new medical graduates.

Plato (429 BC–347 BC): A philosopher and mathematician, Plato was a disciple of Socrates. He founded the Academy in Athens, the first institution for higher learning in the west. According to Plato, knowledge of nature does not require observation; it is attainable through reason alone. However, problems arose as the Greeks began to accept many apparent truths as absolute truths. From earth, it appears that the earth stands still and everything else moves around it!

Aristotle (384 BC–322 BC): Aristotle was an exceptionally talented Greek philosopher-scientist, a pupil of Plato and the tutor of Alexander the Great. Although Plato's main interests were philosophy and logic, Aristotle dabbled in many areas of science along with philosophy. He wrote extensively on natural sciences; some even considers Aristotle as the 'first scientist' in the world. He set a pattern about earth and the universe that was to last for about 2000 years to come. For him, a large stationary earth is at the centre of the universe, and the sun, the moon, and the stars orbiting circularly around the earth. He also contributed considerably to the knowledge of biology. He classified animal species and arranged them into hierarchies. Aristotle formulated the now infamous *Scala Naturae* (the great chain of beings) to explain the diversities of living things. Accordingly, a rat was born as a rat and frog as a frog; all creatures remained unchanged as they were created as per the wisdom of God. This theory was accepted immutable as it was consistent with religious dogma, and no scientists dared to offer alternative explanations until the nineteenth century.

Euclid (325 BC–270 BC): The knowledge of geometry developed by Egyptians was perfected and presented in an axiomatic form by Euclid around 300 BC. Many of the geometric principles could be presented in a deductive system as with many theorems, which could follow logically from a few self-evident axioms or obvious truths. This must have made a deep impression on the Greek mind. Eventually, the notion that the behaviour of nature could be described using logic alone began to prevail. This is the essence of *rationalist philosophy.*

Archimedes (287 BC–212 BC): Archimedes contributed much for the advancement of science. He is famous for the floatation principles in physics. One important principle is named after him, the *Archimedes principle.* He calculated the value of 'pi' (π), and made known the mechanical advantages of lever and pulley systems.

Ptolemy (AD 90–AD 168): Ptolemy was a Greek-Egyptian, who influenced human belief regarding the organization of the universe for many centuries. He developed the now infamous *geocentric theory* of the universe in about AD 140. As Ptolemy's model agreed with the apparent movements of the heavenly bodies and because it was accepted as consistent with religious dogma, nobody dared to question his model for about 1400 years to come.

Galen (AD 129–AD 200): Galen was a Greek anatomist and physician who made invaluable contributions to the knowledge of anatomy and physiology. His theories remained unchallenged until the sixteenth century when new insights into human anatomy and physiology were made by Andreas Vesalius and William Harvey.

The Dark Ages

Logically, one might expect that human scientific endeavours would have grown rapidly following the great discoveries and logical methods introduced by the Greek civilization. What actually ensued should serve as a lesson for our modern scientific and technological world! After the decline of the Greek civilization, there was a decline in the practice of scientific logic. Autocratic political systems did not allow free thought. Although Romans were great administrators and soldiers, they were not scholars. Science and technology did not flourish under their rule as happened during the heydays of Greek civilization. Christianity and Islam, which rapidly spread during the period, also did not help much in the development of science. They accepted most of the knowledge gained by the Greeks as consistent with divine revelation.

The 'Middle Ages' or the 'Dark Ages' of Europe (fifth to fifteenth century) was a sad period for science as a whole (The Middle Ages is the central period of a three-period division of European history—classical, medieval and modern). Religion characterized by *dogmatic beliefs* (preconceived beliefs held to be true no matter what contradictory evidence might exist) wielded tremendous powers over politics and society at that time. Naturally, the rate of new scientific discoveries came to a standstill. This dark age of dogmatism prevailed for over 10 centuries. History shows us the importance of freedom to use scientific methods, if we are to continue to learn about nature. The period was, however, not dark all over the world. For example, **Aryabhatta**, **Varahamihira**, and **Brahamagupth** from the Indian sub-continent contributed much during this period. Important contributions of Indians were the numeral '0' (zero) and decimals. Algebra was developed by the Arabs.

Revival of Science

Until the revival of science by the end of the Middle Ages, science and philosophy were considered together, and both science and philosophy had became bound with theology. From the philosophers of ancient times to the fifteenth century, the contributions to science were very negligible. However, science and technology advanced to boundless levels during the last 500 years, bringing an ever-more comprehensive view of the world. This was possible because of some radical changes that took place in Europe by the end of Middle Ages putting an end to the long period of 'darkness' in science. The period between AD 1500 and AD 1700 is considered as the age of 'Revival of science', and this was instrumental in effecting a change in the western attitude from the medieval to the modern. However, the Church and the emerging *empiricism* were at loggerheads, and the opposition of Church continued for about 200 years. Despite the opposition, new ideas in physics, astronomy, biology, human anatomy, chemistry, and other disciplines rejected many doctrines and beliefs prevalent in the society ultimately ushering in modern science.

In fact, three major inventions that happened during the fifteenth century were responsible for most of these changes. One of these was the invention of *printing*. Printing required the development of paper to replace parchment. Developed originally in China, the technique of papermaking reached Europe in 1200s, and by 1450, papermaking was common. The effective application of printing was achieved by Gutenberg in 1438. The printing press promoted literature and science. Because of the printing press, multiple copies of the Bible could be made available for the public. People began to read religious scriptures, and started questioning the teachings of religious authorities. Two other radical inventions that paved the way for revival were the *compass*, which helped increased trade, travel, and discovery of new continents and islands; and *gunpowder* and *firearms*, which increased the combative ability of humans. The thoughts of **Leonardo da Vinci** (1452–1519), the great Italian artist, acted as a catalyst for scientific advances. According to Da Vinci, 'experience never deceives; it is one's judgment that deceives'.

Alexandre Koyré, a philosopher and historian, in 1939 coined the term *scientific revolution* to describe the radical changes that took place in science culminating the Dark Ages. Many scholars believe that the scientific revolution began with the publication of two classical works in 1543, **Nicolaus Copernicus**' *De revolutionibus orbium coelestium* (*On the Revolutions of the Heavenly Spheres*) and **Andreas Vesalius**' *De humani corporis fabrica* (*On the Fabric of the Human body*).

Some great works, which caused paradigm shifts and revolutionized the whole gamut of science during this revolutionary period are mentioned below:

Nicolaus Copernicus (1473–1543): Copernicus, a Polish astronomer (Original name, Niklas Koppernigk, Nicolaus Coppernicus is Latinised name) advanced the *heliocentric theory*, and showed that the earth revolves around the sun and the sun is the centre of the planetary system. This paved way for a true 'paradigm shift' in spite of objections from the Church. This revolutionary discovery disproved Ptolemian geocentric theory and led to a chain reaction in many fields. In 1543, he wrote and published *De revolutionibus orbium coelestium* (*On the Revolutions of the Heavenly Spheres*), and many consider this work as the starting point for the scientific revolution that followed.

Andreas Vesalius (1514–1564): Vesalius was a Belgian who was able to conduct human dissections, and found that the Greek anatomist, Galen (~129–200 AD), was in error in many details of his descriptions of human anatomy. The works of Vesalius contributed much towards a modern understanding of basic human anatomy, and he is rightly regarded as the 'father of anatomy'. Vesalius published *De humani corporis fabrica* (*On the Fabric of the Human Body*) in 1543, revolutionizing the science of human anatomy.

Galileo Galilee (1546–1642): Galileo, an Italian Astronomer, made the first telescope and founded the science of dynamics. He provided convincing support for the Copernican model of the universe, but had to face the 'consequences' from the religious hierarchy. He was persecuted and had to distance from his own findings.

During the medieval period, many scientists were persecuted for their courage to experiment with nature. A prominent figure is **Giordano Bruno**, a Dominican monk, who was burnt alive in 1600 for his acceptance of the Copernican Model.

Francis Bacon (1561–1626): Francis Bacon championed for 'scientific methods'. He developed ***induction*** as a method of advancing science. His particular philosophy, how to do science, was so influential that historians refer to it as the 'Baconian Philosophy of Science.'

Johanan Kepler (1571–1630): Kepler showed that the planets move in elliptical paths, and they travel most rapidly when near the sun. He is credited with developing the 'laws of elliptical orbits'. Prior to Kepler, circular motion was accepted as a natural and ubiquitous motion of heavenly bodies.

William Harvey (1578–1657): Based on careful observation and experimentation, Harvey could demonstrate that the heart pumps blood, and that the blood circulates constantly through the heart and blood vessels.

Rene Descartes (1596–1650): Descartes was a mathematician, physicist, and philosopher. He is regarded as the father of analytical geometry, the link between algebra and geometry. Descartes was one of the leaders in the *scientific revolution.* He is perhaps best known for the philosophical statement 'I think, therefore I am', found in his famous book, *Discourse on the Method of Rightly Conducting One's Reason and of Seeking Truth in the Sciences* (1637). Descartes was the first to introduce *reductionism* (sometimes also called Cartesian reductionism) to western thinking and philosophy (see Sect. 2.9). He also championed for the *method of deduction.* In the history of science, Rene Descartes and Francis Bacon are regarded as the pioneer thinkers who provided a philosophical framework for natural sciences.

Antonie van Leeuwenhoek (1632–1657): Leeuwenhoek invented the simple microscope. He made glass bead lenses and used them to magnify previously unknown microscopic organisms. He discovered bacteria, protozoa, unicellular algae, and many others. This prompted us to discard supernatural explanations for many diseases, which were actually caused by microorganisms.

The End of Alchemy and Replacement of Phlogiston Theory

Alchemy could be regarded as the precursor of modern chemistry prior to the universal acceptance of the scientific method. Alchemists searched for ways to transform matter especially lead into gold and produce an elixir that would confer eternal life. Alchemy flourished in the early 1600s, and at that time, it was considered equal to medicine and chemistry. Alchemists were interested in mysticism, and they believed that the matter was composed of four elements—earth, air, fire, and water. Although the works of alchemists helped to build up some descriptive knowledge about materials and processes, it did not help to understand the nature of materials. We may be surprised to know that among the prominent alchemists of Europe, the names of Roger Bacon, Saint Thomas Aquinas, Sir Isaac Newton, and Sir Thomas Browne were also included!

Robert Boyle (1627–1691): Boyle is well known for his studies on gases, leading to the formulation of Boyle's law. He contributed greatly to the science of chemistry, and exploded the Aristotelian theory of the four basic elements, earth, air, fire, and water. Robert Boyle was also an alchemist, but by his time, alchemists had rejected most of the occultist beliefs.

Isaac Newton (1642–1727): Newton was able to give meticulous explanations for motion on earth and in space. His three laws of motion and the key concepts of mass, momentum, acceleration, and force revolutionised physics. The influence of Newton spread far beyond physics and astronomy. The ready acceptance of Newtonian physics led many to believe that all of the universe could be explained in precise mathematical terms. Newton's theories and laws reigned undisputed for about 200 years as a scientific and philosophical view of the world. Albert Einstein's theories of relativity did not overthrow the world of Newton, only modified some of its most fundamental concepts.

The Royal Science Society in England was formed in 1654, and scientists like Robert Boyle and Isaac Newton were active members of the Society.

Antoine Lavoisier (1743–1794): Lavoisier refuted the *phlogiston theory* and made advancements in chemistry.

Phlogiston theory is now a discredited seventeenth century hypothesis regarding combustion. According to the theory, combustible substances contain a hypothetical ingredient known as *phlogiston*, which would be liberated through burning leaving a residue. Although the hypothesis was advanced by **J. J. Becher** late in the seventeenth century, the word phlogiston was first used by a German chemist, **Georg Ernst Stahl** early in the eighteenth century. 'Phlogisticated' substances were those that contain phlogiston; and once burned, the 'dephlogisticated' substance would be in its 'true' form, the calx. According to the theory, the ash of the burned material was the 'true' material. Stahl hypothesized that rusting of iron was also a form of burning, where phlogiston was freed; and finally, iron would become ash or calx. This hypothesis received popular support during the eighteenth century until it was refuted by Lavoisier who revealed the true nature of combustion. In unravelling the mystery of burning, Lavoisier established the modern science of chemistry.

Justus von Liebig (1803–1873): Modern agriculture owes a great to the contributions of Liebig, a great German chemist, in understanding plant nutrition. He destroyed the *humus myth* prevalent at that period that soil fertility was largely dependent on the accumulation of humus, believed to be the sole and direct source of plant nutrients. Liebig discovered that crop yield is dependent on a few elements in the soil—notably nitrogen, phosphorus, and potassium. Liebig claimed that by feeding these elements to plants in water-soluble form, one could achieve yields much higher than usual. The *law of the minimum,* postulated by Liebig in 1862, stated that the element in the shortest supply was the limiting factor in crop yield and that all of the elements removed with the crop must be replaced for better growth and yield.

The Beginnings of Modern Biology

Carl Von Linne (1707–1778): Linne, the Swedish naturalist and the author of *Genera Plantarum* and *Species Plantarum*, synthesized the sciences of botany and zoology from the data accumulated over the years by early naturalists and explorers. In fact, Aristotle's work on classification of organisms went unchallenged until 1753 when Linne published the book, *Species Plantarum.* He devised a convenient system of naming plants and animals, the *binomial nomenclature*, which is accepted universally. Linne adopted Latin or Latinised names for naming organisms. He also changed his name to Carolus Linnaeus.

Charles Darwin (1809–1882): The works of Charles Darwin in the nineteenth century paved way for the development of modern biology. Darwin was the first to discover *natural selection,* the biological process in nature, which could account for changes in species over time and ultimately in *speciation* (formation of new species). The first edition of the '*Origin of Species*' was published on 24 November 1859. The '*Origin of Species*' has survived the scrutiny of numerous subsequent studies over the past 150 years, and it has provided us with a natural explanation for the origin of species.

Gregor Johann Mendel (1822–1884): Mendel was an Austrian monk and botanist. In 1866, he formulated the laws of inheritance after experimenting with garden peas and paved the way for genetics. Although Mendel is now regarded as the 'father of genetics', his theory remained unknown to the scientific world until 1900. In 1900, Hugo de Vries and others rediscovered Mendel's contributions marking the beginning of the science of genetics.

Louis Pasteur (1822–1895): In 1862, Louis Pasteur, a French chemist disproved *spontaneous germination* theory, which held that microorganisms could arise spontaneously from inanimate or non-living matter. Until then, it was widely believed that living things could arise spontaneously from non-living, dead, or waste materials, because people saw such materials 'generate' living things such as mould or maggots. Pasteur demonstrated that fermentation is a biological phenomenon. He advanced *germ theory* that diseases are caused by germs.

Robert Koch (1843–1910): Koch, a qualified physician established that an organism could cause a specific disease. For ascertaining the validity of research towards the cause of diseases, he issued certain postulates, the famous *Koch's postulates*.

Science in the Twentieth and Twenty-First Centuries

Science became a great movement during the twentieth century. The *Industrial Revolution* started in the eighteenth century continued during the nineteenth century. The period from 1750 to 1850 witnessed radical changes in manufacturing, mining, transportation, agriculture, and other technologies. All these changes had a deep impact on the social, economic, and cultural conditions of the people. Bernal (1969) stated that in 1896, there were hardly 15,000 people engaged in research. In 1962, it rose to about one million. Similarly, the money spent on research also increased by 2000 fold. In 2021, can you imagine what would be the corresponding figures? It is estimated that in USA alone, there are more than 5 million scientists! China is fast becoming a scientific super power. Although India ranks 4th in terms of engineers and scientists, in terms of quality of research, it is ranked only 22nd! India spends only about 0.5 percent of its GDP on research and development, when the USA and China spend 2.5 percent and 1 percent of their GDP.

In the nineteenth century, scientific advancements were more or less restricted to the West. However, it spread to all the nations in the twentieth century. If we compare the total knowledge gained from time immemorial until the beginning of the twentieth century and that during the twentieth century alone, the latter would still outweigh the former. We usually divide the prehistoric period into the Stone Age, the Bronze Age, and the Iron Age. In a similar fashion, many would like to divide the twentieth century into the atomic age, the space age, the polymer age, the electronics age, the computer age, the biotechnology age, the nanotechnology age, and so on. About 90 percent of the scientists so far seen in the world lived or is living during the twentieth and twenty-first centuries. Atomic physics, quantum mechanics, and astrophysics developed during this period are exploring the macro and micro universe. Molecular biology, biochemistry, and genetic engineering explore the universe at micro levels. Presently, nanotechnology that explores the universe at nano level is catching up.

It is estimated that primitive humans took thousands of years to make a stone knife with sharp edges. At present, new inventions and discoveries are made very frequently. The transformation in various fields, for example, mode of transport from trains to aeroplanes and from there to space shuttles and mode of communication from telegraphy to telephone, radio, television, Internet, and smart mobile phone were very fast. The speed at which new knowledge is utilized to devise technological breakthroughs is astonishingly faster now. Within just five years from learning about nuclear fission, the first atom bomb was dropped. Within nine years, the first atomic power reactor was established. Like these, you can list out several examples of breakthroughs in science and technology during the twentieth century. Out of the several breakthroughs, three are mentioned below, as they were revolutionary landmarks; the theory of relativity by Albert Einstein, the discovery of DNA structure by Watson and Crick, and the green revolution ushered in by Norman E. Borlaug.

Albert Einstein (1879–1955): Einstein came up with theoretical ideas that made radical contributions to the understanding of nature. One of these was the *special theory of relativity* (1905), wherein he considered space and time to be closely linked dimensions. The special theory of relativity is important as it asserts the equivalence of mass and energy—that is, any form of energy has mass, and matter itself is a form of energy. This is expressed in the celebrated equation $E = mc^2$, in which E stands for energy, m for mass, and c for the speed of light. About 10 years later in 1915, Einstein published the theory of *general relativity*. The theory considered the relationship between gravity, time, and space in which Newton's gravitational force is interpreted as a distortion in the geometry of space and time. The relativity theory has been tested repeatedly by checking predictions based on it, but it never failed. The theory of relativity had several unexpected implications revolutionising science as a whole.

Discovery of DNA structure: Another scientific landmark is the discovery of the molecular structure of DNA (deoxyribonucleic acid) in 1953. **James Watson** (b. 1928) and **Francis Crick** (1916–2004) were the major players but others were also involved. In 1951, Watson joined Crick at Cavendish Laboratory in Cambridge, England. At that time, **Rosalind Franklin**, **Maurice Wilkins**, and **Linus Pauling** were also battling to determine the structure of DNA. The experiments of Franklin and Wilkins using X-ray crystallography provided much information about DNA. The article by Watson and Crick proposing the structure of DNA as a double helix was first published in *Nature* in 1953. The double helix model of DNA has led to an explosion of research and deeper understanding about the function of living organisms and the mechanisms of heredity and evolution. The discovery further provided a massive thrust to research in the emerging fields of molecular genetics and biochemistry. For this historic work, Crick, Watson, and Wilkins were jointly awarded the Nobel Prize in 1962.

Norman E. Borlaug (1914–2009): Borlaug was one of the most influential scientists of the twentieth century. He received Nobel Prize for peace in 1970 for averting hunger and famine in the third world countries through the adoption of modern plant breeding techniques.

Cereal-grain yields increased magnificently in many developing countries in the late 1960s because of the works of Borlaug, and the phenomenon came to be known as 'green revolution'. Borlaug, while heading 'Cooperative Wheat Research and Production Programme' in NW Mexico (the precursor to CIMMYT, the International Maize and Wheat Research Centre), produced many high yielding wheat varieties using some new concepts. Borlaug could obtain an efficient plant type of wheat, Norin-10, capable of responding to very high doses of fertilizers and irrigation from Japan (The word 'Norin' in Japanese means 'agriculture and forestry', and varieties officially released and registered with the Japanese Ministry of agriculture and forestry are designated as 'Norin', and distinguished by the numerals following it). Borlaug succeeded in developing broadly adapted, shortstemmed, disease-tolerant, high-yielding wheat cultivars containing Norin-10 genes, and the improved seeds produced by him revolutionised Mexican wheat production. The mean productivity of wheat was catapulted from less than 700 kg/ha in 1925–30 to more than 2400 kg/ha during 1965–69. This breakthrough created hope, and the green revolution spread to India, Pakistan, and many other developing countries by 1960s to 1970s. This silent 'bloodless' revolution helped averting famine in India and Pakistan, thus earning the 1970 Nobel Peace Prize for Norman E. Borlaug.

Borlaug championed for increasing productivity and total crop yields, and he argued that it would curb deforestation. He believed that increasing per hectare productivity of agriculture can help control deforestation by preventing further conversion of forests for farming. In the present milieu of rising global food demand, if you want to follow traditional low-yielding farming methods, the world population should decrease through voluntary means or by mass starvations; otherwise, conversion of more forests for farming is indispensable. Sometimes referred to as *Borlaug hypothesis*, it is argued that intensive high-yielding techniques are highly essential for saving planetary ecosystems from destruction.

Lessons from the History of Science

Before concluding this section on history of research, we should also note a disturbing but persistent trend of poor reception to thoughts from persons who champion for new ideas contrary to prevailing dogmas. What actually happened to great thinkers such as Socrates, Galileo, and Bruno are examples. When the telescope was invented, many natural philosophers and astronomers were reluctant to test their theories by looking through the new instrument fearing persecutions from the religious hierarchy. The situation has changed drastically now, but still there are problems. For example, even after piling up numerous evidences in support of Darwin's theory of evolution, many people still cling to old beliefs. Similarly, superstitions and pseudoscience are on a come back trail. Even educated people are succumbing to pseudoscientific notions, both old and new (Sect. 2.8). Post-modernist thinking and explanations are creating more confusion (Sect. 2.11). There are no sure ways to change this state of affairs forthwith. Creating a critical or scientific attitude is the best way out. Along with this, people should also imbibe the spirit of '*scientific temper*'. Scientific temper is an attitude everyone should cultivate, which involves the application of logic and reasoning in decision-making by avoiding bias and preconceived notions.

1.6 Induction and Deduction in Research

One of the main objectives of scientific research is to find out facts and establish reliable theories and laws based on them, which can be used to explain and predict phenomena or events. For arriving at correct conclusions, a systematic approach to *argument* or *reasoning* is essential. The researchers should use appropriate *logic* to prove their points. Before the modern idea of research emerged, the term the old philosophers used to call research was *logical reasoning*. Therefore, it is natural that some of the basic characteristics of logic have been carried over into presentday research. Thus, the inductive and deductive methods of reasoning became part of research.

The first systematic approach to reasoning commonly attributed to the Greeks was the *method of deduction*. Deductive reasoning is the process of reasoning from a general assumption to a specific application. In this case, the major premise is based on a theory, rule, law, principle, or general understanding. On the contrary, when the major premise of an argument is based on observation or experience, we would say that it is based on the *method of induction*. Often, induction is explained as moving from the specific to the general, while deduction starts with the general and ends with the specific (see Sect. 14.4 for more examples).

Consider the following logical reasoning using the method of deduction:

All A's are B's.

All B's are C's.

Therefore, all A's are C's.

The above reasoning consists of three statements; a major premise, a minor premise, and a conclusion that derive from the premises. The question is simple; what is the relationship between the two classes A and C, if you know already the relation of both A and C to the third class B? Deductive reasoning is supported by deductive logic; for example, note the following general propositions and specific propositions made out of them through deductive logic.

General proposition: All crows are black (already known).

Specific proposition: This bird is a crow; therefore, it is black.

General proposition: All pesticides are toxic (already known).

Specific proposition: This is a pesticide, therefore, it is toxic.

Deductive method contributed much for the development of primitive science. This method was highly successful in mathematics, but the problem was that it did not succeed well in exploring the universe, and it was not all useful for arriving at new facts. For this, we have to use the method of induction. A researcher can arrive at the general laws with the help of specific propositions through inductive arguments. In fact, the widespread acceptance of induction as a method of advancement of science happened just about 450 years ago only, when **Francis Bacon** (AD 1561–1626) championed for its use. The experiments of **Galileo Galilee** and **Tycho Brahe** during that period also gave fillip to inductive methods. Since 1600, inductive method became widespread, and it has been highly successful in exploring nature.

Induction normally involves generalization from the behaviour of a few samples to that of a population. For example, after smelling a number of jasmine flowers, you could make an inductive statement: 'All jasmine flowers are scented'. According to inductive logic, if a situation or condition is true in *all observed cases*, then the situation or condition must be true *in all cases*. Therefore, after completing a series of experiments that support the inductive proposition, you can affirm that the proposition holds good in all cases. However, when the sample size is small, the generalized statement may not be reliable. Induction involves a bottom up approach; starts with specific observations, detect patterns and regularities, formulate a tentative hypothesis, and finally develop some general conclusions, which may eventually become a theory or law according to the strength of evidences.

Both inductive and deductive methods should be considered as different ways of approaching the same objective. In fact, a combination of induction and deduction is practiced in science now. The observations made through induction are further verified deductively through applications to new situations. The scientist proposes a *hypothesis* through induction and then tries to deduce the probability that it is false through empirical evidences. This is what is commonly known as the *hypothetico-deductive method* or *inductive-deductive method*. This hypothetico-deductive method is now recognized as equivalent to scientific method or experimental method (see also Sect. 1.8).

1.7 Hypothesis, Theory, Law, Fact, and Others

In science disciplines, *hypothesis, theory, law,* and *fact* are four very different types of reasonable or scientifically acceptable statements offered to explain natural phenomena.

A supposition, an assumption, or a guess that is not supported by observation is called a *conjecture*. This is not more than a simple guess or assumption on a phenomenon or something. However, if the conjecture is formulated based on some observation, it is called a *hypothesis*. In fact, a hypothesis is a contention or idea that is still in the process of active testing; it may or may not be correct. Once the predictive phase has been carried out and there are enough experimental evidences to support the hypothesis, it would be called a *theory*. In other words, a theory is a corroborated hypothesis, which denotes an explanation that has been confirmed sufficiently for acceptance, but less firmly established than a *law*.

When a long held theory could not explain some new data or phenomenon, researchers working in that field attempt to construct a new hypothesis. This effort becomes more and more complicated as our knowledge increases, because the new theory should be able to explain both new and old data regarding the phenomenon. Changes in our knowledge paradigm become inevitable because of new theories. In science, the testing and improving or occasional discarding of theories are a never-ending process. Remember that there is only one truth, and truth does not change, but theories and laws are held to be true as far as science can ascertain. The theories that have survived many confirmations and not falsified by convincing and repeatable experimental evidences may eventually get the status of a '*law*'. As far as science is concerned, we have the highest confidence in a 'law'.

There are very few laws in biological sciences except some like Mendel's inheritance laws. In biological sciences, in practice, there is no clear distinction between a theory and law. For example, it is still the 'theory' of evolution and not the 'law' of evolution, although it does not mean that biologists consider it is inferior to a law. Evolution is a fact as it has been observed in numerous situations. In spite of the accumulated evidences, evolution is still considered as a theory! This is mainly because biologists are dealing with living organisms, and living organisms are so diverse that they exhibit exceptional variations. Therefore, generalizations in biology are probabilistic as there can be some exceptions. This problem of 'laws' is more extensive in social sciences. In economics, although we hear the names of certain laws, for example, the 'law of diminishing returns' and the 'law of equi-marginal utility', majority are just theories. Some famous laws in science are law of floatation, Boyle's law of gases, conservation laws, Ohm's law, and the four laws of thermodynamics.

In science, the word '*fact*' has special meaning. A fact is something that has been repeatedly confirmed; and for all practical purposes, it is accepted as true or factual. Fact is an empirically verifiable observation. Scientific knowledge is used to understand the universe and unravel underlying facts. It is important to realize that there is no absolute truth in science, and a fact now may be modified or later discarded. As far as science is concerned, a fact is what most people agree to be true. However, just because everyone agrees, a fact may not necessarily be true! They could be our best guesses at that time. A 'fact' is a fact only because we believe it. As facts are evident with the instruments available to us, including our senses, we are not sure whether our senses are being deceived. Along with theories and laws, facts may also be falsified by new observations. For example, the 'fact' that heavier objects fall faster than lighter ones was falsified by Galileo by conducting an experiment after climbing the Pisa tower. There are several examples like this.

If we follow scientific method, all 'results' must be considered provisional, and this must include the so-called 'laws' and 'facts'. Newton's 'law of gravitation' is a famous example of a law that has been found to be only partially correct. Within most scientific disciplines, the elevation of some 'theory' to the status of 'law' usually takes place after a very long time during which the 'theory' is used, tested, and verified. In science, theories are accepted when proved that correct predictions can be made utilizing them. In addition, theories, which are simpler and more mathematically elegant, tend to be accepted over theories which make similar predictions but which are more complex (the *principle of parsimony*, see Sect. 2.7). Some popular theories that have been disproved off include phlogiston theory, spontaneous generation theory, Lamarckism, and the geocentric theory of the universe.

Auxiliary Hypothesis and Ad Hoc Hypothesis

In science, observation of phenomena can lead to hypotheses. The hypothesis implies certain definite observations. We can think of confirmation of our inductive hypothesis in symbols as:

$$A \rightarrow B; \quad B; \quad \therefore A$$

(*A* implies *B* (the hypothesis); *B* is observed; therefore, *A* must also be true or present).

However, if we do not observe what we expect to observe, there is no other option but to reject the hypothesis using the method of deduction.

$$A \rightarrow B; \quad -B; \quad \therefore -A$$

(*A* implies *B* (the hypothesis); *B* is not observed; therefore, *A* must not be true or present).

Sometimes, the expectation of a certain observation is based on the inductive hypothesis created to test its effectiveness in concurrence with some background assumptions. These background assumptions or additional assumptions used to infer a prediction are called *auxiliary hypotheses*. When the observations do not agree with expectations, it means that at least one of the assumptions or hypotheses that lead to expect a given observation is false. It may be the hypothesis we want to test, or it may be one of our auxiliary hypotheses. A classic example is the discovery of planet Neptune. In the early part of the nineteenth century, astronomers observed some mysterious perturbations in the orbit of Uranus. They advanced several explanations. One suggestion was that the law of 'universal gravitation' by Isaac Newton (1642–1726) was failing at great distances. Another probability was that an unknown object was influencing the motion of Uranus. They also put up an auxiliary hypothesis to save Newton's law considering the fluctuations in Uranus's orbit. The auxiliary hypothesis was that another undiscovered object was influencing the orbit of Uranus. Taking the perturbations in the orbit of Uranus as a basis, astronomers calculated the size and position of the unknown object that might be responsible for such fluctuations. Consequent to these observations and explanations, they could identify it as a new planet, and named it Neptune. Understand that auxiliary hypotheses can also be verified; in the example, it was possible by a visual confirmation, and they could save Newton's law from refutation. Please note that only those auxiliary hypotheses that are amenable to an independent investigation are useful. However, auxiliary hypotheses that cannot be verified are worthless, often termed ad hoc hypotheses.

An 'ad hoc hypothesis' is a hypothesis simply added to an existing hypothesis or theory in order to save it from being falsified. The meaning of 'ad hoc' is 'to this', indicating that something is for a specific, limited purpose, for example, an ad hoc committee is a temporary one to deal with a particular situation. Ad hoc hypotheses are common in pseudoscience and religion, which are usually created to explain away evidences that contradict some favourite ideas. Remember that much of scientific understanding relies on the modification of existing theories, but these modifications are distinguished from ad hoc hypotheses lacking independent support, which are adopted to save a theory from refutation. Such hypotheses are not good for science because any hypothesis or theory can be rescued from refutation in this way.

It is common among ordinary people to save many of their long held beliefs from being refuted by concocting some ad hoc hypotheses. When faced with some contradictions in their argument, they try to prove how the contradiction will disappear, if some new assumption is taken into account. See a simple example of conversation between two school going girls, one is suffering from cold.

Nisha: If you take *Thulsi* water for some days, you will be freed of cold.

Sruthi: I tried that for three days, and still got a cold.

Nisha: Did you take *Thulsi* water everyday?

Sruthi: Yes.

Nisha: Well, I am sure something went wrong with your *Thulsi* preparation.

In this case, the protagonist, Nisha is suggesting a simple remedy out of her beliefs, but to her dismay, it is not working. The only reliable option for her is to prove that Sruthi's *Thulsi* water (a decoction made from sacred basil) is not good to act as a medicine for cold by producing reliable evidence. If Nisha cannot do so, her attempt to rescue her hypothesis (that *Thulsi* water heals colds) will not succeed, and we should simply reject her ad hoc hypothesis that there were problems in its preparation!

Axiom and Theorem

An *axiom* is a statement that is accepted as true without proof, because it is conspicuous or self-evident. For example, 'Things that are equal to the same thing are equal to one another' and 'The whole is always greater than any of its parts' are two axioms. From a single basic axiom, you can prove many things through the deductive method as Euclid (c. 300 BC), a Greek mathematician, did. Euclid collected and systematized all the geometrical theorems known during his time. He arranged them in such a way that each theorem could be proved with the help of previous theorems. The word '*theorem*', which is often used in physics and mathematics, is a formula or principle inferred or to be inferred from other formula or principle. Note that a theory is different from a theorem. The theory is a statement of physical events and cannot be proved from basic axioms, at the same time, theorem is a statement of mathematical fact that logically follows from a set of axioms.

Model

In science, a *model* is a system of postulates or inferences presented as a conceptual description of a system. Sometimes, an equation or formula can be used to estimate the effect of a physical process on some other process. In certain cases, simple charts, diagrams, or three-dimensional figures are also used. *Empirical models* are a prominent group of models frequently used for problem solving. An empirical model is based on observation or experiment, and not derived from theory. A model so established can be used to fit the observed facts or data, which allows the prediction of what will happen in certain situations, because what has happened before in those situation is known. The reliability of such methods depends on the database created through experience.

1.8 The Scientific Method

When dealing with science, everyone is supposed to follow certain rules and processes to make it a reliable source of information. **William Whewell** (1794–1866) coined the phrase *hypothetico-deductive* method for the formalized strategy of scientific investigation integrating both inductive and deductive approaches. The method has been incredibly successful in investigating nature. As *hypothetico-deductive* type of investigations is widespread in science, it is often referred to as the *scientific method*. It is also sometimes called *inductive-deductive* method. Accordingly, a scientific investigation is initiated by forming a hypothesis through induction in such a way that it could possibly be falsified by an experiment. Popper's views on falsification (Sect. 2.6) are also taken into consideration in the modern scientific method.

Scientific method is a broad term for the diverse processes through which the sciences are built up. It depends on observation, measurement, prediction, experimentation, or verification, thus differentiating science from other fields of knowledge. The scientific method is based on evidences rather than beliefs or arguments. This character distinguishes science from faith or authority. The scientific method is often described as comprising the following main actions:

- Make observations or gather information
- Develop a hypothesis
- Predict results
- Design an experiment
- Conduct the experiment and collect data
- Evaluation and conclusion
- Acceptance, modification, or rejection of the hypothesis.

The scientific method can be used to solve everyday problems as well! For example, suppose that when you turned the ignition key, your car failed to start. You can guess many reasons and form many tentative hypotheses like: exhausted fuel, weak battery, problems in the carburettor, and so on. You can verify each of these 'hypotheses'. You can check the fuel tank, ask some friends to push the vehicle while you start the engine, or something similar. Only after such preliminary tests, you may decide to consult a mechanic.

A major advantage of scientific method is that it is unprejudiced. You do not have to believe a particular scientist; but you can redo the experiment and decide whether your results are true or false. The conclusions will hold irrespective of the state of mind, the religious persuasion, the state of consciousness of the investigator, or the subject of the investigation. However, this is not the case with various pseudoscientific theories or beliefs. A hypothesis is accepted as a theory based on the results obtained through experiments or observations, which anyone can reproduce. Repeatability of results obtained through scientific method is another advantage. You may not always conduct experiments; in certain cases, mere observations are sufficient. For example, for studying the stars and other celestial objects, it is impossible to perform experiments; therefore, information is gathered from natural observations. Various stages or steps involved in scientific methods are indicated below:

1. **Make observations or gather information**

Scientists observe something in nature and explore how, why, and when something occurs. From observations, they begin to detect certain patterns. A scientist should look critically and avoid all sources of bias in observation. Information about a defined problem may be gathered in many ways; from books and journal articles in a library, from information on the Internet, from discussions with experts and colleagues who are also interested in the problem, and by preliminary observations by the scientist.

2. **Develop a hypothesis**

Scientists offer possible explanations for the phenomenon under study with the observations or information gathered. They form hypotheses using inductive logic. The most important aspect of an explanation is that it must be '*falsifiable*' (see Sect. 2.6).

Explanations should also satisfy the '*principle of parsimony*' or '*Ockham's razor*' (see Sect. 2.8); that is, the hypothesis is expected to contain the least possible number of unproven assumptions. Please note that the 'principle of parsimony' is not an invariable aspect of the scientific method, but only a rule of thumb for quickly evaluating which hypotheses are likely to be fruitful. Select the best option based on your observation and past experiences.

3. **Prediction of results**

The prediction is a formal way to put a hypothesis to test. Suppose you want to find out whether giving protein rich feed increases milk yield in cattle, you can make a hypothesis and do a prediction like this:

> If protein rich feed increases milk yield in cattle, then all these cows fed on protein rich diet will give more milk yield than the control group.

If this prediction holds, then you should accept the hypothesis. After the experiment, if you find that this prediction does not hold well, you will reject your hypothesis.

4. **Design an experiment to falsify the hypothesis**

Every hypothesis must be tested by performing appropriate experiments and evaluating the results. Plan the experiments with suitable research designs and replications. When you publish research articles, you must describe the methodology adopted so that anyone can reproduce or verify experiment. *Reproducibility is an important aspect of experiments.*

5. **Conduct the experiment and collect data**

Physical and natural sciences rely heavily on numerical data and on replicable experimentation to measure and calculate results. Usually, two groups—the control group and the experimental group—are identified for the test. Identification of relevant variables for the experiment is a major task. For identifying and deciding on the variables, you can rely on your experience, but discussing with an expert is also good. Social science disciplines such as sociology, instead of experiments, usually rely on surveys and use descriptions and inferences to arrive at conclusions.

6. **Evaluation and conclusion**

Once the experiment is over, subject the data to proper analysis, summarize the results, and arrive at conclusions. You should also go back to your observations, data, and original hypothesis for consistency. If you did not prove your hypothesis, nothing to worry, you have achieved something in another sense! Unsuccessful experiments provide information that can lead to answers by eliminating other options and also save someone the trouble of repeating your experiments.

Sometimes, you may also get some ideas for solving similar problems. Always remember that research builds on the work of others.

7. **Accept, modify, or reject the hypothesis**

The last stage of the scientific method is acceptance, modification, or rejection of the hypothesis. If the data supports the hypothesis, it is accepted. If the data does not support the hypothesis, then we reject it. Sometimes, the data may partly support the hypothesis. Then we modify the hypothesis and repeat the whole process by a new set of experiments.

Remember that we can never prove any theory to be 'absolutely true' by following the scientific method; we can only be sure that it fits well with the evidences so far collected. For example, if a conclusive deviation or exception is found on a later date, it would be accepted as enough proof to disprove the theory. The custom in science is to use the word '*corroborated*' instead of '*proved*' when a hypothesis or theory is tested. The concept of attaining absolute truth is difficult in science; therefore, we recognize some uncertainty as part of science. Still, we should admit that most scientific knowledge is durable. Prevailing theories or ideas are not rejected completely, but suitably modified based on new evidences.

Experimental Research and Scientific Method

As discussed above, scientific method is a practical, but formalized methodology of learning about the world. A pertinent question is relevant here: Are there any differences between 'experimental research' and 'scientific method'? Although these are used almost in the same sense, some distinction can be made. Scientific method is broader in outlook than experimental research. For a more formal, systematic, and intensive process of conducting a scientific investigation after formulating a hypothesis, we generally use the term 'experimental research'. Experimental research is usually conducted in a formal setting as that of an academic institution.

When we simply say scientific method, it involves an informal application of problem identification, hypothesis formulation, observation, analysis, and conclusion. For example, you could reach a conclusion why your seeds did not germinate or why your personal computer went wrong by employing simple scientific methods. However, as the processes involved here are not structured as in a research problem, it cannot be called 'experimental research' in the scientific sense. Experimental research must be a formal and structured investigation in the laboratory or field with the objective of new inventions and discoveries, formulation of new theories and laws, and revision of accepted theories and laws in the light of new scientific information.

1.9 Research Methodology and Research Methods

We often use the terms research methods and research methodology. *Research methods* mean the methods and procedures a researcher employs to accomplish a research task. Research methods provide precise and detailed procedures of how to start, implement, and complete a research project. These include the research techniques, data collection methods, statistical techniques for the analysis of data, evaluation of research results, etc. At the same time, research methodology is used in a wider perspective. *Research methodology* deals with the general approaches or guidelines for conducting research. It is the science and philosophy behind research methods. Research methodology can be defined as the systematic study of the research process starting from the planning process to reporting the results.

'Research methods' constitute only a part of the wider field of 'research methodology'. Research methods are more important during the implementation phase of a research project or experiment, whereas, research methodology is relevant from the planning stage itself. For a researcher, there are many methods and techniques, which can be selected for completing the tasks. The researcher should know not only the theory and practice of these methods or techniques but also needs to know which of these methods are relevant and how these are selected. In other words, research methodology is about the research methods and the logic behind in selecting a particular method and how the researcher proceeds and concludes the research project. Research methodology encompasses the strategies we use during planning, implementation, and reporting stages of a research project. Certainly, writing research reports and speaking about research are also part of the wider field of research methodology.

1.10 The Research Process

The research process starts from identifying a research problem and completes by the publication of research results. It often begins from an idea or from an unsolved problem. It can also be a continuation of previous research done by the researcher or others. Identifying the problem and comparing it with similar ones is the first task. If this step is successfully completed, then you have to get the approval of concerned authorities to proceed with it. This step is crucial as the research requires funding, and funds would be released only if it is approved by the competent authorities. The authorities approve the projects based on the merit of the problem. After getting necessary approval and funding, you may proceed with it by selecting and applying the most appropriate research methods. You have to exercise extreme caution and vigil during this step because a wrong decision can ruin the entire effort. Once the research work is completed, the next step is to analyse the data you have gathered, and come out with definite conclusions. If you have succeeded in solving the problem, you should inform others, especially those who have some interest or stake in the problem, about your findings and how you have solved the problem. In other words, it is the duty of the researcher to prepare a report of the results and communicate it in a proper way. A research student first executes this in the form of a dissertation or thesis, and then as research papers or presentations in symposia or conferences.

All the activities mentioned above form part of the research process. In short, the research process consists of a series of steps necessary to carry out the research effectively. It consists of closely related but often overlapped activities. At each operational step in the research process, the researcher chooses from a basket of methods, procedures, techniques, or models to accomplish the research objectives. For the smooth flow of the research process and its successful completion, a thorough knowledge on various tenets of research methodology is crucial. The following are the important steps in the research process.

1. Identify the research problem
2. Review of literature
3. Develop the objectives
4. Decide the research design
5. Formulate the research protocol
6. Get approval from competent authorities
7. Conduct the research work and collect data
8. Analysis of data
9. Interpretation of data
10. Preparation of the thesis/report
11. Presentation of results
12. Publication of reports.

Helpful hints have been provided in the forthcoming chapters to finish successfully all the above steps of the research process.

1.11 Academic Research

Before concluding this chapter, we should also examine some of the shortcomings of academic research, the kind of research conducted by students in academic institutions. Postgraduate courses such as Masters, M.Phil.s, and Ph.D.s are concerned with learning how to do research. Depending upon the courses, students may undertake various research projects, small or big, as a part of their academic programmes. It is mandatory for M.Phil. and Ph.D. degrees to have original research projects as a part of the programme; and students are expected to submit thesis or dissertation in the form of a report containing the results and interpretations. '*Original research*' means that the researcher will be able to make a unique contribution to knowledge in that particular branch of study. For technical subjects such as agriculture, engineering, and medicine, M Phil courses are not common; therefore, it is stipulated that at the Masters degree level itself, they undertake original research and present the report in thesis form.

Presently, for Masters programmes in traditional subjects too, short-term project works and submission of project reports or dissertation have been made obligatory in many universities. However, such project works normally do not require original research. The research element in these Masters programmes is designed to make use of some of the methods of research. However, the main emphasis is on the learning process; therefore, more importance is given to understand the methodology.

A big problem with most academic research is that motivation for research is lacking. The initial motivation may be simply the practical need to satisfy the academic requirement. Naturally, most of such studies do not make a significant contribution to knowledge. From this kind of academic research, significant results cannot be expected because of constraints such as time limit, lack of finance, and lack of experience. The student's interest may be to prepare a thesis somehow and get a degree rather than continue a career in research. Therefore, relatively safe research is done. At the doctoral level, however, some significant contribution is expected as the time spent on research would be higher than Masters. In spite of shortcomings, when academic research is done under the close supervision of a research adviser or guide who is sincere and actively involved in research, students' thesis may make an important contribution to science.

For research students, formative years in research are very important, because during these stages, they learn the basics and other nuances of research and the ethics involved. Students should be motivated to do meaningful research, and only sound techniques in tune with research ethics should be followed (see Chap. 23). Never take a casual attitude to students' research work; and under any circumstances, do not ask them to breech any of the accepted norms of ethics. Advisors and guides should understand that the students always look upon them as role models!

Bibliography

Bernal, J. D. 1969. Science in History, 4 volumes (4th Ed.), Pelican Books.
Best, J.W. and Kahn, J.V. 1993. Research in Education (7th Ed., Indian Reprint, 2004). Prentice-Hall of India, New Delhi, 435p.
Carey, S.S. 2011. A Beginner's Guide to Scientific Method (4th Ed.). Wadsworth, Boston, 160p.
Cohen, M. N. 1977. The Food Crisis in Prehistory. Yale University Press, New Haven, 351p.
Gower, B. 1997. Scientific Method: An Historical and Philosophical Introduction. Routledge, London, 276p.
Gray, D.E. 2014. Doing Research in the Real World (3rd Ed.). Sage Publications, London, 729p.
Holmes, D., Moody, P., and Dine, D. 2010. Research Methods for the Bioscience (2nd Ed.). Oxford University Press, Oxford, 457p.
Hughes, W. and Lavery, J. 2004. Critical Thinking: An Introduction to the Basic Skills. Broadview Press, Canada, 403 p.
Jarrard, R.D. 2001. Scientific Methods [E-book]. Available: http://www.emotionalcompetency.com/sci/sm_all.pdf. 236p. [Accessed 10 June 2018].
Kerlinger, F.N. 1986. Foundations of Behavioural Research (3rd Ed.). Holt, Rinehart and Winston, USA, 667p.
Leedy, P.D. and Ormrod, J.L. 2010. Practical Research: Planning and Design (9th Ed.), Pearson Education, New Jersey, 360p.
Marder, M.P. 2011. Research Methods for Science. Cambridge University Press, 227p.
Moore, J.A. 1993. Science as a Way of Knowing: The Foundations of Modern Biology. Harvard University Press, Cambridge, 530p.
Popper, K.R. 1959. The Logic of Scientific Discovery (2004 Reprint), Routledge, London, 513p.
Raman, V.V. 2008. The scientific enterprise 1. Science: some definitions and views. Resonance 13(6): 596–605.
Ross, S. 1962. Scientist: The story of a word. Ann. Sci. 18(2): 65–68.
Salkind, N.J. 2008. Exploring Research (7th Ed.). Prentice Hall, New Jersey, 316p.
Trochim, W.M.K. 2003. Research Methods (2nd Ed.). Bizantra, New Delhi, 363p.
Walliman, N. 2011. Your Research Project: Designing and Planning Your Work (3rd Ed.). Sage Publications, New Delhi, 378p.

Chapter 2
Philosophy of Research

There are in fact two things, science and opinion; the former begets knowledge, the latter ignorance.
Hippocrates (460 BC–377 BC)

Most of the knowledge that we call science has been generated during the past 500 years. It took a long while to determine how the world is better investigated. One earlier method was to just talk about it or argue as old philosophers did. To cite an example, Aristotle, one of the great Greek philosophers, believed that women have fewer teeth than men have, and provided long arguments in support of his belief. If he had bothered to count the teeth of any one of his wives, he would not have committed this 'deductive blunder'! This kind of logic is unreliable. Simple arguments will not prove anything. To ascertain whether a statement is correct, it requires proofs.

We now accept that science follows certain rules and processes. However, it took a while for the universal acceptance of these rules, and people simply relied on other methods for knowing things such as the methods of tenacity, authority, or intuition despite their weaknesses. Presently, there are better approaches to know the universe and its various phenomena, and one approach is to use the 'method of science', that is, do experiments and make careful observations. In earlier times, there were no distinctions between various scientific disciplines as we see now. Everything was 'philosophy'! Even the word 'science' is relatively a new term. However, science has a philosophical nature, and a mission for the 'philosopher of science' is to understand that nature. The nature of scientific investigations and the knowledge generated from such studies are the main concerns. The concepts and techniques that have been developed within philosophy are useful for science too for a better understanding and reasoning.

2.1 Philosophy and Science

Before understanding the philosophy of research or science, one should have some idea about the word '*philosophy*' itself. The word philosophy is difficult to define,

C. G. Thomas, *Research Methodology and Scientific Writing*,
https://doi.org/10.1007/978-3-030-64865-7_2

and it is a 'philosophical question' indeed as what philosophy is! Common people generally use the word 'philosophy' to indicate one's perception or outlook on life, for example: 'my philosophy of life', or basic principles behind something, for example: 'my philosophy of teaching'. In academic circles, however, philosophy is accepted as a distinct discipline just like any other disciplines such as physics or economics. Philosophy is the study of the nature, meaning, and justification of principles or beliefs about various things in the universe. Instead of using empirical methods, philosophers use logic for their studies and present solutions based on arguments. The concepts such as existence, goodness, knowledge, and beauty are considered part of philosophy.

The meaning of the word 'philosophy' is 'love of wisdom'; 'philo' comes from the Greek word '*Philein*', meaning to love and 'sophy' comes from the Greek word '*Sophia'*, meaning wisdom. Earlier, 'sophist' was the word used to denote a 'wise man' or 'thinker', but later, everybody began to use the term 'philosopher' instead of 'sophist'.

In earlier times, the scope of philosophy was all abstract intellectual endeavours including the scientific disciplines such as physics and biology. Some well-known philosophers such as **Aristotle** and **Rene Descartes** made significant contributions to science as well. Similarly, famous scientists such as **Newton** and **Einstein** have also contributed valuable philosophical insights. Even up to early modern times, the present day 'scientists' were referred to as 'natural philosophers', that is, philosophers who study nature; and the common term for science was 'natural philosophy'. You may recall that Isaac Newton's classic work, a landmark in the history of modern science, was titled '*Mathematical Principles of Natural Philosophy*'. Similarly, the highest degree in science in most centres of higher learning is still Ph.D., standing for *Philosophiae Doctoris* (Doctor of Philosophy), meaning 'teacher of philosophy' (the Latin word, *Doctoris* means teacher). During the eighteenth century, the word 'science' began to be used in the sense of any systematic field of study distancing itself from philosophy. After the establishment of the 'British Association for the Advancement of Science' in 1831, the word became more popular. Around that time, **William Whewell**, a British philosopher of science introduced the word 'scientist' to describe a 'practitioner of science'.

Over the years, the scope of philosophy became smaller, as different sciences developed into independent disciplines. Probably, mathematics and physics were the disciplines that parted company much earlier. Biology and chemistry parted ways in the eighteenth and nineteenth centuries. Psychology, economics, sociology, and linguistics were once the fiefdoms of philosophers, but now, they have only a weaker connection with philosophy. When all these disciplines developed on their own strength, 'philosophy' remained as a separate discipline focusing on all abstract, non-experimental, and intellectual endeavours of human beings.

There could be a 'philosophy' of everything taught or studied! For example, *'philosophy of science'* is the branch of philosophy that explores what constitutes science and scientific methods. It is mostly concerned with the motivation for scientific approaches, the means for determining the validity of information, the formulation and use of the scientific method, and the types of reasoning used to arrive at conclusions. Science philosophers use concepts and techniques that have been developed within the main field of philosophy such as empiricism and rationalism. Philosophy of science also includes, as sub-disciplines, the philosophies of scientific disciplines as 'philosophy of physics', 'philosophy of biology', and 'philosophy of sociology'. The purpose of these sub-branches of philosophy is to help interpret the 'philosophical' aspects of research work in these disciplines.

Philosophers of science often distinguish scientific disciplines as 'hard' and 'soft' sciences based on the approaches used in explaining those subjects. For example, in disciplines such as physics and chemistry, it is relatively easy to keep the conditions under control during experiments. On the other hand, in disciplines such as agriculture and medicine where living beings are involved, it is difficult to maintain the conditions consistently when experiments are conducted. In social sciences such as sociology and psychology, it is still more difficult to decide a method for measuring the experimental outcome in an objective way. Therefore, science disciplines such as physics and chemistry are often referred to as 'hard sciences', while others such as sociology and psychology are referred to as 'soft sciences'. Objective measurements are easier in hard sciences but more difficult in soft sciences.

2.2 Epistemology

The term 'epistemology' comes from the Greek word *Epistêmê*, meaning knowledge. In philosophy, *epistemology* is the critical study of knowledge dealing with the origin, nature, scope, and limits of human knowledge. The major concern of epistemology is how we know things, an important issue for researchers. Typical epistemological issues in scientific research are the definition of knowledge, the sources and criteria for knowledge claims, the kinds of knowledge possible, the validity of knowledge claims, and the relationship between the knowledge claims and objects of knowledge. As these issues are involved in epistemology, researchers should be concerned with their knowledge claims. Research makes knowledge claims; therefore, researchers should be prepared to bring out clearly the basis for such claims.

Science uses experience and observation as the basis of knowledge claims. Research that uses experience and observation is using an epistemology called *empiricism* (Sect. 2.3). However, empiricism is not the only way of research. For example, in mathematics, one uses a process of *logical reasoning* usually from axioms. In such disciplines, researchers use an epistemology known as *rationalism* (Sect. 2.4). However, scientists might use more than one epistemology to substantiate their research findings. In other words, logical reasoning or rationalism is also part of research.

The knowledge we acquire may be a priori knowledge, the knowledge known independently of experience. It is non-empirical or arrived at beforehand. A posteriori knowledge is knowledge known by experience; that is, it is empirical or arrived at afterwards. Certain scientific disciplines such as physics treat all knowledge as empirical, while some disciplines such as mathematics and logic are exceptions.

Rationalists suppose that knowledge is primarily attained by a priori processes or is inherent (see also Sect. 1.1).

Some distinction has to be made between '*epistemology*' and '*methodology*'. As far as the philosophy of science is concerned, methodology is also concerned with how we acquire knowledge, but it is much more practical in nature. It pays attention to the particular methods or strategies we use to understand our world better. To put it differently, we can say that epistemology involves the philosophy of how we come to know the world, and methodology involves the practice.

2.3 Empiricism

Empiricism or dependence on evidence is a major principle in the philosophy of science that all knowledge is the result of our experiences (*Empeira* is the Greek word for experience). *Empiricists* believe that truth comes only from direct experience. Justification through evidences is needed to believe something. They contest intuition, rationality, authority, and other such concepts. Empiricists particularly oppose *rationalism*, saying that ideas or thoughts alone cannot be taken for granted without empirical evidences. Empiricism is at the centre of the *scientific method*, which says that acceptance of theories should be based on experiments and observations.

Francis Bacon (1561–1626), an English philosopher and political leader, is considered as the father of 'empiricism' as he argued vehemently for scientific method. Bacon presented his philosophical thoughts through his famous book *Novum Organum* (New Instrument) appeared in 1620. He projected 'induction' as the logic of experimentation and 'deduction' as the logic of argumentation (see Sect. 1.6 and Chap. 14). Bacon deplored 'appeal to authority' and vociferous arguments, and criticized Aristotelian logic prevalent during that time as ineffective for the discovery of new theories and laws. Bacon campaigned for a new scientific method based on inductive generalization from observation and experimentation, and he was the first to formulate rules of inductive inference. Bacon's 'theory of induction' or 'theory of method' was quite different from that of Descartes' 'theory of deduction'. Bacon believed that scientists should observe without any preconceptions, and then, they could generalize based on these observations using inductive logic. In this way, one would finally turn up at the most fundamental and comprehensive laws of nature.

The term '*empirical*' means that the hypotheses formulated are testable using observations or experiments. All the scientific statements such as a hypothesis, theory, law, fact, or principle are derived from our experiences or observations. The word 'empirical' is also used synonymous with 'experimental'. In scientific research, empirical evidence is important because it is the evidence that anybody can experience. It is repeatable, and therefore, empirical evidence can be checked by anybody when knowledge claims are made by an individual. Sometimes, the term '*semi-empirical*' or '*quasi-empirical*' is also used to describe methods that make use of natural observations or previous experimental results to describe aspects of the scientific method, which are not amenable to refutation by experiments.

Scientists collect evidences through various means and take measurements in situations ranging from natural settings such as a forest to completely contrived ones such as a green house or a laboratory. For recording observations, scientists use their own senses, devices that enhance those senses such as microscopes, and certain instruments for noting characteristics quite different from what humans can sense such as electricity. In certain cases such as oceanic phenomena, earthquakes, migration of birds, and plant succession, they need to observe the events passively only. Certain situations, however, demand an active investigation, for example, mining into the earth's crust to study its nature or managing treatments in a field or laboratory.

While employing empirical methods, researchers try to control conditions to obtain better evidences. They may, for example, control temperature and light, change the concentration of chemicals by varying just one condition at a time, and hope to identify its exclusive effects on what happens. However, control of conditions is impossible or impractical in several studies as in studying stars, may be unethical as in studying people, or liable to deform the natural set up as in studying wild animals in captivity. In such situations, observations need to be made from a wide range of naturally occurring conditions. As great importance is placed on evidences, it is important to develop better instruments and techniques for collecting them.

2.4 Rationalism

The practice of logical reasoning is often denoted by the term *rationalism*. According to rationalists, truth can be discovered through reason and rational thought. Logic and mathematics are classical rational disciplines. Although many of us use them regularly, most often, the truth of our assertions is open to question. It is believed that classical Greek philosophers laid the foundations of logical thinking.

According to rationalists, the world is *deterministic* and for every phenomenon, cause and effect relationship is applicable. The notion that the processes of life are not explicable by the laws of natural sciences such as physics and chemistry alone and that life is in part self-determining is called *determinism* or *vitalism*. This idea of determinism originated with the Greeks essentially assumes that the universe is orderly and rational and is governed by predictable laws. They also assume that the cause and effect can be understood through sufficient understanding and thought. A priori (prior to experience) or rational insight is also a source of knowledge. On the other hand, the notion that holds natural processes to be mechanically determined and capable of complete examination by the laws of natural sciences such as physics and chemistry is *mechanism*; for them the world is *mechanistic*.

All religions are deterministic while science is mechanistic.

Rationalism is often compared with empiricism. Broadly speaking, these approaches need not be mutually exclusive, a scientist or philosopher can be a rationalist and empiricist at the same time. Ideas or thoughts come to us through induction and deduction (Sect. 1.6). What is important here is the real source of human knowledge and the most authentic technique to substantiate what we think or what we know.

Among lay people, emotional thinking, hopeful thinking, and wishful thinking are much more common than logical thinking as they are far easier and more amiable to human nature. However, emotion, hope, or wish cannot be evidences; therefore, scientists and critical thinkers should learn to think logically and accept only logical evidences. Children learn to think logically by the traditional methods of reading, writing, and mathematics. Science can be considered as a fourth method of logical thinking. In spite of all these, it is unfortunate that most people never learn to think logically! Because of this mentality, the society accepts many unfounded arguments and statements without posing any questions often leading to disastrous consequences.

2.5 Skepticism

Another important idea in philosophy of science is *skepticism*, the regular habit of questioning beliefs and conclusions. A scientist should also be a skeptic and should question the veracity and reliability of the knowledge claims made by others and the knowledge they already have. The word skepticism comes from the Greek *Skeptesthai,* meaning 'to examine'. In this etymological sense, scientists must be true skeptics and vigilant enough to examine the evidences, arguments, and reasons for their beliefs. Scientists should not succumb to self-deception and deception of themselves by fellow scientists, students, or others. A scientific skeptic always questions the reliability of knowledge claims by subjecting them to a systematic investigation. If we consider the stringency of scientific method, science itself is a structured kind of skepticism! You must also note that a scientific skeptic need not always be a scientist who conducts experiments, but can be anyone who accepts only those claims that are possibly true based on falsifiable hypotheses and testing.

Skeptics consider beliefs cautiously, and they are prepared to change their attitudes taking into account new evidences or valid reasons. It is quite natural that efforts to defy conventional and long accepted notions face strong resistance. However, with the passage of time and the acquisition of new knowledge, older worldviews (paradigms) would often have to be given up. You may be knowing that many theories, once considered scientific truths, were overthrown and became obsolete theories. Although skeptics have open minds, they resist believing something without adequate evidence or reason. Scientists treat new ideas also with the same skepticism. There would be strong opposition against introducing extremely new ideas. We constantly hear many fantastic, bizarre, and outrageous claims pertaining to several fields. Pseudoscientific claims are rampant in many fields such as medicine and agriculture. We must have some method of deciding what to believe weaning away pseudoscientific notions. The scientific method that uses critical thinking is the only method available for this purpose.

Several claims and practices are prevalent in society that scientists must see as unlikely to be true on scientific grounds. Some examples of such claims are health claims surrounding certain foods, certain alternate systems of medicines such as Reiki, existence of supernatural entities such as ghosts, the existence of ESP (extra sensory perception), telekinesis, psychic powers such as clairvoyance, clairaudience, premonition, and telepathy, visits of aliens, UFOs, astrology, creationism, intelligent design, dowsing, conspiracy theories, biodynamic farming, zero budget natural farming, and many others.

2.6 Falsifiability

According to the philosophy of science, *verifiability* means that a statement such as hypothesis, theory, law, or principle is empirically verifiable by conducting experiments or observations. It was the norm in science until **Karl Raimund Popper** (1902–1994) came up with his doctrine of *falsifiability* in 1930s. Many consider Popper as the foremost philosopher of science in the twentieth century. Popper asserted that no hypothesis, theory, or law could be considered scientific, if it does not permit the probability of an opposing or contrary case. Until then, the challenge of separating science from pseudoscience has been an uphill task for the scientists. Popper could not agree to the 'scientific' theories of **Sigmund Freud**, **Alfred Adler**, and many others, especially the manner in which their theories seemed to account for any observation, and thus verified each time! After Karl Popper, 'verifiability' was replaced by 'falsifiability'. The school of thought that highlights the significance of falsifiability as a philosophical principle is *falsificationism* in contrast to *verificationism* prevalent earlier.

The principle of 'falsifiabilty' or 'refutability' states that in order to be scientific, a scientific statement or an assertion (i.e. hypothesis, theory, law, or principle) must be falsifiable or refutable. If we say that some assertion is falsifiable, it does not mean that it should be proved false; rather it means that if the assertion is false, then this could be shown by observation or experiment. It only indicates that the empirical statement must admit logical counterexamples. For example, after observing several crows, we can come to an inductive conclusion that all crows are black! However, the assertion: 'All crows are black', could easily be falsified, if we succeed in observing a coloured crow! As of now, nobody has spotted a coloured crow. However, we cannot just rule out the possibility of finding a red, white, or some other coloured crow in the future! In contrast, the hypothesis: 'There are goblins on earth but they will disappear whenever anyone looks for them', is not falsifiable. These goblins are deliberately made in such a way that no one could ever see them! On the other hand, the statement that, 'There are no goblins on earth', is scientific; we can disprove this statement by presenting a goblin before everybody! Arguments on evil spirits, snowmen, UFOs, astrological predictions, etc., are other examples of 'hypotheses' or 'theories' that cannot be falsified.

Karl Popper rejected classical empiricism and called his approach '*critical rationalism*'. A critical rationalist believes that there is a reality independent of our thinking about it. A 'critical rationalist' recognizes that all observations are fallible and have error; and therefore, all theories are revisable. In other words, the critical rationalist is 'critical' of our ability to know reality with certainty. As all measurements are fallible, the critical realist stresses the need to use *triangulation* by taking observations and measurements through multiple means.

Popper's new approach to science, falsifiability, gained much attention worldwide among philosophers of science. His work, *The Logic of Scientific Discovery* (1959) is a classic in the field. Popper's scientific work was influenced by his study of Albert Einstein's theory of relativity, which he used to demonstrate the difference between a truly scientific theory and a pseudo-scientific theory. Karl Popper believed that scientific theories such as that of Einstein could be easily falsified by simple experiments. Remember that Einstein's theory of relativity made predictions against Newton's theory. According to Einstein's theory, light would be deflected when it passes close to a strong gravitational field. At that time, it was thought that light should be unaffected, but during a total solar eclipse, stars could be seen contrary to expectations, which ought to have been concealed by the sun. These experiments could have created results that challenged the theory of Einstein. Therefore, the theory was, and still is, refutable or falsifiable! According to Popper, this criterion of falsifiability and the practice of using experiments not to verify but to refute scientific theories must be the basis of true science. This was sharply in contrast to the commonly held view at that time that science was based only on empiricism through inductive reasoning and experimental verification. This principle of falsifiability has been accepted for hypothesis formation and testing in modern science.

2.7 The Principle of Parsimony

The *principle of parsimony* or the *Ockham's razor* is another prominent concept in the philosophy of science. **William of Ockham** (1285–1349) was a Franciscan monk and philosopher from Ockham, England. According to him, for explaining various phenomena, one should use the simplest description as possible. The principle affirms that the validation of any phenomenon should make as few assumptions as possible, eliminating those that make no difference in the observable predictions of the hypothesis or theory. In other words, all other things being equal, when multiple theories or explanations are possible, the simplest one is to be preferred. The simplest explanation is usually the best. The principle is highly useful to select the best from hypotheses that predict the same thing.

The principle of parsimony is more often taken today as a rule of thumb that directs economy, parsimony, or simplicity in scientific theories. Parsimony is also a factor in statistics and models. Often, mathematical models with the lowest number of parameters are favoured, as each parameter introduced into the model adds some uncertainty to it.

In science, the principle of parsimony is generally used for evaluating hypotheses and selecting the best possible ones. When competing hypotheses are identical in other respects, the principle recommends selection of the hypothesis with the least assumptions and postulates for existing data. Scientists are supposed to choose the simplest explanation consistent with the data available at a given time. However, bear in mind that these simplest explanations often succumb to complications, as new data are made available.

2.8 Pseudoscience

Along with the rise in popularity of science, pseudoscience and pseudoscientific claims are also on the rise. Gullible people often succumb to such claims. The meaning of 'pseudo' is false. A body of knowledge claiming to be both factual and scientific but lacking empirical evidence, consistency with existing well-established scientific theories and laws, experimental accessibility, and falsifiability can be considered as pseudoscience. Pseudoscience does not adhere to the philosophy and methods of science. However, by frequently using many scientific icons such as scientific terms, descriptions, and designations, the proponents of pseudoscientific ideas try to give their false claims an aura of true science. Some people, knowingly or unknowingly, promote pseudoscience because of several reasons. It may be because of ignorance about the nature of science and the scientific method or intentional fraud for financial or some other benefits. You may find many types of pseudo-scientific theories claimed to have the backing of apparent evidences, but when scrutinized closely, they are just statements of faith! Astrology is a typical example.

The significant difference between scientific theories and pseudoscientific theories is that the former theories are treated as falsifiable whereas the latter are immunized against falsification by various strategies designed to make them immune to refutation. For example, astrological predictions are so vaguely expressed that whatever happens could be interpreted as confirming them! However, according to the principle of falsifiability, repeated confirmation of a particular theory does not guarantee that it is a good theory. It may be merely reflecting the ingenuity of the scientist or persons concerned in finding appropriate ad hoc hypotheses (Sect. 1.7) using vague concepts or reinterpreting concepts arbitrarily according to the situations.

The boundary between pseudoscience and real science is often thin and vague. If we can test the claim and subject it to scrutiny, it may be real science. However, if the assertion is at variance with established theories or research results, it can be considered as pseudoscience. Some examples of pseudoscience are astrology, dowsing, and biodynamic farming. On the other hand, the fields such as 'Acupuncture' may be categorized as ***protoscience*** instead of branding it as pseudoscience. Many expect that when these fields are experimentally examined, the results unequivocally prove the supremacy of science! Similar is the case with Homeopathy or Ayurveda, as many of the principles underlying the above systems of medicine do not obey the principle of falsifiability.

Any hypothesis, theory, principle, law, or doctrine must follow the principle of falsifiability and the principle of parsimony to be accepted as scientific. Otherwise, these must be categorized as pseudoscience. Similarly, if the assertion is without supporting evidences, contradicts with experimentally established facts, or fails to provide an experimental possibility of reproducible results, there is no other option but to dismiss it as pseudoscience.

Understand that certain systems of thought that rely on 'divine' or 'inspired' knowledge, for example: revelation, theology, or spirituality, are not considered pseudoscience or science. They are simply faiths or beliefs. Most scientists are also believers in some faith or religion. However, if somebody claims that some religious practice or belief is scientific, it is a serious matter, for it should also obey the above rules including falsifiability.

2.9 Reductionism

Reductionism means a bottom up approach to understand complex natural phenomena by 'reducing' them to their fundamental parts and studying their interactions under controlled conditions. Scientific reductionism is most often used to describe the notion that everything in the universe can be broken into smaller and smaller parts and that the whole can be understood in this way. This includes many things such as objects, theories, phenomena, and explanations. The idea of reductionism was first introduced by **Rene Descartes** (hence, sometimes also referred as *Cartesian reductionism*) in his famous book *Discourse on the Method of Rightly Conducting One's Reason and of Seeking Truth in the Sciences* (1637). Reductionist approaches help us to break complex systems down into their components, and each piece can be studied individually by way of disciplinary and sub-disciplinary approaches. If we know the parts, the dynamics of the whole system can be derived. Reduction as a method is highly successful in disciplines such as physics, chemistry, and biology. However, a major problem with this approach is that scientists often fail to revert back to see how their findings fit in with totality.

Probably, **Fritjof Capra** might be the most prominent and vocal critic of Descartes' reductionist views (Capra 1975). Influenced by Fritjof Capra, and probably due to some confusion over the philosophy of science, many critics among environmentalists, anti-imperialist movements, and feminists have started criticizing reductionism (e.g., Shiva 1988; 1991). According to 'deep ecologists' and 'eco-feminists', reductionism is opening the way to merciless exploitation of nature. This is not a correct interpretation as reductionist thinking and methods are the bases for many of the well-developed areas of modern science. As explained earlier, reductionism seeks the explanation of the whole by looking at the relationships between the component parts that can be experimentally tested, thus eliminating the requirement of assumptions on extra forces such as consciousness, vital force, and others (Nanda 1997, 2004). A significant degree of reductionism is required for the growth of medical and agricultural sciences, otherwise, it is impossible to determine significant versus non-significant measurements.

The rejection of reductionist ideas is often called *holism*. Holism asserts that things have certain properties as a whole, which cannot be explained based on its components parts. The word 'holism' is from Greek, meaning *all*, *whole*, *entire*, or *total*. A key principle of holism is: 'The whole is different from the sum of its parts'. Contrary to reductionism, holism inquires about a whole system by probing the system as a whole in its entirety instead of dismantling it and studying the parts.

The supporters of holism assert that the properties of a given system such as physical, biological, chemical, social, and economic cannot be explained based on its component parts alone. For them, the system as a whole determines how the parts behave. Although the idea of holism has ancient roots, introduction of the term 'holism' is credited to the South African diplomat Jan Smuts. He made a mention of holism in his 1926 book, *Holism and Evolution* (Smuts 1926). The term *holistic* evolved into the term *wholistic* in the late 1990s to clarify the concept even further. Schools with wholistic learning styles and wholistic medicine that considers the mind, body, and spirit in diagnosis and treatment are examples. You may also find wholistic models in other fields such as agriculture. However, most of the claims of supporters of 'only wholism' are unsubstantiated. Post-modernists reject reductionism and embrace wholism. Many of them do not have much faith in replicated experiments following the principles of experimentation as they simply reject reductionism. People who are opposed to science often use the words 'reductionism' and 'reductionist' to condemn whatever they dislike about modern science. This is the main tradeoff between public funded research and activist run research.

In fact, science requires both reductionism and holism. We usually dismantle complex systems into various component parts employing reductionist approaches to get first hand information about the system. In most cases, this also allows scientists to put the pieces together again by way of holism. Take the simple example of sugar. If it is reduced to atoms of carbon, hydrogen, and oxygen, it may not have the qualities of sugar. However, you can have original qualities of sugar up to the molecular level. Another important case is living beings. You may require a reductionist approach for studying about various organs and the functioning of living beings. However, unlike machines, you cannot dismantle a living being and re-assemble it into its original living state again based on our present level of knowledge!

Reforms in education call for inter-disciplinary convergence, and for bringing together the multiple specialties of a particular discipline, intra-disciplinary convergence. In fact, reductionism focuses on the properties of parts; and holism, on the relationship between them. Put together, reductionism and holism stand out as supplementary rather than conflicting ideas. However, the sad part of the whole issue is that inter-disciplinary convergence is impossible in the present disciplinary organization of academic institutions in India. It is now a fashion to form new watertight departments, faculties, and even universities based on false claims. This kind of *greedy reductionism* is not good for the growth of science. Arguably, some situations call for holistic thinking, but those requiring a closer look at the parts of the system are better suited for reductionism. For example, watershed management is an area where both reductionist and holistic approaches are employed. We approach watershed management holistically in a multidisciplinary platform, but when going for individual components, we also employ some amount of reductionism.

2.10 Paradigm and Paradigm Shift

The word *paradigm* has originated from the Greek word *Paradeigmia* meaning pattern or example. Paradigm is used in the sense of a 'world view', a way of looking at the world. A paradigm is essentially our perspective or outlook on a situation. **Thomas Samuel Kuhn** (1922–1996), a well-known science philosopher of the twentieth century, used the word 'paradigm' to describe a set of practices in science. He wrote eloquently on the history of science and developed important ideas and postulates in the philosophy of science.

Kuhn postulated that the practice of science go through three phases. The first phase, which undergoes only once, is the *pre-scientific phase*, wherein there is no consensus on any competing theories. During this phase, several incompatible and incomplete theories may surface. Subsequently, one of these theories succeeds leading to the second phase, the *normal science*. A scientist working within this phase behaves according to the dominant theory. Kuhn called the dominant theory, which other scientists begin to follow as a *paradigm*. Scientists try to elaborate, expand, and further justify the paradigm within this phase of normal science. However, in course of time, problems may surface calling for amendment to the theory incorporating accumulated experimental evidences, which might seem to contradict the original theory. When the situation worsens as the current explanatory theory fails to explain certain phenomenon, someone may put forward a complete replacement or redefinition of the theory. According to Kuhn, this is a *paradigm shift*, which brings in the third phase, the *revolutionary science*. Kuhn says that all scientific disciplines go through these paradigm shifts several times, as new theories replace the old ones.

Before Copernicus, almost all were accepting the long held geocentric theory of Ptolemy that the sun revolves around the earth. However, Copernicus challenged this concept and came up with the revolutionary heliocentric theory that the earth revolves around the sun and not the other way. Despite oppositions, slowly, people began to accept this new theory. In fact, Copernicus' theory ushered in a new period of 'normal science'. Although some modifications were added by Kepler and Newton, they adhered to the new paradigm. The latest examples of refinements to the paradigm are the acceptance of Einstein's general relativity to replace Newton's account of gravity in the 1920–30s and Suess and Wegener's plate tectonics in the 1960s.

Thomas Kuhn says that a scientific revolution takes place when research workers stumble upon certain anomalies, which are not explainable based on the current paradigm within which they work. When new findings are made, which cannot be reconciled with the existing paradigm; and when these findings are confirmed repeatedly and independently by other scientists, the scientific community develop a new paradigm in line with the new evidences. This is where religion and science differs. The most important merit of scientific method is that those who swear in its name are prepared to change their stand based on new findings and genuine explanations. On the other hand, in religion, those who cling to the existing beliefs are not willing to change, and they will not allow the belief system amenable to logic or questioning.

Paradigm shifts have some common characteristics. In the modern world, many things are changing continuously, and everybody has to adjust to that change, signifying that paradigm shifts are essential part of life. At times, paradigm shifts can create problems too. The society needs stability for survival, and continuous shifts in our paradigms would make our lives very difficult. We cannot abandon a paradigm until we have one to put in its place. This is mainly because our paradigm is what allows us to function and it is very difficult to work without a paradigm. Similarly, a long period is often needed to effect a paradigm shift. Paradigm shifts do not happen instantly but requires years or even decades to bring about a new paradigm.

Due to the unrelenting pursuits of adventurous scientists, paradigm shifts continue to occur in the field of science. The Renaissance period (fourteenth to seventeenth century) of European history saw many unique paradigm shifts. A few renowned examples, which forced the human kind to think differently, are noted below:

- 'Heliocentric theory' (earth revolves around the sun and the sun is at the centre of the planetary system): It replaced the 'geocentric theory' of Ptolemy and paved the way for a chain reaction in many fields.
- 'Germ theory' by Pasteur (diseases are caused by germs): It replaced the 'spontaneous generation theory' accepted as immutable for a long period.
- 'Theory of evolution' by Charles Darwin: It replaced 'creationism' (God created life in the form we now find) bringing in a chain reaction in biology and religion.
- 'Theory of general relativity' by Albert Einstein: This brought down previous ideas of space and time.

Before concluding this section, a warning about the general use of the terms 'paradigm' and 'paradigm shift'! These are two widely abused terms in English, which has attained the status of clichés. Because of the indiscriminate usage of the terms to represent any type of change, the charm and punch of the words have lost. The phrase 'paradigm shift' has escaped from the Kuhnian context, and many simply use it in a general sense for a radical change in personal beliefs, complex systems, or organizations. Avoid using these terms in such out of context situations.

2.11 Post-modernism

Modernism and *post-modernism* are western phenomena related to philosophical, economic, scientific, and social facets of life emerged in the seventeenth to eighteenth centuries. 'Enlightenment' is at the core of modernism, a general attitude advanced during the renaissance movement, of attaining truth, justice, and happiness through knowledge and reason. Science emerged from the older philosophy to the present modern science because of modernism. Modern movements occurred in arts, architecture, sculpture, painting, and almost all walks of life. However, some intellectuals criticized this movement; for them, modernism was excessively restrictive. As a reactionary movement, post-modernism emerged in the 1960s embracing freedom and diversity of styles with the objective of rejecting and superseding 'enlightenment' and 'modernism'.

There are many followers for post modernism in a wide variety of disciplines or fields such as arts, architecture, music, film, literature, sociology, agriculture, health, communication, fashion, and technology. The term 'post-modernism' is applied to several movements in the above disciplines that reacted against modern movements, typically marked by revival of traditional elements and techniques.

Post-modern thinking can be found in science too. Post-modernists often blame modern science for all the environmental problems of the world. They campaign for traditional knowledge and local practices, most of which are not led by rational and scientific criteria. Post-modernists assert that modern science is just one culture-bound way of looking at nature! According to them, the content of all knowledge is socially constructed; even the 'facts' and 'laws' of modern science are Western constructions reflecting dominant interests and cultural biases of Western societies. In a post-modernist's world, truth becomes relative, and often knowledge is replaced by interpretation.

Many prominent intellectuals in India have also jumped on to the bandwagon, and started arguing for developing 'local science' or suggesting 'alternative sciences'! This perspective has numerous sympathizers among the environmental and feminist movements. Their criticism is not limited to uses or abuses of science; they question the very basic content and methodology of science, especially 'reductionism'. It is unfortunate that post-modernist attacks on science is a blessing for all religions as they are relieved of the demand to review their philosophies in tune with the progresses made in understanding the universe.

Post-modern ideas can be seen in almost all spheres of human activity. **Ivan Illich** questioned the role of educational system with his deschooling theory (*Deschooling Society*, 1971). He also called for de-medicalization (*Limits to Medicine: Medical Nemesis*, 1976) questioning the role of the medical system and medicines. The book by **Fritjof Capra**, '*The Tao of Physics: An Exploration of the Parallels between Modern Physics and Eastern Mysticism*', first published in 1975 was a landmark in the whole history of post-modernism. In this book, Capra makes an assertion that both physics and metaphysics are leading to the same system of knowledge! Capra exhorted the society to abandon conventional linear thoughts and the mechanistic view of Descartes. He was highly critical of Descartes' reductionism. There were many takers for the interpretation of modern physics and metaphysics by Capra, but these are still outside the realm of mainstream science.

Post-modernism related to agriculture actually began with the publication of the book '*Silent Spring*' by **Rachel Carson** (Carson 1962). In the 1970s, philosophical ideas such as lifeboat ethics (Hardin 1974), deep ecology (Næss 1989), bioregionalism (Berg and Dasmann 1977) and ecofeminism (Mies and Shiva 1993) influenced alternate farming movements such as natural farming, organic farming, and several of their variants. Initially, **James Lovelock**'s Gaia Hypothesis also contributed much (Lovelock 1979). Although these global movements emerged as holistic ideas concerned with pollution and the environment, eventually, these paved way for the consolidation of green politics and activist campaigns against modernism.

Taking a cue from post-modernism, some so-called environmentalists in India extended their criticisms to modern agricultural movements such as green revolution, and came up with 'alternate' models based on post-modern thoughts such as deep ecology. Most post-modern followers insist on accepting and promoting ethnic or traditional practices in agriculture without subjecting them to empirical tests. Although the post-modern ideas emanated from India in agriculture are shared and promoted in the West as radical and revolutionary, authors like Nanda (1997, 2004) and DeGregori (2004) came out with sharp criticisms for such anti-science rhetoric. They say, for example, the call of post-modern intellectuals to return to traditional agriculture might help the rich land-owning farmers, but devastate the lives of the peasants and the poor.

Post-modern ideas are creating much confusion among intellectuals and common people alike. The problem is whether we can have a post-modern outlook on natural sciences with 'alternate sciences' based on traditions or simple naivety as happened in other fields such as art and literature. As Nanda (2004) remarked, while there could be different styles in art, literature, mythology, architecture, and the like, science cannot have different styles, and the criteria of scientific inquiry is applicable uniformly to all, cutting across national and cultural differences. Science is universal, and by following the much-acclaimed scientific method, it allows everyone to learn more about the universe. It also corrects itself through socially established ways of subjecting existing knowledge to empirical tests. As post-modernism does not permit to have a universalistic view of science, it should not have any role in science!

Bibliography

Berg, P. and Dasmann, R. 1977. Reinhabiting California. The Ecologist 7(10): 399–401.

Best, J. W. and Kahn, J.V. 1993. Research in Education (7th Ed., Indian Reprint, 2004). Prentice–Hall of India, New Delhi, 435p.

Bird, A. 1998. Philosophy of Science, UCL Press, London, 313p.

Capra, F. 1975. The Tao of Physics: An Exploration of the Parallels Between Modern Physics and Eastern Mysticism. Shambhala Publications, Berkely, California, 330p.

Carson, R. 1962. Silent Spring. Houghton Mifflin, Boston, 368p.

DeGregori, T.R. 2004. Origins of the Organic Debate. Iowa State University Press, Ames, 211p.

Gower, B. 1997. Scientific Method: An Historical and Philosophical Introduction, Routledge, London, 276p.

Hardin, G. 1974. Living on a lifeboat. BioScience 24 (10): 561–568.

Jarrard, R.D. 2001. Scientific Methods [E-book]. http://www.emotionalcompetency. com/sci/sm_all.pdf. 236p. [Accessed 10 June 2018].

Kerlinger, F.N. 1986. Foundations of Behavioural Research (3rd Ed.). Holt, Rinehart and Winston, USA, 667p.

Kuhn, T. 1962.The Structure of Scientific Revolutions, Chicago University Press, 173p.

Losee, J. 2001. A Historical Introduction to the Philosophy of Science (4th Ed.), Oxford University Press, New York, 328p.
Lovelock, J. E.1979. Gaia: A New Look at Life on Earth. Oxford University Press, New York, 157p.
Mies, M. and Shiva, V. 1993. Ecofeminism. Zed Books, London, 336p.
Næss, A. 1989. Ecology, Community and Lifestyle: Outline of an Ecosophy. Cambridge University Press, 233p.
Nanda, M. 1997. History is what hurts: A materialist feminist perspective on the green revolution and its ecofeminist critics. In: Hennessy, R. and Ingraham, C. (eds.), Materialist Feminism: A Reader in Class, Difference, and Women's Lives. Routledge, New York, pp. 364–394.
Nanda, M. 2004. Prophets Facing Backward: Postmodernism, Science, and Hindu Nationalism. Permanent Black, New Delhi, 309p.
O'Hear, A. 1989. An Introduction to the Philosophy of Science. Oxford University Press, New York, 239p.
Popper, K.R. 1959. The Logic of Scientific Discovery (2004 Reprint). Routledge, London, 513 p.
Ross, S. 1962. Scientist: the story of a word. Ann. Sci. 18(2): 65–68.
Shiva, V. 1988. Staying Alive: Women Ecology and Survival in India. Kali for Women, New Delhi, 215p.
Shiva, V. 1991. The Violence of the Green Revolution: Third World Agriculture, Ecology, and Politics. Zed Books, London, 265p.
Smuts, J.C. 1926. Holism and Evolution. The Macmillan Company, 362p.
Walliman, N. 2011. Your Research Project: Designing and Planning Your Work (3rd Ed.). Sage Publications, New Delhi, 378p.

Chapter 3
Approaches to Research

None of my inventions came by accident. I see a worthwhile need to be met and I make trial after trial until it comes. What it boils down to is 1 per cent inspiration and 99 per cent perspiration.
Thomas Alva Edison (1847–1931)

New knowledge arises in many ways. Many of the earlier findings made by our ancestors were accidental or by trial and error. Chance discoveries played a crucial role in the development of science and technology in prehistoric times. Primitive humans learned many things through observation. Observation of wild fire and the consequent making of fire, domestication of animals and plants, and use of cooked food are typical examples. The life of primitive humans was centred on hunting, fishing, and gathering food from nature. Because greater part of the day was spent in search of food, they could not devote much time for intellectual or cultural pursuits. However, after the origin of agriculture, they were freed of the burden of finding food throughout the day, and slowly began to dabble in leisure activities, some involving intellectual efforts. The transition from shifting cultivation to settled agriculture further increased time for leisure, and the history tells us that humans created knowledge through the processes of both deduction and induction (Sect. 1.6). In earlier days, there were no organized research activities as we see now. In fact, the method of induction was accepted as a tool in science only after Francis Bacon (1561–1626).

Presently, scientists employ many well-established research methods to study the unknown. Intuition and chance discoveries also play a part. In agriculture, research seeks answers to key questions, which could lead to significant improvements in existing agricultural practices and finally production. Research in any scientific discipline is done with similar motives. In this chapter, an overview of various research methods relevant to science and technology is attempted.

C. G. Thomas, *Research Methodology and Scientific Writing*,
https://doi.org/10.1007/978-3-030-64865-7_3

3.1 Intuition in Research

In Chap. 1, we discussed the importance of intuition as one of the methods of acquiring knowledge. Sometimes, new ideas originate unexpectedly from an observation or incident or while doing an experiment. This talent of individuals, the sudden stimulation to problem-solving by a new perspective, is usually denoted by the term *intuition*. Intuition is also known by terms such as insight, revelation, inspiration, enlightenment, and sudden comprehension.

Traditionally, intuition is considered as a method of acquiring knowledge, wherein, reason is considered as the criterion of truth rather than experience. It is also called a priori method of learning things, as reasoning is done from what is prior or before. A priori knowledge is considered true, as it is obtained through the method of deduction. An idea rises suddenly into consciousness as a solution to a problem in which one particular individual is intensely interested, something like a premonition or 'sixth sense'. A typical feature of intuition is the thrill or ecstasy it generates. For example, Archimedes could not control his excitement when he found the answer to the puzzle of specific gravity while he was in the bathtub; he screamed, 'Eureka'!

Scientists often make use of intuitive ideas. However, the usefulness of insight may be varied; we cannot simply dismiss it as irrational. For example, some researchers get ideas instantly and conduct experiments to test their ideas. Some others try to gather data from various sources until an answer emerges. In practice, intuitive ideas most often occur to those individuals who are specialists in specific areas than generalists. However, bear in mind that hard work has no substitutes as Thomas Edison rightly remarked, 'None of my inventions came by accident. I see a worthwhile need to be met and I make trial after trial until it comes. What it boils down to is *one per cent inspiration and ninety-nine per cent perspiration*'. Thus originated the popular saying attributed to Edison, 'Science is 99 percent perspiration and 1 percent inspiration'.

Favourable Conditions for Intuition

Insight or intuition may not happen always. Certain circumstances favour achievement of insight faster. The following five conditions are conducive for intuition (Jarrard 2001).

Define the Problems Correctly

You should have clarity in identifying the problem. If the problem is clear to you, the chances of success through intuition are better. Sometimes, describing the problem to others may help, because it forces the researcher to define the problem in simple terms.

Accumulate All the Needed Facts

Collect all the relevant information related to the problem, and try to understand the information so collected with a critical mind. Often, conducting a literature search may prompt insights, because it compels us to go through the data and its interpretation in a logical sequence.

Desire a Solution

Sometimes, we may have a personal stake in a problem, which can help or hinder intuition. An interest in solving the problem can promote concentration and imagination. The interest pulls us to pursue a task and persist with it despite difficulties and disappointments. At the same time, intense desire for a solution can be counterproductive, if it leads to disturbing worry and anxiety.

Discuss with Others

Discussing the ideas with others may help in intuitive ideas. As they have different perspectives, we may benefit from those perspectives directly, or we may be able to combine our knowledge with theirs. When carried out in a relaxed and friendly manner, these discussions are most likely to encourage insight.

Temporary Abandonment of the Problem

Another option is to simply relax and abandon the work temporarily whenever you encounter anomalous data in your experiment or observation. These can create a favourable environment for the insight to happen.

Barriers to Intuition

Intuition must happen in a free flowing atmosphere. Some factors can act as barriers to intuition. Sometimes, we may not be able to recognize the impediments. The following are some obvious barriers, which prevent intuition or insight to happen in a free and uninterrupted atmosphere (Jarrard 2001).

Distractions

Both pleasant and unpleasant distractions can be unfavourable to insight. For example, unpleasant distractions such as domestic worries, anxiety, and fatigue destroy the receptivity needed for insight. Similarly, pleasant distractions such as excitement or preoccupation with something other than the instant problem can also be inimical to insight. A distraction-free environment is necessary not only for insight but also for the success of research as a whole.

Interruptions

Probably, interruptions are the most troublesome type of barriers to intuition. Total immersion in the problem is essential. Try to build up an interruption-free environment in the work place and the laboratory.

Conditioned Thinking

Most scientists always think in a particular direction; that is, they become conditioned, and may not notice new developments contrary to their way of thinking. Conditioned thinking is a major problem of established scientists. Often, this can prevent a person from adopting a new perspective that may be needed to solve a problem. There are some considered methods to break free from conditioned thinking (Beveridge 1957).

Take a temporary break: Often, taking a temporary break from the problem and resuming it after a while helps. Temporary abandonment of the problem allows the established notion to weaken, probably permitting a new way of thinking.

Consult others: It is a good idea to discuss the issue with your colleagues and friends. Discussing the riddle with others can provide a new perspective. Sometimes, they can provide you the needed insight!

Read relevant research articles: Reading relevant research articles related to the problem and gathering information from different sources help you to free from conditioned thinking.

3.2 Unconventional Ways to Intuition

Intuition or insight can happen in several unconventional ways too. The usefulness of these unconventional sources may vary depending upon the situations. Serendipity, unexpected results, borrowing from other disciplines, and contribution of amateurs are some unconventional alternatives to traditional insight.

Serendipity

Sometimes, researchers make some wonderful discoveries or inventions simply by accident or luck. The term *serendipity* is often used to denote chance discoveries, especially while looking for something else. The term is actually derived from the title of a fairy tale '*The Three Princes of Serendip*', the heroes of which were always making wonderful discoveries. Probably, pharmacology and chemistry are the two major fields where serendipity is more common. Chance discoveries can play a prominent role in research. Chemical synthesis of urea in 1828 by Friedrich Woehler is an example. He was trying to produce ammonium cyanate by mixing potassium cyanate and ammonium chloride. In the process, he obtained urea, the first organic chemical compound to be synthesized! Another celebrated example is the discovery of X-rays by Wilhelm Roentgen. In 1895, Roentgen was doing some experiments on the effects of cathode rays at the University of Wurzburg. He noted that certain rays were emitted when a current was passed through a covered discharge tube, which illuminated a barium-covered screen. He used photographic plate to capture the image of various objects placed in the path of the rays. Through further experiments, he confirmed that X-rays were produced by the impact of cathode rays on objects. Roentgen named these rays, 'X-rays' because of their unknown properties. In 1901, Roentgen received the Nobel Prize for the accidental discovery of X-rays.

Serendipity is only a chance, and you cannot rely on it alone for finding answers to scientific questions or for solving problems. For this, you have to undertake research in an established way. One can seek insight but cannot seek chance discoveries! Be alert for any unexpected results, but resist the temptation to rationalize away or discard them. However, one requires considerable background knowledge to identify the unexpected result and to appraise its significance properly. Probably, this is the reason why most of the accidental discoveries occurred in the field of specialization of scientists who made them.

Unexpected Results

Unexpected results including serendipity can happen in many ways. Austin (2003) described four types of chances, from Chance I to Chance IV. Although these four kinds of chances are closely related to each other, they are extremely different in the type of intelligence, receptivity, and participation involved.

Chance I: Usual serendipitous discoverers by accident are examples of Chance I type. An example is Roentgen's discovery of X-rays mentioned already. This is a case of sheer luck that may provide a great opportunity to anyone to continue research. In life sciences, the unexpected finding of a plant, insect, or microorganism may open up great opportunities. It is easy to identify the discovery when it happens but you cannot do anything to increase its incidence! In the examples mentioned, chance occurred naturally as a matter of probability without anybody's intervention.

Chance II: This kind of chance favours those who have a persistent curiosity about many things coupled with an enthusiasm to experiment and explore. If you are busy doing something, the chances of luck are greater. Heath (1985) narrated the interesting case of Paul Ehrlich. Ehrlich believed in the possibility of a chemical cure for syphilis and kept on testing compounds. Finally, he succeeded with Salvarsan, the 606th compound he tested!

Chance III: In this type, chance presents only a slight indication, and ordinary people may overlook the potential opportunities. However, a researcher who has sufficient background knowledge with a prepared mind may grasp the significance and continue with it. For example, when Alexander Fleming in 1929 noticed that a strange fungus was growing on his culture dish, he isolated and purified it, leading to the chance discovery of penicillin, the first antibiotic. Its practical significance, however, was revealed only in 1939 by the works of Florey and Chain.

Chance IV: In this case, the researcher may try ideas not in conformity with established thinking. It is often due to a combination of determination and imaginative thinking. Chance IV type cannot be considered as mere luck but good intuition.

Among the four chances discussed above, the first three–characterized by luck, exploration, and sagacity–are variations of serendipity. Although chance I and II depend upon accidental discoveries by mere luck, chance III is not passive; it comes only to those who think and act. Unexpected results can foster insight and breakthroughs, but most often we encounter disappointments. It is easy to become distracted and dragged in a new direction by an unexpected result, and the whole effort would be wasted. Do not forget what Louis Pasteur (1822–1895) said, 'In the field of experimentation, chance favours only the prepared mind'.

Borrowing from Other Disciplines

In certain cases, borrowing from another field or discipline is effective in achieving the needed insight. You can easily extrapolate and try the principle, technique, relationship, or equipment from one discipline to another discipline. Often, many inventions and discoveries are motivated by the sudden change in our competence to make a new type of measurement or to make a more accurate measurement than before. For example, the inventors of the electron microscope, laser, CT scan, and many others could never have imagined the fields that these devices would be used. In most cases, the person who is familiar with a problem recognizes the applicability of a new instrument or technique developed by scientists in some other disciplines.

It is true that science is developing at an unbridled speed, becoming more and more specialized, but most researchers are seldom aware of technical or conceptual developments in other disciplines or specializations. Consequently, a potential application may go unnoticed for many years. In fact, popular magazines and newspapers carry news of articles of such breakthroughs for experts in unrelated fields (only if they read them!). However, the problem is that not all the new scientific advances may be recognized by the media. Interaction with experts from other fields and attending their conferences will be of great help in this regard.

Contributions of Amateurs

There were many breakthroughs brought out by amateurs or those with scanty experience of the relevant evidence. In certain fields, you will see extraordinary contributions from amateurs. As pointed out by Jarrard (2001), the 'amateur' need not be a newcomer to science, but might be an experienced scientist who has just changed research areas. The new entrant may bring a technique or concept from the original discipline to a different field. A similar phenomenon is breakthroughs made by young scientists. Most revolutionary breakthroughs in science are led by the younger generation. Although the youngsters may have more energy than the older generation, they may have less efficiency and less knowledge. The higher discovery rate among the newcomers is because of their greater flexibility of thought devoid of conditioned thinking and assumptions. What is needed is a little grooming by the elders.

3.3 Variables in Research

Research is done in different ways depending upon the characters involved and the purpose. In experimental research, the researcher manipulates a character under highly controlled conditions to see if that causes any changes in a second character. Often, the investigator directly manipulates a character or some characters keeping all other characters constant as far as possible. The way to establish a cause and effect relationship is to isolate and eliminate all the factors that might be responsible for a particular outcome and test only one factor that the researcher directly want to measure. The particular factor the researcher wants to test is a ***variable***. Then, the question arises; what is a variable in research? Simply speaking, anything that can vary is a variable. For research purpose, *a variable can be defined as a character, condition, or concept that can take different values, and hence measurable.* In agriculture, cultivars of crops, yield, breeds of animals, fertilizers, pesticides, etc. are variables as they can take different values. In health and social science research, race, gender, age, agreement, attitude, programme, treatment, medicines, diets, blood pressure, body weight, etc. are examples of variables. Variables are important in both experimental and non-experimental research.

Often, variables are classified as *dependent, independent,* and *extraneous variables* based on the causal relationships between them.

Independent and Dependent Variables

Independent and dependent variables are important when exploring cause and effect relationships. In experimental research, the *independent variable,* also known as experimental variable or predictor variable is the one manipulated by the investigator, whereas the *dependent variable* or outcome variable is measured for reaction due to the independent variable, which may be a treatment, programme, or cause. In an experiment, an independent variable would be the apparent cause and the dependent variable the supposed effect. When we manipulate independent variables, some changes occur in characters such as yield, attitude, or blood pressure, which are our dependent variables. For example, crop yield (an effect, which is a dependent variable) depends on many factors or causes such as rainfall, irrigation, and agronomic practices (independent variables responsible for the effect).

When you conduct research, you may find that many dependent variables are easy to observe, but for some, special equipment is needed for correctly measuring them. For example, it is very easy to measure the height of crop plants, but measuring photosynthesis is difficult, and you need special apparatus and cumbersome procedures. Therefore, in certain situations, you may try to find out a relation between the characters by predicting the difficult to observe variables from easily observable ones through regression and model building.

In certain studies, researchers may not be able to control and manipulate an independent variable. For example, the variables such as gender, age, education, body weight, and ethnicity cannot be manipulated, which are the *status variables* or *attribute variables*, reflecting the characteristics or status of the study population. These attributes have already been determined and cannot be altered. Although researchers are unable to control or manipulate status variables, they can treat them as independent variables and study their effects using appropriate research designs. At the same time, those independent variables that can be manipulated, changed, or controlled are called *active variables* or *treatment variables.*

An *attribute* is a specific value on a variable. For example, the variable gender has 2 attributes, male and female; and a common variable in social research, 'agreement' might be defined to have 5 attributes, 1 = strongly disagree; 2 = disagree; 3 = neutral; 4 = agree; and 5 = strongly agree.

Independent variables may also differ depending upon their nature of values. They can take *numerical values* or *text values.*

Examples of variables with numerical values:

Fertilizer doses (kg/ha)	0, 60, 90, 120
Seed rate (kg/ha)	100, 150, 200

Examples of variables with text values:

Gender	Male, female
Crops	Wheat, rice, sorghum, and maize
Cultivars of rice	Jaya, Jyothi, and Uma

Qualitative and Quantitative Variables
When we consider the nature of variables, we may sometimes come across attributes that cannot be measured. For example, qualitative characters such as religion, colour, and gender cannot be measured. These are the *qualitative variables*, also known as *categorical variables*. At the same time, *quantitative* variables are measurable, for example, height, weight, and length. However, certain qualitative variables can act as quantitative variables too. For example, although seed dormancy is a qualitative character, if we take the number of days taken for germination after the dormancy is broken, it becomes a quantitative character.

For the purpose of measurements, qualitative variables are categorized as *nominal* and *ordinal* variables, while quantitative variables are classified as continuous and discrete variables.

Nominal and Ordinal Variables
A *nominal variable* may have one, two, or more sub-categories depending upon the extent of variation; for example, the variable 'educational status' can have the sub-categories as non-literate, literate, matriculate, graduate, postgraduate, and doctorate; and the variable 'gender' can have males and females as subcategories. On the other hand, *ordinal variables* have no absolute values, but these can be arranged in either ascending or descending order according to the magnitude of variations. As ordinal variables have no absolute values, the real differences between adjacent ranks may not be equal. Ranking spaces them equally, although they may not actually be equally spaced. For example, the variable 'agreement' can have five subcategories as 1-strongly disagree, 2-disagree, 3-neutral, 4-agree, and 5-strongly agree.

Continuous and Discrete Variables
Quantitative approaches to research involve the study of continuous variables or discreet variables. In such studies, numerical data are collected to explain, predict, or control the phenomena of interest. *Continuous variables* have continuity in their measurement, and therefore, take any value. For example, take the case of 'plant height'. It may range from 10 to 11 cm as 10.01, 10.1, 10.312, 10.5, 10.88, 10.9, and 10.99: However, *discrete variables* take only discrete or pre-assigned values. For example, the 'size of a family' can take only integer values such as 3, 4, 5, and 6.

The nature of characters specifies whether the character is continuous or discrete. Sometimes, certain discrete variables are treated as continuous variable, for example, population when expressed in lakhs, millions, crores, or billions. The population of India according to 2011 census is 1,210,193,422. However, after approximation, it is possible to write 121.02 crores or 1.21 billion.

Continuous variables are often further classified as *interval* and *ratio* variables. An interval variable is comparable to an ordinal variable mentioned previously, but the intervals in an interval variable are spaced equally (ordinal variables have no absolute values and the intervals are not equally spaced). In the case of variables of this kind, the intervals can be measured numerically along a range. For example, the difference between 3 °C and 4 °C is same as the difference between 10 °C and 11 °C. The interval is same, and hence, the difference between the characters is same.

A ratio variable, in addition to the equal interval properties of an interval variable, has two additional features. It has a true zero, meaning a fixed starting point. Therefore, it is possible to indicate the complete absence of a property. For example, temperature measured in Kelvin is a ratio variable as 0 K (absolute zero) indicates that there is no temperature whatsoever. On the contrary, 0 °C does not mean that there is no temperature at all. In other words, temperature measured in degrees Celsius is not a ratio variable but an interval variable. Another feature is that the numerals of the ratio scale have the qualities of real numbers and can be added, subtracted, multiplied, and divided and expressed in ratio relationships. Height, weight, area, income, and age are examples of ratio variables (See also scales of measurement in Sect. 6.2).

Extraneous Variables

When you consider variables as dependent and independent, a third group of variables, which cannot be fitted in either of them may also surface. For example, consider two variables, irrigation and yield. It is common knowledge that yield is affected by not only irrigation but also many other factors. Not all the other factors or variables that interfere with yield can be categorized as dependent or independent variables. Yield can be written as a function of several inputs.

$$\text{Yield} = f(\text{inputs}).$$

In the example, the inputs might be fertilizers, irrigation, or some others. However, the dependent variable (yield) is influenced by not only the quantity of inputs supplied (independent variable), but also some other factors such as soil moisture status, soil aeration, depth of soil, drainage, acidity, temperature, sunlight, shading, and incidence of pest and diseases. Such uncontrollable factors are called *extraneous variables* or *confounding variables.* Extraneous variables are those uncontrollable variables that may have a significant influence upon the results of a study. For example, when there is a yield increase upon the application of fertilizers, a part alone is explained by the effect of N, P, and K fertilizers; and unexplained variability occurs due to variability in other factors.

Uncontrolled variations are common in biological experiments. When conducting experiments, you may suspect uncontrolled variations on several occasions when none of the initial precautions and safeguards taken help to correct the error. Such problems can arise whenever an experiment gives surprising, unexpected results. In a perfect experiment, the investigator must be able to hold all the extraneous variables constant except for a single independent variable. This is an ideal situation only, and in practice, it is difficult to ensure such an ideal situation. Although it is impossible to eliminate all extraneous variables, by following the principles of experimentation and including a 'control', the experimenter can limit the influence of external variables to the minimum.

3.4 Formal Approaches to Research

Intuition or insight is a part of the research process. However, one cannot always wait for intuition. If it occurs, it is good, and the researcher's job becomes easier. Even after getting the idea through intuition, researchers have to proceed further. They get ideas for research from many other sources—literature, experience, observation, seniors, colleagues, and many others related to the problem (Sect. 7.2). However, to test the idea through scientific methods, it is essential to conduct research in an established way. Research studies may take various forms according to the subject and the nature of the problem. An appropriate research design must be used to achieve this objective. Different types of research questions demand different types of research designs.

Research studies can be classified based on several considerations as basic or applied; descriptive or analytical; quantitative or qualitative; conceptual or empirical; experimental or non-experimental; retrospective or prospective; and cross-sectional or longitudinal.

Basic and Applied Research

According to the purpose or application of research results, research can be broadly classified into two—basic research and applied research. *Basic research,* also called *fundamental research* or *pure research*, has no immediate application; whereas *applied search* has immediate application. Basic research is taken up with the main objective of enhancement of knowledge. However, applied research is for practical benefits.

Basic Research

Basic research is fundamental to most of the scientific advances ever made. It is very unpredictable, and often, there is no initial connection between the research and its application. Basic research is primarily concerned with the development of theory. In basic research, we study the processes that are universal in their application to scientific knowledge. For example, in biology, basic research is conducted to increase the understanding of fundamental life processes. Basic research is usually mentioned under two heads, pure basic research and strategic basic research.

Pure basic research: Experimental or theoretical works carried out mainly to get new knowledge without a specific application at the moment come under the purview of pure basic research. Pure basic research is undertaken without looking for long term economic or social benefits but just for the advancement of knowledge.

Strategic basic research: Strategic basic research also involves experimental or theoretical works taken up mainly to obtain new knowledge without a precise application in view; however, a major difference is that it can be directed into particular broad areas in the hope of useful discoveries or solving problems.

Applied Research

Applied research is normally concerned with the application of theory to the direct solution of various problems. It is also original work but undertaken primarily to acquire knowledge with a specific practical application in view. In agriculture, most of the experiments conducted are applied in nature with the objective of solving specific problems or improving existing practices. Mainly, there are two types of applied research.

Evaluation research: In evaluation research, data are collected, evaluated, and decisions are made based on the findings. It involves the systematic process of gathering data on the variables of interest to make decisions of relevance. Examples include evaluation trials of herbicides, insecticides, and fungicides; comparisons of a number of cultivars; alternative manurial practices; and systems of management.

Developmental research: It is also sometimes referred as *research and development (R&D)* and has the major function of developing effective products for use. In technological disciplines, its main objective is extension of the results of applied research towards finished products or processes. In social sciences, developmental research involves the process of transferring knowledge gained through research into operational programmes, for example, farmer training materials, teacher-training materials, and student learning materials.

Descriptive and Analytical Research

Descriptive or *exploratory research* attempts to determine, describe, or identify the state of affairs, as it exists at present, while *analytical research* attempts to establish why the state of affairs is that way or how it came to be like that. Analytical research studies are used to describe associations and analyse them for possible cause and effect. Experimental studies, quasi-experiments, and correlational studies are analytical studies. In analytical research, the researcher has to use information already available or generated and analyse these critically to provide answers.

Descriptive studies are commonly employed when not much is known about a particular phenomenon. The researcher observes, describes, and documents various aspects of the phenomenon under consideration through the method of *observation* or *survey*. No attempt, however, is made to manipulate variables or look for cause and effect relationships. These are ideal to describe a situation or area of interest factually and accurately, to find out the frequency with which it occurs, and to categorize the information. The results provide the knowledge base for potential hypothesis formation leading to detailed studies, mainly of the experimental type. These designs are useful as evaluation and decision-making tools of administrators and policymakers, as they provide a wealth of information to base their decisions.

Descriptive research can be followed in both natural sciences and social sciences. In natural sciences, this approach is used for accurately describing nature as in morphology, anatomy, and astronomy by observation through the naked eye, through microscope, or telescope. Systematic or taxonomic studies are also descriptive studies that involve naming and classifying organisms. The results from descriptive studies provide the knowledge base for the generation of potential hypotheses, which may lead to further studies especially of the experiment type.

In social sciences, descriptive research is used to describe the characteristics of an existing phenomenon through methods such as *census, survey,* or *observation* (Sects. 6.4 and 6.5). Examples of descriptive research in social sciences include public opinion surveys, fact-finding surveys, job descriptions, surveys of literature, documentary analyses, anecdotal records, critical incident reports, test score analyses, survey of traditional plant protection measures, adoption pattern of new cultivars, and survey of involvement of women in decision-making in farms.

Quantitative and Qualitative Research

Research designs can be classified as quantitative or qualitative based on the nature of data collected. In *quantitative research*, the knowledge generated is based on collection of numerical data and analysis. In general, quantitative research is confirmatory and deductive in nature. Data analysis is mainly statistical. Remember that deductive reasoning is the process in which the investigator begins with an established theory, where concepts have already been reduced into variables, and then gathers evidences to test whether the theory is supported. Most often, quantitative designs are about quantifying relationships between the independent variable and the dependent variables. On the other hand, *qualitative research* involves the collection of data in narrative form to understand the particular phenomenon or process. Qualitative research relies primarily on qualitative measures, which include any measure where the data are not recorded in numerical form. They are exploratory and inductive in nature, and begin with the assumption that reality is subjective and not objective. In situations where not much is known about a particular phenomenon or concept, it is customary to use a qualitative design first. Once concepts and themes are identified, they can be further tested using a quantitative approach.

In most qualitative research methods, the data analysis includes coding of data and production of a verbal record. *Historical research* and *ethnographic research* (Sects. 4.2 and 4.3) are examples of qualitative research. *Participatory learning and action* (PLA), which evolved as an alternate approach to data collection to replace the time-consuming socio-economic survey, is also qualitative (Sect. 4.11).

Conceptual and Empirical Research

Conceptual research is often associated with some abstract concepts, ideas, or theory. The researcher split down a theory or concept into its components to get a better understanding of the deeper issues concerned with the concept. Although conceptual analysis is a useful method, it should be used in combination with other methods of analysis to produce reasonable results. Conceptual research is the preferred method in social sciences and philosophy, and philosophers and thinkers often use it to develop new theories or concepts and to reinterpret existing ones.

In empirical research, on the other hand, data are collected through observation and experiments. The researcher form a working hypothesis or guess based on observed facts capable of being verified by further observation or experiment. The researcher then works to get enough facts or data to prove or disprove the hypothesis, using a suitable design and manipulating the persons or the materials concerned. Empirical research is appropriate when you want to prove that certain variables affect other variables in some way. Evidences gathered through empirical studies are considered the most powerful support possible for a given hypothesis. Because observation and experimentation are employed for obtaining verifiable results, empirical research is commonly used in scientific studies

Experimental and Non-experimental Research

Quantitative research designs are either *non-experimental* or *experimental*. For quantifying the relationships between independent variable and dependent variable, quantitative designs are used. This relationship is studied in two ways; in the first, the introduced treatment is understood to be the cause of change, and therefore, wait for sufficient time to produce the change. In the second case, the researcher after observing a phenomenon tries to understand its cause. In this case, the researcher starts from the effects to outcomes and determine causation. If a quantitative study can be conducted starting from the cause to the effects, it is called *experimental research*, whereas, if we start from the effects to trace the cause, it is *non-experimental research*. In the first case, the independent variable can be observed, introduced, controlled, or manipulated by the investigator, whereas in the second case, this is not possible, as the assumed cause has already occurred. Therefore, the researcher retrospectively links the outcomes to the causes.

Non-experimental research is a common term where independent variables cannot be manipulated, and therefore, cannot be experimentally studied. These designs are used to describe, differentiate, or examine associations between or among variables, groups, or situations. Manipulation of variables, random assignment, and comparison groups are normally absent in non-experimental studies; the research worker just observes what happens naturally. Non-experimental designs are taken up because of several reasons. These might be the only way out in a situation wherein the variables cannot be manipulated. For example, experiments involving human subjects need care, and there are several restrictions on conducting experiments involving humans. Because of ethical compulsions, certain variables should not be manipulated. In some other instances, independent variables have already taken place, therefore, manipulation is impossible.

Non-experimental designs can be classified based on the timing of data collection as *cross-sectional* or *longitudinal* (Sects. 4.4 and 4.5). In a cross-sectional study, data on the recognized variables are collected one point in time, and the relationships between them are decided. In a longitudinal study, however, data are collected at different points over time. Sometimes, non-experimental studies are classified based on the timing of the event being studied as *prospective* or *retrospective* (see below). In a typical retrospective study, an incident, event, or phenomenon recognized at the present is linked to causes or variables in the past. At the same time, in a prospective study, the causes and variables related to some phenomenon or events identified in the present are linked to possible outcomes in the future.

An *experimental research* is one where the independent variable can be directly manipulated by the investigator, keeping other variables constant as far as possible. The way to establish a cause and effect relationship is to isolate and eliminate all the factors that might be responsible for a particular outcome and test only one that you directly want to measure. A *semi-experiment* or *quasi-experiment* design has the properties of both experimental and non-experimental designs; part of the study may be experimental and the other part non-experimental. Another group of experiments, which cannot be classified as true experiments or quasi-experiments are *ex post facto* research. *Ex post facto* experiments are usually adopted when human subjects in real-life situations are involved, and the investigator comes to the scene after the event occurred, that is, 'after the fact'. More details on experimental research are given in Chap. 5.

Prospective and Retrospective Research

As discussed above, a non-experimental study may be *prospective* or *retrospective*. If the researcher gathers data at a particular time in the present and traces the differences into the past, it is a 'retrospective' study. In contrast, if the investigator starts to observe in the present and wait for the results in the future, it is a 'prospective' study. A prospective study looks forward in time. For example, you can select a group of subjects and watch them for several years. A prospective study is slow, and normally, you have to wait for years to accumulate sufficient data to draw any strong conclusions. A prospective study watches for outcomes such as the development of a disease during the study period and relates this to other factors such as a suspected cause or risk. In a prospective study, usually a 'cohort' of subjects are studied and observed over a long period (Sect. 4.6).

In a retrospective study, the researcher tries to connect the events into the past, and examines exposures to suspected causes or risk with reference to an outcome that has been established before the start of the study. For example, we may observe an epidemic and try to figure out the reasons for the epidemic. As the subjects are affected by the disease, we just note down various effects on humans and environment and go for tabulation at the end. Nevertheless, take special care to avoid sources of bias and errors due to extraneous factors, which are more common in retrospective studies than in prospective studies. *Case-control studies*, a form of *ex post* facto studies, are usually retrospective studies (Sect. 5.12).

Some studies can be considered as *retrospective–prospective* studies. These focus on past trends in an incident or phenomenon and study its future trends. The study usually starts by collecting retrospective data from existing records before introducing some intervention. The study population is then observed for a specific period to determine the effects of the intervention. *Time series studies* come under this group (Sect. 5.11).

Bibliography

Austin, J. H. 2003. Chase, Chance, and Creativity: The Lucky Art of Novelty (2nd Ed.). The MIT Press, Cambridge, 267p.

Best, J.W. and Kahn, J.V. 1993. Research in Education (7th Ed., Indian Reprint, 2004). Prentice–Hall of India, New Delhi, 435p.

Beveridge, W. I. B. 1957. The Art of Scientific Investigation (3rd Ed.), Vintage Books, New York, 239p.

Bryman, A. 2008. Social Research Methods (3rd Ed.). Oxford University Press, Oxford, 748p.

Fowkes, F.G.R., and Fulton, P.M. 1991. Critical appraisal of published research: introductory guidelines. Br. Med. J. 302: 1136–1140.

Heath, D.A. 1985. Research: Why do it? In: Hawkins, C. and Sorgi, M. (eds.), Research: How to Plan, Speak and Write about It (4th Indian Reprint, 1993). Narosa Publishing House, New Delhi, pp. 1–11.

Holmes, D., Moody, P., and Dine, D. 2006. Research Methods for the Bioscience. Oxford University Press, Oxford, 381p.

Jarrard, R.D. 2001. Scientific Methods [E-book]. Available: http://www.emotionalcompetency.com/sci/sm_all.pdf.236p. [Accessed 10 June 2018]

Kerlinger, F.N. 1986. Foundations of Behavioural Research (3rd Ed.). Holt, Rinehart and Winston, USA, 667p.

Kothari, C.R. 2004. Research Methodology: Methods and Techniques (2nd Ed.). New Age International, New Delhi, 401p.

Kumar, R. 2011. Research Methodology: A Step-by step Guide for Beginners (3rd Ed.). Sage Publications India, New Delhi, 415p.

Leedy, P.D. and Ormrod, J.L. 2010. Practical Research: Planning and Design (9th Ed.), Pearson Education, New Jersey, 360p.

Mann, C.J. 2003. Observational research methods. Research design II: cohort, cross sectional, and case-control studies. Emerg. Med. J. 20:54–60.

Marder, M.P. 2011. Research Methods for Science. Cambridge University Press, 227p.

Trochim, W.M.K. 2003. Research Methods (2nd Ed.). Bizantra, New Delhi, 363p.

Wellington, J. and Szczerbinski, M. 2008. Research Methods for the Social Sciences. Continuum International Publishing Group, London, 234p.

Chapter 4
Major Research Methods

Washing one's hands of the conflict between the powerful and the powerless means to side with the powerful, not to be neutral.
Paulo Freire (1921–1997)

When planning for research, we select appropriate methods and designs to fulfill its objectives. Each discipline of science has its own ways of doing research. Social sciences mainly make use of qualitative and non-experimental methods although experimental methods are also adopted occasionally. Some of these qualitative and non-experimental methods are employed by natural scientists too. From a practical standpoint, depending upon the disciplines and nature of study, researchers choose a research design and plan a project for achieving the objectives. Researchers can choose any of the numerous qualitative or quantitative methods as they wish, but depending upon the circumstances, time, resources, and the type of variable involved.

In this chapter, selected non-experimental methods—natural observation, historical research, ethnographic research, cross-sectional study, longitudinal study, cohort study, case study, correlational research, model building, action research, and participatory learning and action (PLA) have been discussed. However, this is not intended to give a detailed account of all the non-experimental methods employed by social scientists. Fortunately, there are numerous books on research methodology for social sciences, and for additional information, readers may consult any of such books as Kerlinger (1986), Best and Kahn (1993), Burns (2000), or Bryman (2008). The methods that come under the category of experimental methods including quasi-experimental methods and *ex post facto* studies have been discussed in Chap. 5.

4.1 Natural Observation

Observation of nature is fundamental to the study of science. In situations where experimentation is difficult or impossible, researchers usually resort to natural observation. The observations are recorded from its natural setting. The investigator dispassionately observes and records some behaviour, process, or phenomenon in

C. G. Thomas, *Research Methodology and Scientific Writing*,
https://doi.org/10.1007/978-3-030-64865-7_4

its natural surroundings. In social sciences, health, and ecology, natural observation, sometimes also called *natural experiments* or *observational studies* are common. In earlier times, observation dominated in almost all scientific disciplines. In the history of science, there are several good examples of natural observations. We know that **Charles Darwin** could recognize several patterns in nature through observation, which allowed him to understand the process of biological evolution through natural selection.

Observation has been a successful method in physical sciences such as astronomy, geology, oceanography, and meteorology. In geology, you may find highly experimental fields such as experimental geochemistry and largely observational fields such as palaeontology. Astronomy usually relies on experiments of this type. It is impractical to do experiments involving celestial bodies. In many cases, using the data collected by observing natural phenomena, we can test our hypothesis. In social sciences, sometimes we venture out to observe humans or animals in real-life situations to study their activities and responses. In natural sciences, this may involve observing an animal or groups of animals, a plant or plant community, or some physical phenomena. In health science-related areas such as human anatomy, observation is the prime method used to describe the construction of the body. In ecology, observation and comparison have always been of paramount importance although experimental methods have also been incorporated.

A major strength of natural observation is that it allows researchers to observe behaviour in natural setting in which it normally occurs. However, as this is a descriptive method and the results are obtained without the controlled conditions of a laboratory, conclusions about cause and effect relationships have to be drawn cautiously. Researchers may have to wait patiently to observe the behaviour or phenomenon of interest. The difficulty of observing behaviour or a process passively without disruption and the difficulty of making observations amenable to statistical analysis are other limitations.

4.2 Historical Research

Historical research is involved with the study of past events. It relates past events to one another or to current events. *Historiography* or historical research endeavours to find out the nature of events that have happened in the past. A historian tries to reconstruct the past usually based on a hypothesis. However, it is very difficult to use scientific method in historical research. The purpose of scientific research is prediction; but historians cannot generalize on past events and predict. Most past events were unplanned, many uncontrolled variables were present, and the influence of one or two individuals might have been crucial. Therefore, typical experimental methods with ‘randomization’ and ‘control’ are not possible in history. Historians cannot control the conditions of observation or manipulate variables as the events had already occurred. Often, they have to depend upon the reported observations of other witnesses of doubtful competence; most of them may not be alive. Still, the historian states a problem, formulates hypothesis, gathers and analyses primary data, tests hypothesis as consistent or inconsistent with available evidences and arrives at conclusions.

Historians use both primary and secondary sources of data for arriving at definite conclusions on the hypotheses they formed. Primary sources for historical research include eyewitness accounts, remains or relics, and oral testimonies. Documents containing eyewitness accounts and the records kept and written by actual participants of events for future generations are good sources of information. Remains or relics are objects associated with a person, group, or period. These include materials such as fossils, skeletons, tools, weapons, utensils, clothes, paintings, and coins, which may provide clear evidence about the past. Oral testimony is an account of a witness or participant. Secondary sources such as history books, reports, and encyclopaedias are also used, although they are less reliable.

4.3 Ethnographic Research

Ethnographic research is about the study of current events rather than past events involving people. The meaning of *ethnography* is 'writing about people'. Ethnography encompasses any study of a group of people for describing their socio-cultural activities and patterns. It involves the collection of descriptive narrative data on many variables, which cannot be numerically recorded over an extended period in a naturalistic setting.

Several methods are used in ethnographic research, but the most common approach is *participant observation*, where the researcher lives with the subjects being observed as an active participant and records extensive field notes. Participant observation is a common method for collection of qualitative data wherein the researcher becomes a participant in the situation being observed and observes the pattern or behaviour. The researcher takes part in the daily activities of people, reconstructing their interactions and activities in field notes taken on the spot or as soon as possible after their occurrence. However, in order to succeed, the researcher should know how to become a participant in a situation or context and her/his specific roles to perform as a researcher. The researcher should also be proficient in preparing and keeping field notes and analysing and interpreting field data. They must be accepted as a natural part of the culture. Participant observation often requires long periods, months, or even years of intensive work.

You can use *direct observation* also as a method of data collection in ethnographic research. In direct observation, the researcher does not become a participant but watches the events with a dispassionate perspective. However, the observer should try to be as unobtrusive as possible to avoid bias in observations. Researchers can use modern techniques such as hidden camera, recorders, and video. Direct observation will not take as much time as that of participant observation.

4.4 Cross-Sectional Study

Cross-sectional studies are common in social sciences, health sciences, and in ecology. Sometimes, these are also called *prevalence studies*, *one-shot studies*, or *status studies*. A cross-sectional study can be considered as a survey of the prevalence of a phenomenon, situation, problem, risk factors, or other characteristics by taking a cross section of a defined population at one particular time. Measurements are made at a single instance only. A cross-sectional study is 'cross-sectional' in respect of both the study population and the time of investigation. It helps to have a picture of a population at a certain time, allowing conclusions across a wide population to be drawn.

In most cross-sectional studies, data are collected using the survey method through questionnaires or interviews (Fig. 4.1). First, decide on the research question or hypothesis, identify the study population, select a sample, and contact the respondents for information. For example, if you want to study the incidence of goitre in a population, scientists can look at a wide range of ages, ethnicities, and social backgrounds. If a significant number of subjects from a certain social background are found to have the disease, then the researcher can investigate further. Some topics, which are ideal for a cross-sectional study, are unemployment status, incidence of a particular disease such as cancer, and the attitude of society towards wetland conversion. Market research often uses opinion polls, a classic example of cross-sectional study. In many cross-sectional studies, the samples used are convenience samples (Sect. 7.1) as drawing a true sample is difficult and costly. Ideally, the sample must be representative of age, gender, and social class structure of the population being studied. The study must be truly cross-sectional.

Fig. 4.1 Cross-sectional study using an interview schedule

A major advantage of cross-sectional studies is that subjects are not deliberately exposed or treated, and hence, there are seldom ethical difficulties. This type of study is relatively inexpensive and quick as only one group is used and data are collected only once. Lesser resources are required to conduct the study, as the studies involve no follow up. Cross-sectional studies are the best ways to determine prevalence of something, and therefore are useful for identifying associations, which can be more rigorously studied later using a cohort study (Sect. 4.6) or randomized controlled trial (Sect. 5.9). Nonetheless, it may be difficult to differentiate cause and effect from simple association. There can be a number of probable explanations for the effects. Similarly, cross-sectional studies are not appropriate for studying rare events or conditions because we cannot be sure of rare conditions in study samples.

4.5 Longitudinal Study

An observational study performed over a period of years to determine the pattern of change in a phenomenon, situation, problem, attitude, or other similar attributes in relation to time is a *longitudinal study*. In contrast to cross-sectional research, where measurements are made on a single instance, in a longitudinal study, measurements are made over a period. A longitudinal study can be compared to a series of repetitive cross-sectional studies.

A longitudinal study may be retrospective or prospective. In a *longitudinal retrospective* study, the investigator studies the present and past events, while in a *longitudinal prospective* study, the investigator follows the subjects for future events. Longitudinal studies allow scientists, especially social and medical scientists, to study long-term effects of a process, programme, medication, or event in a human population. These are also useful when you need to collect data on a continuous basis. In agriculture or ecology, longitudinal studies can be planned in many situations such as plant growth analysis, plant succession studies, soil erosion studies, and river flow analysis.

In a longitudinal study, the researcher contacts the study population many times for a longer period at regular intervals to gather necessary information. There are no fixed intervals; the interval can vary depending upon the nature of the study. Sometimes, the intervals may be as short as a week or longer than a year. For data collection, although the same population is used, it need not be from the same individuals.

A major disadvantage of longitudinal studies with human subjects is the *conditioning effect*. This situation can occur, if the same respondents are contacted frequently. Soon, they begin to know what is expected of them, and without careful thought, they may respond to questions in a conditioned manner, or sometimes, they may lose interest in the inquiry! A way out is to change the respondents from the same population, if possible.

4.6 Cohort Study

The word '*cohort*' is from the Roman cohorts, the soldiers who marched in groups together into battle. In a cohort study, a particular group having a common trait within a subgroup of population (the cohort) is observed over a period. The common traits can be age, marriage, disease, height, weight, education, etc.

You can see several examples of specific cohorts in real life. You may see groups of people who are affected with a particular condition such as a disease, medical condition, pollution, natural disaster, and so on. People who have taken a particular drug also form a cohort. This is the most appropriate method for studying the prevalence and background of a condition. Humans, animals, or even plants living near a certain polluting source are also examples of cohorts. A cohort study can probe even further by dividing a cohort into sub-groups, for example, a cohort of alcoholics could be sub-divided, with one group suffering from diabetes and another from high blood pressure.

Cohort studies can be retrospective or prospective. However, most cohort studies are planned as prospective studies. In a *retrospective cohort* study, phenomena that have already happened are considered, while in a *prospective cohort* study, the study starts from the present. The retrospective cohort study is historical in nature and often uses data already collected for other purposes. The researcher collects and evaluates historical data to ascertain the effects of variables. A retrospective study is comparatively less costly, and it is an effective method for studying health and environment related risks due to pollutants, toxins, and drugs. Although retrospective studies give results rapidly, validity may be a problem. Complete removal of confounding variables from historical records and interviews is impossible. In a prospective cohort study, certain variable are observed over a period for their effects. Initially, a group of individuals is chosen who are free from the effects of the variable being tested. The subjects in the group are observed to see whether they develop the effects over a period.

In single cohort studies, those people who are free from the effects are used as internal controls. Where two cohorts are used, one group is exposed to or treated with the variable of interest and the other is not treated, thereby acting as an external control. To maintain validity, all the subjects included in the study must be initially free from the effects of the variable being tested. For example, an investigation into the effects of some occupational disease must ensure that all the volunteers are free from the disease.

You can easily differentiate a 'prospective cohort' study from a 'longitudinal' study in that in the former, subjects are brought together based on some common experience, and they are observed for a specific period at regular intervals to check which group develops the outcome of interest. If you take a classical example of smoking, you can watch smokers and non-smokers for a definite period to ascertain whether the two groups respond differently with respect to a particular outcome. However, in a longitudinal study, what is important is the time in which subjects are followed. After selecting the individual subjects, they are observed continuously over a specific period. For example, if you want to check the effectiveness of a drug that claims to reduce cholesterol in blood, you need to follow the same subject over a significant period.

In health research, cohort studies are often taken up in situations where randomized controlled trials (Sect. 5.9) are unethical to take up. Research on risk factors is mainly done using cohorts. For example, you cannot deliberately expose people in the name of experiment to insecticide spraying or cigarette smoking. By using cohorts, we get an idea on the influence of each variable on the likelihood of developing the phenomena or events. You must note, however, that a prospective cohort study may be inappropriate for rare events.

4.7 Case Study

Case studies are useful in both natural and social sciences. These are usually taken up when researchers want to get a detailed background view of a particular case or phenomenon. The case may be a person, plant, group, process, disease, event, community, or any other similar unit. These were found to be appropriate to explore intensively the background, status, and environmental interactions of a given social unit, individual, group, institution, or community. Case studies are common to identify the social and background factors that might have caused certain strange manners in an individual.

A case study is characterized by three features: (1). it is focused on a selected case, (2). the researcher has the desire for in-depth understanding of an issue; and (3). data can be collected in many different ways such as observations, interviews, and secondary data review.

A major characteristic of this method is the combination of different sources of information, for example: family history, personal data, medical records, interviews, observations, and other similar sources. This approach is based on the assumption that the case under study is typical of cases of a similar kind so that generalizations can be made, which are applicable to comparable type of cases.

Case studies are frequently used to evaluate something engaging both qualitative and quantitative methods. Agricultural and life scientists might use this method to study a crop, animal, or a single instance of some physical phenomenon. Case studies are also helpful when researchers are unable to undertake experimental studies. Often, it provides in-depth information about what is going on, and gives direction to a more comprehensive study later.

Some limitations must also be noted. Primarily, this is a descriptive method, not an analytical or explanatory one. Case studies normally involve a single case or just a few only, and therefore, may not be representative of the population. Because case studies often rely on information provided by different people, chances of leaving important details are high. In addition, as most of the information gathered is retrospective data or recollections of past events, the problems inherent to memory might also be expected.

4.8 Correlational Research

Correlational research is typically cross-sectional; it is often used to examine whether changes in one or more variables are related to changes in other variables. The nature, degree, magnitude, and strength of relationships between two or more quantifiable variables can be investigated through correlational research. If two variables are correlated, we may be able to use the relationship to predict the value of one variable over the other. Often, the outcome of a correlational study is useful for formulating hypotheses, which could be later tested in experimental and quasi-experimental studies. However, it is important to remember that just because there is a significant relationship between two variables; it does not mean that one variable causes the other. Remember correlation can be used for prediction of outcomes but not for predicting causes!

Although correlation is often described as a method of research in its own right, it is more of a mathematical technique for summarizing and relating data. In fact, for most researchers, it is a common statistical tool. Correlation, regression, and curve fitting are often tools used by researchers for prediction based on relationships.

Sometimes, a distinction is made between different types of correlational studies. A *predictive study* is done simply to learn which variables from a set is the best to predict the dependent variable. The objective here is simply to maximize prediction. A second type of study is *relationship*. With relationship studies, the goal is to understand those variables that are theoretically related to a dependent variable. The greatest limitation of correlation study is that it only shows that two variables are related in a systematic way, but neither proves nor disproves that the relationship is a cause and effect relationship.

4.9 Model Building

Occasionally, an equation or formula can be used to estimate the effect of a physical process on some other process. We usually call all such problem solving without doing experiments as '*models*'. There are several kinds and types of models, but simple models based on regression are frequently used. *Empirical models* are a prominent group of models based on observed facts, which allow prediction of what would happen in certain conditions, because we already know what has happened before in those conditions. The reliability of such methods depends on the database created through experience. An empirical model may be a simple relationship or a complex multiple equation. Most of the present-day models tend to become increasingly complicated and consequently heavily dependent on computer programmes to solve the problem.

The Universal Soil Loss Equation (USLE) is an empirical model designed to predict long-term average soil losses from a land in specific cropping and management systems. Various factors affecting soil losses through erosion such as rainfall erosivity, soil erodibility, gradient of slope, length of slope, vegetative cover, and soil conservation practices are considered in the equation to predict expected soil losses for a particular area. Similarly, several empirical models have been developed for estimating consumptive use of crops from routinely measured meteorological data. Many of them have been determined and tested locally. You can select an appropriate empirical formula such as Penman–Monteith to calculate the reference crop evapotranspiration (ETo), and using the crop-coefficient, crop evapotranspiration (ETc) can be estimated. Model fitting is popular in other research areas too such as ecology, physiology, economics, and engineering.

4.10 Action Research

Action research is a process of solving practical problems through the application of the scientific method involving the collaboration and cooperation of all stakeholders—researchers, extension workers, and beneficiaries in teams. Action research encompasses two activities, *action* and *research*, and 'research' is the means for 'action', for example, solving a problem or issue. It is promoted and put into practice by many development agencies and community organizations around the world by applying innovative ideas, techniques, and technologies in solving issues and problems leading to socio-economic development. In simple terms, action research is 'learning by doing'; a group of stakeholders identify a problem, plan and undertake something to solve it, and evaluate the success of their efforts.

Action research is mainly undertaken to develop new skills or new approaches and to solve issues and problems with direct application to any applied setting. A problem situation is diagnosed first, remedial action is immediately planned, implemented, and effects monitored. It is a powerful tool to tide over some contingency situations. For example, if some epidemic broke out and there are no proven remedies, we would not get enough time to form a hypothesis, validate it through experiments, and then put to practise. What is required is immediate action. In consultation with the experts and extension workers in the field, some ad hoc remedies are formulated and action research is conducted. Action research has four basic characteristics (Burns 2000).

- ***It is situational***: A problem is diagnosed in a specific context, and attempts are made to solve it in that context.
- ***It is collaborative***: Action research is a collaborative effort with researchers, extension workers, practitioners, farmers, or non-professionals.
- ***It is participatory***: All the team members take part directly in implementing the research work.
- ***It is self-evaluative***: Monitoring and evaluation are continuously done. Modifications are evaluated within the ongoing situation to improve practice.

Action research differs from other types of research in several features. Its focus is to involve the people concerned also along with researchers. Those who take up action research are not immediately concerned with adding more 'knowledge' to the already existing knowledge base. However, in course of time, it may contribute simultaneously to basic knowledge and to social action. Instead of validating theories independently and then applied to practice, validation is done through practice. In other words, in action research, research and practice go together. As the focus of action research is in solving real problems, it is used in real situations as against experimental research, which is mostly done in contrived or artificial situations of a laboratory or field. It is the ideal method when the methodology and settings need the involvement of people or a change must take place immediately.

4.11 Participatory Learning and Action

Participatory research methods have evolved for assessing development issues such as natural resources in a locality, environmental concerns, and gender issues in local communities in a relatively short period. Although these are the main domains of social scientists, scientists from other fields should also be aware of some of these techniques as most developmental projects (e.g. watershed development) are usually done by a multidisciplinary team.

The main objective of any participatory methods (PMs) is to enable ordinary people to play an active and influential part in decisions which affect their lives. This is possible by empowering the local community by including its members in all the stages of research process. In this process, the researchers or the extension workers play a catalytic and supportive role. They will not dominate but act as facilitators, working to involve stakeholders in every aspect of the research process. The techniques enable the officials or extension agents (*the outsiders*) to use the criteria and choices of local people (*the insiders*) for data collection and understanding the local environment with clear local priorities. In comparison to the traditional survey or questionnaire methods followed in cross-sectional and longitudinal studies, where participation is low to none, participatory methods empower the local people to achieve their development goals using the expertise of outsiders.

Participatory Methods: RRA, PRA, PAR, and PLA

Among the different types of participatory methods, *Rapid Rural Appraisal* (RRA) was the first one to emerge, which focused on how outsiders (officials and development agencies) could quickly learn from local people about their situations, requirements, and challenges. It was spearheaded by Robert Chambers at the University of Sussex, England. In the late 1970s, RRA began to be used by development practitioners in poor and backward countries. RRA, as the name suggests, is a rapid method to find out quickly the required information necessary for project formulation. In due course, RRA wanted to be more participatory in the collection of information by involving local people in data-gathering and analysis through the use of popular education methods such as mapping, transect walks, and ranking. Later, as the emphasis shifted from quick collection of data to the involvement of end-users and learning from experiences, RRA gave rise to *Participatory Rural Appraisal* (PRA), with the focus on facilitation, empowerment, and sustainable action.

Participatory Rural Appraisal (PRA) is an approach for shared learning between local people and outsiders, but the term 'rural' in it is somewhat misleading. PRA techniques are equally applicable to urban settings. It is not limited to project appraisal only but useful throughout the project cycle and for research studies. In fact, the term PRA is one of many labels for similar participatory assessment approaches, the methodologies of which overlap considerably. In the meantime, the differences between RRA and PRA became clear. The major difference lies in who leads the appraisal. If the learning process is mostly done by outsiders such as extension workers, it is called RRA. On the other hand, if it is a continuous research and action process by the local community, it is called PRA.

Another participatory approach is *Participatory Action Research (PAR)*, which is actually an activist approach to research. It has its origins in the work of radical social scientists from Latin America, Paulo Freire (1921–1997) and Orlando Fals-Borda (1925–2008). The concept of *empowerment* and *participation* became popular after Paulo Feire put forward the concept of *conscientisation* in his influential work, the *Pedagogy of the Oppressed.* Paulo Feire believed that through participatory and empowering approaches, rural people who are currently poor and oppressed, would progressively transform their environment by their own methods.

For RRA, PRA, or PAR, the techniques used are almost same, therefore, a new term *Participatory Learning and Action* (PLA) has been suggested as an umbrella term encompassing a variety of participatory approaches. PLA has been used widely in natural resource management (watershed management, soil and water conservation, forestry, wild life, biodiversity, fisheries, etc.), village planning, community health, food security, and poverty eradication programmes.

Features of PLA

Participatory learning and action (PLA) tools have several advantageous features compared to the traditional methods of data collection.

Participation: PLA is designed to equip local people not only as sources of information but also as partners with the PLA team in gathering and analyzing the information. It ensures participation by the local communities, and they are given an opportunity to describe how they do things, what they know, and what they want.

Focused learning: In PLA, primary data are collected with a data optimization approach. The focus is not on learning everything but on learning what is necessary. PLA must be efficient in terms of both time and money, and it intends to gather just enough information to make the necessary recommendations and decisions. This feature is also mentioned as 'optimal ignorance' or 'minimal ignorance'.

Teamwork: For generating PLA data, informal interaction and teamwork among the stakeholders and the facilitator are necessary. A PLA can be conducted successfully by a balanced team of local people representing different socioeconomic groups, culture, gender, and age and a few outsiders consisting of a mix of subject matter specialists.

Flexibility: PLA is highly flexible and does not provide any particular plans for its practitioners. The combination of techniques appropriate in a particular context will be determined based on the size and skills of the PLA team, the time and resources available, and the nature and location of the proposed work.

Speed: PLA is relatively quick because it does not attempt to collect a statistically valid sample. The total time needed depends upon factors such as the size of the community, social structure, and the overall purpose of PLA.

Triangulation: It is very difficult to conduct statistical analysis on the data gathered through PLA because of its largely qualitative nature and relatively small sample size. Therefore, the method of triangulation is generally employed for cross-checking and cross-validating information for accuracy. Please recall that in trigonometry, one can indirectly but precisely measure the location of a point by appropriate sightings from two other points. Similarly, we can apply this method of triangulation to social measurements. If we make use of more than one measuring techniques producing agreeable results, we can have more faith in the information collected.

In PLA, triangulation is achieved by using different tools to gather information on the same issue, for example: maps, transects, and trend lines to examine environmental changes. A similar case is by listening to different people with different points of view about the same topic, for example: woman/man, young/old, poor/wealthy, etc.

On the spot analysis: As PLA tools depend on schematic pictures, maps and diagrams, information is immediately available for analysis. On the spot analysis allows for gaps to be filled and inconsistencies to be checked on a regular basis. In other words, information generated can be reviewed, analyzed, and added continually throughout the process, thereby, allowing team members to modify questions and review the focus of the study, if needed.

Limitations of PLA

PLA cannot be considered as an error-free technique as it involves some risks and limitations. Before trying the methods, the researchers should be aware of limitations.

The risk of raising expectations: This is a major problem of PLA techniques. Although this may not be entirely unavoidable, it can be minimized with careful and repeated clarification of the purpose of the PLA and the role of the team in this exercise. It is essential that the PLA team should be frank and open from the beginning about the flow of resources.

Quality of the work and the perceptions produced: There is a tendency to use PLA as a standard survey method to gather quantitative data. Besides, incorrect practices such as using large samples and questionnaires instead of semi-structured interviews could greatly compromise the quality of the work and the insights produced.

The tendency to use too many different techniques: If the PLA team is not adequately trained in the methodology, they may have a tendency to use too many different techniques, some of which may not be relevant to the topic under consideration.

Inadequate time for preparation: Often, sufficient time may not be available for the team to relax with the local people, listen to them, and learn about the more sensitive issues under consideration. In many cases, the team may miss the views of the poorest and less communicative members of the communities they interacted.

The credibility of the PLA findings: Conversion of PLA insights into a standard evaluation report poses considerable challenges, and individuals unfamiliar with participatory research methods may raise questions about the credibility of the PLA findings.

Revealing failure or need for change: In areas where development activities are being implemented, the findings from PLA may reveal negative impacts, failures, or needs for improvement. It is the responsibility of PLA team to communicate the findings to the implementing agency. However, this needs care as they should not feel that the team is trying to persuade them to change a programme.

Learning about illegal activities: The villagers may tell the PLA team the illegal activities taking place among them, for example, illicit brewing of liquor, drug trafficking, cultivation of Ganja, illegal sand mining, conversion of wet lands, unauthorized collection of fuel wood, etc. The right way to handle such information depends on the openness and responsiveness of the authorities. However, the team should not use such information in any way that might result in serious problems to the villagers who participated in the appraisal.

Stirring up conflicts: Sometimes, instead of harmony, the exercise may expose deep differences and conflicts among various groups. When such conflicts are visible or likely to occur, the PLA team either stop the appraisal or use conflict resolution methods. They should learn to work with the conflicts and solve them amicably.

Agency domination: The development agency who implements the programme should be ready to relinquish control and leave the space to the local people to decide themselves. If the agency ignores the priority issues raised by the villagers and addresses only those already on its agenda, then the most important results of the PLA process would be neglected. They should be ready to accept the unpredictable outcomes. If the agenda of the agency is something else, never use participatory appraisal lest it may raise local people's expectations.

4.12 PLA Tools

From the basket of PLA tools, the most appropriate for the project context must be selected. The central part of any PLA is *semi-structured interviewing*. It can be done as *individual interviews*, *natural group interviews*, *focus group discussions*, *community meetings,* and *brainstorming*. Another major tool is *transect walk*, a kind of exploratory or observational walk. Social transects and land use transects are

Fig. 4.2 View of participatory mapping, a participatory learning and action tool

common. Several diagrammatic techniques are frequently used to stimulate discussions and record the results (Fig. 4.2). Some of the key diagrammatic techniques are *mapping techniques* (resource maps, ITK maps, historical maps, and institutional maps), *ranking exercises* (problem ranking, preference ranking, matrix ranking, and wealth ranking) and *trend analysis* (seasonal calendars and daily activity charts). Plenty of literature (including on line sources) is now available on PLA. For more details on PLA, you can consult any of such sources as Chambers (1994), Pretty et al. (1995), and Narayanasamy (2009). An overview of most common PLA tools is provided in this chapter.

Semi-structured Interviews

An *interview* is a verbal interchange, often face to face, in which an interviewer tries to seek information, beliefs, impressions, or opinions from another person. For eliciting opinions or perceptions from a person or a group, interview is an ideal choice. Interviews can be structured, semi-structured, or unstructured (Sect. 7.5). Semi-structured interview is an important PLA tool.

The person who conducts semi-structured interview uses a guide list with some broad questions for extracting opinions of participants. The guide list normally includes broad and open-ended questions to be answered in a free flowing, conversational, relaxed, and informal setting. The interviewer is free to rephrase questions and to ask probing questions for added details. Family details, family income, gender issues, use of natural resources, occupation, details on farming, community living, and many other topics can be effectively investigated by this technique. Interviewers must be able to make relatively brief notes during the interview. Immediately after the interview, they should fill out the summary form and editing for completeness and errors.

Practice and experience are needed for using semi-structured interviews. Interviewers should have good communication and summarizing skills, and must be aware of the general topics to be covered in the interview. Semi-structured interview can be adopted for individuals (individual interviews) and groups (group interviews).

Individual Interviews

Individual interviews may involve all the individuals in the locality or randomly selected individuals, the key informants. A *key informant* is an individual, who as a result of his/her knowledge, previous experience, or social status in the community, has access to information such as perceptions about the functioning of society, their problems, and needs. By conducting *key informant interview* involving individuals, good amount of information can be extracted from key informants. However, as the selection of key informants is not random, the issue of bias can occur. Care must also be taken to separate the informants' bias towards various issues.

Group Interviews

Group interviews can take many forms depending on the purpose they serve, the structure of the questions, the role of the interviewer, and the circumstances under which the group is convened. *Focus group discussion* and *community interviews*, and *interviews with natural groups* are useful for planning development projects.

Focus group discussion: A planned semi-structured discussion with a small group of homogeneous people with similar background and experience is called focus group discussion or focus group interview. Examples of such groups include beneficiaries of a pond, shareholders in an irrigation system, and users of a public service. A small list of open-ended questions is used to focus the discussion. Participants need to study the questions posed by the interviewers, give their own comments, listen to what others in the group say, and respond to their observations. For using the method, you may prepare a broad interview outline first. Then, fix the number of focus groups based on the size of the community. In a small community, 2 groups of 6–12 persons each and representing key categories, for example: men and women, small and marginal farmers, or wealthy and poor, may be sufficient.

Two facilitators are generally required for conducting focus group discussion. The interviewer acts as a group facilitator introducing the subject, guiding the discussion, cross-checking each other's comments, and encouraging all members to express their opinions. A second facilitator acts as a rapporteur. The rapporteur should be able to write rapidly to capture people's expressions as accurately as possible. Before starting the discussion, explain the purpose of the session to the group. Each person should get minimum one chance to provide ideas. If somebody dominates the discussion, it would become necessary to pose questions directly to less talkative participants. As with semi-structured interviews, the facilitator can use a variety of probing questions to extract ideas and to keep the talk focused. If possible, record the session using a voice or video recorder with the permission of the group. The discussion may yield good information, views, and attitudes. Immediately after the session, carefully review the notes and recordings.

Focus group discussion may be conducted with different groups to identify trends in perceptions and opinions. A main advantage of this technique is that participant interaction helps to eliminate false or extreme views, thus providing a quality control mechanism.

Community interviews: The entire community can take part in community interviews. These are conducted as public meetings in which the community as a whole is consulted. A set of close-ended questions are asked during community interviews. The interviewer asks a question, and the group interact to have agreement around an answer. Community interviews will provide salient information on how well a project is working. A major flaw is that some high status community members may hijack the opportunity and try to have agreement on their own preferences.

Interview with natural groups: Any group of individuals with some common interests form natural groups. Examples of typical natural groups are farmers working in their fields, farm labourers, mothers fetching their children from school, people queuing for a bus, traders, and customers at the market, and patients waiting in a health centre. The facilitator can start casual conversations with these groups of people. Interviews with people met during transect or observational walks and other forms of participant observation sessions will also help. The effort elicits important hints about local views on the issues of interest. It is also helpful to have initial contacts and rapport with local people.

Although the technique is effective, good communication skills are needed to get the most out of this technique. To avoid embarrassing situations, the facilitator must have some previous understanding of local etiquette. Sometimes, the answers may be exaggerated or coloured to suit the expectations of the outsider. Therefore, conduct a countercheck of the perceptions collected through this technique.

Brainstorming

Brainstorming is an ideal tool to gather divergent ideas on a specific topic from a group of stakeholders or experts. Multiple ideas generated and the ensuing discussion allow the members to explore and compare an assortment of possible solutions.

A brainstorming session starts with some opening and non-leading general statements. It should not overstate a particular point of view. The facilitator introduces the topic and then writes the key questions on the blackboard or on a flipchart. Participants are then asked to provide short answers, comments, or ideas. They can give ideas written on cards using a few key words. These are then pinned to a canvas or wall. In the beginning itself, point out the importance of generating new ideas, with the impression that 'all ideas are good ideas'. If someone does not agree with someone else's point, a better idea should come out rather than arguing or disagreeing. Accept only additional contributions during the brainstorming. Defer those to the discussion afterwards. Always encourage fresh ideas rather than repetitions of earlier items. Each participant is allowed to express his/her view. However, talkative participants need to be restrained and encourage silent participants to come out with ideas.

Keep the brainstorming relatively short, about 10–30 min is sufficient to obtain most of the ideas on a specific topic without exhausting the participants. After the exercise, review the results with the groups. Remove all the duplicate items and form clusters of similar ideas. Highlight differences of opinion, if any, and discuss those until a list of clearly described ideas is achieved. Finally, summarize the results and keep them for future reference. For the brainstorming to be successful, the facilitator should have sufficient experience in dealing with group dynamics.

Transect Walk

A transect walk involves a kind of exploratory or observational walk undertaken by a team along with local people. The observational walk is commonly called *transect walk* as it is structured along a straight line cutting across the terrain in a specific way. These kinds of 'walks' help to verify the information provided on maps through direct observation and in discussions with people. The main disadvantage is that it may take much time. Two broad categories of transect walks are *social* and *land-use transects*.

Social transects concentrate on housing, infrastructure, and amenities; religious features; cultural features and behaviours; economic activities; and skills and occupations of people. *Land-use transects* focus on environmental and agricultural features such as cultivated lands, crops grown, fallow lands, forests, waste lands, soil erosion problems, streams, water bodies, and soil types. It provides much information about farming practices, cropping patterns, irrigation facilities, and agro-ecosystem of the locality.

Before planning a transect walk, collect some preliminary details about the area. Gather enough details about the boundaries and land characteristics. Determine the routes of transect walk based on these details. Decide the issues to focus on and the information that needs to be gathered. Arrange the teams along with villagers who are willing to walk. It is better to select elderly and experienced villagers (key informants) and village youths. Explain the purpose of the walk to them beforehand and assign various responsibilities such as asking questions, recording information, drawing quick sketches, and collecting materials like unknown plants and plant products to different team members. Sometimes, you may be able to gather all the details with a single transect walk. If the area is large or with undulating terrains, more transect walks have to be planned. Often, details are gathered through '*across transect walks*' but verified through '*along transect walks*'.

Diagrammatical Techniques

PLA involves several diagrammatical techniques to stimulate debate and record the results. Initially, many of these are not drawn on paper but on the ground with sticks, stones, seeds, and other local materials and then transferred to paper for a permanent record. However, in a community of educated villagers, the mapping techniques can be directly done on paper itself. Although visual-based techniques are important tools for enhancing a shared understanding between outsiders and insiders, they may hide important differences of opinion and perspective when drawn in group settings. Similarly, they may not reveal information on culture and beliefs adequately. Therefore, these are to be crosschecked by other techniques such as interviews and direct observation.

Diagrammatic techniques are usually grouped into *mapping techniques*, *ranking exercises*, and *trend analysis*.

Mapping Techniques

Participatory maps are helpful for understanding the physical characteristics of the area, the socio-economic conditions of the community, and perceptions of participants about their community. Mapping starts with collective discussions among groups of community members, and then drawing maps of their perceptions about the geographical distribution of environmental, demographic, social, and economic features in their locality. In certain cases, purchased maps, aerial photographs, or basic drawings on paper can be used as a basis for the participatory exercise.

Mapping exercises are helpful for providing a broad overview of the situation. These are useful to observe links, patterns, and inter-relationships in their territory. Another advantage is that even illiterate individuals can participate in the exercise. However, as bias and superficial treatment can influence the outcome, take care to complement mapping exercises with information obtained through other participatory tools.

Commonly attempted participatory maps include resource maps, indigenous technical knowledge (ITK) maps, timelines, and institutional maps.

Resource mapping: Participatory resource mapping is a tool that helps to understand the resource base of a community and its working using these resources. Resource maps can be drawn with the help of local people. For example, maps can be drawn of a village, a small watershed, an irrigation scheme, or command area. You need not be worried about having a perfect map; what is required is to obtain information on their perceptions of resources. Resource maps usually include the following information:

- Land use (location, crops grown)
- Assets (houses, public buildings)
- Infrastructure (roads, rail, others)
- Water resources (wells, rivers, springs, dams, canals)
- Irrigation and drainage system (irrigation canals, drains, water logging)
- Agro-ecological zones (soils, slopes, elevations)
- Communication facilities
- Supply and services (marketing, processing, and related facilities)
- Financial facilities, public utilities, recreational facilities, etc.

Organize a meeting for the whole community in consultation with local leaders and extension workers. Make sure that it is scheduled at a time when both women and men can attend and that all socio-economic groups have been invited. Resource maps can be prepared on a suitable open space on the ground or on large sheets of paper, if all the participants are literate. If the map is to be drawn on the ground, materials such as sticks, pebbles, leaves, sawdust, flour, or any other local material can be used. If you are working on paper, be ready with large sheets of paper and marker pens or pencils.

Before the start of the mapping exercise, discuss the subject of the mapping exercise and agree on the graphic symbols to be used. Allow the participants to choose their own symbols. Entrust a participant the responsibility of plotting symbols according to the suggestions of the group. Begin the process of mapping by indicating a central and important landmark. Prompt the participants to draw other landmarks on the map. Allow sufficient time to the group to discuss different opinions and perceptions. Once the map is finalized, ask the participants to interpret the overall picture and make corrections, if needed. Persuade them to identify the main problems revealed by the map. Finally, preserve the map for future use.

Indigenous technical knowledge (ITK) map: Local people may be practicing several successful indigenous technologies in agriculture, forestry, animal husbandry, and fisheries. Most of these would be proven methods, practices, and techniques, generated by their own and their ancestor's experience. Experienced villagers must be identified and discussion conducted to find out ITKs. Probe the reasons behind the practices, method of utilization, advantages and disadvantages, and note them. It is a good idea to collect samples, if available, and preserve them.

Timelines: In timelines or historical maps, sequences of changes occurred in the community are recorded. This mapping exercise is helpful to introduce the time dimension in environmental appraisal. It also gives an idea about environmental degradation and population changes. For example, it can generate discussion on issues such as increased soil erosion, population growth, climatic changes, and environmental degradation. The timeline technique is also useful to indicate the occurrence of droughts, fertilizer usage, years in which major crop failures took place, and shift in agricultural practices. It is important for the outsiders to take stock of this historical knowledge and the underlying experiences to understand local decision-making processes.

Timelines may be drawn through a participatory discussion. Start discussion by asking questions to the elderly key informants first. Ask them to review the major events and changes that took place over the years. Encourage some of them to write down the events in chronological order. However, do not insist too much on specific years or dates. Compare the changes and identify the trends.

Institutional maps: Another type of mapping exercise is institutional map, sometimes called *Venn* or *Chapatti diagram.* A Venn diagram can be used to look at relationships within the institutions or relationships between the community and other organizations. It shows the perceptions of participants about access to resources or of social restrictions. This technique can be done either as part of a group discussion or can be undertaken by individuals to illustrate the different perspectives they have on the organization.

You may be knowing that a Venn diagram is a tool used in mathematical set theory. Venn diagrams in participatory mapping mimic these mathematical diagrams. In a Venn diagram, a set of parameters is represented through a circle. The area of intersection of two circles represents common items occurring in each set. Thus, two partially overlapped circles are used to indicate the presence of common factors. Often, cut-outs of circles with different dimensions are used. Use larger cut outs to represent larger organizations. The distance maintained between two circles representing two organizations represent their closeness in relationships. Venn diagrams help the villagers to visualize the organizational interrelationship.

Ranking Exercises

Four types of ranking techniques are commonly used in participatory evaluation—problem ranking, preference ranking, matrix ranking, and wealth ranking.

Problem Ranking

The objective of problem ranking is to obtain perceptions of local people on their important problems. It is usually done with groups of men and women involving all the socio-economic groups. During the discussion, the participants list out their problems. Rank the problems according to importance and select the most important ones. Prioritize these problems by discussing the causes and effects with them. Another technique called *pair-wise ranking* is also common. In this technique, cards representing different problems are used. The facilitator prepare 'problem cards' and shows two at a time, asking them to respond, 'Which is the bigger problem?' The participants give the answers after comparing, and the results are noted. After counting the number of times each problem 'wins', these are arranged appropriately, and finally ranking is done.

Preference Ranking

In preference ranking, participants assess different options before them using their own criteria. It is almost similar to problem ranking but used to compare the preferences. For identifying the preferences, scores can be assigned. For example, the problems of water availability can be compared on a 5-point scale, by giving 5 for the most important and 1 for the least important. The usual procedure is mentioned here. Use a blackboard or flipchart, and draw a priority-setting background in a tabular format. Include the criteria in columns and possible actions in rows. Then, the group must evaluate possible actions and list them in the left column of the table. Distribute the scoring cards among the members so that each member gets one set of cards for each of the scoring criteria. The number of cards in each criteria set must be equal to the number of actions being ranked, for example, if there are four actions, then each criteria set should contain a number from 1 to 4. To assess the 'feasibility' of a specific action, the participant must display the appropriate card. After all the criteria were evaluated, note down the scores in the table. Add the individual scores and report the sum in the right column of the table.

To complete the process of preference ranking, all the participants are expected to study the final results and make comments on them. Explain to them clearly that the scores are meant to assist in decision-making. Ask the participants to make decisions considering both the trends revealed by the total scores and the suggestions from the discussion.

Matrix Ranking

In *matrix ranking* or *matrix scoring*, a group of options is evaluated by applying different criteria and assigning scores to each criterion. It uses a matrix with options along the horizontal axis and the elicited criteria along the vertical axis. This technique enables the facilitators to find out the reasons for the preferences and priorities in various practices such as crops, cultivars, tree species, soil and water conservation measures, irrigation methods, fertilizers, plant protection measures, and other technologies.

Matrix scoring can reveal interesting differences among group members. Start discussion on a particular technology; for example, if it is improved cultivars of rice, list all the cultivars grown by them. List the reasons in short phrases. Draw up a table of matrix by placing the reasons in rows and technologies in columns. Then, ask the villagers to rank the cultivars of their choice for each season. Indicate the cultivar best suited for the first season, second season, and so on, until all the cultivars are covered. If five technologies are considered for ranking, the best one for the specific reason is to be given a score of 5, the next best 4, the next 3, the next 2, and the last 1. Calculate the total score for each technology over all the reasons mentioned. Repeat the process for all participants. Based on the total score, make inferences on adoption of technologies and use these inferences for future planning strategies.

Wealth Ranking

In wealth ranking, people are ranked and placed on different steps of the social ladder based on their wealth. The objective is to have a first-hand information about the stratification of local society into rich, middle, and poor categories as viewed by people themselves. The exercise helps officials concerned to have an idea about the differences in wealth in every neighbourhood leading to an overall understanding of the socio-economic conditions of the population in the territory. This exercise is also helpful for selecting the beneficiaries for various schemes. The most common version of this technique involves a focus group discussion involving members of a locality. The facilitator introduces the procedure using local terms for wealth and poverty and encourages them to first discuss how they define these terms and how they would describe a poor or rich household. Using these criteria, cards are made and actual ranking is done by sorting cards.

Wealth ranking may be difficult in urban and semi-urban areas, where people tend to be less familiar with their neighbours than in rural settings. Moreover, in some communities, relative wealth ranking is a very sensitive topic, and therefore, it is better to conduct this technique in a private setting allowing participants to talk freely.

Trend Analysis

Trend analysis shows the changes in the past few years regarding the daily life pattern of people in a locality. It can be in the form of a line graph, which indicates the relationship between two types of data. Trend analysis can be done for the items such as the change in crop yield (increased or decreased), cost of cultivation of crops, wasteland area, farm produce prices, consumption of fertilizers and pesticides, land use changes, cropping pattern, population growth, coverage of improved cultivars, tree species grown, fuel used, flow of income (increase or decrease), changes in credit, and status of transport and communication facilities. *Seasonal calendars* and *daily activity charts* are two common trend analysis techniques.

Seasonal Calendars

Seasonal calendars or *seasonal diagrams* are tools, which help to explore changes taking place over the period of a year. In such calendars, attempts are made to determine the seasonal activities as understood and practiced by the villagers. Seasonal behaviour of features such as labour shortage, pest and disease incidence, harvesting of crops, ploughing activities, sowing of crops, intercultural operations, input requirements, irrigation requirement, weather elements such as temperature and rainfall, soil moisture status, and fuel and fodder availability can be assessed based on seasonal calendars.

Seasonal calendars are often compiled based on semi-structured interview and group discussions. Calendars can be drawn on a large piece of paper, traced in the sand, or on a concrete floor using stones or seeds. A series of seasonal variables can be included in one calendar to give an overview of the situation throughout the year. Important seasonal events such as festivals can also be shown.

Draw a horizontal line all the way across the top of the paper explaining that the line represents a year. Ascertain from the group how they divide the whole year as months or seasons. Accordingly, the participants have to mark the seasons along the top of the horizontal line. It is convenient to start the calendar by discussing the rainfall pattern. The participants may put stones, grains, sticks, or leaves under each month of the calendar according to the relative quantities of rainfall. It is better to scale it from 0 to 10, no stones meaning no rain, and 10 means maximum rain. The seasonal calendar can be made on paper too, by replacing the stones with dots in each month with a marker. After completing the seasonal calendar for rainfall, other aspects of interests can be covered.

Daily Activity Charts

Daily activity charts or *daily activity clocks* show graphically how the community members spend their day. The diagrams also make it easy to compare the daily activities of different groups of people such as women versus men, employed versus unemployed, rich versus poor, married women versus widows and young versus old. Daily activity charts show who works for longer hours, who concentrates on a small number of activities, who must divide their time for a multitude of activities, and who has the most leisure time. Just as a seasonal calendar shows the busiest times of the year, a daily activity chart shows the busiest times of the day, and therefore will be useful in monitoring changes in the time use of project beneficiaries and whether these vary by gender. It is a useful tool for gender analysis.

For conducting the exercise, organize separate focus groups of women and men making sure that each group includes people from different socio-economic groups. Explain the purpose of the exercise. Ask women and men groups to produce their own charts. They should first focus on the activities of the previous day. Illustrate all kind of activities such as agriculture work, livestock care, childcare, cooking, and sleep. Ignore slight variations in daily activities of persons, but go for daily activities typical of the group.

Gender Analysis

The term *gender* refers to socially constructed and culturally determined differences attributed to men and women and their roles in the household, community, or the society. These differences and roles change over time and vary widely within and between cultures. At the same time, the word *sex* is used to mean the real biological differences between a male and a female. A systematic study to document and understand the roles, needs, and priorities of women and man and their relations within a given context is called *gender analysis.*

Most of the PLA tools can be used for gender analysis too. Semi-structured interviews and daily activity charts are some typical tools. However, it is important to access and record the data in disaggregated and specific terms with respect to men and women. This helps us to understand the gender division of labour, how they use natural resources, rely on them, and have access to alternatives. Gender analysis enables us to explore and highlight the underlying inequalities in the relationship between men and women by asking fundamental questions such as *who* does or uses *what*, *how*, and *why*, and *who* benefits, and *how much*. When conducting discussions or interviews, both men and women should be allowed to provide their answers and their views. If necessary, separate meetings shall be conducted for men and women, because in some cultures, women are reluctant to attend meetings and to speak their minds. In such cases, a woman facilitator should assist the discussion and help women to find out what they wish to discuss and how. Some examples of questions are:

- Who owns the land? Who migrates?
- Who has access to what resources and their benefits—finance, equipment, land, natural products, etc.
- Who uses which natural resources, and for what?
- Who is paying the price?
- Who decides? How?
- What is getting better for the women or men?
- What is getting worse for the women or man?

In certain cases, we also need to include some additional questions such as: Which men? Which women?

Gender roles and responsibilities are not only structured by the socio-economic position of a household in terms of caste and class but also by other factors such as age. For example, women from lower caste and poor families shoulder most of the burden of domestic work. In contrast, upper caste and rich women may not bear the same burden of work than their hapless counterparts do. However, they are generally more restricted in social mobility. Similarly, age gives some more freedom for the women as older women cutting across caste or class categories enjoy in social mobility.

The Socio-economic and Gender Analysis (SEAGA) publications of the Food and Agriculture Organization (FAO) offer practical tools and methods for integrating socio-economic and gender issues at different levels and within different technical areas. Their main concern is to promote gender awareness, especially when meeting development challenges. The 'Field Level Handbook' (FAO 2001) is very useful for conducting gender analysis. It provides practical information on how to undertake socio-economic and gender analysis at the field level.

Organizing PLA

Adequate preparations should be made in advance for organizing PLAs. A typical PLA activity should be carried out by a multi-disciplinary team with a team leader and subject matter specialists according to the subjects to be explored. Usually, a team of people works for two to three weeks on discussions, analyses, and fieldwork. Logistical arrangements include accommodation, arrangements for food, transportation, funds to purchase refreshments for community meetings during the PLA, video or audio recorder, and supplies such as chart paper and marker pens. You can use portable computers too. Training of new team members is required. Appraisal results are influenced by the length of time allowed to conduct the exercise, scheduling and assignment of report writing, critical analysis of data, and recommendations made out of them.

A conducive environment must be created before attempting participatory learning. Meet the villagers with an open mind. They must be told about you and your intentions. Seek good rapport with the people. Identify the villagers who are willing to share their experiences.

PLA begins with a semi-structured approach. However, allow inclusion of unexpected or unrelated issues or topics that may come up during the discussions. A PLA covering relatively few topics in a small area normally takes between 10 to 30 days. However, a PLA with a wider scope over a larger area can take several months. Immediately after the fieldwork period, reports must be written. Try to make available a preliminary report within a week of the fieldwork, and the final report should be made available to all participants and the local institutions that were involved.

Among different PLA techniques, correct sequencing is important. The PLA team must decide the most appropriate techniques to elicit the desired information. Depending on the topic under investigation, the techniques can be combined in different ways. Mapping exercises are good to begin with as they involve several people, which stimulate much discussion and enthusiasm. These also provide the appraisal team an overview of the area. After the maps are drawn, it is a good idea to proceed with transect walk, along with some of the people who have constructed the map. As these local persons know the area clearly, it gives much information. Seasonal calendars and timelines can reveal changes and trends throughout a single year or over several years. Ranking techniques such as wealth ranking that have some sensitivity in divulging information, may be done only after establishing good rapport with the people.

Brainstorming serves as a good introduction for more structured and focused discussions at a later stage. Preference ranking is a good opener at the beginning of a group interview and helps focus the discussion. Subsequently, individual interviews can be conducted on the preferences of local people already spelt out and the reasons for these differences. At the end of the whole exercise, some clear cut ideas should emerge.

4.13 Some Special Research Methods

In agricultural and life sciences, research need not always involve experiments based on hypothesis formation and testing. It may be simply descriptive as in anatomy, morphology, or ecology. The approaches we adopt are dependent upon the nature of the discipline and particular topic. The following are some additional types of research employed for specific purposes.

Prototypes and Pilot Studies

Prototypes and pilot studies are two special forms of scientific troubleshooting. When designing a new apparatus or equipment, it is better to make a *prototype* first. Similarly, when beginning a novel type of experiment, it is recommended to conduct a *pilot study*. The prototype is a routine step in technology or applied science to act as a bridge between theory and practical application. Two scales of prototype are often used, *laboratory prototype* and *pilot plant*. A prototype is considered essential whenever a company or institution substantially modify existing equipment design. It is also useful when a researcher assemble equipment based on published works and scanty details.

The pilot study is the procedural analogue to a prototype. Pilot studies can give valuable information on whether a detailed experiment will succeed and how one should cope up with uncontrollable variables.

Directed Research

A study taken up by a researcher in response to a request from somebody to explore a specific problem is called directed research or targeted research.

Clinical Research

Clinical research addresses important queries related to the normal functioning of the human body and diseases using human subjects such as new treatments, new medicines, new techniques to use existing treatments, improved screening techniques, and new diagnostic techniques. Clinical trials are conducted based on a plan, usually called a ***protocol***. The protocol describes what types of patients may be enrolled for the study, schedules of tests and procedures, drugs, dosages, length of study, and measurable outcomes.

On Farm Research

It is a general term for evaluation, refinement, or development of farmer-oriented need based-research under the existing biophysical and socio-economic conditions of resource poor farmers. On farm research is usually designed to take the study results from the experimental field or laboratory to the farmer's fields.

Verification Farm Trials

Verification farm trials reflect a high level of confidence on the part of the researcher that a new technology is technically feasible and effective. It is often a straight comparison of new technology and current farmer's practice, usually done in large plots. Although it is a farmer managed practice, implementation is done jointly by the farmers, researchers, and extension workers.

Farmer Participatory Research

Farmer participatory research (FPR) is a kind of agricultural research system that ensures participation of farmers by including indigenous knowledge in the research process.

Operations Research

It is the use of scientific methods for decisions regarding human–machine systems involving repetitive operations. Operations research deals with strategic and tactical problems of various operations. It is also sometimes called *operational research* or *management science.* Operations research is used as a scientific approach to problem solving for executive management. *Operational research project* (ORP) is the application of specific methods, tools, and techniques to operations of systems for optimum solution to the problem.

Meta-Analysis

Meta-analysis is a statistical procedure for combining the results from several previously made independent studies. This helps to improve the dependability of the results. Meta-analysis gives a more objective appraisal of the collected evidences by different researchers than a traditional review of literature on the problem. It is also helpful to provide a more precise estimate of a treatment effect and explain heterogeneity between the results of individual studies. However, beware of biased analysis in which relevant studies are purposefully excluded or inadequate studies are included.

Bibliography

Barton, T., Borrini-Feyerabend, G., de Sherbinin, A., and Warren.P. 1997. Our People, Our Resources. IUCN (International Union for the Conservation of Nature and Natural Resources), Gland, Switzerland, 259p. Available: https://portals.iucn.org/library/ sites/ library/ files/documents/1997-038.pdf [Accessed 12 June 2018].

Best, J.W. and Kahn, J.V. 1993. Research in Education (7th Ed., Indian Reprint, 2004). Prentice–Hall of India, New Delhi, 435p.

Bryman, A. 2008. Social Research Methods (3rd Ed.). Oxford University Press, Oxford, 748p.

Burns, R.B. 2000. Introduction to Research Methods (4th Ed.). Sage Publications, London, 613p.

Chambers, R. 1994. The origins and practice of participatory rural appraisal. World Dev. 22 (7): 953–969. https://doi.org/10.1016/0305-750X(94)90141-4 [Accessed 10 May 2018].

FAO [Food and Agriculture Organization of the United Nations].2001. Field Level Handbook, SEAGA Socio–Economic and Gender Analysis Programme. FAO, Rome, 135p. Available: http:// www.fao.org/docrep/012/ak214e/ak214e00.pdf [Accessed 02 Jan. 2018].

Fathalla, M.F. and Fathalla, M.M.F. 2004. A Practical Guide for Health Researchers. WHO Regional Publications Eastern Mediterranean Series 30. World Health Organization Regional Office for the Eastern Mediterranean, Cairo, 232p.

Fowkes, F.G.R., and Fulton, P.M. 1991. Critical appraisal of published research: introductory guidelines. Br. Med. J. 302: 1136–1140.

Gray, D.E. 2014. Doing Research in the Real World (3rd Ed.). Sage Publications, London, 729p.

Holmes, D., Moody, P., and Dine, D. 2006. Research Methods for the Bioscience. Oxford University Press, Oxford, 381p.

Kerlinger, F.N. 1986. Foundations of Behavioural Research (3rd Ed.). Holt, Rinehart and Winston, USA, 667p.

Kothari, C.R. 2004. Research Methodology: Methods and Techniques (2nd Ed.). New Age International, New Delhi, 401p.

Kumar, R. 2011. Research Methodology: A Step-by step Guide for Beginners (3rd Ed.). Sage Publications India, New Delhi, 415p.

Mann, C.J. 2003. Observational research methods. Research design II: cohort, cross sectional and case-control studies. Emerg. Med. J. 20:54–60.

Marder, M.P. 2011. Research Methods for Science. Cambridge University Press, 227p.

Narayanasamy, N. 2009. Participatory Rural Appraisal: Principles, Methods and Application. Sage Publishers, New Delhi, 363p.

Pretty, J.N., Guijt, I., Thompson, J., and Scoones, I. 1995. Participatory Learning and Action: A Trainers Guide. International Institute for Environment and Development, London, 270p.

Trochim, W.M.K. 2003. Research Methods (2nd Ed.). Bizantra, New Delhi, 363p.

Wellington, J. and Szczerbinski, M. 2008. Research Methods for the Social Sciences. Continuum International Publishing Group, London, 234p.

Chapter 5
Experimental Research

> *The strongest arguments prove nothing so long as the conclusions are not verified by experience. Experimental science is the queen of sciences and the goal of all speculation.*
> *Roger Bacon* (1214–1294)

Experiments are part of the scientific method that helps to decide the fate of two or more competing hypotheses or explanations on a phenomenon. The term 'experiment' arises from Latin, Experiri, which means, 'to try'. The knowledge accrues from experiments differs from other types of knowledge in that it is always shaped upon observation or experience. In other words, experiments generate empirical knowledge. In fact, the emphasis on experimentation in the sixteenth and seventeenth centuries for establishing causal relationships for various phenomena happening in nature heralded the resurgence of modern science from its roots in ancient philosophy spearheaded by great Greek philosophers such as Aristotle.

When somebody mentions the term 'experiment', you might remember the chemistry or physics laboratory and the experiments conducted there. These experiments are conducted in highly controlled situations, and you always get repeatable results no matter how many times you repeat the experiment. In a laboratory, the researcher can control most of the variables that could affect the running of the experiment and its outcome. However, experiments need not always be conducted in laboratories! For wider external validity, experiments can be planned and conducted in natural settings, and such experiments conducted within a natural setting are usually called ***field experiments***. In fact, we are extending the classical laboratory methods to field or real situations with some safeguards against uncontrollable variations as in most agricultural or health experiments.

An experiment can be defined as a test designed and conducted by deliberately changing some variables in a system with the intention of falsifying a hypothesis involving selected variables. It involves the creation of a simulated situation in which the researcher manipulates one or more independent variables and measures the outcome variables. Carefully designed and implemented experiments continue to be one of the most powerful methods of science for establishing causal relationships. Experimental research differs from other research approaches, as it has greater control over the objects of its study. When you conduct experimental research, you are not

C. G. Thomas, *Research Methodology and Scientific Writing*,
https://doi.org/10.1007/978-3-030-64865-7_5

going to merely describe a condition, determine the status of something, or record past events as in other non-experimental methods described in the previous chapter. Instead of limiting your search to observe and describe what exists, you intentionally manipulate some variables to make sure how and why a particular event occurs. In other words, you are trying to establish cause and effect relationships between events or phenomena by forming hypotheses.

5.1 Cause and Effect Relationships

For establishing true cause and effect relationships, conducting experiments is the easiest and definite method. There are two major variables of interest in an experiment—the *'cause'* and the *'effect'*, and you directly manipulate causal variables, keeping other variables constant as far as possible. For establishing cause and effect relationships, you have to isolate and eliminate all the factors that might be responsible for a particular outcome and test only one that you directly want to measure. The researcher can control certain causes (such as a fertilizer schedule, growth regulator, or hybridisation process), but certain causes occur naturally or outside the researcher's influence (such as a flood, drought, or pest attack). The effect is the outcome that you wish to study. You must ensure that your manipulation of the independent variable is the only one having an effect on the dependent variable. Researchers do this by holding all other variables (that might affect the dependent variable) constant. This highly controlled procedure ensures that the observed changes in the dependent variable were caused by the manipulations alone. For example, take the case of comparing the effects of a cause of interest to us (e.g. fertilizer schedule) relative to other conditions (e.g. another fertilizer schedule or no fertilizers). A key factor in a causal study like the above is how you decide what experimental units (treatment plots) receive your fertilizer schedule, and what are placed in an alternative condition.

Experimental methods are widely used in agriculture and life sciences including health sciences for determining cause and effect relationships. The relationship or question is generally expressed as a statement of *hypothesis* that must be subjected to falsification through experimentation. In a typical experiment, the experimental area is divided into *experimental plots* or *experimental units.* Note that the term 'plot' is synonymous with 'experimental unit'. It refers to the unit of material to which a treatment is applied. It can be a single leaf, a whole plant, a pot, or an area of ground containing many plants, a person, group of persons, and so on. The objects under comparison called *treatments,* for example: cultivars, seed rates, fertilizers, medicines, or whatever under comparison, are assigned to each unit.

Sometimes, an independent variable of specific experimental interest (treatment) is referred to as a *factor*. Normally, the term is used when an experiment involves more than one variable. *Level* refers to the degree or intensity of a factor. There can be several levels for a factor including a zero level. Another term often used with experiments is *blocks,* which refers to categories of subjects within a treatment group.

Blocking is common in agricultural field experiments to take care of heterogeneity in experimental area (see Sect. 5.4). Blocks are also possible in other kind of experiments, for example, in a health study, we might divide the human subjects into old aged, middle aged, and youngsters forming three blocks.

Unlike physical sciences, variability is a characteristic of biological materials. There would be differences between experimental plots, even if you assign the same treatment to all the plots, and there would be substantial differences in yields. However, the differences may show a pattern. Often, neighbouring experimental units produce almost similar yields than distant plots. A major problem of such experiments is non-uniformity of experimental plots. There may also be systematic trends or local periodic variations across a field. Therefore, it is important to determine whether differences between experimental units are due to real treatment effects or because of unaccounted variations. By applying proper statistical tools, we can tide over this difficulty.

The limitations of experimental research must also be noted. The first and foremost limitation is that it can only be used when the conditions are appropriate for manipulating the variables. Sometimes, certain practical and ethical issues may prevent us from adopting this method. Experiments involving human beings and animals need stringent scrutiny in terms of ethics (see Chap. 23). For example, a psychologist might want to know, if parent's method of disciplining their children has an effect on how their children behave. It would be neither practical nor ethical to ask the parents to discipline their children in a certain way just to see if that affects their children's behaviour! Other examples include the effect of smoking, drinking alcohol, pesticide effects, and consumption of certain food items on the incidence of certain diseases. A second limitation is the artificiality of experimental conditions. The experimental studies are usually undertaken in laboratory or research farms in a highly controlled environment, and as the conditions are artificial, it is very difficult to apply the results in the less controlled and complex field situations. This issue is related to the validity of experimental findings (see Sect. 5.3). A third limitation is time. Experiments, in general, require more time and resources to complete the tests.

Experiments are of different types. Before examining various types, we must have an overview of some general concepts of experimentation such as hypothesis formation, errors, validity, and control.

5.2 Hypothesis in Experiments

Hypothesis formation is an important step in experimental approach. A *hypothesis* is a possible or probable answer to a specific question, which describes what you expect to happen in your study in definite terms. A hypothesis may or may not be correct; it may be still in the process of active testing. However, not all research studies have hypotheses. Sometimes, a study is designed to be exploratory or descriptive as in anatomy, ecology, or astronomy. In such studies, we may not be able to state a hypothesis in a formal way, as the purpose of the study is to explore some particular area of interest more thoroughly. Probably, the results obtained from such studies may be helpful to develop some specific hypothesis or prediction for future research.

For an experiment, we usually formulate two hypothesis statements—one that shows our assumption and the other that shows all the other probable outcomes with respect to the relationship set out in the hypothesis. For example, take two variables A and B. We can assume two types of relationships between the two variables, one positively and the other negatively. If we say that A and B are related, we should also take into account the other possible alternative that A and B are not related. The hypothesis we support is called the *research hypothesis* (H_1). It is a positive statement. For example:

> Irrigation during the critical stages of rice ensures higher grain yield and savings in water compared to continuous irrigation.

For conducting statistical analysis, you formulate the *null hypothesis* or statistical analysis (H_0) in a negative form to describe the possible situation, if the original research hypothesis is false. It is written as a negative statement for testing purpose only, for example:

> There is no significant increase in grain yield or savings in water when irrigation is done during the critical stages of rice compared to continuous irrigation.

Null hypothesis is important for statistical analysis. It assumes that there is no difference between the mean of the sample and the mean of the population, and whatever you notice is only a random sampling error. Alternate hypothesis is a statement about the population parameter or parameters, which gives an alternative to the null hypothesis, within the range of pertinent values of the parameter. That is, if Ho is rejected, what hypothesis is to be accepted and vice versa.

The philosophical basis for the null hypothesis is related to the doctrine of *falsification* by Karl Popper (see Sect. 2.6), wherein he argued that science makes progresses by testing and falsifying hypotheses. Rejection of H_0 is equivalent to falsifying it, and therefore, provides support for the research hypothesis (H_1) as the only alternative.

You can never prove a null hypothesis, as any one result that differs from the hypothetical result is sufficient to disprove it. It can, therefore, be either disproved and rejected or accepted within the limits of sampling errors. The data from the experiments are subjected to statistical analysis. The probability of occurrence of the null hypothesis is calculated. If the probability is lower than the level of significance, you reject the null hypothesis. The null hypothesis is tested for rejection under the assumption that it is true, as distinct from the research or alternate hypothesis. Rejection of null hypothesis suggests that you have enough evidences to support your research hypothesis.

The rejection or acceptance of a null hypothesis is based on some level of significance. In most field experiments, 5 percent ($p = 0.05$) level of significance is often used as a standard for rejection (in certain cases, you may even go for 1 percent ($p = 0.01$) level of significance). When you reject a null hypothesis at 5 percent level of confidence, it indicates that a difference in means as large as that between the experimental and control groups would have resulted from sampling error in less than 5 out of 100 replications of the experiment. In this case, the probability that the difference is due to the experimental treatment rather than to sampling error is 95 percent.

A research hypothesis is a working assumption. It is a tentative explanation for an observation, certain behaviour, phenomenon, or event, which has occurred or will occur. The hypothesis gives an idea about the researcher's expectations pertaining to the relationship between the variables in the study. It is formulated after preliminary observations or following a thorough review of related literature, or both, but before the execution of the study. The researcher collects enough data that either support the hypothesis or do not support it. If the hypothesis works, the scientist can continue further studies. A good hypothesis must be:

- *Testable*: You must be able to design an experiment or make an observation to determine the validity of the hypothesis.
- *Falsifiable*: The possibility exists that the hypothesis could be false (Sect. 2.6).
- *Simple*: Hypotheses that are simpler tend to be accepted over others that make similar predictions but are more complex (principle of parsimony, Sect. 2.8).

5.3 Validity of Experiments

A major limitation of experiments is that the conclusions and inferences made out of the experiment should have a wide range of validity or authenticity applicable to other settings and regions. An experiment is valid only if observed changes in the dependent variable are actually due to the independent variable and the results of the experiment apply to the real world outside the experimental setting such as a laboratory. If the experiment in question fails to achieve these conditions, it is of no value. The first condition is related to *internal validity*, and the second is *external validity*. Campbell and Stanley (1963) started the debate on validity threats and the factors responsible for the threats.

An experiment is said to have high internal validity when the treatment factors that have been manipulated (independent variables) actually have a genuine effect on the observed effects (dependent variables). In other words, the experimental treatment alone should make the difference rather than other extraneous factors. A study with high internal validity permits conclusions on true cause and effect relationships. After the experiment is completed, you are likely to make some claims that your research findings have implications for other settings and at other times. When you do so, you should examine the external validity of these claims, that is, the extent to which the findings of the study can be generalized. In fact, external validity is another name for generalizability. You would achieve nothing, if your observed relationships between the variables were valid only in the experimental setting!

While planning and conducting experiments, we may confront several threats to internal validity and external validity. We may not be confident enough to say that a relationship exists between the independent and dependent variables, if threats to internal validity are high. Similarly, our confidence in stating whether the results of the study are applicable to other settings and groups is also compromised, if the threats to external validity are not under control.

Threats to Internal Validity
The factors that affect internal validity of a research design are: (1) history, (2) maturation, (3) testing effects, (4) instrumentation, (5) sampling bias, (6) attrition, and (7) selection bias.

History
While the experiment is under implementation, if some external events other than the treatments such as a drought, flood, or pest incidence occur, these can influence the results, and internal validity is at stake. These kinds of events during the period of experimentation are usually denoted by the term 'history'. By including a control group from the same population as the experimental group sharing the same history, we can reduce the effects of history largely. The threat from history is relatively less in experiments of short durations.

Maturation
Maturation refers to changes that can occur in the subjects for over a period influencing the results of a research study. Unlike history, effects of maturation are not the results of specific events but because of ageing, changes in physical, intellectual, or emotional characteristics occurring naturally over time, or similar changes in the experimental subjects. These are more problematic with respect to human subjects, because during the experimental period, physical or psychological changes taking place within the subjects can jeopardise the conclusions. As in the case of dealing with the effects of history, the effects of maturation can be overcome by using a control group chosen from the population from which the experimental group is drawn. The effects of maturation are likely to be less in short duration studies.

Testing Effects
This threat to internal validity is due to the effects of a previous test upon the performance after a second test. For example, exposure to a pre-test usually influences observations from a post-test. This threat is more obvious in experiments involving human beings. If the intervening time between pre-test and post-test is short, testing can become a serious threat to internal validity. In most cases, the use of a research design that does not include a pre-test can eliminate testing as a potential threat to internal validity. However, if pre-treatment data are required, the only option is to use data collection techniques without revealing the actual intention to participants.

Instrumentation
Changes in the calibration of instruments, questionnaires, or interviewing technique can create problems to internal validity. Standardized instruments, data collection procedures, and the training of observers are some of the measures, which reduce problems from instrumentation. Most threats from instrumentation can be eliminated by specifying the measurement procedures. Occasionally, the instruments being used

for the experiment have to be recalibrated, if their readings become suspect. For example, voltage fluctuations and temperature variations can play havoc with your data. However, by keeping these factors constant during the course of investigation, these can be prevented from influencing other variables. This *stabilization* approach is the best for variables that have a disruptive influence, for example, voltage or temperature variations. In many instances, you can use a *standard sample* for managing simultaneously uncontrollable or unknown variables.

Sampling Bias
This threat occurs when the assignment of subjects to experimental groups is done in a way likely to prejudice outcomes, for example, when individuals are selected for an intervention or treatment based on extreme scores on a pre-test. Avoiding the use of extreme scorers and using random sampling will minimize the threat due to sampling bias.

Attrition
Attrition or mortality is the differential loss of individuals from treatment or comparison groups, which may bias the results. Loss of plants, animals, and human subjects come under attrition threats. In experiments with human subjects, this is often a problem when research participants are volunteers. They may drop out of the study, if they find that it is consuming too much of their time or the task is too arduous. To reduce mortality threats, an experimenter need to select large groups and must make certain that they are true representative of the population.

Selection Bias
Selection bias pertains to the possibility that differences exist between individuals in treatment and control groups at the start of the study and that those differences may affect the results. For example, one group may differ from another group in age, ability, gender, ethnic composition, or other ways. If the differences in features of a group affect the result of a study, these amount to a serious threat to internal validity. Internal validly threats due to selection can be minimized by adopting random selection and random assignment of subjects and blocking.

Threats to External Validity
As discussed earlier, external validity means the extent to which a particular experimental result can be generalized assertively to other settings and groups, that is, the 'real-world' larger than the group selected for the study. External validity is under threat whenever conditions inherent in the research design are such that the generalizability of the results is limited. In general, you can reduce threats due to external validity by taking various precautions to ensure that the selected sample, the environment, and the background are representative of the population.

Threats to external validity of results may be related to the *population*, the extent to which a sample is representative of the population from which it was selected; or to the *ecology,* that is, the characteristics of the context or setting of the study are representative of the context and setting to which the results are to be extended.

Accordingly, two forms of external validity—*population validity* and *ecological validity*—are usually mentioned.

Population Validity
The question of population validity arises when the selected sample is not a true representative of the population. The more representative the sample, the more confident we would be in generalizing from the sample to the population. Another question is how extensively the results can be applied. When a particular research result works across many different kinds of settings—even those not represented in the sample— external validity is high.

Ecological Validity
We may have created a set of environmental conditions for the research work and the results may be valid only in those conditions. The extent to which the results of a study can be simulated from the background environmental conditions of the study to other environmental conditions is often called *ecological validity*. Threats to ecological validity aggravate when the environmental background of the study is not representative of the population.

The following are some types of ecological validity: (1) effect of testing, (2) multiple-treatment interference, (3) selection-treatment interaction, (4) reactive effects of experimental arrangements, (5) experimenter effects, and (6) specificity of variables.

Effect of Testing
This can happen mostly in the case of human subjects. The administration of a test, especially pre-test, may affect the performance of the participants in a research study. If this occurs, it means that the performance of the subjects being studied may be different from what it might have been. Therefore, the results may not be generalizable to situations where pre-testing will not occur. This threat is similar to the internal validity threat of 'testing', and the solution is similar. The use of research designs that do not include pre-tests can help eliminate this potential threat.

Multiple-Treatment Interference
Multiple-treatment interference can happen in situations where individuals participating in a study receive more than one treatment. In such research studies, the effects of multiple treatments may interact and show some effects contrary to our expectations. When the researcher applies several treatments, it may not be easy to understand the effectiveness of each treatment separately, probably, only the combination of the treatments works effectively.

When the same experimental subjects receive two or more treatments simultaneously, there may be a carryover effect between treatments such that the results cannot be generalized to single treatments. When there is sufficient reason to believe that there will be multiple treatment interference, the researcher should select a design where only one treatment is assigned to each subject. In case such a design is not workable in the given conditions, the researcher may take steps to measure the effects of all pertinent treatments and integrate them into a multiple treatment design such as a factorial design.

Selection-Treatment Interaction

This can happen when selected treatment groups or control groups are more or less sensitive to the treatment prior to initiating the treatment. This is most likely to occur when the treatment and control groups are not randomly selected or not matched. There is a possibility that some characteristic of the participants for the study interacts with some aspect of the treatment. Such characteristics include prior experiences, personality factors, or any other traits that might interact with the effect of the treatment. For the results to be validly generalized to a larger population, the sample must possess the same traits, characteristics, experiences, and so on as the population.

Selection-treatment interaction is analogous to the internal validity threat of selection, and the measures to take care of this threat are alike. The external validity threat due to selection-treatment interaction can be reduced by following measures such as matching, blocking, random selection, and random allocation of treatments. When random selection or random assignment is not practical, statistical techniques such as analysis of covariance can obviate most differences due to measurable attributes of the individual, thus minimizing threat from selection-treatment interaction.

Reactive Effects of Experimental Arrangements

Sometimes, reactive effects of experimental arrangements may pose threats to external validity. Reactive effects mean the awareness of participants or volunteers that they are participating in a research study. Because of that awareness, their response or performance would differ from what it would have been otherwise. The reactive effects may be due to (1) the novelty of the experimental treatment, for example, a new technique or machine (*novelty effect*); (2) the belief on the part of participants that they are participating in an experiment and receiving some special treatment (*Hawthorne effect*); (3) the participant's belief in the effectiveness of the treatment (*placebo effect*); or (4) the participants in the control group view themselves in competition with the treatment group and so changes their behaviour (*John Henry effect*). For details on these reactive effects, see Sect. 6.4.

Experimenter Effects

In certain situations, the experimenter may inadvertently influence the performance of treatments in a study. This is known as experimenter effects. The experimenter effects may be passive, for example, the gender, race, or personal attributes of the researcher affecting participant's performance; or active, if the researcher is directly involved. This usually happens, if the researcher communicated their expectations to the participants in a manner affecting their performance. *Rosenthal effect* is the most common experimenter expectancy effect (Sect. 6.4). *Personal bias* can also influence the results adversely.

If some form of experimenter effect is anticipated, use *blind studies* to minimize such threats to external validity (Sect. 5.10). In a blind study, instead of directly collecting data or make observations, the researcher trains an ordinary observer and she/he collect the data on behalf of the researcher. Make sure that the person collecting the data or making the observations does not have any idea about the purpose of the study and ignorant about the specific treatments.

Specificity of Variables

Specificity of variables is also important to ensure external validity. If the variables included in a study are not adequately described and operationally defined, the ability to generalize the results of the study is threatened. This is also useful to allow another researcher to replicate the study, if required. Similarly, ensure that for the description and definition of variables, you employ only reliable and valid devices and equipment.

When designing and conducting experiments, you may confront several threats to both internal and external validities similar to the above. In many cases, validity threats cannot be evaded fully. The mere coincidence of a threat to validity does not mean that the experimental results are erroneous or confusing. However, a prior knowledge on validity threats enables the researcher to take appropriate actions to reduce their influence.

5.4 Principles of Experimentation

As discussed in Sect. 5.3, both internal and external validities are serious issues, which require careful consideration while designing experiments. The influential ideas for experimental designs applicable to laboratory and field experiments in agriculture, medicine, other life sciences, and social sciences can be traced back to Sir Ronald A. Fisher (1890–1962). Fisher was a geneticist, who joined Rothamsted Experimental Station in 1919. He was asked to do something on the accumulated data over the years from agricultural experiments. He systematically introduced the principles of experimentation—*randomization, replication,* and *local control* or *blocking* to take care of validity issues and broadened insights into the applicability of analysis of variance. The publication of Fisher's *Statistical Methods for Research Workers* in 1925 and *The Design of Experiments* in 1935 gradually led to the acceptance of the three principles of experimentation worldwide. Therefore, one can say that the experimental designs have their origin in agricultural research. Before Fisher's pioneering work came to limelight, most researchers used systematic schemes rather than randomization to assign participants to the levels of a treatment; often leading to possible bias and inaccurate interpretation of the data.

Design of experiments is also used in many other fields of application in addition to agriculture, for example, medicine, psychology, and sociology. Since majority of present-day experimental designs originated from agricultural experiments, it is not surprising to see several terms of agriculture such as treatment, yield, plot, and block in experimental designs related to other disciplines.

Fisher developed the concept of randomization based on small samples by consulting William Sealy Gosset (known popularly as 'Student'). The three principles that Fisher vigorously championed—randomization, replication, and local control—remain the foundation of good experimental design still now, and they are considered as the major principles of experimentation to increase both internal and external validity.

Randomization

In experiments, assigning plots or subjects to various treatments must be done in such a way that treatment and control groups are comparable in all respects except for the treatment. For example, if a researcher selectively assigns treatments to participants, personal bias can creep in. Therefore, randomisation of treatments is an important principle to avoid personal bias, and it ensures that each treatment have an equal chance of being assigned to any experimental plot. Randomization is also helpful to deal with uncontrolled variables. When you resort to random selection of samples, you can easily translate the biasing effects of uncontrolled variables to random unbiased errors. For example, 'time' can influence an experiment as equipment may show minor changes because of the period factor. Here again, by randomizing the order of sample measurements, the researcher can evade temporal influences of this kind from potentially biasing the outcome.

Groups can be created in two ways: through *matching* and *randomization.* The subjects may be matched according to characteristics such as family lineage, age, weight, health condition, soil fertility levels, or some other characteristics. This involves distributing these subjects or crop plants so that each subject in one group exactly matches characteristics of subjects in other groups. However, it would be difficult to have equal groupings; therefore, randomization is preferred to matching. Random assignment makes sure that the experimental groups are probabilistically alike at the time of assignment. Randomization can be done with a table of random numbers or by drawing lots.

In fact, randomization is based on the statistical principle of *normal distribution.* If you subject population data to normal distribution, most individuals will fall within the middle range of values for a given characteristic with progressively fewer ones towards either extreme (remember the bell curve of normal distribution). Differences between groups will average out and become more comparable.

Replication

Experiments should be repeated more than once to increase reliability. The process of repeating the treatment more than once is known as *replication* (a coined term from *repetition* and *duplication*). Replication increases statistical accuracy and precision. The treatment must be replicated at least twice to estimate experimental error. However, too many replications are costly, and beyond a certain limit, additional advantage due to precision is not significant if you consider the additional cost involved. The number of replications needed depends on the magnitude of experimental error likely to be obtained in the experiment and the degree of precision wanted. The experimental error can be kept to the minimum by providing more *degrees of freedom* (Sect. 6.9) for the experimental error.

Generally, we follow a rule of thumb to determine the number of replications. Choose the number of replications to provide not less than 10 degrees of freedom for estimating experimental error (Gomez 1972). However, Panse and Sukhatme (1985) suggested 12 degrees of freedom as the lower limit. In most experiments, we set the limit at 12 degrees of freedom for error. By following this thumb rule, with five treatments in a randomized block design, a minimum of four replications are needed (degrees of freedom is an interesting concept in the analysis of variance). The number of degrees of freedom in a distribution is the number of values that are independent of each other, which cannot be deduced from each other (see Sect. 6.9).

Blocking/Local Control

Ronald Fisher observed that by dividing the experimental fields into different blocks, and then conducting experiments in these blocks, the data gathered and inferences drawn would be more reliable. This process of *blocking* or *local control* helps the researcher to separate out differences due to confounding variables and by excluding them from the estimates of error effects.

Blocks refer to categories of subjects within a treatment group. In agricultural field experiments, variations between experimental plots can be reduced largely through blocking. When the blocks differ widely, there will be substantial reduction in experimental error. However, plots within a block should be homogenous. When the fertility pattern of an experimental field is known, blocks can be oriented in such a way that soil differences between blocks are maximized and those within blocks are minimized. When fertility patterns are not known, avoid using long and narrow blocks. Use compact blocks as nearly square, since closer plots can be more alike than those that are farther apart.

We can have blocks in experiments related to human beings too. Suppose we want to test two medicines. We might divide the subjects into three blocks—old aged, middle aged, and young patients; and further divide the groups into a group treated with medicine A and another treated with medicine B. This arrangement allows us to find out how the treatment affects different age groups. In the example, we might observe that medicine B out performs medicine A except for older patients where medicine A outperforms medicine B. This phenomenon is due to *interaction* between the treatment (the medicine) and subject characteristics (age). Interaction between treatment factors can occur in factorial trials when there are more than two factors.

5.5 'Control' in Experiments

Normally, an experiment demonstrates that an event occurs after a certain treatment is given, and that the event does not occur in the absence of the treatment. By dividing the entire experimental units into two groups, the 'control group' and the 'experimental group', you can confirm the effect of a variable, even if many unknown variables interfere. In such experiments, a '*control group*' is useful, especially when one must deal with several uncontrolled variables. The two groups should be as similar as possible except for the one aspect whose effect is being tested. In this context, you should particularly note that a 'controlled experiment' is not an experiment under control; it simply means that the experiment is designed with the 'experiment' and the 'control' groups.

In agriculture, you resort to controlled experiments when there are a number of alternative treatments. These are tested along with a standard treatment (control). Each treatment is applied to each experimental unit, and then several observations are made on each unit. Enough safeguards are taken to separate out experimental and random variations from true variations. Experiments conducted by observing adequate precautions and stipulations help the researcher to separate out differences between the treatments from the uncontrolled variation that is assumed to present. When planning for experiments, the treatments to be included are a point of concern. The researcher should know very well why these particular quantities for the treatments, and why these combinations alone and not others have been fixed.

Take sufficient precautions so that the two samples—experimental and control—are identical in all important respects. The dependent variable is measured in both groups for an identical period before the treatment. The treatment is then introduced into the experimental group only, and the dependent variable is measured in both groups for an identical period after introducing the treatment. The treatment effect is determined by measuring differences in the dependent variable in the control group and experimental group.

Many researchers believe that for all kind of experiments, a 'control' must be included. Consider seriously this question and make sure whether a control is actually needed. 'Control' need not be insisted in all the experiments, but examine each case individually and decide whether a control is essential. Although most scientists vouch for the need for a control, understand that there are experiments where a control can be harmful! For example, if there is already an effective cure for a disease, it is unethical to conduct an experiment comparing a new cure with 'control' (no treatment). In this case, the treatments ought to be the new cure and the one currently in use, instead of a 'no treatment' control. Similarly, in a trial with some new pesticides, if there is 'no treatment' control, then the pest may multiply on those plots and spread to other fields.

5.6 Classification of Experiments

Hypothesis testing experiments are generally classified as (1) true experiments, (2) quasi-experiments, and (3) ex post facto experiments based on the applicability of randomization and manipulation of treatments.

True Experiments

True experiments are characterized by three main features:

1. Randomly formed groups
2. Manipulation of treatments
3. Measurement of dependent variables.

Random assignment is the most feasible practical strategy to ensure that groups are statistically equivalent. Note that random assignment is different from random sampling. Random sampling is one method for selecting subjects to participate in a study. At the same time, random assignment is a method for assigning subjects such as people, animals, or plants to a treatment group, but not a method for selecting the study subjects. A true experiment requires randomly formed groups; however, one could use convenience sampling to select study participants.

The second requirement of a true experiment is manipulation of treatments, which means that the investigator can determine who receives what treatment. If the researcher decides who gets what, then manipulation occurred. The third requirement is that groups are compared and the dependent variables are measured. In most experiments, at least two groups are compared by measuring the dependent variable.

In agriculture and life sciences, true experiments are often grouped under three categories—laboratory experiments, pot culture experiments, and field experiments.

Laboratory Experiments

Laboratory experiments are carried out under controlled conditions, and chances of errors are comparatively less. In such experiments, one or more independent variables are deliberately varied and measurements on the dependent variables are taken in a contrived, artificial setting, which could eliminate all the possible extraneous variables. Laboratory experiments are commonly employed for basic studies in disciplines such as physics, chemistry, biochemistry, physiology, biotechnology, and other science disciplines (Fig. 5.1). However, laboratory experiments with crops are rare in agriculture as growing of crops under artificial conditions has only limited practical value.

Pot Culture Experiments

These are conducted under controlled conditions but not so rigidly as in laboratories. These have greater variations than laboratory experiments but are extremely useful in solving fundamental problems in agriculture. Certain disciplines such as plant pathology, soil science, and plant physiology employ pot culture experiments in a big way (Fig. 5.2). A negative result obtained in pot culture trials would, in all probability, be negative in field trails also, while a positive result need not necessarily be so. Pot

Fig. 5.1 Laboratory experiments. Both traditional and modern methods are used in the laboratory. Seen left is a typical chemistry experiment (**a**). In the right, a student is working with an Atomic Absorption Spectrophotometer in a chemistry laboratory (**b**)

culture studies alone or in conjunction with laboratory studies are common for basic and some applied studies.

Field Experiments

Field experiments are routine in applied sciences in which the independent variables are manipulated and measurements on the dependent variable are made on experimental units or plots in their natural setting. Once there were not many takers for

Fig. 5.2 A typical pot culture experiment in plant pathology

Fig. 5.3 A view of a field experiment with rice

field experiments, and many scientists considered them as ineffective as a scientific instrument of research. However, in the recent decades, field experiments have received large attention from scientists and statisticians. By the use of modern techniques and application of appropriate statistical methods, field experiments have the potential to show significant results. In agriculture, field experiments are widely used (Fig. 5.3). Heterogeneity experienced in the fields can be eliminated largely by suitable experimental designs, layouts, and statistical analysis. For more details on field experiments, see Sect. 5.8.

Quasi-experiments

Unlike true experiments, in a quasi-experiment, groups of subjects are formed not based on random assignment but by some other methods. For all practical purposes, it is just like a true experimental study with the only difference of missing randomly assigned groups. When using human subjects, it is often impossible to do random assignment, especially when they are part of intact groups such as schools, colleges, communities, neighbourhoods, or hospitals. Although groups might be reasonably similar in a practical sense, using data from intact groups limits the conclusions that can be drawn regarding treatment effects (more details in Sect. 5.12).

Ex Post Facto Experiments

The meaning of ex post facto is 'after the fact'. The phrase ex post facto is used in situations wherein the researcher devises an experiment to examine the effect of a treatment after it has already taken place. The causal event of interest has already happened without the manipulation of treatments by the researcher. These kinds of experiments are useful when human subjects in true situations are involved, and the researcher comes to the scene only 'after the fact'. Most ex *post facto* studies are observational studies as no manipulation of treatment is possible. In such situations, the investigator begins the investigation by identifying the outcome variables and then tries to identify possible reasons, that is, independent variables responsible for the outcome. As an example, take the case of two groups of patients in a hospital, one treated with a particular medicine and the other with an alternate medicine. If the researcher is deliberately selecting the patients and giving the medicines randomly, then it would be a true experiment. On the other hand, if the researcher has no role in medication and she/he just takes observations after the medicines were given by somebody else, it would be an ex post facto study.

Quasi-experimental research and ex post-facto are commonly used in health and social research. More information on research designs with quasi-experimental and ex post-facto are provided in later sections in this chapter.

Classification of Experiments Based on Timing of Observation

Experimental studies can also be grouped based on whether observations are measured before the introduction of the treatment and whether there is a control group. Accordingly, you have the after-only designs, the before-and-after designs, after-only designs with control, and the before-and-after designs with control.

The After-Only Design Without Control

In 'after-only' experimental designs without control or *post-test only without control*, data are collected after the subjects have been exposed to the independent variable. In this design, baseline data are created depending on the respondents' recall of the situation before the intervention or from existing records. Changes in the dependent variable are reported as the difference between the baseline data and the observations made 'after' the intervention. This type of design is adopted when you do not have the time to take 'before' observations directly. For example, if you want to study the impact of a campaign by the government, which must be implemented as a contingency measure, probably you may employ this method. In this case, baseline data are created from the records, and any change in trends is assumed to be because of the new programme. The main problem with after-only designs is that they do not have any control over other confounding factors that could have affected the post-test observations.

The After-Only Design With Control Group

In this experimental design, the experimental groups and the control groups are measured and compared only after the implementation of an intervention. The 'after-only design with control group' is also sometimes called '*post-test only control group*' design, wherein subjects are randomly assigned to either a control group or an experimental group. The groups are not pre-tested. One group is exposed to a treatment or a series of different treatments, and then both groups are post-tested. This design assumes that the control and experimental groups are equivalent except for the randomly assigned intervention. It is assumed that the difference between the groups is the effect of independent variable. Between group differences are measured and subjected to statistical analysis to determine treatment effects.

Most of the comparative experiments being taken up in agriculture, especially in agronomy and horticulture, fall under this category.

The Before-and-After Design Without Control

This is also known as the *pre-test/post-test design without control.* In this type, a single experimental group is chosen and the impact of dependent variable on it is measured before introducing the treatments. After introducing the treatments, the dependent variable is again measured and the difference between the pretest values and post-test values are estimated. The before-and-after design is an upgraded version than the after-only design, because the researcher is considering the probable effect of the independent variable by observing differences between the observations on the dependent variable before and after the test. This kind of study normally includes two sets of cross-sectional data collection points on the same population to find out the change in the phenomenon or variable between two points in time, the second being undertaken after a certain period. The difference between the two sets of data collection points with respect to the dependent variable is considered the impact of the programme.

The before-and-after design without control is considered as the most suitable design for measuring the impact of a programme or activity. It is also one of the most common designs in evaluation studies. We get the overall impact of introducing the treatments from the differences in observation before the treatment and after the treatment. However, the main difficulty with such a design is the chances of interference from extraneous variables on treatment effects especially if the interval between observations is too long.

The Before-and-After Design With Control Group

In this type of experimental studies (also called *pre-test/post-test control group design*), two test groups are established first—an experimental group, which would be exposed to the independent variable and a control group, which would not be subjected to the variable under study. Both groups are pre-tested. After exposing the experimental group to a treatment or different treatments, both groups are post-tested. The before-and-after design with control is designed based on the assumption that the confounding factors, if any, would affect equally the experimental and control groups, and therefore, any differences in the data collected from the two groups can be credited to the impact of the treatment alone.

5.7 Experimental Design

Experimental designs are concerned with the planning and conduct of experiments including the system of assigning the selected treatments to the experimental plots and analysing the generated data to obtain valid conclusions. A good experimental design allows valid comparisons between treatments. In the case of agricultural field experiments, soil heterogeneity is a main source of variation. The selected experimental design must balance the requirements and limitations of the problem in which one works so that the experiment can provide the best conclusion about the hypothesis being tested. An experimental design recognizes the independent, dependent, and confounding or external variables, and specifies the manner in which the principles of experimentation— randomization, replication, and blocking—and statistical aspects of an experiment are taken up. The main objective of an experimental design is to establish a causal connection between the independent and dependent variables. Another major goal is to extract the maximum information with the minimum utilization of resources.

Single Factor Experiments

The experiments wherein only a single factor differs while all other factors are maintained uniform are called *single factor experiments*. The treatments selected include different levels of a single factor. For example, suppose that you want to compare a group of crop cultivars for higher yield. In cultivar trials, the single variable factor is the cultivar and the factor levels (i.e. treatments) are different cultivars. In such trials, only the cultivar varies from experimental plots. Other management practices will be uniformly applied to the plots. Fertilizer trials, where several rates of a single fertiliser element are tested; insecticide trials, where several insecticides are tested; plant population trials, where several plant densities are tested, etc., are examples of single factor experiments.

For dealing with single factor experiments, two groups of experimental designs are used: complete block designs and incomplete block designs. *Complete block designs* are suited for experiments with a small number of treatments and is characterized by blocks, each of which contains at least one set of treatments. *Completely randomised design* (CRD), *randomised block design* (RBD), and *Latin square design*s (LSD) are the three most commonly used complete block designs. Complete block designs are often used for simple experiments with relatively small number of treatments. The analysis of data is simple, and handling of missing data, if any, is easy with this kind of designs.

In a *completely randomized design (CRD)*, the treatments are allocated completely at random so that each experimental unit has the same probability of getting any one treatment. It is assumed that the differences, if any observed, between experimental units receiving the same treatment are due to experimental errors. This error must be kept to the minimum. Therefore, CRD can be used for experiments with homogenous experimental units such as laboratory experiments and pot culture experiments, where near uniformity can be ensured and extraneous factor effects are relatively easy to manage. For field experiments in agriculture, however, CRD is not normally used because of field heterogeneity. In clinical or epidemiological studies, the experimental units are human subjects, and an experiment with completely randomized design is usually referred to as *randomized controlled trials (RCTs)* (see Sect. 5.10).

Randomized complete block design (RCB) is a widely used design in field experiments in agriculture because of its simplicity and flexibility. A unique feature of RCB is the presence of blocks of equal size each of which contains all the treatments.

Latin square design (LSD) is appropriate where the differences in experimental units are in two directions, for example, differences in fertility gradients by length and breadth. In such cases, blocks are to be considered in both directions following a Latin square design. The design gets its name from an ancient conundrum regarding the methods in arranging Latin letters in a square matrix so that each letter appears once in each row and once in each column.

Incomplete block designs are suited for experiments with a large number of treatments and are characterized by blocks, each of which contains only a fraction of the treatments under test. An example is varietal tests, which usually involve many varieties. *Lattice* and *group balanced block designs* are the common incomplete block designs.

Multiple Factor Experiments

Plants and animals are simultaneously exposed to many growth and environmental factors during their lifetime. The response of an organism to a particular factor may be different with the level of other factors. If the effect of one factor changes as the level of the other factor changes, we can say these two factors are interacting. When the response to a particular factor is expected to differ under the influence of another factor, you choose a two-factor experiment. If more factors are involved, you sometimes go for multiple factor experiments. Factorial experiments are used to handle two or more factors simultaneously. The experiments wherein the treatments comprise of all probable combinations of the selected levels of two or more factors are called *factorial experiments.*

Factorial experiments can also be divided into *complete block designs* (factorial RBD and factorial LSD), *incomplete block designs* (balanced lattice and confounding), and *split plot designs* (split plot, split-split plot, and strip plot).

In experiments involving a combination of treatments where certain treatments require larger plots size than others, split plot design may be used. A split-split design is used when there are three factors, each of which differs in their requirement of plot sizes and accuracy.

When the factors included in the study need large plots for effective application of treatments, a strip plot design will be more efficient. Similarly, if too many treatments were to be tested in a factorial scheme, it would generally be more efficient to adopt a *confounded design* than a simple randomized block design.

In certain situations, *fractional factorial design* is adopted. When the number of factors or levels of factors are large, it would become cumbersome to test all the complete set of factorial combinations simultaneously in a single experiment. In such a situation, it would be worthwhile to test only a fraction of the total number of combinations. For example, an experiment with three factors at five levels would mean that there are 243 treatment combinations in a single replication! A large or impractical experiment like this may not be needed if our concern is to estimate only lower order effects under the assumption of absence or negligible higher order effects. This technique of eliciting useful information by observing only a part of the complete factorial is known as *fractional factorial design.* In these experiments also, confounding is necessary to reduce block size. Ordinarily, only higher order interactions are confounded. It is an ideal design to study the effects of several major and micronutrients together on crops.

A detailed treatment on design of experiments and the application of statistics in experiments are beyond the scope of this book. Fortunately, there are several books, which give excellent insight into these aspects. For more information on design of experiments and application of statistical procedures, refer to Cox (1958), Hill (1971), Snedecor and Cochran (1980), Gomez and Gomez (1984), Panse and Sukhatme (1985), or similar books.

5.8 Field Experiments

A field experiment is a true experimental study performed outside the laboratory, that is, in the real-world situations such as farms, forests, grasslands, ponds, polluting sources, rivers, watersheds, factories, hospitals, and so on. For example, an ecologist studying an ecosystem will not be able to move the entire system into the laboratory for conducting an experiment. In agriculture, field experiments along with control are widely used, as the crop growing conditions cannot be recreated in the laboratory. Although field experiments are carried out in a more or less realistic situation such as a farm, the degree of control over the independent variables is comparatively less unlike a laboratory experiment wherein all the conditions for the experiment will be under the control of the investigator. Therefore, the necessity of a 'control' group normally does not arise.

Although once considered ineffective as a scientific instrument of research, the world began to note and recognize the potential of field experiments by the classical works of Sir Ronald Fisher in 1920s and 1930s, who introduced the principles of experimentation and analysis of variance. During the last five to six decades, they have been identified with unlimited possibilities by the application of appropriate statistical methods. Before Fisher, although soil heterogeneity was known to exist, it was either ignored or attempts were made to even out by various means. At present, the presence of soil heterogeneity is fully recognized, and its effects are eliminated by suitable experimental designs, layouts, and statistical analysis.

In field experiments, the researcher manipulates one or more independent variables in field situations taking all precautions to have control over the environment. However, as everything is not under the control of the investigator, it is recommended to have a 'control group' in field experiments. In such experiments, you begin the experiment by creating two or more treatment groups, which are almost equivalent so that measurements on characters should be similar among the groups, and that the groups should be able to react almost uniformly to a particular treatment. This equivalency between groups can be ensured by applying the principles of experimentation (randomization, replication, and local control), which take into account the variation between treatment groups and between the individuals in each group. Once equivalent groups have been formed, the researcher tries to treat them identically except for the one variable that she/he wishes to isolate. A major advantage

Fig. 5.4 Researchers can use modern techniques to record observations from the field. In the photograph, plant canopy analyser is used to record various canopy features

of a field experiment is that it is practical and amenable to experimentation without artificially introducing confounding variables.

A field experiment follows all the steps of the scientific method starting from the formulation of a research hypothesis. In scientific disciplines such as agriculture, biology, ecology, and geology, field experiments are long considered as a sound experimental practice following various steps of the scientific method. Researchers can also use modern techniques to record observations (Fig. 5.4). However, a major concern is the cost of field studies, as they tend to be very expensive compared to a laboratory study.

5.9 Field Experiments in Agriculture: Some Practical Considerations

In agriculture, especially those involving crops, field experiments are routine. The following are some practical issues, which are to be considered while planning and conducting field experiments in agriculture.

Experimental Design

Special attention is needed in selecting appropriate designs for the experiments. The most suitable design depends largely on the number and nature of the proposed treatments. For pot culture experiments, *completely randomized design* without blocking can be conveniently used. *Randomized block design* with local control becomes necessary when soil heterogeneity or differences in experimental materials are suspected. However, if the fertility gradient is in two directions, by length and breadth, then blocks are to be considered in both directions following a *Latin square design*. If the number of treatments is large, and a uniform area within the block cannot be attained, an incomplete block design like *lattice design* may be used. For example, if many genotypes are to be tested, it is desirable to use an incomplete block design.

In experiments involving a combination of treatments, wherein one of the factors requires a large-sized plot as in irrigation, it might be advisable to adopt a *split plot design* with irrigation treatments assigned to main plots. If both the factors need large plots for effective application of treatments like irrigation and land configuration treatments, a *strip plot design* will be more efficient. Similarly, if many treatments are to be tested in a factorial scheme, it is necessary to adopt a *confounded design* than a simple randomized block design. Because of its simplicity and flexibility, simple randomized block design is the most preferred design for research workers in agriculture. With the usually adopted plot sizes ranging from 20 to 40 m^2, randomized block design can be adopted safely for experiments with up to 20 treatments without appreciable loss of efficiency (Panse and Sukhatme 1985).

Number of replications

Experiments should be repeated more than once to increase reliability. The greater the number of replications, the lesser would be the errors due to chance. Fixing the number of replications for an experiment is also a problem for novice researchers. Replications needed for a field experiment is dependant on factors such as the natural variability of the experimental materials, the experimental design adopted, the number of treatments, and the desired degree of precision.

Generally, we follow a rule of thumb to determine the number of replications. Choose the number of replications to provide not less than 10 degrees of freedom (Gomez 1972) or 12 degrees of freedom as suggested by Panse and Sukhatme (1985) for estimating experimental error. For example, as the error is based on $(t - 1)(r - 1)$ degrees of freedom in a randomized block design, if the treatments to be tried is 10, then a minimum of 3 replications are needed (for an explanation on degrees of freedom, see Sect. 6.9).

Size and Dimension of Plots

Size of plots is another crucial factor, which affects experimental error. Smaller plots are unreliable, but too large plots are unmanageable and misuse resources and labour. For the same plot area, the perimeter dimension of a square plot is lesser compared to a rectangular plot, and therefore, only lesser number of plants are exposed to border effects in square plots. While laying out plots, the length and width of plots must be measured correctly, as even a small error in plot dimensions can greatly affect the experimental results. This is especially so when we project the yield from a small net plot area to a large area, for example, a minor error in projecting the yield from 12 m^2 to one hectare might be substantial.

Fix the plot dimensions after careful consideration of planting methods and spacing. For a broadcasted crop, a plot of 20 m^2 (5 m × 4 m) might be alright but when row to row spacing and plant to plant spacing are involved, the length and width dimensions should be fixed as multiples of spacing. For example, if the spacing given is 60 cm × 30 cm, it means that row-to-row spacing is 60 cm and plant-to-plant spacing is 30 cm. Instead of fixing 5 m × 4 m dimensions for the plot, it may be slightly varied and take 4.8 and 4.2 m so that we will get 8 rows and 14 plants per row, leaving exactly half the spacing at the borders (30 and 15 cm). If you take 5 m × 4 m, then you will get 8 rows and 13 plants only, leaving wider borders of 50 and 20 cm, thus not only wasting land but also complicating border effects.

For crops like rice, the size of plots generally ranges from 8m^2 to 25 m^2, depending on the nature of treatment and uniformity of plots. For field crops like rice, decide the size of plots in such a way that a minimum net area of 5 m^2 is available for harvesting, leaving border rows and rows for destructive sampling. Normally, fertilizer trials require larger plots than a variety trail. An irrigation trial requires still larger plots to take care of buffer strips between plots. Similarly, for those experiments involving pesticides, the width of the plots should be decided by the coverage of the sprayer being used.

Selection of Treatments

Treatments for the experiment are to be fixed after carefully considering many aspects. Before going for experimentation in a detailed way, it is often advisable to try the treatments in an observation trial. The preliminary tests often reveal possible difficulties that might be involved in the application of the treatments or the unsuitability of some treatments under field conditions. The treatments must be selected in such a way as to achieve the objectives and to increase the precision of the experiment. This is especially important when studying the effects of inputs in increasing doses such as herbicides, fertilizers, water, fungicides, or insecticides. In such experiments, it is more beneficial to determine how the experimental units respond to increasing rates of treatments than to know whether different rates are significantly different. The major objective of these kinds of experiments should be to determine a response curve with a range of rates going well beyond the optimum. For determining the shape of the response curve, too many treatments are not required; but four or five well-structured rates are enough. For example, if the optimum rate of nitrogen is expected to be around 100, then suitable treatments for the trial might be 0, 50,100, and 150 or 0, 40,80,120 and 160 kg/ha This indicates that rates in equal or multiple increments within the expected range of the response are most efficient in establishing equations for a rate: response curve.

Border Effect and Harvest Area

In agricultural field experiments, an 'experimental plot' is a unit of land area on which random assignment of treatments is made. The whole unit of land receiving a treatment (gross plot size) may not be the harvest area (net plot size). Harvest area is decided excluding the 'border plants'. The phenomenon in which the yield or other characteristics of crops in the border areas of a plot differ from those of the central portions is called *border effect*. Border effects can occur because of several reasons. Consider two examples, one a varietal trial and the other a fertilizer trial. In a varietal trial, border plants of vigorous varieties usually gain by competition with plants of less vigorous neighbouring plants, an advantage not available to inner plants. Thus, if we are not accounting border effects, the difference between the performances of the varieties in the field is liable to be overestimated. Apart from the possibility of bias, the border effects would aggravate error variances by increasing heterogeneity among plants. In fertilizer trials, chances are that the nutrients at different doses supplied to various plots might seep up to the border plants in the unfertilized plots, which ultimately affect the response of border plants to the fertilizers applied. In the case of trials involving irrigation, more precautions to avoid border effects are needed. In such trials, as lateral movement of irrigation water is to be expected, double bunds and buffer areas between the plots are required in addition to excluding border plants from observations.

The size, shape, and orientation of plots should be in such a way that experimental error is reduced to the minimum. In varietal yield trials involving crops like rice, where inter-varietal competition is expected, plots with at least six rows should be used leaving out one row on each side of the plot as border, so that four centre rows are available for harvest. One or two plants at both ends of each row should also be excluded as border plants.

Destructive Sampling

Some experiments involve the measurement of several plant characters. In certain cases, it may also involve destruction of plants as in the case of dry matter production. In weed management trials, periodic monitoring of dry matter production of weeds is common by uprooting weeds. Uprooting of weeds enable crops in the weed uprooted area to be vigorous and more competitive. You may follow some techniques to avoid plant competition near the spot from where uprooting is done. Leaving an area in the centre of the plot for final harvesting, and using the surrounding areas for sampling is a common method. Another option is to set up an area entirely for destructive sampling lengthwise. For example, if you are taking a plot size of 6 m $\times$ 4 m, provide 2 m lengthwise entirely to conduct sampling.

Experimental Site

Make sure that the site selected for the experiment should be uniform in all respects to the maximum possible. By following efficient blocking, we can take care of heterogeneity to a certain extent. We often observe that the quantitative characters measured from neighbouring plots sown simultaneously with the same cultivar and treatment are not uniform and differ mildly or widely depending on several factors. The most obvious and probably the most important features that magnify the differences are related to soil heterogeneity. The following are the features that amplify soil differences (Gomez and Gomez 1984).

Slope: Slope of the land affects soil uniformity. If the experimental site is not uniform, soluble nutrients may move to lower slopes along with runoff water. Differences in slope may create differences in soil texture and soil depth too. Obviously, an ideal experimental site is one that has no slope. In the absence of such levelled lands, we can go for lands with uniform and gentle slope having predictable fertility gradients using efficient blocking along the contours.

Previous cropping history: Previous crops and various treatments given to them must be taken into consideration. For example, various management practices such as fertilizer schedules, plant protection, and crop rotation are sources of additional soil heterogeneity. Therefore, it is necessary to standardize the field to obtain uniform fertility by some measures. A commonly adopted practice is to sow the area with a green manure crop or cover crop for at least one season, and to equalize the nutrient contents of soil before conducting an experiment. Another method is to assess the fertility gradient by uniformity trials and proper block arrangement.

Grading: Levelling or grading the field is a normal operation in farms for various purposes such as irrigation and drainage. It is done by removing top soil from elevated areas and spreading and filling crevices and uneven areas. However, for an experiment, such areas are not ideal as this operation causes an uneven depth of surface soil and at times exposes the subsoil. As far as possible, avoid such sites. Nevertheless, if there is no option except the present site, assess the pattern of soil heterogeneity through uniformity trials and follow blocking or adjustments through analysis of covariance techniques.

Presence of large trees, buildings, and other structures: These structures influence the surrounding areas by shading. Large trees with their massive root system offer stiff root competition. Sometimes, shade may not be a problem but top soil is removed for construction and subsoil is exposed. Avoid such problem sites.

Unproductive site: Avoid waste and unproductive lands for experimentation. A field with very poor or problem soils should not be selected unless it is an experiment designed specifically to evaluate such problem conditions.

Sowing

Sowing and transplanting are important operations to ensure uniformity. In general, over sowing seeds by 25–50 percent may be required to ensure that enough seedlings emerge to establish the required plant population. Use seeds from the same seed lots. The seeds should be sown on the same day completing all the formalities in all the plots. If there is difficulty to cover all the plots in one day, cover at least all the plots of one replication at a time. For uniform germination of seeds, compact the soil around the seeds immediately after sowing and ensure soil moisture.

Thinning and Gap Filling

Excess plants in the experimental plots should be removed within a reasonable time. The removal of excess plants is usually called *thinning*. Thinning should be done within 6–10 days after seedling emergence to avoid unevenness in the growth of the crop. While removing excess seedlings, take care to avoid disturbance to the remaining plants. Seedlings should be left as equally spaced as possible. A common practice is to lay a 2 m pole or stick along the row and remove the seedlings in excess of the calculated numbers. The most vigorous seedlings of similar size are to be retained. Diseased, damaged, and weak seedlings may be removed while thinning. As far as possible, the same person should be engaged to thin all plots in a replication to reduce bias. Thinning should be done within the shortest time in the whole replication in a day. Normally, transplanting seedlings in vacant places is not recommended because the transplanted plants are usually weaker and less productive than the normal plants. A researcher has two alternatives in dealing with missing plants. One is to manipulate the thinning process so that the areas adjacent to the vacant space have more plants or to ignore the gaps at the seedlings stage and adjust the number of plants at harvest or adjust plot size at harvest.

Manures and Fertilizers

Manures and fertilizers required for each plot should be spread uniformly so that each plant in the experimental plot gets its due share. Plants in a plot may behave differently if these are not applied uniformly. Subdivide the experimental plot into smaller units to enable uniform application to smaller areas. Applying fertilizers to each row of plot separately can also be considered. The usual procedure is to weigh or measure the fertilizers required per row, and uniformly spreading it on each row by hand.

Cultural Operations

All the common cultural operations including ploughing, seedbed preparation, levelling, weeding, earthing up, and plant protection must be done uniformly to all plots, covering all the plots in one replication at a time. Differences in cultural operations can aggravate experimental errors. Sufficient protection from animals, birds, and intruders must be given. Damaged and disturbed plots should not be included for observation, but considered as missing plots at the time of statistical analysis.

Off-Type Plants

Sometimes, you may notice a few exceptionally tall or vigorous plants in the plots. These off-type plants may have come from the seeds of a previous crop or may be natural mutants. You cannot simply ignore these plants as they may already have affected surrounding plants. Their removal creates missing hills or gaps in the row and thus affects the surrounding plants. Therefore, they are normally allowed to mature, but just before harvest, such plants are counted and removed. The yield of the plot can then be computed as given by Gomez and Gomez (1984):

$$Y = \left(\frac{a+b}{a}\right)y$$

where Y is corrected plot yield; y is actual grain yield from normal plants in the plot; a is normal plants in the plot; and b is number of off type plants.

For example, take the case of a plot with a total plant stand of 100, in which 4 plants are off types. Suppose that after removing the 4 off types, the remaining 96 plants yielded a harvest of 2 kg. The corrected plot yield is 2.083 kg. This correction is applied based on an assumption that the competition effects offered by the off-types to the surrounding plants are similar to those of the normal plants.

Error in Data Collection

Data collection is an important aspect, which determines the credibility and repeatability of the experiment. Follow good data collection practices. Try to review the data collected immediately after completing the measurement to detect any anomalous data, for example, unusually high or unusually low readings. The doubtful figures need to be checked again by taking observations from the specific plot. Always keep the sample until the analysis of data is completed. Another source of error is the differences between individuals in taking the measurements. Keep the number of persons involved in data collection to the minimum. Similarly, allow the same individual to evaluate all the plots in one replication. It is important to double check row dimensions and plot size as plots are harvested, and avoid mislabelling data by counter-checking to reduce errors.

Competition Effects

Competition between plants for various resources is common in crop communities. Competition between crops and weeds for the same resources such as solar energy, soil nutrients, and moisture should be kept to the minimum. Competition effects between crop plants within a plot (intraspecific competition) should also be considered and kept uniform. Make sure that the plant response represents the actual conditions being tested. Experimental and sampling errors are likely to increase with variation among plants within a plot. Some commonly encountered competition effects are noted below:

Varietal competition: This is a case of intraspecific competition, which is most obvious when the experiment involves different cultivars or genotypes.

Different cultivars or genotypes sown in adjacent plots may be subjected to different environments depending upon their location relative to adjacent plots. In such trials, the plants near the borders generally experience the effects of varietal competition.

Nutrient competition: When neighbouring plots receive different rates of fertilizers, plants that receive a higher rate tend to be more vigorous and more competitive. Another source of competition is seepage of water from fertilized plots, which may spread nutrients to the root zone of the adjacent plots. Ultimately, this may favourably influence the plots receiving smaller doses of fertilizers.

Pathways or alleys: It is common to see vigorous and luxuriously growing crops along the pathways and alleys in a farm. This happens mainly because of less competition along pathways and alleys than those in the centre of the plot.

Missing Hills or Plants: Missing hills or plants are common problems for farm researchers. It is very difficult to ensure full stand for all plots in an experiment even for the vigilant researcher who have taken all precautions. Poor germination, bird damage, insect and disease attacks, physical mutilation, and similar causes may be responsible for the death of a few hills or plants in a plot. Plants near to the spots of missing plants perform better than the normal plants because of less competition compared to other plants creating more problems. Discard all the plants immediately adjacent to a missing hill, and harvest only those that are fully competitive. While computing total yield, this loss of plants must be accounted.

Records to be Maintained by Researchers

Researchers have to maintain certain records pertaining to their research projects for their own reference and for future use by the institution. In research institutions, transfers are routine, and it is possible that a scientist after starting a project is transferred. This has to be entrusted to his/her substitute or to some others in the institution. A smooth handing over of charge is possible only if you have maintained the records properly. The following are the mandatory records a researcher has to maintain.

Basic record: Usually, it is a bound register made of good quality paper. This contains all the relevant details of an experiment and the observations, both raw and analysed. Basic record should contain all details of the project such as name of the project, names of principal investigators, co-principal investigators, or associates, date of start, date of completion, project code, administrative and technical sanction details, if any, and a brief write up on technical programme including lay out in the case of long field experiments. Season-wise tabulated data of all the characters should be noted in the basic record.

Observation book/field book: Researchers enter the data obtained from observations first in this book, which will subsequently be transferred to the basic record. This should contain details of the treatments such as crops grown, laboratory works, lay out, cultivation details of crops, animals used, and/or human subjects. Design the field book in such a way that all data are entered directly by pencil, so that all statistical analysis of the data can be done directly from it or only one data transcription is needed for all analysis. This will save time, money, and the chances for errors. To assess the experimental data properly, chronological notes should be recorded in the field book related to emergence, growth of the crop, incidence of pests and disease, differential behaviour of treatments, abnormal events, rainfall data, labour utilization, and other scheduled events. It should also contain the field lay out plan.

Project file: A copy of all the correspondence made with the funding agency, controlling institution, and others regarding the project must be kept in a project file. Always keep a copy of periodical research reports and audit reports, publications, etc., related to the project. Project files, field books, and basic records of completed projects should be kept as permanent records. The principal investigators of the scheme have the responsibility to maintain these records. The heads of research institutions should also a play a regulatory role. They should occasionally verify the registers. They should also ensure that only updated documents are handed over to the next scientist on his/her relief from the institution by way of transfer or retirement. When the PI is on leave, she/he should entrust the works to the Co-PI or associates so that the project continues smoothly.

5.10 Randomized Controlled Trial

A clinical or epidemiological experiment involving human subjects with completely randomized design is usually referred to as *randomized controlled trial* (RCT). RCTs are generally employed in testing the safety and efficacy of a medicine or drug, medical device, surgery, or some other treatment. In an RCT, subjects are allocated into an experimental group and a control group through a random method. The researcher studies two or more interventions on selected individuals who receive them in random order. The experimental group receives the treatment or medication to be compared, and the control group commonly receives a *placebo* (e.g. vitamin tablets or sugar pill) or no intervention at all. Those who takes part in a randomized controlled trial is often called a *volunteer*, *participant*, or *subject*.

Randomization in RCTs ensures that the subjects in the groups are similar as far as possible with respect to all the factors that might affect the outcome. Whenever it is found ethical and practical, a randomized controlled design should be considered in all such clinical studies. The RCTs designed for clinical research are also called *randomized clinical trials*, and sometimes, *randomized controlled clinical trials*. Although RCTs are mainly used for clinical trials, these types of trials can be employed in certain other research areas too such as education, nutrition, and social work.

The researcher has to take some extra precautions when clinical or nutritional experiments involving human subjects are designed, especially against external variables such as the *placebo effect*. Placebo is an innocuous medication prescribed for the mental relief of a patient than for its actual effect such as vitamin tablets or sugar pills, which creates a psychological effect. Placebo effect is the improvement in the condition of a sick person in response to placebo treatment. In many cases, it has been established that the mere illusion of receiving a treatment can play an important role in a patient's recovery. Placebo effect is noticed in many other cases such as dietary studies also.

Blind Studies

Blind studies are usually planned to eliminate the influence of placebo effect, especially when the effectiveness of a drug or food item is evaluated. In such blind studies, the human subjects do not know whether they are receiving real or fake treatments. Compared to an ordinary RCT, blind studies make sure that the responses by the volunteer are due to the treatment alone and not a response to the understanding that they are being treated. A blind experiment is said to be *single blind* when the researcher is aware of the situation but not the participants. Normally, a single blind RCT is enough to handle the issue of placebo effect. However, a *double blind* test may become necessary, if we suspect placebo effect and *Rosenthal effect*, the influence that a researcher can exert on the outcome of a research investigation (Sect. 6.4). In a double bind test, both the researcher and the participants are completely 'blind', which means that they may not know which individuals are in the control group or in the experimental group until all the observations are taken. For conducting a double blind test, the investigator has to make use of trained helpers.

When a new processed food product is made, it is common to evaluate its taste and acceptability using organoleptic tests. A psychological barrier can function here, if the ingredients or the method of preparation are revealed to the volunteers, who come forward to taste the products and give score values. A similar psychological effect with respect to the quality of several farm products is prevalent among lay people. For instance, most people believe that 'organic foods' are of supreme quality compared to the farm products produced following integrated approaches. The mere act of knowing that a particular product is 'organic' can induce a placebo effect; therefore, when you conduct a test, it can suppress actual values. Therefore, to verify or falsify such claims, conduct single or double blind tests involving organoleptic tests.

5.11 Paired Design

In a true experiment, random assignment ensures that experimental subjects are treated alike. In certain cases, however, random assignment alone will not save the experiment from problems arising out of differences between subjects. Consider that a researcher is testing the effectiveness of a new medicine to treat a specific ailment. The researcher doubts that the age or some other character of the patient might be related to the effectiveness of the drug. If you are planning a randomized controlled trial (RCT), you would be randomly assigning the subjects to the treatment and control, and possibly age can become a confounding variable affecting the experimental outcome. For example, after administering the medicine to the subjects consisted mainly of young people, if you observed no difference between the treatment and control, you cannot conclude decisively whether the drug is ineffective or whether the ineffectiveness was because of the age of the subjects consisted mainly of youth.

In situations similar to the above, when there is an extraneous variable that the researcher believes might become a confounding variable, a *paired design* can be used. This is also called *matched pairs design* as the pairs are formed based on some matching variable. A matched pair design is a special case of a randomized block design when the experiment has only two treatments. This design can be used when the experimental units or subjects can be grouped into blocks of size two such that within a block, the two units are as homogenous as possible. Pairing or grouping should be based on a character that is likely to influence the actual character under the study (the matching variable). In the case of human subjects, each pair can be matched based on gender, age, health condition, height, weight, interests, income, race, religion, IQ, and so on depending on the character under study. The researcher after measuring the matching variable creates groups of subjects who have the same value or level of that matching variable.

Occasionally, you may confront a situation wherein there are very few subjects with identical values on the matching variable. For example, if you have subjects with the following ages: 72, 65, 60, 56, 55, 49, 44, 41, …, etc., you cannot go for identical matched pairs. However, you could take the two older people together, make a pair, and randomly assign them to the two conditions; then take the next two in terms of age and similarly complete the process. Although the pairs are not identical in terms of their ages, they are almost similar in age. Within each pair, subjects are randomly assigned to different treatments. The number of experimental units must be an even number. After pairing, the treatments are allotted randomly to the two units in each pair. However, to have minimum of 10 degrees of freedom for error, it is advisable to have 11 or more pairs of units for the experiment. In other words, a minimum of 22 units are required to run the experiment.

In fact, paired design is a 'pre-test/post-test design with control'. For obtaining perfectly matched pairs, you should have reliable and valid procedures for pretesting. Matched pairs design is common in health and nutrition studies and those involving animals. Analysis of data can be done as in the case of a paired *t*-test or as RBD. Examples include comparing the milk yields of cows after giving a particular feed, the performance of farmers after receiving a particular training, and success of a drug in treating a disease related to some specific features. Pre- and post-test data and matching are essential in all these cases.

Paired design has some limitations. The researcher has to begin from measuring the matching variable followed by creating matched pairs and then assigns the selected subjects to experimental conditions. Considering these difficulties, researchers may prefer to use an RCT rather than a matched-groups design. However, if the researcher is contemplating a study that involves higher costs, more time, special subjects that are hard to find, and if there is an obvious extraneous variable such as gender and age related to the dependent variable, then possibly, it is meaningful to adopt the matched-groups design.

5.12 Quasi-experiments

Quasi-experiments or semi-experiments are almost like true experiments except that they lack probabilistic equivalency between groups. For all practical purposes, it is just like a true experimental study with the only difference of missing randomly assigned groups. Quasi-experiments are common in the area of health and veterinary research and social sciences in situations where it is not possible to create a truly controlled group because of ethical reasons or when randomization is not possible. In experiments involving human subjects, often we find it very difficult to adopt random assignment, especially when they are part of intact groups such as schools, colleges, communities, neighbourhoods, or hospitals. Sometimes, the groups may be reasonably similar; but utilizing data from such intact groups restricts our conclusions on treatment effects.

Participants are assigned to groups based on some characteristics or quality these bring to the study, for example, differences in sex, age, class, or type of job. Assignments to groups have already taken place before the experiment begins.

There are four major forms of quasi-experimental designs—crossover designs, time series designs, non-equivalent control groups designs, and natural experiments.

Crossover Design

In most experiments, the experimental units that include animals, persons, or crops remain on the treatment from the start of the experiment until the end. These are often called *continuous trials*. For example, RCT is a continuous trial. In a *crossover trial*, however, each experimental unit will receive consecutively two or more treatments during the course of the experiment. In such trials, the period of comparison is divided into a number of sub-periods. A crossover study can be considered as a longitudinal study in which the experimental subjects receive a series of various treatments. A crossover design is also called *counterbalanced*, *switchover*, or *rotation design*. The treatments are administered to the groups in a different order, but the number of groups should be equal to the number of treatments. For example, if there are four different treatments, there should be four groups.

Crossover designs are employed commonly in many scientific disciplines such as animal husbandry, health, psychology, education, and pharmaceutical science. Randomized, controlled, crossover clinical experiments can be useful for health and drug-related trials. In a crossover clinical trial, each subject is randomly assigned to a sequence of treatments, of which one treatment may be a control or placebo.

In a crossover design, the subjects are divided into two equal groups (if there are only two treatments) and one-half of the participants are randomly assigned to start with the control or placebo and the other half with the treatment. After a period, enough to allow any treatment effect to vanish, the treatments are crossed over. In the basic crossover trial involving only two treatments (A and B), each experimental unit receives both treatments in either of the sequences A-B or B-A.

The experimental units—animals or human beings—available for the experiment should be allocated to the two sequence groups at random.

In a usual continuous trial with persons or animals, it is common to place them on a standard diet or treatment, prior to their random allocation to the experimental treatments. A crossover trial has the advantage that since two or more treatments are compared on the same experimental unit, the between experimental unit variation does not enter into the experimental error. In other words, analysis of covariance to take account of variations between subjects is not needed, and the standardization period plays a minor role only. In each period, we have both the treatments; hence, comparisons between treatments are free of period effects. Similarly, as each unit receives both treatments, comparison of treatments is within unit, thus removing between-unit variation from the treatment differences. As each subject included in the study serves as both an experimental subject and a control, the crossover design helps to reduce the number of subjects required.

Time Series Design

A time series is simply a sequence of measurements taken at various points in time. A research design is often called *time series design* when observations of the same variables are taken at different points in time for studying social and other trends.

Because of this stress on trends, such designs are also called *trend designs.* Time series designs are distinguishable from cross-sectional designs where measurements are taken only once. This group of quasi-experimental designs involves repeated measurement of a group; and often, the experimental treatment is induced between two of the measures.

In time series designs, repeated measurement of the dependent variable is done before, during, and after the introduction of the independent variable in the experiment. Therefore, time series design can be considered as a detailed version of the 'before-and-after design'. In a time series design, the subjects are repeatedly pre-tested and post-tested rather than being tested once at the beginning and a second time at the end of the treatment. For example, we could measure the attitude of trainees of a particular training programme each day throughout the training period, and we could see how their attitude changes over time.

Two forms of time series design are generally in use. The *interrupted time series design* studies one group of the same participants for a particular period, whereas the *multiple time series design* studies two or more groups of the same participants for a particular period. Measurements are made at multiple occasions over time before and after an intervention to detect whether the interference has significant effects.

Some problems of this design must also be noted. As there are no randomly drawn and assigned experimental and control groups, it is not possible to point out changes in the dependent variable directly to the effects of the experimental treatment. For example, individual groups taking part in a time series design may improve their performances from pre-testing to post-testing. Therefore, it must be ascertained whether it is the treatment or some other event that produced the change. A variety of time series designs are used in health science, some of which can be used in agricultural science too. Time series designs are usually used in epidemiological research. A single population group of defined size is observed over a certain period during which preventive or curative measures are applied, and measurements are made at specified intervals.

Non-equivalent Group Design

Non-equivalent group design is similar to 'before-and-after experiment with control', but it lacks the key feature of the randomized experiment—random assignment. In a non-equivalent design, we try to use intact groups, which are similar as the treatment and control groups. In community-based research, we might use two similar communities. Although we try to select groups that are as similar as possible, we can never be sure whether the groups are comparable. Most often, the groups may not be equivalent, that is, the groups may be different prior to the study, and hence the name 'non-equivalent groups design'. It is a known fact that if there were differences between the groups before the start of the experiment, it would affect the outcome of the study. The differences between the treatment group and control group prior to the study are used to take account of the effects of selection differences.

A non-equivalent group design is desirable when the subjects are allowed to self-select themselves into different treatments when a researcher assigns subjects to different treatments or when different treatments are implemented in different sites.

The groups are generally chosen from clustered units such as classrooms, farmers groups, and groups of patients. Groups are assessed at exactly two points in time, one before and another one after the treatments are implemented. Differences on the post-treatment observation are used to assess the effects of the treatment.

Natural Experiments

A quasi-experiment is sometimes called a 'natural experiment' when it involves testing a hypothesis based on data gathered by observing a natural system. The variables are not controlled or manipulated in such experiments. By the term 'natural experiment', we mean events, interventions, or policies, which are not under the control of researchers, but amenable to observation and recording of data. We may not be able to design a true experiment with natural phenomena; but we can collect data by simply observing a phenomenon for testing a hypothesis concerned with it. Natural experiments are observational studies, and these may not have a 'control' in the traditional sense of a randomized experiment. 'Natural experiments' are almost similar to 'natural observation' already described in Sect. 4.1; the only difference is that the natural experiment is done to test a hypothesis while natural observation may be simply a descriptive study.

Natural experiments are accepted as study designs in many disciplines wherever controlled experimentation is difficult. For example, astronomy usually relies on natural experiments of this type, as it is impractical to do experiments involving celestial bodies. In social sciences, this usually entails observing humans or animals in real-life situations.

In life sciences, natural experiments involve observing an animal or groups of animals, a plant or plant community, or some physical phenomena. Natural experiments are also feasible in health science when a randomized controlled trial is difficult to conduct because of ethical issues. A typical example of a natural experiment is to relate the quality of water in a particular locality and the occurrence of diseases like cholera.

5.13 Ex Post Facto Experiments

Ex post facto studies resemble a typical experiment because groups are compared; however, a key difference is there, no manipulation of independent variables. The causal event of interest has already happened without the manipulation of treatments by the researcher. In such experiments, the 'effect' becomes the dependent variable and the probable 'cause' becomes the independent variable. Ex post facto experiments are useful when human subjects in true situations are involved, and the researcher comes to the scene only 'after the fact'. Most ex post facto studies are observational studies as no manipulation of treatment is possible. In such situations, the investigator begins the investigation by identifying the outcome variables and then tries to identify possible reasons, i.e. independent variables responsible for the outcome. Ex post facto studies are often criticized because of the post hoc fallacy (see Chap. 14).

Causal-comparative studies and case-control studies are examples of ex post facto studies.

Causal-Comparative Studies

In a causal-comparative study, the researcher attempts to trace an effect that has occurred without any manipulation to its possible causes. Some researchers consider causal-comparative research as a type of descriptive research only because it describes conditions that already exist. Actually, it is a retrospective study, as it starts with effects and investigates back to its probable causes. Without any manipulation, the independent variables have already occurred, and therefore, the same degree of controls as in an experimental study is not possible. Because of this ex post facto nature, interpretation of results requires caution. Nevertheless, certain control procedures such as *matching* can be adopted to improve the reliability of results and consequent interpretations. Normally, a causal-comparative research is planned to approximate the conditions of a true experiment in a situation that does not allow control or manipulations of relevant variables.

In a causal-comparative study, the researcher selects two groups of subjects, the *experimental group* (who have subjected themselves to the independent variable assumed to be the cause) and *comparison group* to act as control (hence the name 'causal-comparative' experiment). The two groups may differ in two ways. One group may possess a characteristic that the other does not possess or each group has the characteristic but to differing degrees. Make it a point to define the independent variable differentiating the groups clearly.

Note that in causal-comparative research, random samples are chosen from two already existing populations, unlike a single population as in experimental research. However, the aim is to have groups that are as similar as possible on all relevant variables except the independent variable as in experimental studies.

An example for the application of causal-comparative experiment is narrated here. Suppose you want to determine the relation of aerial spraying of pesticide X (as the independent variable) on the occurrence of cancer (as the dependent variable) in humans. You cannot conduct an experiment with humans dividing them into groups and by asking them to be exposed to aerial spraying of pesticide X! It is unethical and impractical. In such situations, you can use the 'causal-comparative method'. You can enlist several people who were exposed to pesticide X spraying in the past. In fact, they have already applied the independent variable in your question, that is, aerial spraying of pesticide X and exposed to it. To test your hypothesis, you must gather data on two groups of people, (1) an 'experimental' group of people who have subjected themselves to your independent variable assumed to be the cause (exposed to aerial spraying of pesticide X), and (2) a 'comparison' group (to act as control) of similar people with the same background who were not exposed to aerial spraying.

After collecting sufficient data, a statistical test is conducted to decide whether the people exposed to pesticide X spraying scored significantly higher than those who were not exposed (the comparison group). By comparing the incidence of cancer in the two samples, you can check whether there are significantly more cases of cancer in the experimental group. If you find no significant difference between the two groups of people on the incidence of cancer, then you accept the null hypothesis and reject the research hypothesis. On the contrary, if there are significantly more cases of cancer in the experimental group, you may infer that your study supports the alternate hypothesis and conclude that exposure to aerial spraying of pesticide X increases the occurrence of cancer.

You must apply caution in attributing cause and effect relationships as above and come to a firm conclusion without doing further studies. In this kind of observational studies, it is very difficult to rule out all causes other than the one being investigated. For example, the supposed cause of an observed effect may be a third variable. Understand that a causal-comparative study lacks randomization, no manipulation of variables, and true control is absent. The investigator cannot randomly allocate subjects to treatment groups because they are already in those groups. The groups may be different on some other important variables too, for example, gender, age, or health, in addition to the identified independent variable.

Causal-comparative research is prone to 'post hoc fallacy', the conclusion that because two factors go together one must be the cause and the other the effect (see also Chap. 14 item No. 39). Post hoc fallacy occurs when a conclusion presumes a causal relationship without sufficient grounds. For example, linking cancer with pollution, pesticide application, fast foods, or other such causes after doing simply a study as above requires caution. This fallacy is named after a Latin phrase post hoc *ergo propter hoc*, meaning 'after this, therefore because of this', which suggests that one event caused another event simply because it took place first. It is likely that other factors have caused the second event. As there is a danger of confusing symptoms with causes, these kinds of studies should test not just one hypothesis but other logical alternatives or competing hypotheses too before rushing for a post hoc fallacy. Therefore, the results of any ex post facto study should not be taken for granted but to be taken as a guide only for deeper and more rigorous research. However, these kinds of studies facilitate decision-making and provide guidance for further studies.

Case-Control Studies

A causal-comparative study when used for epidemiological studies particularly for studying infrequent events with 'matching' is usually called *case-control study*. Sometimes, a case-control study is also referred to as a *case-referent* or *case comparison* study. In a typical case-control study, the characteristics of a group of people with a specified medical condition such as a disease, risk factor, a treatment, or an intervention (*cases*) are matched with a control group of people without the condition but are otherwise similar (*controls*) (hence the name, case-control study). Case-control studies are usually retrospective in nature, and the major objective is to verify retrospectively the exposure to the risk factor of interest from each of the two groups of subjects—cases and controls. In fact, these types of studies are ideal to estimate odds. The magnitude to which each participant was previously exposed to the independent variable can be easily identified. For example, in the case of a disease like cancer, the investigator looks backwards in time for exposures that might have caused the disease. By examining medical case records and through interviews, the researcher can build up the history of the subject's 'case'. Case-control studies are useful for studying conditions with long intervals between exposure and outcome. Statistical analysis is also possible to find out whether there are any significant differences between the groups, and it allows the researcher to arrive at a firm conclusion.

The first step in conducting a case-control study is to decide on the research question to be answered. Based on the researchable question, formulate a hypothesis and then decide what will be your observations, and how they are measured. The next step is to specify the features of the study group, and decide how to form a valid control group. After having finalized these aspects, the 'case' group and 'control' group are compared often by conducting interviews. Medical records are also used to build up history of the subject's life. Sometimes, blood tests or other tests are also conducted after drawing samples.

In situations where the outcome is rare, a case-control study may be the only possible option to get enough information from a few subjects. Similarly, in situations where there is relatively a long period between an exposure and the disease, case-control studies are ideal. As some of the subjects have been intentionally chosen because they have the disease in question, case-control studies are more cost efficient when compared to cohort or cross-sectional studies.

The main weakness of case-control study, as in the case of causal comparative, is that as no intervention is attempted and no effort is made to alter the course of the condition, it is purely an observational study, and therefore, does not provide the same level of evidence as randomized controlled trials. Another problem is *sampling bias*. Ideally, you should use random sampling, and the cases chosen should be a random sample of all the subjects with the disease or problem. Most often, however, you have no other choice but to study a sample of those subjects whom you are able to recruit (convenience sampling). There are issues in selecting the controls too. To enable the controls to represent the same population as the cases, *pair wise matching* of participants is often followed. For each subject in a group, the researcher finds a match in the other group with the same or very near score on the control variable. If a volunteer does not have a suitable match, s/he is eliminated from the study. The resulting matched groups are very similar with respect to the recognized extraneous variable.

Case-control studies are very poor in establishing cause and effect relationships, and it is very difficult to rule out all causes other than the one being investigated. A case-control study usually depends on retrospective data, and because such studies rely on memory, the possibility of *recall bias* cannot be ruled out. Recall bias is the inclination of people with a disease or condition to recall events better than the people who do not have that condition. In certain cases, recall bias can be eliminated by using data recorded for other purposes before the outcome had occurred. The success of this strategy, however, is limited by the availability and reliability of the data collected. Because of the problems of matching, recall bias, and the presence of confounding variables, utmost care is needed while evaluating results.

Case-control studies are also prone to the 'post hoc fallacy', as discussed earlier in the case of causal-comparative studies, the conclusion that because two factors go together one must be the cause and the other the effect. Because there is a danger of confusing symptoms with causes, case-control study should test not just one hypothesis but other logical alternatives or competing hypotheses too before rushing for a post hoc fallacy. Therefore, the results of any case-control study should not be taken for granted but to be taken as a guide only for deeper and more rigorous research. They are, of course, useful for generating hypotheses. The hypotheses thus formed can be tested more thoroughly by other methods such as randomised controlled trials.

5.14 Requirements of a Good Experiment

A well-planned and properly laid out experiment is important for success. We can ensure success in our research endeavour, if we are able to meet the requirements such as the absence of systematic error, precision, wide range of validity, simplicity, and proper statistical analysis (Cox 1958).

Absence of Systematic Error

Chances of experimental error are higher, if you treat all experimental units or plots alike. Random errors are out of your bounds but you can take enough safe-guards to 'control' or reduce systematic error to the minimum. In other words, the units receiving one treatment should show only random differences from units receiving other treatments including control, and they must be allowed to respond independently of one another.

Blocking, proper plot techniques, and proper data analysis are the three techniques followed in agricultural experiments for the control of systematic error. In field experiments, where substantial variation is expected within an experimental field, experimental errors can be reduced by employing proper *blocking*, also called local control (Sect. 5.4). *Proper plot techniques* comprises of measures to reduce soil heterogeneity, competition effects, and mechanical errors. In experiments where blocking is not done effectively, *proper data analysis* can help to control error greatly. *Analysis of covariance* can be adopted in such cases (Sect. 6.10). Suppose you want to lay out an experiment in a coconut garden where the yield levels of coconut trees differ considerably. Using the initial yield as the covariate, final yield after the experiment can be adjusted to the values that would have been attained had all experimental trees started with the same yield.

Precision

If there are no systematic errors and the estimates of treatments differ from its true value only by random errors, the experiment has more precision. You measure random errors usually by *standard error*. Standard error should be sufficiently small. If it is large, the experiment itself is useless. The precision of any experiment depends on the intrinsic variability of the experimental material, the accuracy of the experimental work, the number of experimental units, and the design of the experiment. Therefore, these aspects have to be considered carefully for better precision.

Range of Validity

The conclusions and inferences made out of the experiment should have a wide range of external validity applicable to other settings and regions. As described in Sect. 5.3, necessary precautions should be taken to avoid or reduce threats to both internal validity and external validity. In agricultural research, most experiments show a great deal of *internal validity* indicating that they are valid in highly controlled conditions; but *external validity* may be low when the experimental results are applied to actual field situations. Therefore, experimental designs in agriculture focus more on the problems of external validity.

Simplicity

The experiment should be simple in design and analysis, for example, simplicity of design, simplicity of analysis, and simple analytical tools. Select a suitable design so that it's lay out, observations, and statistical analysis can be done easily. In the case of field experiments, follow simple designs like CRD, RBD, or simple factorial combinations. Because of the problems in getting large area for experiments and huge costs involved, the modern trend is to avoid complicated designs. Similarly, include only those methods and techniques that you can do with the available resources and facilities.

Proper Statistical Analysis

The data collected during the course of investigation must be subjected to appropriate statistical treatment. Without making artificial assumptions, a proper statistical analysis of the results should be possible. For this, the investigator should have a basic knowledge on statistics; otherwise, seek the help of a statistician.

Bibliography

Best, J.W. and Kahn, J.V. 1993. Research in Education (7th Ed., Indian Reprint, 2004). Prentice–Hall of India, New Delhi, 435p.

Campbell, D. and Stanley, J. 1963. Experimental and quasi-experimental designs for research. In: Gage, N.L., Handbook of Research on Teaching. Rand McNally, Chicago, pp. 171–247.

Chandel, S.R.S. 1991. A Handbook of Agricultural Statistics. Achal Prakashan Mandir, Kanpur, 560p.

Cox, D.R. 1958. Planning of Experiments. John Wiley & Sons, New York, 308p.

Fathalla, M.F. and Fathalla, M.M.F. 2004. A Practical Guide for Health Researchers. WHO Regional Publications Eastern Mediterranean Series 30. World Health Organization Regional Office for the Eastern Mediterranean, Cairo, 232p.

Fowkes, F.G.R., and Fulton, P.M. 1991. Critical appraisal of published research: Introductory guidelines. Br. Med. J. 302: 1136–1140.

Gall, M.D., Borg, W.R., and Gall, J.P. 1996. Education Research: An Introduction (6th Ed.). Longman, New York, 788p.

Gomez, K.A. 1972. Techniques for Field Experiments with Rice. International Rice Research Institute, Manila, Philippines, 46p.

Gomez, K.A. and Gomez, A.A. 1984. Statistical Procedures for Agricultural Research (2nd Ed.). John Wiley & Sons, New York, 680p.

Hill, A.B. 1971. Principles of Medical Statistics (9th Ed.). Oxford University Press, New York, 390p.

Holmes, D., Moody, P., and Dine, D. 2010. Research Methods for the Bioscience (2nd Ed.). Oxford University Press, Oxford, 457p.

Kerlinger, F.N. 1986. Foundations of Behavioural Research (3rd Ed.). Holt, Rinehart and Winston, USA. 667p.

Kirk, R.E. 2012. Experimental Design: Procedures for the Behavioural Sciences (4th Ed.). Sage Publications, 1072p.

Kothari, C.R. 2004. Research Methodology: Methods and Techniques (2nd Ed.). New Age International, New Delhi, 401p.

Kumar, R. 2011. Research Methodology: A Step-by step Guide for Beginners (3rd Ed.). Sage Publications India, New Delhi, 415p.

Leedy, P.D. and Ormrod, J.L. 2010. Practical Research: Planning and Design (9th Ed.), Pearson Education, New Jersey, 360p.

Marder, M.P. 2011. Research Methods for Science. Cambridge University Press, 227p.

Panse, V.G. and Sukhatme, P.V. 1985. Statistical Methods for Agricultural Workers (4th Ed., revised: Sukhatme, P.V. and Amble, V. N.). ICAR, New Delhi, 359p.

Ross, S.M. and Morrison, G.R. 2004. Experimental research methods. In: Jonassen, D.H. (ed.), Handbook of Research for Educational Communications and Technology (2nd Ed.). Lawrence Erlbaum Associates, New Jersey, pp. 10211043.

Snedecor, G.W. and Cochran, W.G. 1980. Statistical Methods (7th Ed.). Iowa State University Press, Ames, Iowa, 507p.

Chapter 6
Collection and Analysis of Data

We really haven't got any great amount of data on the subject, and without data how can we reach any definite conclusions?
Thomas Alva Edison (1847–1931)

Once the project proposal is approved and the source of funding is finalized, you have to act quickly and start the project. Do not wait for all the formalities to start the project; otherwise, you may not be able to complete the project within the stipulated time. Once you have started and the project is underway, periodic monitoring is essential for the smooth running of the project in the right direction. Gathering data meticulously through surveys or observations is essential for the success of a project. You need to prepare and send progress reports and other reports asked by the funding agency or the institution in time. In the reports, you need to present only the refined data through neatly prepared tables and figures after subjecting the collected raw data to appropriate statistical analysis. If you feel that you have generated some meaningful research findings, try to present them in relevant conferences or symposia. Once the project is completed, you should publish research articles, technical bulletins, or books based on these results. If you are fortunate to get some breakthrough, you should apply for patents in time.

In the case of thesis projects by students, the mechanism in operation is almost similar to a normal research project as mentioned above. However, the supervisory guide or the chairperson of the advisory committee has a greater role in properly guiding the student in data collection and analysis. An overview of data collection techniques and analysis is provided in this chapter.

6.1 Sampling

In research studies involving many individuals or items, it may not be possible to observe and record data on all of them. Often, observations may be confined to representative items only. The larger group is called a *population,* and the representative

C. G. Thomas, *Research Methodology and Scientific Writing*,
https://doi.org/10.1007/978-3-030-64865-7_6

portion from which information is collected is called a *sample.* The procedure of selecting a representative sample from a big mass of population is called *sampling.*

A population is a group of individuals having one or more characteristics in common. In some cases, you may attempt to obtain information from all the elements of a population or complete enumeration of a population as in a *census*. A researcher, however, attempts to conduct a census, only if the population is sufficiently small. When the population is large, it is impossible to collect information from all the members of a population, and the researcher has to plan for some shortcuts. A widely accepted procedure is to collect information from representative samples of the population. If this subset is a true representative of the overall population and exhibits similar characteristics to any randomly chosen division of the population, then the generalization or conclusion may have applicability to behaviour of the entire population.

The process of sampling enables us to draw generalizations based on careful observations. A measured value based on sample data is called a *statistic.* You can use this statistic to estimate the characters of a population. A population value inferred from a statistic is a *parameter*. The following are the merits of sampling over complete enumeration or census of a population.

- ***Reduced cost***: If the data are secured from only a small fraction of the population, naturally, expenditure is less.
- ***Greater speed***: Sample data can be collected and summarized more quickly than with complete enumeration.
- ***Greater scope for accuracy***: Studies that rely on sampling have more careful supervision than a complete enumeration.
- ***Degree of precision***: Information generated through sample collection is prone to some uncertainties. A major problem is the occurrence of sampling errors, as only a part of the population is measured in sampling. Although there will not be any sampling errors in complete enumeration, possibilities of non-sampling errors are greater. If you want to have high degree of precision, take large samples and use superior instruments for measurements.

Sampling and Non-sampling Errors

In sampling, instead of studying a whole population, you study a fraction of the population, the sample, and infer the situation based on that sample. This is actually an inductive process (See induction and deduction in Sect. 1.6). Inferences are made by observing a small section of people, animals, or plants and extrapolate them for the whole population, which they represent. Note that representative sampling is a prerequisite for successfully averaging out random errors and avoiding systematic errors. The quality and usefulness of your inference depends on how best is the representative sample.

The error occurring because of the likely faults in the sampling process is called *sampling error.* Sampling error is the extent to which a sample drawn from a population differs from the original population. Errors can also occur due to other reasons like errors in measurement, investigator bias during data processing, and interpretation. Such errors are called *non-sampling errors*. It includes personal error also, for example, things hidden or forgotten. When the population is large and sample size is small, chances of errors are more. The degree of precision can be increased by taking larger samples and by using superior instruments for measurements.

In census, sampling errors are absent, only non-sampling errors; while in sampling, both types of errors occur. When the sample size increases, sampling error decreases, but non-sampling error increases. When the sample size becomes equal to the population, there is no sampling error as there is no difference between the population and sample.

Sampling Methods

Sampling methods are classified as *probability sampling* and *non-probability sampling*. In probability sampling, each constituent of the population has a known probability of being selected. Because of this character, sampling error can be estimated—a major advantage of probability sampling. When extrapolating data from samples to that of population, values are presented plus or minus the sampling error. Common probability methods include *simple random sampling, systematic random sampling*, and *stratified random sampling*.

In non-probability sampling, samples are chosen from the population using some non-random procedures. Non-probability sampling methods include *convenience sampling, quota sampling, judgment sampling,* and *snowball sampling.* A major disadvantage of non-probability sampling is that the extent to which the sample differs from the population remains unknown, and therefore, it is very difficult to estimate sampling error. Non-probability sampling methods are generally used for qualitative studies.

Simple Random Sampling

For small and almost uniform populations, we generally go for simple random sampling. As the name indicates, it is the simplest form of probability sampling where individual units are selected at random. When you attempt sampling, ensure that every constituent of the population has an equal chance of being selected. You have to fix up the number of samples based on the size of population, and every element of population has the same probability of being selected. Data collection will be made from the representative samples only and not from the entire population. For example, when you conduct an experiment with crops, you may not measure the heights of all the crop plants but only samples selected through simple random sampling.

Techniques of Simple Random Sampling

Perfect randomness is an important factor in sampling because many of the common statistical techniques used to process results are valid on truly generated random data only. The method of selecting a sample at random depends to some extent on the size and nature of the population. The following are some of the commonly used methods to select samples:

- ***By drawing lots***: This lottery method is a commonly used procedure for random sampling especially when the population is small. Names of all the items or individuals in a population are written on pieces of paper, and random draws are made.
- ***By selective-service numbers***: In this method, the units are arranged in a particular order, for example, alphabetically, numerically, or geographically. Then, you use selective-service numbers for selecting the sample. For example, every 10th, 20th, or Nth unit can be selected.
- ***Random numbers***: A most commonly followed approach for scientific experiments. Usually, random numbers prepared by Fischer and Yates or Tippett are used to draw samples. Presently, computer generated random numbers are also in use.
- ***Grid system***: Grid system is mainly employed for studying an area, especially in ecology and geography using quadrats. A quadrat is a small plot to isolate a standard unit of area for studying the distribution of an item over a large area, and particularly suitable for plants, slow-moving animals such as snails and millipedes, and some aquatic organisms. The actual size of the quadrat is decided by the habitat being sampled and by the purpose of the study. Usually, square quadrats with sides 0.5 m are used in plant sampling from experimental plots. In the simplest form of using quadrats, the quadrat is thrown to fall at 'random' within the site. However, this is unsatisfactory in most cases as personal bias can creep into the throwing act, that is, it may not be truly random. Moreover, this 'throwing' would be possible only in cases where quadrats of small size are used. Normally, it may not be possible to throw anything larger than a 1 m × 1 m quadrat. Therefore, a better method to use quadrat is to map the entire area and then to lay a numbered grid over the map. A random number table is then used to select the grids for observations.

Systematic Random Sampling
Sometimes, a variation of simple random sampling—*systematic random sampling*—is adopted if the entire population is finite or can be listed. It is also called an Nth name selection technique. First, the required sample size is determined, suppose it is 50 from a population of 1000. Then, the number of intervals is found out by dividing the population by the sample size. One would select the first item by choosing a randomly selected item, and then, every 40th (Nth record) item is selected from a list of population until the sample of 50 items is completed. For example, if the first sample is the 6th item, subsequent samples will be 46th, 86th, 126th, and so on until we get 50 samples. This kind of systematic sampling is as good as random sampling.

Stratified Random Sampling
Stratified random sampling is a commonly used probability method when large samples are involved. In such cases, it is considered superior to random sampling as it helps to reduce sampling error. Stratified sampling is adopted usually for sampling heterogeneous populations. The population is divided into homogeneous subgroups or strata, and from each subgroup or stratum, a random sample is drawn. Before

going for sampling, the researcher has to identify relevant strata with some common characteristics and their actual representation in the population. Afterwards, for each stratum, subjects are selected in proportion to its frequency in the population using random sampling procedure.

Multistage Sampling

It is a type of stratified sampling suitable for infinite populations, where a list of members is absent, or when the individuals are living in widely scattered groups. This is also called *cluster* or *area sampling*. The population is first divided into different stages, and random samples are drawn. Initially, the population is divided into first stage sampling units, from which a random sample is selected. This sample is then divided into second stage units, and again a sample is selected. In this way, a random sample is selected at each stage. There must be at least two stages in this type of sampling.

Convenience Sampling

Convenience sampling is a non-probability sampling technique, which is also called *accidental sampling*. It is most often used in descriptive research where the concern of researchers is to get an inexpensive estimate of facts. As the name implies, selection of samples is based on convenience. This non-probability method is often used during preliminary stages of research to get a rough estimate of the results.

Judgment Sampling

Judgment sampling or *purposive sampling* is another commonly used non-probability method. The researcher selects the sample based on some judgment, especially when the entire population is inaccessible. This is actually an extension of convenience sampling. Suppose that an investigator has to take samples from several districts. The researcher may decide to select samples from one representative district only instead of several districts, having convinced that the chosen district is truly representative of all the districts to be sampled. Although bias can occur in judgment sampling, you can still have good representation of population, if you can do it objectively.

Quota Sampling

Quota sampling is another non-probability sampling method almost similar to stratified sampling. In quota sampling too, the researcher first identifies the strata and their proportions as they are represented in the population. However, after selecting the stratum, samples are drawn from each stratum using the procedure of convenience or judgment sampling unlike stratified sampling, where each stratum is filled by random sampling.

Snowball Sampling

Snowball sampling is a special non-probability sampling method for situations where the desired sample characteristic is rare. In certain occasions, locating experimental subjects may be very difficult or costly. In such cases, the researcher can try referrals from initial subjects to generate additional subjects. Although this technique substantially reduces difficulties of researchers in locating samples and cut the cost on searching, this may increase sampling bias.

6.2 Scales of Measurement

Researchers need to measure several characteristics of the variables under study. It is essential to have a numerical method for describing observations. Four scales of measurement are commonly used depending upon the nature of variables; (1) the nominal or classificatory scale, (2) the ordinal or ranking scale, (3) the interval scale, and (4) the ratio scale.

Nominal Scale

Nominal scale (also called *classificatory scale* or *categorical scale*) is the most elementary method of quantification of observation, and the least precise. This scaling is done for nominal variables (see Sect. 3.3). A nominal scale describes differences between characters by assigning them to categories. By using nominal scale, you can classify animate beings, inanimate objects, or events into a number of mutually exclusive categories, so that each member of the subgroups has some characteristics in common. For example, the variable 'gender' can have two categories, males or females; and considering 'educational qualification', six categories can be suggested, non-literate, literate, matriculate, graduate, post-graduate, and doctorate. See the fictitious data in Table 6.1.

Using nominal scale, you can only allot individuals to a category, but cannot rank them. Sometimes, code values are given for processing data, for example, 1 for female and 2 for male. Individuals having a single value are alike and those with different values are different. The labels tell us that the categories are quantitatively different from each other. However, they have no quantitative significance, implying that they

Table 6.1 Educational level of people in a village in Kerala (fictitious data)

	Male	Female	Total
Non-literate	5	7	12
Literate	80	52	132
Matriculate	78	86	164
Graduate	46	42	88
Post-graduate	26	22	48
Doctorate	3	2	5

Table 6.2 Level of adoption of IPM in rice by farmers

Rank	Indication
0	No adoption
1	Slight adoption
2	Partial adoption
3	Moderate adoption
4	Majority adoption
5	Full adoption

cannot be added, subtracted, multiplied, or divided. Probably, the only arithmetic operation possible with nominal scales is counting.

Ordinal Scale

With a nominal scale, the researcher may only be able to indicate that certain things differ as they fit into certain categories. However, with an ordinal scale (also called *ranking scale*); one may be able to assert the amount or degree of their differences. For ordinal variables (see Sect. 3.3), ordinal scaling is done. Note that they will have all the properties of a nominal scale; but in addition, it is possible to rank the subgroups in a certain order. They can be arranged either in ascending or descending ranks according to the magnitude of variation, but actual differences between adjacent ranks may not be equal, as they have no absolute values. All the properties of nominal scale can be applied to ordinal scores too, but not vice versa. For ordinal scale, frequency can be identified. As in the case of nominal scores, in ordinal scale too, individuals with the same scores are treated alike. After tabulating and ranking the scores, these can be subjected to analysis. Consider the example given in Table 6.2. In the example, ranks were allotted based on the degree of adoption of integrated pest management (IPM) in rice by farmers. The lowest score of '0' was given to 'no adoption' and the highest rank of '5' was given to 'full adoption'.

In a similar way, if you want to study the level of infestation of a disease in a crop, scores can be given by visually observing the disease infestation and assessing its severity. The scores can be arranged either in increasing or in decreasing order. If 0–5 is the range of scores to be given for the level of intensity of infestation, you can give score 0 for no infestation and 5 for the severest infestation. The scores would be 0- no infestation, 1—very slight infestation, 2—slight infestation, 3—moderate infestation, 4—severe infestation, and 5—very severe infestation. You can rank the units as 5 > 4> 3 > 2> 1 > 0.

Interval Scale

Although the ordinal scale is an improvement over the nominal scale, it is not without problems. As the real differences between adjacent ranks are not equal with ordinal scale, the differences between two values do not have any specific meaning. If our intention is to explain such differences, we must go for interval scales. Interval level scores are in a position higher than ordinal scores as the differences or intervals convey some more meaning. All the characters of ordinal values are applicable in interval level scores too, but additionally, interval scale uses a measurement unit, which allow the responses to be placed at equally spaced intervals in relation to the spread of the variable. For example, the difference between 3 and 4 is same as the difference between 10 and 11. Because the interval is same in both cases, the differences between the characters are also the same. Interval scale is applicable for interval variables coming under continuous variables (see Sect. 3.3).

Most quantitative measurements are amenable to interval-level measurement. The interval scale has a starting point and a terminating point divided into equally spaced intervals. Nevertheless, the starting and terminating points and the number of intervals between them are arbitrary and vary from scale to scale. A common example of interval scale is temperature measurement scales, Celsius and Fahrenheit, which start with different points of origin. The point of origin for the same natural phenomenon, the freezing point of water, is 0 on the Celsius scale and 32 on Fahrenheit scale. Each degree or interval is a measurement of temperature; however, as the starting point and terminating points are arbitrary, they are not absolute. Therefore, saying that 100 °C is twice as hot as 50 °C is wrong. The main limitation of interval scale is the absence of a true zero, and you cannot measure the complete absence of a feature using this scale. Because it is a relative scale, no mathematical operation can be performed on its reading.

Ratio Scale

A ratio scale, in addition to having equal interval properties of an interval scale, has two additional features. It has a true zero, meaning a fixed starting point. Therefore, it is possible to indicate the complete absence of a property. Another feature is that the numerals of the ratio scale have the qualities of real numbers and can be added, subtracted, multiplied, and divided, and expressed in ratio relationships. Measurement of height, weight, area, income, and age are examples of this scale. Ratio scale is applicable for ratio variables coming under continuous variables (see Sect. 3.3).

In physical and natural sciences, variables are mostly expressed in ratio scale. However, behavioural sciences such as sociology and psychology are generally limited to describing variables in ordinal scale, nominal scale, or interval scale warranting the use of nonparametric tests.

6.3 Collection of Data

Data collection is an important feature of the whole process of research. Measured facts used as a basis for reasoning, calculation, or decision-making are called *data* (Although *datum* is singular and *data* is plural, in technical writing, the plural form *data* is always used). When you collect data, your intention is to make inferences based on the collected data. Sometimes, the information you are trying to find out is already available in records but needs to be extracted. However, this may not be the case always, and in most situations, data have to be generated by doing some research. According to the broad approaches to information gathering, data are categorized as *primary data* and *secondary data*. When the data are collected by direct observation or survey, they are called primary data. It gives first-hand information to you. If the data are collected from already published books, census reports, journals, theses, project reports, published statistics, and similar documents, they are called secondary data.

Observation and *survey* are the two major primary data collection procedures employed by researchers. The choice depends upon the purpose of the study, nature of the problem, the resources available, and the skills of the researcher. Surveys do not involve direct observation; rather, inferences about behaviour or situations are made from data collected through interviews or questionnaires. Survey methods are used widely in social sciences and management to assess prevalence, attitudes, and opinions on different subjects, especially in non-experimental studies such as cross-sectional studies (see Chap. 4). More details on observation and survey are given in the following two sections.

6.4 Observation

Observation of nature is the fundamental basis of science. It is a purposeful and selective way of watching and recording an interaction or phenomena as it occurs. When you conduct an experiment, you observe and record information on several features. For example, if it is a growth analysis study of plants, you will observe and record number of leaves, leaf area, total dry weight, stem weight, height, root weight, and so on. Similarly, when you are conducting a titration study with chemicals or designing a device and testing its functions, you are making observations. In social sciences too, observation is made in several situations such as the behaviour of a group, personality traits of an individual, and functions of a worker.

Types of Observation

Observations can be divided into three types, direct observation, natural observation, and participant observation based on the way in which these are collected.

Direct Observation

Direct observation is the norm in natural sciences such as physics, chemistry, biology, and most applied sciences. It can also be used in social sciences. The researcher watches the events with a dispassionate perspective. In all kinds of experiments, researchers note down observations directly as the variables show signs of changes. However, the observer should try to be as unobtrusive as possible to avoid bias in observations. Researchers can use modern techniques such as microscopes, hidden cameras, recorders, and videos to record observations.

Natural Observation

In natural observation, researchers passively observe and record the behaviour, phenomena, or some other events in its natural background. The researcher will not in any way interfere with the system. This kind of observation has been a successful method in such physical sciences as astronomy, geology, oceanography, and meteorology. In social sciences, humans or animals are observed as they go about their activities in real life situations. In natural sciences, this may involve observing plants, animals, or some physical phenomena in natural settings. In health science-related areas such as human anatomy, observation is the primary method used to describe the construction of the body. In ecology, observation has always been of importance although experimental methods have also been utilized (For more details see Sect. 4.1).

Participant Observation

Participant observation is one of the most common methods for qualitative data collection. In participant observation, researchers live with the subjects being observed as an active participant and record extensive field notes. They take part in the daily activities of people, reconstructing their interactions and activities in field notes taken on the spot or as soon as possible after their occurrence. Compared to direct observation, participant observation often requires long periods, months, or even years of intensive work. Time is a major constraint with participant observation as a data collection method.

Problems with Observation

In research, observations must be taken with utmost care. The observer should be neutral, and his/her emotions or bias should not under any circumstances influence the data. A few common problems with observation include personal bias, placebo effect, Hawthorne effect, Rosenthal effect, John Henry effect, and novelty effect.

Personal Bias

In the so-called 'hard' sciences like physics and chemistry, possibilities of personal bias are less. However, chances of observer bias are greater in agriculture, health, and social sciences, which are considered 'soft'. If the observer is biased, s/he can introduce biased observations and may come up with a wrong conclusion. Sponsored researches from corporate firms or projects funded by sources with dubious agenda need greater scrutiny in this regard. These are intentional biases.

Personal bias of a lesser degree may also occur when selecting plots for a particular treatment or locating the plots. In field experiments, unintentional biases can be eliminated by adopting the three principles of experimentation—randomization, replication, and local control.

Placebo Effect

Placebo is an innocuous medication prescribed for the mental relief of a patient than for its actual effect. This creates a psychological effect, commonly called the placebo effect. In many cases, it has been established that for the patients receiving treatment, the mere illusion of receiving treatment can play an important role in their recovery from an illness. Placebo effect is noticed in many other cases also, for example, those involving dietary studies. When human subjects are involved in experiments, special safeguards against placebo effect must be taken. *Blind studies* are planned to isolate the placebo effect (see Sect. 5.10). In such blind studies, the human subjects do not know whether they are receiving real or fake treatment, and we can be sure that any effects on the participants are due to the treatment itself and not a reaction to the awareness that they are being treated. A blind experiment is said to be *single blind* when the researcher knows the situation well but not the participants. A blind experiment may be designed as *double blind*, in which both the participants and the researcher are completely 'blind', which means that they may not know which individuals are in the control group or in the experimental group until the observations are taken.

Rosenthal Effect

In support of the need for double blind experiments, the *Rosenthal effect* is often quoted. Robert Rosenthal, an acclaimed psychologist, has demonstrated the 'experimenter expectancy effects', the influence that a researcher can put forth on the outcome of a research inquiry. Unless stringent safeguards are taken to minimize experimenter bias, the results may conform to experimenters' expectations. Later, this effect came to be called 'Rosenthal effect' in honour of Robert Rosenthal.

Rosenthal and Fode (1963) demonstrated the experimenter expectancy effect in the laboratory on maze-running performance of test rats with the help of some undergraduate students. The students were told that they had two groups of test rats, a 'bright' strain and a 'dull' strain specially bred for the purpose. In fact, the two strains were not different and genetically identical. The researchers had misled the students for the purpose of the study, and the students' expectations of the rats had prompted them to try different methods of treatment, which had affected the learning ability of rats. After some time, the relative performances of the groups were compared. The results showed that the 'bright' rats learned the mazes more quickly than the 'dull' rats. It appears that students had unconsciously influenced the performance of their rats, based on what they had been told. Actually, the influences that they exerted on their test animals were unintentional and unconscious.

These experiments highlighted the threat of experimenter expectancies to scientific research and stressed the need for properly trained researchers and the adoption of *double blinding* as noted above to avoid Rosenthal effect.

Hawthorne Effect

Hawthorne effect can occur when human subjects are involved in experimentation. When individuals participating in an experiment become aware that they are being observed, they may change their behaviour, which may be positive or negative. When a change in the behaviour of persons or group is attributed to their being observed, it is known as Hawthorne effect. The name is based on a 1924 study of productivity at the Hawthorne factory in Cicero, Illinois, wherein a series of research studies were conducted on workers. Researchers increased light intensity and found that the productivity of workers increased. When they reduced the light intensity, to their surprise, they found that productivity of workers still increased. This suggested that the fact of being studied increased productivity rather than the increase or decrease of intensity of lights. Hawthorne effect must be taken into consideration when assessing the effectiveness of any intervention in a natural setting.

John Henry Effect

John Henry effect is an experimental bias, which can occur in experiments involving human subjects by the reactive behaviour of the control group. The term *John Henry effect* was coined from the behaviour of a legendary American steel driver (a person who works by hammering and chiselling in the construction of tunnels for railway tracks) John Henry in the 1870s. The legend says that when John Henry heard his output was being compared with that of a steam drill, he worked very hard to outperform the machine. He won the competition but died in victory with hammer in his hand. This phenomenon has been observed in several similar experiments. In an experiment, if the control group participants come to know of their status, they may work hard to overcome the disadvantage of being in the control group. A true treatment effect may be masked because of this John Henry effect.

Novelty Effect

This is also observed in experiments with human subjects. The novelty effect is the tendency of human subjects to improve their performance initially when something new is introduced. This may not be due to any real improvement in learning but in response to the enthusiasm generated by the novelty factor. For example, a treatment may work simply because it is new and the subjects respond to this uniqueness rather than the actual treatment. After a while, however, as the novelty wears off, the response may decline.

6.5 Surveys

The survey method, a common tool for data collection in social sciences for descriptive studies, does not involve direct observation by a researcher. Rather, inferences about behaviour or situations are made from data collected through interviews or questionnaires. Surveys are highly useful when researchers are interested in collecting data on characteristics of human behaviour, which are often not amenable to direct observation. Surveys are widely used in social sciences such as sociology, psychology, economics, and management to assess attitudes, opinions, or prevalence of something in different subjects. In agricultural and medical sciences too, these are used especially for those studies involving farmers and patients, respectively. Cross-sectional studies and ex-post-facto studies generally use survey methods for data collection.

Some limitations of surveys must be noted. A major limitation of survey is that it depends heavily on a self-report mode of data collection, and therefore, problems such as poor memory, non-cooperation, purposeful cheating by way of providing wrong responses, vague responses, or imprecise responses because of confusion in questions can contribute to inaccuracies in collected data. Furthermore, as this method is descriptive and not explanatory or analytical, it is incapable of offering any insights into cause and effect relationships.

Many kinds of survey procedures are used by researchers. Surveys can be divided into three broad categories: *attitude surveys, questionnaires,* and *interviews.*

Attitude Surveys

Attitudes are predispositions to react positively or negatively to some social objects. In attitude surveys, also called *opinion surveys*, standardized questionnaires indicating the degree of favourability towards an object are used to measure attitudes or opinions. It is mainly based on the individual's agreement or disagreement with a number of statements relevant to the attitude object. An attitude scale usually consists of the belief component of the theoretical attitude in a number of statements. Many attitude scales were developed by researchers, for example, *Thurstone scale* (Thurstone 1928) and the *Likert method* (Likert 1932). The Thurstone technique of scaled values is a method to facilitate interval scale measurement. The researcher selects a large number of evaluative statements, usually 20 or more, about a particular group, institution, idea, or practice. The statements should be unambiguous and cover a full range of attitudes towards the topic; and after the set of statements are made, a panel of judges are asked to rate each statement, either agree or disagree, and finally analysis is done. However, in surveys with questionnaires, Likert scale is the most widely used scale. A Likert item is a statement, which the respondent is asked to evaluate according to any kind of subjective or objective criteria. Often, the level of agreement or disagreement is measured on 5- or 7-point scale response levels. For example, if 5 is taken, 2 are subjective degrees of affirmation, 2 are subjective degrees of negation, and 1 neutral. The format of a typical 5-point scale Likert item is (1) strongly disagree, (2) disagree, (3) neutral (neither agree nor disagree), (4) agree, and (5) strongly agree.

Questionnaires

A *questionnaire* consists of a series of pre-determined questions that can be self-administered, administered by mail, or asked by interviewers. The information is collected usually through mail by sending a proforma. The respondent is asked to fill it, and there is no interaction between the investigator and the respondent. Hence, simple questions only are included in questionnaires. When the questionnaire is administered by interview, it is often called an *interview schedule* or simply *schedule.* Here, investigators ask questions directly to the respondents using the schedule. They frame a set of questions, and while asking questions, they can make some variations in the mode of asking questions.

Questionnaires are used in research based on a fundamental and important assumption that the respondents are willing to give truthful answers. They must also be able to give responses freely. Interviews or questionnaires commonly include three types of questions, (1) closed questions or forced-choice questions, (2) open-ended questions, and (3) scale items.

Forced-choice items: Forced-choice items or closed items allow the respondent to choose from two or more fixed alternatives. Most frequently used are the dichotomous items, which offer two alternatives only, for example: 'yes/no' or 'agree/disagree'. Sometimes, a third alternative is also given as 'undecided' or 'don't know'. The alternatives offered should cover all the possibilities.

Open-ended items: The respondents are free to answer anything to the questions offered to them. The answers may be a short note or short essay depending on the nature of question. No restraints are imposed on their answer to the question.

Scale items: As already mentioned, the scale is a set of verbal items to which the respondent answers by indicating degrees of agreement or disagreement. Individual responses are located on a scale of fixed alternatives, for example: 'strongly disagree' to 'strongly agree'.

Interviews

Interview is a personal form of research compared to a questionnaire. For seeking opinions or impressions from a person or group of persons, interview is an ideal choice. An interview is a verbal interchange, often face to face, in which an interviewer tries to elicit information, beliefs, or opinions from another person. In interviews, the interviewer works directly with the respondent. If required, the interviewer can probe further by asking follow up questions. Interviews can be conducted in many ways; it may be a structured interview, unstructured interview, or semi-structured interview based on the interview schedule used and the context. Quantitative analysis is possible only from the numerical data generated out of structured interviews. If you need only qualitative information, then either semi-structured or unstructured interviews can be adopted.

Structured Interviews

In structured interviews, the interviewer uses a pre-prepared questionnaire (an interview schedule) and conducts the interview as such without much variation. A major advantage of structured interview is that the quantitative data generated are amenable to statistical analysis.

Unstructured Interviews

This takes the form of a conversation between the informant and the researcher. There is no standardized list of questions, and it is a free flowing conversation in a natural setting. In an unstructured way, it focuses on the respondent's perception or opinion on various issues. This is also called *open-ended interviewing* or *in-depth interviewing*.

Semi-structured Interviews

Semi-structured interview makes use of a guide list with some broad questions or issues, which are to be discussed for possible investigation during an interview. The list normally includes broad and open-ended questions to be answered in a free flowing, conversational, relaxed, and informal setting. The interviewer is left free to rephrase the questions and to ask probing questions for added details. Semi-structured interview is one of the main tools used in 'participatory learning and action' (PLA), an umbrella term for a variety of participatory approaches to development (see Sects. 4.11 and 4.12).

6.6 Possible Errors in Research

Any difference between the measurements you made and the true value is an *error.* It is easy to obtain closely reproducible results from experiments in 'hard' disciplines such as physics and chemistry, provided your experimental techniques are sound and the equipment are working correctly. In such disciplines, it is relatively easy to keep the conditions under control during experiments, and chances of experimental errors are relatively less; and in most cases, when you repeat the experiments you get concurrent values. On the other hand, experimental errors are more common in biological experiments as the variables involved in the study are human beings, animals, plants, soil, or climate, which are dynamic. Fortunately, the reality of such fluctuations in 'soft' disciplines such as agriculture and health sciences has been recognized and measures to deal with them have been standardized by statisticians.

Consider a simple experiment of comparing two cultivars of rice. You assigned cv. A to Plot 1 and cv. B to plot 2. After the harvest, you found that the yield of cv. A in plot 1 was higher than cv.B in plot 2, and concluded that cv.A is superior to cv.B. However, the conclusion may be entirely wrong because of a false assumption! It is assumed that any difference between the yields of the two plots is caused by the cultivars alone and nothing else. In fact, the two plots were not uniform and differed on many counts, for example, soil structure, texture, water-holding capacity, pH, fertility gradient, and so on. There might also be differences in pest and disease occurrence, bird damage, etc. Therefore, you should separate out treatment differences from other sources of variation. This kind of variation observed between experimental plots when they are treated alike is called *experimental error*. These errors, caused mainly by uncontrollable factors, must be accounted for while reaching conclusions. Based on the magnitude of experimental error, one can decide whether a difference observed in the experiment is due to genuine reasons or just due to chance alone. Therefore, plan the experiment in such a way as to have an estimate of experimental error.

Experimental errors can hide the effect that the researcher is trying to investigate. Errors should not be confused with personal mistakes or faults, which can be rectified by taking appropriate corrective measures. Personal mistakes can be eliminated largely by checking one's work frequently. Scientists should be careful to observe and correct their own mistakes. Even if all the precautions are taken, errors can creep in, and such errors are natural variations, fluctuations, or deviations that are present in the data or a set of observations.

Unavoidable errors are a great problem in both laboratory and field experiments. The researcher should be aware of the possible effects of errors on experimental data and consequent conclusions made out of it. It is impossible to compare an experimental result to a theoretical prediction, compare two experimental results, or evaluate whether an apparent correlation is real without knowledge of the possible errors. There should be some mechanism to estimate the errors.

Random and Systematic Errors

Unrecognized variations in experiments can be of two types, *random errors* and *systematic errors*. Random errors are produced by a combination of uncontrollable and unknown variables. If these errors occur as negative and positive deviations from true values, and if they have a mean of zero, they do not normally bias mean values. However, they may affect the accuracy of measurements from replications. At the same time, if positive and negative values of errors average to a nonzero value, such errors are called *systematic errors*. Experimental units receiving a treatment should show only random differences from units receiving other treatments including the control. These units should be allowed to respond independently of one another.

Systematic Errors

While conducting an experiment, we must make sure that there are no systematic differences between experimental units receiving a treatment from those receiving another treatment. At times, systematic errors may be more complicated. A *constant systematic error* affects accuracy of values, but it will not influence the correctness of measurements. At the same time, a *variable systematic error* affects both accuracy and correctness of measurements. Systematic errors occur mainly because of two reasons, *experimental unit (plot) variations* and *faulty techniques*. It is assumed that any difference between the values of two experimental units (plots) is caused by the treatments alone and nothing else. In field experiments, chances of plot variations due to soil heterogeneity are big problems. Soil heterogeneity occurs mainly because of texture, fertility gradients, pH variations, and erosion hazards. There might also be differences in pest and disease occurrence, bird damage, etc. These variations can also occur when the 'plot' consists of animals or even human beings. Soil heterogeneity is tackled largely by random arrangement of plots, replications, and suitable layouts.

Faulty techniques include inaccurate scales and measures, faulty instruments, wrong methods, or personal bias. For example, a frequent source of systematic errors is calibration errors with equipment and instruments. If the calibration is faulty, the values may be too high or low. This kind of errors can be eliminated by the skill, care, and experience of the researchers. Researchers well versed in research methodology and instrumentation seldom commit this kind of mistakes.

Random Errors

Random errors are so called as these can occur due to chance alone, which cannot be attributed to any specific cause. These errors may happen quite unexpectedly despite all the precautions taken. Because they are governed by probability, statistical methods are used to estimate the error due to random variation. Random errors can be averaged out by making many replicates or repeat measurements. When the number of replicate measurements is more, it permits us to predict the proper value with greater confidence. An estimate of random error helps us to settle on whether an observed difference is genuine or due to chance alone. Therefore, every experiment should have provisions to estimate errors due to random variations.

Most experiments tackle the issue of reducing errors using the principles of experimentation—randomization, replication, and local control—and by representative sampling. A *representative sample* is a small division of the population, displaying the same features as that of the original population. Non-representative sampling is a frequent pitfall, which can be checked by taking adequate precautions. If your sample is representative of the overall population, it exhibits similar characteristics to any other randomly chosen sample of the population, and your generalization may have applicability to the behaviour of the population as a whole.

6.7 Analysis of Data

The data recorded in observation books or basic records have to be tabulated well. For describing and analysing the results, a basic understanding on statistics and its application in agricultural and life sciences is necessary. In agricultural university system, students are exposed to one or two courses on statistics at the undergraduate level itself. In addition, students majoring in those disciplines such as agronomy, horticulture, plant breeding, and other subjects, which involve extensive field trials, should compulsorily take a course on design of experiments and analysis at the Masters level.

A brief introduction to some of the most widely used concepts of statistics, both descriptive and inferential statistics as applied to agriculture and other life sciences, is attempted here in simple language excluding technical jargons to the maximum extent possible. For more information on design of experiments and application of statistical procedures in agriculture and life sciences, refer to Hill (1971), Snedecor and Cochran (1980), Gomez and Gomez (1984), Panse and Sukhatme (1985), Holmes et al. (2010), or similar books. Remember that for applying statistics, you must have done a better experiment as Ernest Rutherford (1871–1937), Nobel Prize winner for chemistry in 1908, quipped: ‘If your experiment needs statistics, then you ought to have done a better experiment.’

6.8 Descriptive Statistics

During the course of investigation, you may be accumulating voluminous data, which need to be summarized and presented in a clear accurate format for easy comprehension. For this purpose of summarizing a data set, you make use of *descriptive statistics* or *deductive statistics*. Using descriptive statistics, you can reduce a set of numbers into simple values and describe the main features of a data set quantitatively. These include calculation of measures of central tendency such as mean, median, and mode; dispersion measures such as standard deviation and range; association measures such as correlation and regression; frequency distribution tables; cross-tabulation tables; and graphs and figures. Normally, a combination of tables, graphs, charts, and a discussion is used to summarize and present the data set. A major drawback of descriptive statistical analysis is that it limits generalization to the particular group of individuals only. Therefore, no conclusions are possible beyond this group, and you cannot assume any similarity to those outside the group.

There are three general types of descriptive statistics used to describe data, the measures of central tendency, the measures of spread or dispersion, and the measures of association.

Measures of Central Tendency

Measures of central tendency are used to describe the central position of a frequency distribution for a group of data. For describing this central position, the *mean, mode,* and *median* are generally used. Among these, the most commonly used measure of central tendency is the arithmetic mean. If all the data fall in a 'normal distribution', the mean, the median, and the mode coincide.

Measures of Spread or Dispersion

Measures of spread help us to have some idea about the variation of measured data around the mean. A number of statistics are used to describe this spread including the *range, percentiles, quartiles, mean deviation, variance*, and *standard deviation.*

Among these, standard deviation is the most popular one among researchers.

Measures of Association

If you want to see whether there is a relation between two sets of observed data, you can make use of the techniques of *correlation* and *regression.* The relationship between two paired variables can be estimated using correlation. You can estimate this even by using natural data generated without doing an experiment. It provides a measure of the degree of association between the variables or the goodness of fit of a prescribed relationship of the data. Correlation can be negative or positive. The extent of relationship can be quantitatively represented by the *coefficient of correlation.* A perfect positive correlation is $+1.00$, and a prefect negative correlation gives a value of -1.00. On the other hand, regression analysis describes the effect of one or more independent variables on a single dependant variable by expressing it as a function of the independent variable. The relationship may be simple, if only one variable is involved, and multiple, if more than two variables are involved. It may be linear, if the form of the relationship is linear. Sometimes, the relationship may be curvilinear or hyperbolic, that is nonlinear. Regression equations or models can be derived and used to predict the effect of independent variables.

Other Descriptive Statistical Tools

Other common descriptive statistical tools include *percentages, proportions, ratios,* and *rates.* Frequency distribution tables, cross-tabulation tables, graphs, and figures are also made according to the type of observation and data collected.

6.9 Inferential Statistics

As already mentioned, you make use of descriptive statistics or deductive statistics for summarizing a data set. When you are summarizing and describing the data using descriptive statistical tools, you may also want to examine whether the estimates made in the study can be generalized and extrapolated beyond the samples under study. It is also essential to ascertain whether the differences or associations found can be explained by chance, or whether there are any real differences. *Inferential statistics* or *inductive statistics* offer the tools to answer these questions. The tools of inferential statistics are used to make claims about the populations to which the samples belong. The main objective of using inferential statistics is to arrive at conclusions about populations based on observations on representative samples. We make use of observed data from samples to make inferences about population parameters. Representative samples are drawn from population, and using the sample statistics, population parameters are estimated. However, remember that as sample data give only representative figures, they cannot be perfect estimators of populations. Different procedures are followed to make sure that the inferences made are appropriate and rational.

As discussed previously, the properties of populations such as the mean or standard deviation are called *parameters* as they represent the whole population. It is very difficult to have access to the whole population in which one have interest. However, one may have access to data from a fraction of a population, a sample. For example, you might be interested to know the fruit yield from mango trees in a particular region. As it is difficult to estimate fruit yield data from all the trees in the region, you may attempt to measure only a smaller sample of trees, say 200 trees selected at random, to represent the larger population of all mango trees. Samples also have properties such as the mean or standard deviation, which are called ***statistics*** and not parameters. Using the sample statistics, one generalizes the behaviour of populations from which the samples are drawn. However, it is important that the selected sample accurately represent the population avoiding *sampling errors* to the maximum. In general, inferential statistics involve two aspects, (1) estimation of population parameters and (2) testing of statistical hypotheses.

Parametric and Nonparametric Tests

Parametric data are measured data, and parametric statistics assume that all the data are 'normally' distributed. As already defined, a parameter refers to a function of the population. In the 'bell shaped' normal distribution curves, most populations display majority of average values at the central bell part of the curve with extreme values tailing off at each end. If you take the case of height of people, it could be seen that most people are of average height, with a few extremely tall or short cases; that is, heights of people are normally distributed. Although individual values vary, the generality of this type of curves among populations is so strong that the parameter is taken as a constant.

For applying parametric tests, three pre-conditions have to be met: (1) the observations must be independent; (2) the samples must have equal or nearly equal variances;

and (3) the variables involved are expressed in interval and ratio scales. Parametric tests include the *z*-test, the *t*-test, and the analysis of variance (ANOVA).

Sometimes, samples and populations do not behave in the form of a normal bell curve. For example, nominal or ordinal data will not take the shape of a normal curve. They are called *nonparametric data*, which are either counted or ranked. For applying nonparametric tests such as chi-square, the stringent assumption of normally distributed populations is not necessary.

The Null Hypothesis

Formation of null hypothesis is an important step for statistical analysis. A *null hypothesis* (H_0) states that there is no significant difference between the mean of a sample and the mean of a population from which the sample is drawn, and whatever you notice is only a random sampling error (see also Sect. 6.6). Our concern is to know whether visible differences are true differences or whether they are simply due to sampling errors. Null hypothesis is mandatory for statistical analysis, which is written as a negative statement. Along with this, you also formulate the alternate hypothesis or the research hypothesis (H_1) generally as a positive statement. Research hypothesis is a statement about the population parameter or parameters, which gives an alternative to the null hypothesis within the range of pertinent values of the parameter, that is, if H_0 is rejected, what hypothesis is to be accepted and vice versa. Null hypothesis means that even if a difference or an association is observed, it should be assumed that it is due to chance, until it is confirmed by statistical analysis. However, if the statistical test rejects the null hypothesis, the research hypothesis is accepted by elimination.

Importance of the formulation of research hypothesis, especially for experiments, has already been discussed (Sect. 5.2). You start with an assumption that there is no difference or association between the variables studied (null hypothesis, H_0). It is written as a negative hypothesis, for example: 'There is no significant increase in grain yield when balanced plant nutrition is ensured'. The null hypothesis is contrary to the research hypothesis (also termed alternative hypothesis): 'There is an increase in grain yield when balanced plant nutrition is ensured'.

You can never prove a null hypothesis, as any one result that differs from the hypothetical result is sufficient to disprove it. Therefore, it can be either disproved and rejected completely, or accepted within the limits of sampling errors. The data from the experiments are subjected to statistical analysis, and the probability of occurrence of the null hypothesis is calculated. If the probability is lower than the level of significance you have fixed, you reject the null hypothesis. The null hypothesis is tested for rejection under the assumption that it is true, as distinct from the research hypothesis. Rejection of null hypothesis suggests that you have enough evidences to support your research hypothesis.

Type I Error and Type II Error

In most research studies, we draw and examine a sample hoping that it will reveal all the facts about the population accurately. However, because a representative sample only is studied instead of the original population, there is every chance of committing mistakes in the decision for rejecting or accepting the null hypothesis. Take the example of a null hypothesis, which says:

> There is no significant decrease in grain yield when irrigation is limited to critical stages compared to irrigation throughout the growth period.

After analysing the data, you set an arbitrary value for the rejection or acceptance of a null hypothesis based on some *level of significance*; p is the probability that a difference or an association as large as the one observed could have occurred by chance alone. You reject the null hypothesis, if the p value is less than the pre-determined level of statistical significance, usually expressed as a percentage. A result is considered statistically significant, if the p value is less than 5 percent ($p = 0.05$) and is said to be highly significant if p is less than 1 percent ($p - 0.01$).

In most field experiments, 5 percent ($p = 0.05$) level of significance is often used as a standard for rejection of null hypothesis. In certain studies related to health, you may even go for 1 percent ($p = 0.01$) level of significance. Assume that the level of significance at which you are willing to reject or accept this null hypothesis is at 5 percent ($p = 0.05$). This allows for a 5.0 percent probability that the outcome could have occurred by chance. When you reject a null hypothesis at 5 percent level of confidence, it indicates that a difference in means as large as that between the experimental and control groups would have occurred from sampling error in less than 5 out of 100 items of the experiment. Conversely, the probability that the difference in values is due to the treatment instead of sampling error is 95 percent. The rejection of the null hypothesis while it is true is called a *type I error*. At the same time, failing to reject the null hypothesis while it is false is called a *type II error*. The chance of committing a type I error, rejecting the null hypothesis when it is actually true, is represented by α, and the chance of committing type II error, proving an association when none exists, is generally represented by β.

The maximum probability of rejecting the null hypothesis when it is actually true or the chance of committing type I error is known as the *level of significance*. As narrated earlier, these probabilities are generally taken as 0.05 or 0.01 (5% or 1%).

It is essential to ensure that the samples are true representative of the original population, and there should be no chance for committing any type of errors. The two types of errors are unavoidable, and it is not possible to eliminate them simultaneously. When one type of error decreases, the other type increases, and therefore, the best option is to fix the probability of one of them and try to minimize the other. As type I error is more serious, we usually fix the probability of type I error and try to minimize the probability of type II error.

A difference between sample and population is assumed genuine and not due to chance mainly based on three criteria.

1. *The quantum of the difference observed*. It is reasonable to expect that the larger the differences between the means, the more likely it is due to treatments.
2. *The degree of variations in the values*. If the values fluctuate widely with extreme values, most probably the differences in means are due to chance variations.
3. *The sample size studied*. If the sample size is large, it is possible that the conclusion drawn from it will reflect the parameters of the population. Therefore, it is

important to avoid small samples, as there are extreme or rare sample means in the sampling distribution, and you are more likely to get one of them in your experiment.

Degrees of Freedom

When analysing the data from experiments using statistical tools, you usually confront an interesting but important concept, the *degrees of freedom*. The concept of degrees of freedom is based on the notion that if the total items in each set of measurements are fixed, then all the measurements minus one are free to have any value. In other words, the number of degrees of freedom in a distribution is the number of items that are independent of each other, which cannot be assumed from each other.

Suppose you want to add two values such that their sum is 20. In this case, you have a freedom to choose only one value, and let that be 11. Then, the second value will not be based on your choice, but upon the condition that their sum equals 20; that is, the second value automatically must be 9. Therefore, in the example, you can say that there is only one degree of freedom. Take another example, and let us suppose that the sum of four values is equal to 20. Here, you can take three values at your will, and let those values be 2, 7, and 6. The fourth one does not depend upon your will, but depends upon the condition that their sum is to be 20. Therefore, the fourth value must be 5. In other words, one degree of freedom is lost due to the restriction, and you have freedom to choose three values. If there are two restrictions, two degrees of freedom will be lost. If the number of items is N and constraints k, then the degrees of freedom $= N - k$.

The precision of experiments would be reduced if the number of degrees of freedom for error were less. When fixing the number of replications in field experiments, you usually follow a thumb rule to have at least 10 degrees of freedom for estimating experimental error (Gomez 1972). By following this thumb rule, with 5 treatments in a randomized block design, a minimum of 4 replications are needed.

Standard Error

The *standard error* (SE) is a statistical measure about the probability that the sample character will reflect the character of the population from which the sample is drawn. This is actually an estimate of the sampling error. Take the example of a sample mean and the population mean. The sample mean is considered as a probable estimator of the population mean. However, different samples drawn from the same population may have different values of the sample mean. In such cases, researchers would be interested to know how far their sample means are fluctuating. For this purpose, estimation of standard error becomes useful. The standard error depends on mainly two factors, the size of the sample, and the variations of measurements in the sample indicated by the standard deviation. For example, standard error of the mean (SEm) is the standard deviation of all those sample means drawn from the population. The standard error of the mean (SEm) is estimated by sample standard deviation divided by the square root of the sample size assuming statistical independence of the values in the sample.

The standard error should be clearly differentiated from the standard deviation. The standard deviation is a measure of the variability in the particular sample studied. At the same time, the standard error is a measure of the uncertainty in a sample statistic. The standard error, which depends on both the standard deviation and the sample size, is recognition that a sample is unlikely to determine the population value exactly. Often, the ± sign is used to join the SE to an observed mean. Standard errors can be calculated on the difference between two means, on a percentage, on a difference between two percentages, and on a correlation coefficient.

Coefficient of Variation

Coefficient of variation gives an idea about the degree of precision with which the treatments are compared. It can be considered as a good indicator of the dependability of the study as a whole. It provides an idea about the experimental error, which is obtained by dividing the standard deviation with the mean and expressing it as a percentage or proportion of the total. Coefficient of variation is helpful to compare the extent of variation between groups with different means. A higher CV value means that the reliability is low. While doing an ANOVA, it is obtained by dividing the square root of error MS by the grand mean, and multiplying it with 100. According to Gomez and Gomez (1984), for field experiments involving rice yield, acceptable range of CV is 6–8 percent for variety trials, 10–12 percent for fertilizer trials, and 13–15 percent for insecticide and herbicide trials. In any case, acceptable maximum CV should be less than 20 percent in most field experiments.

6.10 Some Common Statistical Tests

Remember that statistics is based on probabilities and not on certainties. Most statistical calculations are based on assumptions. For those studies, which involve complicated designs and analysis, the advice of a statistician should be sought at the planning stage itself. However, it is the researcher and not the statistician who knows the questions to be answered and the nature of data. Therefore, s/he must fully grasp the concepts behind statistical calculations and the limitations. Researchers should also be able to understand the factors, which were taken into consideration to decide on the appropriate test for use and the logic behind the tests. Understand that the type of statistical test depends primarily on whether the data involved are qualitative or quantitative; accordingly, suitable parametric tests or nonparametric tests must be selected.

The *z*-Test and the *t*-Test

The *z*-test and the *t*-test are popular tools to find whether an observed difference between the means of two groups can be considered significant. Here, the word 'significant' has a specific meaning. For example, we know that 10 and 12 are not equal, and the difference is 2. However, 10 and 12 may not be 'significantly' different, if the differences are due to chance. Our concern is how to find out whether the differences are real (significant) and not due to chance (non-significant). The researchers should have some skepticism in this regard, especially when using statistics. We assume that all results are chance results unless shown otherwise (remember how a null hypothesis is written). The statistical procedure for deciding whether the difference under study is significant or non-significant is called the *test of significance*.

The z-test is generally used to ascertain the significance of difference between the mean of a large sample and the hypothesized mean of a population or to compare the difference between the means of two independent large samples (when n is greater than 30). The z-test is also utilized to compare the sample proportion value to a theoretical value of population proportion or to compare the difference in proportions of two independent samples when n is large. For applying z-test accurately, the data should follow normal distribution and standard deviation must be known. The z-value is worked out, and compared with its probable value, obtained from normal table at a specified level of significance (5% or 1%). For example, to compare the score of students of two classes, we may use the z-test, if standard deviation is known. Sometimes, we may want to compare the mean yields of two rice samples, oil yield of two soybean types, oleoresin content of two ginger types, or staple length of two cotton types. Our intention might be to test whether the two samples can be considered as having been drawn from the same normal population. If the samples are large enough to calculate standard deviation, we can use the z-test.

The t-test is the preferred test of significance when small sample sizes are involved (n is less than or equal to 30) and the population variance is unknown. The distribution curves of small sample sizes do not follow strictly the normal distribution and may become flatter at the centre; called a t-distribution. The logic of t-test is that if the difference between two means is large, the variability among the data is small. Similarly, if the sample size is reasonably large, the probability is increased that the difference is not due to chance, but due to a cause. We calculate t-value using a specific formula considering the difference between the two means and the variability among the data. The calculated t-value is then compared with the t-value obtained from the table. The values of t for the chosen level of statistical significance can be found out by consulting published critical value tables. The critical values necessary for rejection of a null hypothesis are higher for small samples at a given level of significance. Each critical t-value for rejection is based upon appropriate degrees of freedom. As the sample sizes increase, the critical t-values necessary for rejection of a null hypothesis decrease and approach the z-values of the normal probability tables.

If the calculated t value is greater than the table t-value, you have to reject the null hypothesis at the chosen level of significance and accept the research hypothesis and proceed further. Conversely, if the calculated t value is less than the table t-value, we have to accept the null hypothesis and reject the research hypothesis. The t-test was developed in 1915 by William Sealy Gosset, a British mathematician. He was working for Guinness breweries of Dublin, Ireland. Gossett's work was to find sampling techniques that would improve the quality of beer-making procedures. As the company rules prohibited publications by its researchers, Gosset used the name 'student' when he published his findings. The t-test is often referred to as the *Student's t-test*.

Note the differences and similarities between the z-test and the t-test carefully. The z-test follows normal distribution while t-test follows Student's t-distribution. A t-test is appropriate when you are handling small samples ($n < 30$) while a z-test is appropriate when you are handling moderate to large samples ($n > 30$). Compared to the z-test, the t-test has more applications, for example, the *paired t-test* for judging the performance of matched pairs. Normally, the z-tests are preferred to t-tests when standard deviations are known. In most situations, however, the population variance may not be known, and we may estimate it by the sample variance, justifying a t-test rather than a z-test. Consequently, the t-tests are more commonly used than the z-tests.

Analysis of Variance

The t-test is generally used to determine whether the means of two random samples are actually different or due to chance or sampling errors. However, when the number of treatment means increase, it would become difficult to use the t-test as such. The analysis of variance (ANOVA) is an effective technique to determine whether the differences between means of more than two samples are statistically significant. As the name indicates, ANOVA uses measures of variance instead of means as in a t test. Although it is possible to use a number of t-tests to determine the significance of the difference between more than two means, it would involve several separate tests. For example, if you want to compare 6 means, 2 at a time, it would involve 15 separate tests (number of pair wise comparisons of N items $= N(N-1)/2$). ANOVA makes it possible to determine whether the 6 means differs significantly with a single test rather than 15 tests. Field experiments involving several treatments are usually subjected to ANOVA.

In ANOVA, one should first use the *F-test* (named by George W. Snedecor in honour of Sir Ronald A. Fisher) to determine whether there exists any real differences between the treatments or they are only the errors of sampling (Snedecor and Cochran 1980). If the null hypothesis has been falsified, and you accepted the research hypothesis, you have to proceed further. You can use a single standard value of the difference of means at a prescribed level of significance—commonly called the *critical difference* (CD) or the *least significant difference* (LSD)—for comparing the observed differences with it. Two treatments are considered significantly different, if their difference exceeds the computed LSD value; otherwise, they are considered to be at par.

The LSD value serves as the boundary line between significant and non-significant differences for any pair of treatment means. The LSD test is simple, and a commonly used procedure for making pair comparisons involving multiple treatments. If any difference between two values is equal to or greater than the LSD, it will be declared as significant. Although LSD test is most appropriate for making planned pair comparisons, it is not valid for comparing all possible pairs of means when the number of treatments is large (Gomez and Gomez 1984). This is because the number of possible pairs of treatments means increases rapidly as the number of treatments increases.

As discussed earlier, it is difficult to study a large number of pair comparisons at $N(N-1)/2$, where N is the number of treatments. For example, with 4 treatments, there would be 6 pairs only, but with 10 treatments, the number of possible pairs would be 45, and with 14 treatments, there would be 91 pairs. Therefore, if the LSD test must be used, use it only when the F-test for treatments effects is significant, and the number of treatments is less than 6. When the experiment is composed of too many treatments (above 6), and if you require the evaluation of all possible pairs of treatments, *Duncan's multiple range test* (DMRT) (Duncan 1955) can be used. Using DMRT, the results are presented using a line notation or alphabet notation based on their ranks. It is customary to use *a* for the treatments with the highest rank, *b* for the second rank, *c* for the third rank, and so on. Give a footnote below the table of treatment means as 'In a column, means followed by common letters do not differ significantly at 5 percent level by DMRT'.

A full discussion on ANOVA or various statistical tests is beyond the scope of this book. Readers may refer to any of the standard books listed in bibliography.

Transformation of Data

In normal cases, application of F-test in the analysis of variance is valid only when some assumptions regarding the data are correct. The following are the assumptions:

- Treatment effects and environmental effects are additive.
- Experimental errors are independent; they have common variance; and all follow normal distribution.

In situations of violation of assumptions as above, you transform the data using an appropriate method to take care of variance heterogeneity. The original data are converted to a new scale giving a new data set, which satisfies the condition of homogeneity of variance. Here, as we are applying a common transformation scale to all observations, it does not alter or distort the comparative values between treatments; and therefore, the comparisons are valid. Selection of suitable transformation depends on the specific type of relationship between the variance and mean.

Logarithmic Transformation

This is most suitable for data where the standard deviation is proportional to the mean or where the effects are multiplicative. Examples include the number of insect pests observed per unit area and number of egg masses per plant or unit area. Usually, simple logarithm of the values ($\log X$) is used. However, if the data involves small values (less than 10), go for $\log(x + 1)$ values and use them.

Square Root Transformation

You generally use square root transformation for data such as the counts of a number of items like the number of infested plants in a plot, number of insects caught in a light trap, and number of weeds per plot. It is also used for percentage data where the range is between 0 and 30 percent or between 70 and 100 percent . Usually, square roots of the data are used ($X^{½}$). However, if the data set include small values (less than 10), especially when zeros are present, use $(x + 0.5)^{½}$ for transformation.

Arc Sine Transformation

Angular transformation is suitable for proportions, data obtained from a count, and data expressed as fractions or percentages. For arc sine transformation, usually a table prepared by Fisher and Yates is used. Percentage data can also be subjected to angular transformation. However, for percentage data lying between 30 and 70 percent, no transformation is required. For percentage data lying within the range of either 0–30 or 70–100 percent (but not both), the square root transmission is most ideal. For percentage data that do not follow the ranges noted, arc sine transmission could be used.

Analysis of Covariance

In statistics, ***covariance*** is used to assess the extent to which two variables change together and the strength of relationship between them. Analysis of covariance (ANCOVA) is a method that includes additional variables—*covariates*—into statistical analysis. This allows researchers to account for variations associated not with treatments variables themselves, but with one or more covariates. In certain experiments, ANCOVA comes to our rescue to eliminate initial differences on several variables between the experiment and the control groups. Suppose you want to lay out an experiment in a coconut garden where the yield levels of coconut trees differ considerably. Using the initial yield as the covariate, final yield after the experiment can be adjusted to the values that would have been attained had all experimental trees started with the same yield. This means that you have to take some ancillary observations to improve precision.

Another example is the layout of a field experiment in a land where you suspect non-uniformity due to fertility or slope variations. In such cases, you can collect ancillary observations from a *uniformity trial* conducted over the experimental plots before the layout of actual experimental trial. Then, this yield data from the uniformity trial is used as an index of variation from plot to plot because of factors such as soil heterogeneity. Even if you adopt local control or blocking effectively, plot-to-plot variations can play spoilsports. The observation from uniformity trial can be utilized in estimating these variations to reduce the error affecting the treatment comparisons. Another common example is plant breeding trials, where yield differences may be noticed due to the failure of some of the seeds to germinate or subsequent death of some already germinated plants. By recording plant numbers and yield from plots, this effect can be accounted. In all these cases, ANCOVA is an ideal tool.

There are several other applications for ANCOVA. Suppose that you have taken all the precautions to design and conduct an experiment. Even then, observations for certain plots might be lost or even affected by accidental causes such as pest attack or prowling by cattle that it would not be proper to consider these as normal experimental observations and have to be discarded creating a problem for the researcher. The plots or data omitted because of such extraneous causes are called *missing plots*.

Adjustments of missing or incomplete observations can be done through ANCOVA using Barlett's technique for missing plots (Panse and Sukhatme 1985).

Validation in Qualitative Research

Qualitative data such as the nominal and ordinal types are usually subjected to nonparametric tests. *Chi-square test* is a prominent test applied to qualitative data. *Triangulation* is another method for validating qualitative data. Other major nonparametric tests include the *Mann–Whitney test* and the *Kruskal–Wallis test*. You may find appropriate nonparametric tests according to your requirements from standard books (e.g., Siegel and Castellan 1988; Pett 1997). Chi-square test and triangulation are mentioned below.

Chi-Square Test

The chi-square test is a nonparametric test used for qualitative data (especially of the nominal types) to test hypotheses concerning the frequency distribution of one or more populations. Remember that quantitative data (interval or ratio) are specified along a continuous numerical scale such as yield, plant height, grain number, and nitrogen content; but qualitative data (nominal or ordinal) are concerned with a finite number of discrete classes. The most common qualitative data are those having two classes, which consist of the presence or absence of an attribute such as male or female, success or failure, effective or ineffective, dead or alive. Examples of attributes with more than two classes include cultivars of a crop, colour, and agreement.

In general, researchers collect attribute data when it is impossible to get measured data. For example, take the case of a study comparing the phytotoxic effect of one new herbicide in rice against a standard herbicide. In some plots, the new herbicide may have showed phytotoxicity symptoms; in certain plots, moderate phytotoxicity only; and in some other plots, no phytotoxicity at all. The performance of the two tested herbicides may have been different. It is not possible to explain this finding by chance alone. As discussed earlier, if the differences are large, and if the size of the sample is reasonable, the probability that the findings are due to chance will be less. In harmony with the null hypothesis, you can assume that there is no difference, and calculate the expected frequency for each plot (e.g., extremely phytotoxic, phytotoxic, and no phytotoxicity). Then, you calculate how different the observed results are from the expected results. From this, a chi-square value can be calculated using a special formula. If the computed chi-square value is greater than the corresponding tabular value, the null hypothesis will be rejected and you accept the research hypothesis. In agricultural research, chi-square test is mainly used for testing the significance of proportions, analysis of attribute data, test for homogeneity of variance, and test for goodness of fit. Chi-square test is used widely in plant breeding experiments, for example: to test the significance of deviation of an observed segregation from a theoretical one, simultaneous testing of a number of questions such as single factor ratios, linkage, and heterogeneity.

Triangulation

In the case of data gathered through methods such as participant observation, semi-structured, and unstructured interviews, focus group discussions, and transect walks, it is very difficult to conduct statistical analysis because of their largely qualitative nature and relatively small sample size. Therefore, the method of *triangulation*, a process of cross-checking and cross-validating information for accuracy is generally employed. In trigonometry, one can indirectly measure the location of a point precisely by appropriate sightings from two other points. Similarly, you can apply this method of triangulation to social measurements. The logic is that if you make use of several measuring techniques, all producing agreeable results, you can have some greater faith in measuring the concept of interest to us. It is hoped that alternative ways ensure the validity and reliability of the findings. Triangulation is achieved by using different tools to gather information on the same issue, for example: maps, transects, and trend lines to examine environmental changes; and by listening to different people with different points of view about the same topic, for example: women/man, young/old, poor/wealthy, and so on.

6.11 Popular Statistical Packages

The arrival of computers has revolutionized statistical work, and presently, you can do the calculations quickly with ease. Before finalizing a particular method of analysis, consult standard books on statistics and understand more about common statistical tests and their applications. Several types of software packages are also available, both commercial and non-commercial, for analysing the data in record time. Some organizations have come with free statistical software as a practical alternative to costly commercial packages. Most free statistical software packages are easy to learn and use by means of menu systems. A few popular packages (both priced and free) are mentioned below.

SPSS (Statistical Package for the Social Science)

SPSS is one of the most popular statistical packages, which can perform highly complex statistical data analysis with simple instructions. The software was originally created for the management and statistical analysis of data from social sciences, but now it is used in other disciplines such as agriculture and health sciences as well for a variety of analytical works. The SPSS software was launched in 1968 by SPSS Inc., but was later acquired by IBM in 2009. Although the software was officially renamed as IBM SPSS Statistics, it is still popular as SPSS. Statistical methods usable in the software include inferential statistics such as ANOVA, MANOVA, correlation, regression (linear, nonlinear, logistic, loglinear, etc.), and nonparametric tests; descriptive statistics; and several others such as probit analysis, cluster analysis, factor analysis, geo spatial analysis, and simulation. In addition to statistical analysis, it is highly suitable for management and documentation of data.

SAS

SAS is an integrated system of software for solving many tasks. Previously called Statistical Analysis System, SAS was developed by SAS Institute. The software can be used for data entry and management, graphics designing, multivariate analyses, advanced analytics, predictive analytics, business intelligence and decision support, operations research and project management, and several such applications.

MaxStat

MaxStat is a software package for statistical analysis specifically designed for students and researchers who do not have much knowledge in statistics. The *MaxStat Lite* version is offered free to use for non-commercial purpose. This is useful for smaller projects and theses. The full version, *MaxStat Pro*, offers analysis of larger data sets. The Lite version features descriptive statistics, t-tests, chi-square, ANOVA, nonparametric tests, linear regression, correlation, and basic graphing functions. The Pro version features advanced analyses, more types of graph, word-processing functions, and options for formatting of results. MaxStat supports over 100 commonly used statistical tests.

Stata

Stata is a statistical software package created in 1985 by StataCorp for general use. It is specifically suited for researchers working in the fields of economics, sociology, political science, and health science. Data management, statistical analysis, graphics, simulations, regression, and custom programming are some of the highlights of Stata.

Microsoft Excel

Excel is a spreadsheet application available with 'Microsoft Office' suite, useful for data management and analysis. With the same basic features of all spreadsheet applications, Excel uses a collection of cells arranged into rows and columns to organize and manipulate data. It is useful to perform basic calculations and statistical analysis. They can also display data as charts, histograms, and line graphs.

MSTAT-C

This statistical package was developed under the leadership of Russell Freed of Michigan State University for simplifying and enhancing field and laboratory research and analysis related to agricultural and biological sciences.

CropStat

CropStat is a statistical package developed by International Rice Research Institute (IRRI) for the management and analysis of data from experiments. Although CropStat has been developed mainly to cater to the needs of agricultural workers for analysing the data from field experiments, some programmes can be used for analysis of data from other disciplines too. Please note that the much popular earlier version,

IRRISTAT is not being used today because it was originally written in the BASIC (Beginner's All-purpose Symbolic Instruction Code) language, which is not compatible with 'Windows'. You can download CropStat and some other useful software from the IRRI website, http://bbi.irri.org/products.

Similar to the above, you may find many more free or priced software. **R**, **Jamovi**, **JASP**, **PSPP** are examples of open-source software, and **Minitab, Matlab, Systat, Statistix, AcaStat, Statwing, Statistica**, and **XLStat** are some proprietary products.

6.12 Monitoring Research

It is essential that the research is carried out under the approved norms in science and that the investigators follow only sound practices. Therefore, a system for quality assurance must be followed to ensure that the study is executed, and the data are generated, recorded, and reported in compliance with the approved research programme. Make sure that no data are missing, and that the data are precise and accurate.

The Problem of Missing Data

Careful monitoring during the study can help in reducing the problem of missing data. In addition to missing data, out-of-range values should also be taken care of. Inaccurate and imprecise data are also problems. Reliability of measurements is an important component of assurance of research quality. To ensure the quality of laboratory measurements, standards can be used. In field experiments, the *missing data formula technique* or the *analysis of covariance technique* can be used to cope up the problem of missing values (Gomez and Gomez 1984; Panse and Sukhatme 1985). In the missing data formula technique, an estimate of a single missing observation is made using a formula based on the experimental design used. This estimate is then used to estimate the missing data, and subsequently, the improved data are subjected to ANOVA.

Tabulation and Periodic Reporting

Periodic tabulation of data using descriptive statistics is very useful in the monitoring process. This may reveal aberrant values and missing data. Sometimes, monitoring agencies or funding agencies may demand periodic reports such as quarterly reports, half-yearly reports, and annual reports. These have to be prepared and sent promptly.

Changes in the Project

It is expected that you carry out the project according to the approved protocol. If some changes are to be made in the technical programme, you should get approval from the funding agency. Violation of protocol is viewed seriously, sometimes resulting in even withholding of funds. In the case of a thesis project too, any change in the ORW or technical programme must be discussed in the advisory committee, and get approval from the competent authority.

Scientific Honesty

Data collection and analysis should be done honestly. Collect the data as accurately and carefully as possible without any subjective bias on the part of the researchers. Deliberate fraud and malpractices are condemnable. Fraud involves deliberate deception and may take the form of concoction of data or manipulation of data. A monitoring mechanism should be there in the institution to oversee the works of scientists and research students and to detect fraud and malpractices (see Chap. 23 for a detailed account on ethics in research).

Financial Discipline and Honesty

Research programmes are commonly supported by the government, private agencies, or international donors. The funds have to be used to meet the research expenses as agreed upon in the research proposal. Expenditure has to be documented. Accurate periodic reports and final financial reports are required and should be submitted in time.

Bibliography

Best, J.W. and Kahn, J.V. 1993. Research in Education (7th Ed., Indian Reprint, 2004). Prentice–Hall of India, New Delhi, 435p.

Byron W.M., Brown, J., and Hollander, M. 1977. Statistics: A Biomedical Introduction. John Wiley and Sons, New Jersey, 456p.

Chandel, S.R.S. 1991. A Handbook of Agricultural Statistics. Achal Prakashan Mandir, Kanpur, 560p.

Cox, D.R. 1958. Planning of Experiments. John Wiley & Sons, New York, 308p.

Duncan, C.B. 1955. Multiple range and multiple F-tests. Biometrics 11: 1–42.

Fathalla, M.F. and Fathalla, M.M.F. 2004. A Practical Guide for Health Researchers. WHO Regional Publications Eastern Mediterranean Series 30. World Health Organization Regional Office for the Eastern Mediterranean, Cairo, 232p.

Finney, D.J. 1972. An Introduction to Statistical Science in Agriculture (4th ed.). Blackwell Scientific, 290 p.

Gardner, M.J. and Altman, D.G. 1997. Statistics with Confidence: Confidence Intervals and Statistical Guidelines. BMJ Books, London.

Gomez, K.A. 1972. Techniques for Field Experiments with Rice. International Rice Research Institute, Manila, Philippines, 46p.

Gomez, K.A. and Gomez, A.A. 1984. Statistical Procedures for Agricultural Research (2nd Ed.). John Wiley & Sons, New York, 680p.

Hill, A.B.1971. Principles of Medical Statistics (9th Ed.). Oxford University Press, New York, 390p.

Holmes, D., Moody, P., and Dine, D. 2010. Research Methods for the Bioscience (2nd Ed.). Oxford University Press, Oxford, 457p.

Ireland, C. 2010. Experimental Statistics for Agriculture and Horticulture. CABI Publishing, CAB International, Wallingford, UK, 351p.

Kerlinger, F.N. 1986. Foundations of Behavioural Research (3rd Ed.). Holt, Rinehart and Winston, USA, 667p.

Lester, J.D. and Lester, J.D. Jr. 2009. Writing Research Papers: A Complete Guide (13th Ed.), Longman, 416p.

Likert, R. 1932. A technique for the measurement of attitudes. Arch. Psychol. 140: 1–55.

Marder, M. P. 2011. Research Methods for Science. Cambridge University Press, 227p.

Moser, C.A. and Kalton, G. 1971. Survey Methods in Social Investigation (2nd Ed., 1980 reprint). The English Language Book Society and Heinemann Educational Books, London, 555p.

Panse, V.G. and Sukhatme, P.V. 1985. Statistical Methods for Agricultural Workers (4th Ed., revised: Sukhatme, P.V. and Amble, V. N.). ICAR, New Delhi, 359p.

Pett, M.A. 1997. Nonparametric Statistics for Health Care Research: Statistics for Small Samples and Unusual Distributions. Sage Publications, California, 305p.

Rosenthal, R. and Fode, K. 1963. The effect of experimenter bias on performance of the albino rat. Behav. Sci. 8: 183–189.

Salkind, N.J. 2008. Exploring Research (7th Ed.). Prentice Hall, New Jersey, 316p.

Siegel, S. and Castellan, N. J. 1988. Nonparametric Statistics for the Behavioral Sciences (2nd Ed.) McGraw-Hill, 399p.

Snedecor, G. W. and Cochran, W. G. 1980. Statistical Methods (7th Ed.). Iowa State University Press, Ames, Iowa, 507p.

Stern, R., Coe, R. Allen, E., and Dale, I. 2004. Good Statistical Practice for Natural Resources Research. CABI Publishing, CAB International, Wallingford, UK, 708p.

Thurstone, L.L. 1928. Attitudes can be measured. Am. J. Sociol. 33: 529–54.

Chapter 7
Planning and Writing a Research Proposal

Good fortune is what happens when opportunity meets with planning.
Thomas Alva Edison (1847–1931)

Research is a costly venture. In a research organization inclusive of university and college departments, funds for research are released only after careful scrutiny of the merit of the proposed work. A formal research proposal and its approval are also necessary for thesis works of students. In addition to the in-house funds of the organization, scientists may get funds from external agencies—both national and international. Fortunately, there are several funding agencies to promote research and development in specified areas, usually on project mode, by providing grants to individuals and groups. Therefore, instead of waiting for the meagre resources of parent institutions, scientists can approach a suitable external agency for financing research projects. Your success in obtaining a research grant means that you and your institution are benefited in various ways.

The funds for research normally come in the form of grants from agencies specifically created for the purpose. Most of such donor agencies are government controlled, but grants are available from private donors and international agencies too. The term *grant* denotes a sum of money disbursed by a donor, often a government department, corporation, foundation, or trust, to a recipient such as an individual, non-profit organization, educational institution, or business institution for a specific project or purpose. Unlike *loans* that must be repaid with or without interest, funds provided in the form of grants need not be repaid. However, you have to compete for research grants by presenting your research proposal in the prescribed format, and funds for research are released only after careful scrutiny of the merit of the proposed work. A formal research proposal and its approval are necessary for thesis works of students as well.

Often, there are two kinds of grant applications for research—by an individual scientist or by an organization. In the case of application by an organization, it must be a collective effort under a group leader. Most institutions must have set up a mechanism for approving research projects through a formal research proposal and a pre-funding appraisal. Research proposals to external agencies must additionally

C. G. Thomas, *Research Methodology and Scientific Writing*,
https://doi.org/10.1007/978-3-030-64865-7_7

meet certain specific requirements outlined by them. Remember that proposal writing is a skill, which requires sufficient knowledge in the subject and practice. The final proposal should be neat and tidy, typeset with elegant typefaces, and attractively bound with all the necessary details.

Scientists should be well aware of various funding agencies, and the procedures they must follow for getting financial assistance. Often, enough funds are available for this purpose, but number of applications may be limited. In majority of cases, when the requests for grants are not approved, the ostensible reasons may not be technical inaccuracies, but faults in the application! Many young scientists find the task of writing research proposals for grant more demanding than executing the project. Consider grant writing as a part of scientists' job, and equip yourselves with the necessary skill and competence.

7.1 Research Projects

A collaborative venture carefully planned and managed to achieve some specific objectives involving definite tasks is called a *project*. It involves a combination of inputs such as human resources, equipment, supplies, communication, travel, and sometimes training to produce some desired outputs. A *project proposal* is a formal request submitted to a funding agency to carry out a project. A project proposal must defend each item in the list of things so that a funding agency can decide if it wants to approve a part or the entire request for fund.

Research Projects and Development Projects

Projects can be development projects or research projects. A development project is normally with the objective of improving the situation of something to make life better for the people. National and state governments spend huge sum of money every year on development projects ranging from creation of roads and other infrastructure to providing drinking water and basic amenities. On the other hand, a research project is designed with the objective of improving existing knowledge on something, solving a problem, or developing a new technology. Although research projects may not have direct impact on development immediately, the results from such works may be useful for future development projects.

Request for Proposals and Call for Proposals

Grants for research or development are sanctioned based on some form of 'grant writing', often referred to as a research proposal or project proposal. It is usually submitted in response to a request for proposal (RFP) or call for proposal (CFP) in pre-defined areas. 'Request for proposal', often abbreviated as RFP, is an open invitation made through media by a funding agency to submit a funding request for consideration. The phrase 'request for proposal' is generally used in connection with development projects, and more often, it is 'call for proposals' in the case of research projects.

Certain funding agencies such as the Science and Engineering Research Board in India accept proposals throughout the year. In these cases, there may not be separate 'call for proposals'. However, majority of funding agencies issue 'call for proposals' periodically based on project review and priorities. If you miss that deadline, you have to wait until the following year. Therefore, always be vigilant about the 'call for proposals' by visiting the websites of potential funding agencies.

Letter of Inquiry and Concept Note

Sometimes, instead of a formal research proposal, the funding agency may be inviting a letter of inquiry (LOI) or concept note. This is a better idea to save time for the investigator and the donor. Full proposals take a long time to prepare, and you may not like to spend this much time unless you are sure that your proposal will get due consideration. Donors are also freed of the burden of reading lengthy proposals, which they may not like to fund. In response to an RFP, those who seek grants for their development projects submit LOI. The LOI includes the purpose of the project, the objectives, and financial requirements. A concept note or concept paper is also similar to LOI; but generally, it is a summary version of a full proposal. Most funding agencies provide the format of a concept note, which normally includes the sections as background, brief review of works done, objectives, materials and methods, activities and duration, expected outcome, project associates, and budget outlay.

Ad Hoc Research Project and Competitive Grant Fund

A scientist can submit a research proposal to a potential funding agency, either on his/her own initiative or in response to the RFP or CFP from the agency. Many agencies give grants for ad hoc research projects but some agencies invite applications for grants in competitive grant fund (CGF) mode. Ad hoc research is designed for a specific purpose, which is conducted on a one-off basis (only once and not regularly) as opposed to regular or continuous research, for example, the CGF projects. CGF research projects are different from ad hoc research projects in that they are created to bring about lasting changes in the research structure of a country.

Research Proposal and Research Protocol

The differences between a research protocol and a research proposal must also be clearly noted. A research proposal is a document containing an initial set of ideas for funding research supported by background literature to justify the study. Most funding agencies start project evaluation based on a formal proposal. If the proposal is approved, probably, they ask for a detailed protocol. A research proposal usually contains a small introduction, objectives, brief review of literature supporting the study, data collection procedures, and financial requirements. However, a protocol is a detailed set of activities for the project proposed, which happens only after the proposal stage is over. The protocol acts as a practical guide and timetable of various activities related to the research project. It should also show some foresight into what the investigator is trying to achieve. Therefore, an adequately clear and detailed protocol must be produced, which could be used by all those who are involved in the study. In short, a protocol should act as a script for the research work! Formal protocols are insisted in many disciplines especially clinical studies that require the cooperation and coordination of many volunteers and participants.

Research Proposal and Research Paper
It is interesting to compare a research proposal and a research paper. A research proposal and a research paper are similar in structure because both have some similar components. For example, both should normally have an abstract, introduction, objectives, review of literature, methodology, and conclusion. However, the 'results' and 'discussion', two important parts of a research paper, are missing from a project proposal. If you consider the whole research process as a chain of events, the research proposal forms the first few links in the research chain and the research paper is the last link. In a research proposal, you project the future work you propose to do before a selected audience for their approval; but in a research paper, you report the outcome of past research for the benefit of a wider audience as a part of information sharing. The research proposal not only deals with research but also with human resources and money. Therefore, it should be planned meticulously and lucidly so that it attracts the attention of the funding agencies and the reviewers.

7.2 Finding a Research Problem

How to start the research activity is a formidable challenge for many new researchers. Often, the research process sets in motion from an idea, an unsolved problem, a hypothesis, or a continuation of previous research done by the investigator or somebody else. Even for a seasoned scientist, first thoughts may be hazy and ideas uncertain. The idea might seem impressive, but usually overestimate the resources or underestimate the time. Initially, researchers may have only a general and diffuse notion of a particular problem, but eventually they have to have a clear idea of the nature of problem. Otherwise, they can hardly get very far in solving the problem with a research agenda.

It is impossible to set any rules or directives for finding a research problem. For a novice researcher or research student, this might be the most difficult part. Researchers often find ideas or problems from many sources such as experience, literature, Internet sources, request for proposals, extension workers, colleagues, research guides, patients, and farmers.

Experience: Most often, you carry some research ideas based on your experience. As a professional or teacher, you may confront with several questions or problems pertaining to your specialization. The background, culture, education, and practical experience of researchers influence the ideas and the way in which they think. Once an idea strikes you, write it down immediately, and explore whether it is suitable for a thesis work or general study. Based on your critical evaluation of the problem, develop an outline. Be optimistic about the idea, and work on it earnestly until a proposal emerges out.

Literature search: A usual source for research idea is the literature in your specific fields. This will help you to understand the present status of the subject and to know the problems encountered by fellow scientists in the field. Normally, just looking at the titles of articles and reviewing them will give you an idea for your own study. Recent textbooks and reference books may also be of immense help. As you gain more knowledge in your discipline, it is relatively easy to find new ideas. You can even use the idea of an already published study for a research problem of your own. Repeating a study in a different context to see if the results hold in the new situation is also helpful. In many concluded reports, there will be a separate section on 'future line of work'. This will also provide you many ideas.

Internet search: A general search on the Internet is helpful for getting new ideas or developing your topics further. General search engines such as Google, Bing, Yahoo! and many others will lead you to a wide range of suitable web pages. Wikipedia, the world's largest online encyclopaedia is helpful to have a broad understanding. Databases such as PubMed, PubAg, and CAB Direct are also helpful (for more information on Internet search, see Chap. 9). These sites are rich in scholarly content, and contain abstracts and links to a variety of sources. However, be careful about biased or spurious sites, and depend only on websites sponsored by governments, universities, international agencies such as FAO, established research institutes, and reputed companies. Always avoid sites of lobbyists and advocacy groups.

Daily observations: You can learn many things from nature. If you keenly observe, ideas will be forthcoming. Observations may be active or passive. Active observation involves detecting something and making a note of its significance by relation to something already known. Suppose you have just observed a malformation in a crop. You may try to attribute some reason to this malady, if there is active involvement of your mind. In passive observation, there is no involvement of mind. Active observation is also helpful to relate many problems that come to you for solving. Farmer's field visits and mixing with farmers and extension workers will provide many ideas for research.

Request for proposals: Requests for proposals (RFPs), call for proposals (CFPs), or the notification inviting research proposals on specific topics from donor agencies are also a good source of research ideas. By describing some problems that the funding agency would like to address, these RFPs or CFPs are giving the researcher some clues or ideas for pursuing research. Usually, the RFP describes the problem that they like to address, the contexts in which it operates, the approach they would like to investigate the problem, and the maximum amount they would be willing to pay for such research.

Field problems brought by farmers and extension workers: In agriculture, animal husbandry, and medicine, farmers and extension workers bring to the notice of the research workers problems they could not solve. If the problems are new and need investigation, the researcher can proceed with the idea of problem solving. New diseases and epidemics are such avenues for medical professionals.

New developments: Technological advancements and social developments are creating new problems and opportunities. For example, food security, genetically modified crops, pesticide residues, environmental pollution, water crisis, wetland conservation, increasing popularity of certain crops, global warming, climate change, energy crisis, need for alternate fuels such as biodiesel, and new diseases (e.g. H1N1 swine flu) are some new developments, which warrant immediate research intervention.

Conferences/seminars/lectures/discussions: These will give you an opportunity to mingle with scientists, hear their views, and exchange new ideas. The interaction and the insights from such meetings provide you an excellent opportunity to seek clarifications or further enlightenment on the ideas.

Colleagues and guides: Sometimes, the seniors in your department or your colleagues would be able to lead you to researchable ideas. For young scientists and research students, this is an excellent way to find topics. Try to develop a research idea in consultation with them. For research students, the research guides are the major driving force in selecting and developing the idea into a project plan.

7.3 Analysis of Research Ideas

After you get an idea for a study, you must think about its feasibility critically. You can endeavour to carry out the study in a hassle-free environment, if there are unrestricted resources and enough control over the state of affairs. However, such ideal conditions seldom exist, and researchers are always forced to look for the best they can find in the prevailing situations. Before making a decision on the feasibility of a research project, consider several practical issues. The following points are important and must be considered carefully before finalizing a research project. If you are able to consider these points at the initiation stage of the project itself, then, most of the problems that you may confront later can be solved or eliminated altogether.

Attempt only feasible ideas: Research is done to solve problems. The new idea should be simple, viable, and the objectives clear. For example, everybody now accept that green revolution style agriculture along with the plenty has created a few unintended problems as well. However, to obviate the unintended problems of green revolution, if you are contemplating the 'post-modern' natural farming type of agricultural practises as certain diehard environmental activist swear, the possibility of a food crisis in the near future cannot be ruled out. Sustainable agricultural practices that also take care of environment are the viable alternatives to feed the burgeoning world population, which is set to reach 8.6 billion by 2030, 9.8 billion by 2050, and 11.2 billion by 2100 from the current 7.6 billion (UN 2017).

The idea must be practical: Never venture to take a work that is simply impossible to carry out in the prevailing circumstances. For example, the idea may require using an electron microscope or a difficult assay using high-performance liquid chromatography (HPLC). Unless you have access to these sophisticated techniques, such works could not be completed. Some ideas may be good, but need many workers and several years to complete. The cost would be prohibitive. Therefore, decide on a course of action only after considering these 'practical' aspects.

Novelty of ideas: Often, new ideas get wide acclaim. Originality and novelty are important considerations in proceeding with an idea. Before embarking on an idea, find out whether it is a new problem or duplication of previous work. Beginners may become disillusioned on finding that their ideas have already been investigated. Probably, the joy of getting 'new' ideas may be short lived, when you found after searching the literature that much has been done on this subject. Even if the idea has already been explored, it is worthwhile to think about this further because of some reasons. Sometimes, the methods employed earlier may have been unreliable; it could be meaningful to repeat it in a different environment or locality; or it would be feasible to extend the idea to some other fields. However, take care to avoid simple duplications.

Expected results: While analysing the research idea, consider the expected results too; but do not forget that results are unpredictable. You must understand that any result is worthwhile, even if it is negative! It is also important to consider the prospects of producing research papers based on the study, and ascertain whether such research papers based on the expected results would be accepted for publication by a journal.

Time available: Research is time-consuming; therefore, you should make sure that you have enough time to spare. Normally, a research grant lasts for only three years. In the case of academic research, the problem of time is acute. For a Masters project, never attempt problems that take more than one year. In the case of PhD work too, time is a major constraint. Normally, a full-time doctoral student is expected to complete all the academic activities including research within a period of three years, although extension of time is granted in specific circumstances. Therefore, do not be overambitious; develop only those projects, which could be completed in the specified period. Formulate the technical programme in such a way as to finish within the period stipulated as per academic requirement or as specified by the funding agency.

Opinion of colleagues: Discussing the research idea with your colleagues is usually helpful. Interacting with scientists who are doing similar works also helps. Consult experts in the field and laboratory, especially for new methods and difficult procedures. However, be careful about excessive fear or love of new ideas (neophobia and neophilia!). There is a general tendency among ordinary people and even scientists to oppose and turn down new ideas. Certain people are excessively neophobic, and they are unwilling to try anything new or break from routine.

Statistical treatment: In most universities, as a part of Masters and PhD programmes, basic courses on statistics including design of experiments are taught to equip the students well versed in experimental designs and analysis of data. Still, if the proposed work involves complicated designs and analyses, it is a good idea to consult a statistician and seek advice before the start of the work.

Competence of researchers: Research is a challenging task. You should do a thorough analysis, and decide whether you have the competence to undertake the project. If it involves too much work or is interdisciplinary in nature, collaboration with other scientists become essential. Contemplate how much work you yourself are going to do. You may land in trouble, if you simply dream up an idea in which you do not have enough competence expecting others do all the work!

Possibility of getting grants: This is important, if your institution is cash starved and not capable of funding the proposal. Think whether you are able to mobilize funds for the idea in your mind from external sources. Based on your research problem, identify an appropriate funding agency such as DST, ICAR, UGC, DBT, ICMR, or others (For a list of funding agencies, see Annexure). However, understand that these agencies agree to fund you, only if the ideas conform to their call for proposals.

7.4 Planning a Proposal: Some General Considerations

Once you are very clear about your idea, it is time to write it down in the form of a project proposal. Before you start writing the proposal, consider the following general aspects as well. This is all the more important, if it is for external funding.

Study the Call for Proposals Carefully
Before writing the proposal, make yourself familiar with the funding agency and its programmes. Study the description of various funding programmes, and carefully read the outline of the call for proposals or request for proposals and understand various provisions indicated. If something in the outline is not clear, clarify it from the concerned agency or source. Ascertain whether the funding agency's programme will address the researchable topic that you have in mind. If several programmes are available, identify the most suitable programme for your research. For example, the Department of Science and Technology, Government of India has several programmes. Once you have decided the most suitable programme, study the application form and instructions relevant to the programme carefully.

Start filling in the form after you are thorough with the programme.

Stress on Contemporary Issues
A successful research proposal, especially those written for external funding, should address some of the contemporary issues. If the proposal is able to address these aspects, chances of getting grants are very high. Some current problems, which may attract the attention of donors include climate change, food security, population explosion, solid waste crisis, pollution, energy crisis, lifestyle diseases, genetically modified organisms, and pesticide residues. Give a clear indication in the proposal that if you do the project immediately and come with suitable findings, some current problems can be solved. Convince the funding agency that spending money on this project will not be a waste, and that something worthwhile will come out. Assert that your institution is qualified to take up the work and have the human resources and the infrastructure to do the work. In short, the agency and the reviewers must understand that the researcher is badly in need of support to sort out some important issues.

Good Writing Skills
Often, most researchers may not have the required writing skills to come with a flawless research proposal. However, through hard work, language skills and vocabulary

can be improved; and with a little bit of patience, any individual can master good writing skills. Use forceful words in the proposal to make the points clear and definite. Always use positive language and avoid 'perhaps…', 'may be…', 'probably…', etc. Use short sentences and make them readable. Remember that good written English is simple and clear. Certain writers have a penchant to use passive voice for everything. Do away with this practice, and use the active voice, with the present and future tense as often as possible. Make the project simple to read and understand. Ensure that each sentence should have a meaning leading towards a conclusion. Try to avoid unnecessary words and phrases, clichés, and bureaucratic and wasted words. A proposal prepared using correct words and phrases will create a good impression. For more details on improving writing skills, refer to Chaps. 16–18.

For highlighting certain points, bullets can be used. Supporting or substituting text by maps, charts, photographs, boxes, and graphs are also helpful. If you have to use 'we', always be unambiguous and identify clearly in the proposal which scientist or partner will do what. After giving final changes to the draft, seek the help of a language editor, if you are weak in English grammar.

Time Analysis and Management

A researcher should analyse the time available to him/her for the completion of the work and manage the time effectively. Research requires time, and make sure that you have enough time to spare. Normally, a research grant lasts for three years. For a postgraduate research, time available is still shorter. Underestimation of time is most acute, if you are new to research, or if you do not have the experience of doing a project similar to the planned one earlier. Similarly, you may fail in estimating time required to design and test a new equipment or machine. Usually, one's overall time estimate is much shorter and less realistic when compared to the actual time requirements of individual steps. Most researchers also forget to include time estimates for delays, setbacks, and unexpected problems.

The proposed research may involve several steps. Consider and determine how much time is needed for each step. You must also check the influence of potential intervening problems on time estimates. Most researchers do not give much attention to time analysis and may neglect the broader concerns. You may even claim that since you are so busy in doing many things, you do not have time for planning! Careful planning of an experiment determines its value leading to success. Failure to consider the time constraints creates some of the more common experimental errors and pitfalls.

7.5 Proposal Outline

Most funding agencies issue a specified format for preparing the research proposal, and this should only be used for applying grant. A research proposal should be arranged neatly and legibly by developing the sections according to the format of the agency. A typical format contains the sections as noted below:

- Title page
- Information regarding investigators and location
- Abstract/executive summary
- Introduction
- Objectives/aims
- Methodology (Materials and methods)
- Time schedule of activities and activity milestones
- Ethical considerations
- Facilities available
- Work programme
- Indicators
- Budget
- References
- Curriculum vitae of the applicants
- Institutional endorsement.

Title Page

A separate title page is necessary that contains the title of the research proposal, the applicant's name and address, and the date of submission. Avoid contradictory, ambiguous, or misleading titles; instead, use winning titles. The title is important as it creates the first impression of the project. It should be able to convey the highest possible information in lesser words—usually not more than 12–15 words. It should indicate the area of research and introduce the research question.

Normally, titles are stated in terms of a functional relationship indicating the independent and dependent variables. Make the titles attractive and legible. Titles can be shortened by deleting familiar phrases that give no additional information. A title may become lengthy and awkward, if you add several prepositional phrases to qualify it. Avoid low impact expressions that add nothing to the meaning such as 'Effect of…', 'Studies on…', 'Research on…', 'Response of…', and 'Investigation on…'.

Do not use abbreviations in titles. However, if the abbreviations are more familiar than the words, they stand for, for example: DNA and pH, such abbreviations can be used in titles. If you are not sure about this aspect, use the term in full and its abbreviation in parenthesis.

Titles need to be captivating. Ideally, the title, apart from being catchy, should be such that it gives a clear idea of what the project aims to achieve. Before finalizing the most appealing title, try several attempts, and decide the most pleasing one. An appealing style is to use a colon in the title dividing the title into two parts, a pre-colon part and a post-colon part. Make the pre-colon part short and attractive using appropriate words. However, the post-colon part must be serious, scientific, and convey the needed information. See two examples:

Original title: Studies on ground water recharge through rainwater harvesting

Improved: Water for posterity: Groundwater recharge through rainwater harvesting

Original title: An exploratory study to identify and document useful weeds
Improved: Weeds as resources: Identification and documentation of useful weeds.

Information Regarding the Investigators and Location

You must indicate the names of the principal investigator and co-investigators with full addresses and contact numbers for correspondence. The location of the research and collaborating institutions must also be clearly shown. Usually, the following information shall be included:

Name
Designation
Department
Name of the Institute
Address
Date of birth
Sex (M/F)
Telephone/mobile/e-mail.

Abstract

Most funding agencies ask for an abstract or executive summary of the project proposal. An *executive summary* is intended to provide a brief overview of the proposal. It usually contains a concise account of the proposal along with background information and conclusion. It is intended as an aid to decision-making by the reviewers. An *abstract* is still shorter than an executive summary, and it is prepared as a précis of the document.

Generally, an abstract of a proposal must include the justification for the study, the hypothesis or the research question, and the methods in a brief format. Remember that abstract or executive summary is an important part of the proposal, as it is the first impression of the proposal. The abstract is often written as a brief summary of about 300 words. Certain funding agencies (e.g. DST, Government of India) ask for still a shorter *project summary* not exceeding 150 words. Please note that for students' outline of research work (ORW), abstract or summary is not usually insisted. If you are asked to provide key words, select the most appropriate keywords (usually less than 6) and include them after summary.

Introduction

The 'introduction' part may appear in different forms according to the formats of funding agencies. In some formats, instead of 'introduction', you may be writing 'background'. After reading the introduction, the reviewers should know why you are interested in the project and why you have selected this particular problem. It must analyse a situation stating the problem or need statement and a short review of relevant literature. You should indicate your hypothesis and the assumptions with all the background logic. The rational and justification for taking up the study must be given. The introduction generally includes the project rationale, research hypothesis or research question, brief review of literature, and practical utility.

Project rationale: Present the rationale of your proposed study in simple language, and show its practical utility. As funding agencies want to support important research projects, it is essential to demonstrate the significance of the project so that your study gets high priority. For example, if the research is based on some issues in agriculture, explain what the results will do for farmers' interest or how they will affect productivity, improve food security, conserve environment, generate employment, or save resources. It is important to explain how the implications of your research are not limited to the narrow confines of the research field or laboratory.

State the research hypotheses or research questions: As a part of introduction, you must define the problem. Indicate major independent variables (causes or treatments) and dependent variables (effects that result from causes), and state your hypothesis, if any. Research hypotheses show the relationship between your research variables and expectations (see Sect. 5.2). Research hypotheses are normally found in research proposals involving experiments that compare differences or relationships between independent variables and dependent variables. For exploratory or descriptive studies, you may not have research hypotheses, but research questions only. You must identify and formulate carefully defined research hypotheses or research questions in unambiguous terms.

Brief review of literature: An important step in the formulation of research project, which should be done in the initial stage itself, is proper screening of relevant literature. Include a brief review of the seminal studies and important recent ones. Often, the literature review is incorporated into the introduction section as a part of it. However, certain proposals demand a separate section, which allows a more thorough review of literature.

It is essential to do a thorough literature search to understand what studies have already been done on the subject and to identify any glaring gaps in the literature. It also demonstrates your understanding of the theoretical and practical issues related to your research question. Read and make notes on as much literature as possible. Generate a few keywords to search and locate relevant literature. With the keywords in hand, you can search a database such as CAB Direct or PubMed for articles and other materials relevant to your research topic (refer Chap. 9 for more details on information retrieval). Selectively review the literature related to your research subject; but while writing the proposal, limit their numbers to 6–10 most relevant sources. Include only the most relevant literature and show the sources separately as 'References' (More details on literature review and citation rules are given in Chaps. 10 and 15).

Benefits or practical utility of the proposed research: The introduction often concludes with the benefits or scientific utility of the proposed research. Concerted efforts are needed to convince the reviewers the importance of the proposed work based on scientific considerations. You may also state clearly the benefits of this research to the stakeholders and the society.

Objectives and Aims

For a research proposal, probably 'objectives' is the most important section. It must target the attention of all reviewers, as all of them will go through it. Understand that if the objectives are not important and urgent, financial support may not be forthcoming. Some researchers use 'aims' and 'objectives' interchangeably confusing them to be synonymous. They are for a rude shock when they see that certain funding agencies ask for both 'aims' and 'objectives'! We generally undertake a research project with an overall aim and probably with more than one objective.

Aim is a broad statement of desired outcomes or the general intention of the research project. In most cases, to accomplish our aim, we may conduct more than one experiment with separate objectives. Each experiment will usually examine a small aspect of the overall topic with specific objectives. In general, a research project will be having not more than two or three aims, whereas, it may have a number of objectives consistent with the experiments included.

Objectives are the specified tasks researchers should undertake in response to the research hypothesis or question to accomplish the overall aim of the project. Some funding agencies just ask for 'broad objectives' (aims) and 'specific objectives' instead of aims and objectives probably to avoid confusion. 'Long-term objectives' and 'short-term objectives' are also on similar lines.

While finalizing the aims and objectives, it is a good idea to decide them after calling all the scientists associated with the proposed project (other colleagues can also be invited) for a brainstorming. From the ideas generated, pick up those most relevant to your project. Do not be over-ambitious; include only those objectives which could be managed by you. Similarly, do not include objectives for which your study cannot provide results. Remember that after the completion of the project, the funding agency will check whether all the objectives have been achieved. In short, the objectives should be SMART, that is, specific, measurable, achievable, realistic, and time bound (Challa et al. 2006).

Methodology

Methodology or 'materials and methods' section must describe the tasks needed to complete the research job. This section should contain sufficient information for the reviewers to determine whether the methodology is sound, and you should explain how you would conduct your research in as much detail as possible. The materials required and the proposed activities necessary for successfully completing the project must be described well. Ensure that the methods and experimental designs included are sufficient to meet the objectives. Be specific on the studies intended to undertake and the reasons for including them. It is also essential to describe the possibilities of failure in techniques and feasible approaches to overcome the difficulties.

Keep the description of the methods brief. Describe the evaluation plan of results, statistical tests, and quantitative and qualitative measure of data. Responsibilities of each of the investigators must also clearly be defined and indicated. As a part of the methodology of experiments, you must also state the location, name of the laboratory, and the places where the experiment will be done.

Time Schedule of Activities and Activity Milestones

Certain formats, especially of the development project types, ask for time schedule of activities and activity milestones in a separate section. The timetable should explain what activities would take place at what time of the year during the study. You can use a project management chart to demonstrate the timing of activities. *Milestones* are an important ingredient of a project management chart, which are distinct points in time when specified and measurable targets should have been achieved. You can assess your progress against expectations, and understand whether you will finish the project according to the original plan.

Ethical Considerations

If the study involves human subjects and animals, a section on ethics must be included. The main ethical concerns are whether the research will place the volunteers under undue risk, and whether they are fully informed about the nature of the study. State whether you have taken all the precautions to avoid any breach of ethics. If approval is needed from an ethical committee, it should be mentioned.

Facilities Available

In this section, convince the funding agency that your institution is qualified to do the work. When more institutions are competing for a similar grant, indicate clearly, why your institution is more qualified than others are. You may highlight better research facilities, more qualified scientists, and better institutional support. The strengths of co-investigators of the programme must also be highlighted. Show that the co-principal investigators too have the training and experience to do the research.

Indicators

Again, proposals on development projects may have a section in which the investigators must explain the likely impact of the study through indicators. *Indicators* are the criteria based on which the results or outputs of a project are evaluated; and therefore, these are important to evaluate the merit of the project proposal. Indicators reflect the essential aspects of a research objective in precise terms. By looking at the indicators, the impact of the results can be easily understood. Include only realistic indicators, and ensure that the changes observed are directly attributable to the project interventions. If asked for, both process indicators and impact indicators must be included. You may have to connect the indicators to objectives or activities.

Process Indicators

Process indicators measure the ways in which programme services and goods are provided. Commonly used parameters are:

- What are the expected results?
- How will you achieve the activity milestones?
- What are the steps involved?
- What are the benefits to your institution?
- Mention internal outputs at specified times during the project period.

Impact Indicators

Impact indicators are the criteria on which the impact of the project can be measured. These indicators provide information on the impact of the project on various yardsticks such as socio-economic conditions, productivity, profitability, cost of cultivation, poverty alleviation, food security, and environment protection. The following checklist can be used as impact indicators for a project in agriculture sector:

- Higher yields
- Higher income
- Gender specific impacts
- Use of indigenous knowledge
- Import substitution
- Creation of jobs
- Enhanced community participation
- Human health benefits
- Prevention and control of pests and diseases.

Budget

Budget is an important part of the project proposal, and it must be prepared carefully taking into consideration all the costs involved. Prepare the budget with realistic estimates. Bear in mind that the reviewers are highly knowledgeable on costs, and they may disallow anything which is estimated on the higher side. Include all categories of cost such as salaries/wages (for project associates and helpers), equipment, contingencies, travel, consultants, and supplies. You may spread the cost under both recurring and non-recurring expenses. Yearly expenses, rate of increase in salaries, costs for unforeseen expenditure, etc., must be shown. Make sure that the budget meets all the needs of your methodology. Remember that if you forget something to include at the time of submission, it would be very difficult to account for such an omission later. Normally, funding agencies provide budget proposal formats, and you have to prepare it according to their guidelines. A commonly followed model for preparing the budget is indicated in Table 7.1.

Sometimes, detailed budget estimates for each item as salaries/wages, consumable materials, travel, other costs, and equipment may also be asked to provide along with justification.

References

References are provided to indicate the sources from which the author has obtained information given in the proposal especially those in the 'introduction' and 'review' parts. Include only the most important references; there need be only 6–10, but they must be relevant and preferably recent. References must be typed out accurately following Name-Year system, as it is irritating for the donor of a grant to locate the references properly. For correct citation and listing of references; see Chap. 15.

Curriculum Vitae of the Applicants

Biodata or curriculum vitae (CV) of the principal investigator and the co-principal investigators must be neatly set out. Include sufficient details to show that you are

Table 7.1 A model for budget estimates

	Items	Budget			Total
		1st year	2nd year	3rd year	
A.	Recurring				
	1. Salaries/wages				
	2. Consumables				
	3. Travel				
	4. Other costs				
B.	Non-recurring				
	1. Equipment				
	2. Others				
C	Institutional charges				
	Grand total (A + B + C)				

competent to do the present work. However, avoid unnecessary bragging. Although synonymously used, the terms biodata and CV differ in essential features to be included. Another related term is resume.

Curriculum Vitae (CV) is a Latin phrase meaning 'course of one's life'. A CV is more focused on academic achievements of a person. It is more detailed than a resume, and its size would be as per the requirement. A CV lists out academic and professional qualifications, skills acquired, professional experience including positions held, professional affiliations, achievements and awards, research publications, etc., in chronological order. A CV must highlight the general talent of the person rather than specific skills.

Biodata is the short form for biographical data, and in a biodata, the focus is on personal particulars like date of birth, religion, gender, race, nationality, passport details, permanent and temporary residence, marital status, hobbies, height, weight, and a photograph. The term biodata is mostly used in South Asian countries like India and Pakistan instead of CV when applying for government jobs or for research grants. However, biodata is not common in the international scene where personal information like age, gender, religion are not required to be submitted by candidates.

Resume is a related term. The term 'resume' is from French, meaning 'summary'. In a resume, one summarizes his/her education, skills, and employment especially when applying for a new job. A resume does not list out all details of a profile, but only some specific skills customized to the target job profile. Thus, it may not be longer than one or two pages of size A4. A good resume may contain academic and professional qualifications, professional experiences, awards and achievements, and accomplishments.

Institutional Endorsement

This is an obligatory part of project proposals for external funding. Include a statement from the Head of Institution stating the willingness of the organization to endorse the proposed research project and indicate facilities, if any, that the institute would make available for the research project. Most funding agencies ask for a specified statement from the institution according to their requirement.

7.6 Presentation and Evaluation of Proposals

Writing a research proposal for funding requires patience and time. Do not show undue haste in preparation. Take your own time, and remember that it is not to be done instantly. Read the draft with a critical mind several times. Ask some of your friends to go through the proposal and comment on it. It is also a good idea to show it to a non-technical person such as your spouse or other relatives. Listen to all the comments with patience. Try to catch errors, repetitions, and inconsistencies. Edit the material to shorten, to make it clear, punchy, and appealing before taking the final print out. As Calnan (1976) outlined, a successful proposal must have answered the following questions:

- Why do I want to do this project? (the need)
- What do I want to do? (the plan)
- How am I going to do the project? (the method), and
- What do I need? (the resources).

Make sure that you have followed all this evaluative questions and prepared a fine research proposal.

Sometimes, the funding agency after preliminary appraisal may invite you for a presentation of the project before a reviewing panel. This is an opportunity for you to argue forcibly for your case. Often, based on the presentation you have made, the funding agency may inform their decision. Sometimes, they may ask you to revise the proposal and intimate you the quantum of assistance they are proposing to give you. Based on the suggestions, you have to prepare a modified proposal. If this is approved, the funding agency will send you a sanction letter with details of funding. Once the funding agency approved the project, you have to act quickly and take steps for starting the project without much delay.

Although there are no set rules to sanction a research grant, a funding agency develops certain criteria to be used by the reviewers as they evaluate the merit of the project proposal. The reviewers usually ascertain whether the proposed procedures address the objectives set out in the proposal. Writing style and the arrangement of the proposals into clear component sections are also important. Before you submit the project proposal, try to get an idea of the review criteria used by the donor agency. You should be aware of these criteria, as the way in which you address them will be a part of the assessment of merit of the project. With minor modifications, the same type of criteria is used at all levels of grant review. Usually, grant applications are evaluated using the following five questions (Kendall and Hawkins 1985).

- Is the idea important?
- Can it be done?
- Is the applicant competent to do the work?
- Can it be done within the specified time?
- Are the costs realistic?

After evaluating the project, the funding agency has to take a decision. The following may be the reasons for rejecting the proposal for funding:

- The researchable idea is not important.
- The applicant has not taken sufficient care to plan the project.
- Heavy dependence on others (Co-PI, project associates, personnel from other disciplines and institutes) to do most of the work.
- Budget is not within the limits.

There is no point in getting annoyed or disappointed, if your grant application is rejected. The funding agency may have some reasons. Sometimes, the agency would inform you the reasons. Otherwise, you have to assume the reasons based on the correspondence with the agency and the comments of the reviewers. Reflect on the reasons behind in rejecting the proposal, and rework on it so that you would be able to submit it to the same agency or similar agencies when they again issue notification inviting proposals.

7.7 Students' Outline of Research Work

In most academic institutions, for postgraduate studies such as M.Sc. (Ag.), M.Sc. (Hort.), MVSc., M.Tech., M.Phil., M.D., and Ph.D., research work is a compulsory part of the programme, and preparation of a research proposal and its approval are mandatory. Postgraduate research work conducted in colleges and universities/institutes also needs scrutiny in terms of quality and relevance. Each institution must have developed a mechanism to ensure quality control of students' research projects. Normally, postgraduate students in consultation with their major advisor or guide (allotted by the institution) prepare an 'outline of research work' (ORW) in a specified format, submit the ORW to competent academic bodies; and only if the approval is given, they can start the work and complete the thesis in the stipulated time. Research guides or major advisors have a greater role in formulating students' ORW. They should properly guide the students in finding a research problem and developing it with proper objectives and methodology with a good title.

Universities also prescribe norms for formulation and finalization of ORW. The format of ORW differs widely between institutions. Sometimes, you may be given a specific format in which the proposal has to be written, or you have to develop a format by yourself. An ORW usually has the details such as the name of student, names of major advisor and members of advisory committee, title of thesis, location, introduction, objectives, review of literature, practical/scientific utility, outline of technical programme, main items of observations, facilities (existing and additional facilities required), duration of study, time schedule of study, and financial outlay.

Although funds allotted for PG research is comparatively less, something worthwhile usually emerges because of the active involvement of the student, research guide, and members of advisory committee. Some universities prescribe a uniform ceiling of funding. For example, in Kerala Agricultural University, the ceiling for funding is Rs. 40,000 for a Masters Degree project and Rs. 100,000 for a Ph.D. project.

Some universities ask candidates applying for Ph.D. to submit a research proposal in a topic of their choice as part of the application. The objective of such a proposal is to prove that the candidates have thought out a worthwhile research project in their chosen fields, and that they have the competence and a proper work-plan to complete it. The quality of such research proposals is very important, as badly written proposals are usually rejected (sometimes, their candidature too!).

Often, the student and the major advisor prepare a draft ORW after a thorough literature review and discussing in detail various aspects of the chosen problem. The draft ORW should indicate how you would conduct the study with sufficient details on methodology and various facilities required in terms of land, equipment, and chemicals. If facilities available in another department (in addition to your parent department) are to be utilized, the major advisor may consult the department concerned and ensures possibilities of collaboration. Usually, a member from that department is also included as a member in student's advisory committee. The major advisor should form an advisory committee with experts from his/her discipline and related disciplines depending upon on the interdisciplinary character of the study. The major advisor will act as chairperson of the advisory committee. The draft ORW is first discussed in the students' advisory committee meeting. Once the advisory committee agrees and approves the draft, it should be presented in a seminar in each department or institution. All the faculty members and postgraduate students are expected to attend such seminars. In the seminar, the student shall explain the importance of the problem, scientific utility, and the technical programme. In the light of discussions in such seminars, the draft ORW is revised by the student and the major advisor, and submitted to the head of the institution for final approval.

7.8 Major Funding Agencies

There are several funding agencies or donors for research and development projects, both national and international, but you have to choose the agency depending upon your objective and area of interest.

International Agencies

International funding agencies are usually classified as *bilateral donors* and *multilateral donors*. Bilateral means that aid flows from one government to another government or non-governmental agency. Most bilateral funding agencies are established by governments of developed countries for giving aids to developing and least developed countries. The source of funding usually comes from government funds of those countries. The following are some popular bilateral funding agencies. You can obtain more information on the pattern of funding by visiting their websites.

- ACIAR (Australian Centre for International Agricultural Research): http://aciar.gov.au
- The Belgian Development Cooperation: https://diplomatie.belgium.be/en/policy/development cooperation
- CIDA (Canadian International Development Agency): http://www.acdi-cida.gc.ca
- DANIDA (Danish International Development Assistance): http://um.dk/en/danida-en/
- DFID (Department for International Development, UK or UKAID): http://www.dfid.gov.uk/
- SIDA (Swedish International Development Cooperation Agency): http://www.sida.se/english/
- USAID (United States Agency for International Development): http://www.usaid.gov/.

Multilateral donors are made up of many members. The funds of such agencies are from various sources and mainly contributions from member countries. Among the multilateral donors, the United Nations Organizations (UNO) is the most important. The following three UN agencies fund heavily for agriculture and related works:

- FAO (The Food and Agriculture Organization)
- UNEP (The United Nations Environment Programme)
- UNDP (The United Nations Development programme).

Some other multilateral funding agencies are:

- AFDB (The African Development Bank)
- AFESD (The Arab Fund for Economic and Social Development)
- ADB (The Asian Development Bank)
- EC (The European Commission)
- IADB (The Inter-American Development Bank)
- IFAD (International Fund for Agricultural Development)
- IBRD (International Bank for Reconstruction and Development (World Bank).

There are certain funding agencies that operate with funds from private sources. These agencies get money from the profits of a corporation or company or by gift from a wealthy individual or group. The funding agencies that works based on private funding are usually called *Foundations*. Often, Foundations are created with a big sum of money, and it works utilizing the interests being accrued. Some of the large and popular Foundations funding agriculture-related activities include:

- The Ford Foundation (www.fordfoundation.org)
- The Rockefeller Foundation (www.rockefellerfoundation.org)

- The McArthur Foundation (www.macfound.org)
- The Toyota Foundation (www.toyotafound.or.jp/english)
- The Packard Foundation (www.packard.org)
- The Bill and Melinda Gates foundation (www.gatesfoundation.org).

Indian Funding Agencies

There are several funding agencies in India for supporting research and development projects for specific purposes. In agricultural research, the main government agency responsible for funding is the Indian Council of Agricultural Research (ICAR). The ICAR sponsors and supports different kinds of research schemes, which aim at filling critical gaps in specific scientific fields. The Indian Council of Medical Research (ICMR) is the major organization for research in biomedical sciences. The Department of Science and Technology (DST), Science and Engineering Research Board (SERB), State Councils for Science and Technology, and the Department of Biotechnology (DBT) are the major funding agencies in India, which support science and technology projects. A brief introduction to various funding programmes of these agencies is provided below.

Department of Science and Technology

In the broad area of science and technology in India, the main funding agency is the Department of Science and Technology (DST), Government of India and the unit of the DST at the state level (State Councils for Science and Technology). The Department of Science and Technology supports research projects in many areas of science and engineering, especially those new and emerging areas from individual scientists, public institutions, and voluntary organizations under many schemes and programmes. The site also gives information on various international bilateral programmes. The following are some schemes as given in the website of DST. For more information and proforma for the submission of proposals, visit the website of DST (www.dst.gov.in).

Scientific and Engineering Research

- Mega Facilities for Basic Research
- Innovation in Science Pursuit for Inspired Research (INSPIRE) programme
- R&D Infrastructure (FIST, SAIFs, PURSE)
- Science and Technology of Yoga and Meditation (SATYAM)
- Programme for Science Students
- Swarnajayanti Fellowships
- National Science and Technology Management Information System (NSTMIS)
- Science and Engineering Research Board (SERB)
- Cognitive Science Research Initiative (CSRI)
- Impacting Research Innovation and Technology (IMPRINT)

- VAJRA (Visiting Advanced Joint Research) Faculty Scheme.

Technology Development

- Technology Development and Transfer
- National Good Laboratory Practice Compliance Monitoring Authority (NGCMA)
- Natural Resources Data Management System (NRDMS)
- Climate Change Programme
- Joint programme on Electric Mobility and Technology Foresighting
- Interdisciplinary Cyber Physical Systems (ICPS) Division.

S&T for Socio-economic Programme

- National Council for Science and Technology Communication (NCSTC)
- Science for Equity Empowerment and Development (SEED)
- National Science and Technology Entrepreneurship Development Board (NSTEDB)
- State Science and Technology Programme.

Women Scientists Programmes (WOS)

- Women Scientist Scheme-A (WOS-A): Research in Basic/Applied Science
- Women Scientist Scheme-B (WOS-B): S&T interventions for Societal Benefit
- Women Scientist Scheme-C (WOS-C): Internship in Intellectual Property Rights (IPRs) for the Self-Employment.

Technology Missions Division

- Water Technology Initiative Programme
- Clean Energy Research Initiative
- Nano Science and Technology Mission
- National Super Computing Mission

Science and Engineering Research Board (SERB)

In 2008, Government of India through an Act of Parliament approved the establishment of the Science and Engineering Research Board (SERB) as an autonomous research-funding agency free from bureaucratic controls. Most of the activities of the erstwhile Science and Engineering Research Council (SERC) under the Department of Science and Technology were handed over to the newly formed SERB. SERB supports research in frontier areas of science and engineering. The following are some schemes and programmes of SERB.

Core Research Grant (CRG)

The Core Research Grant (CRG) scheme provides core research support to active researchers to undertake research and development in frontier areas of science and technology such as life sciences, physical sciences, chemical sciences, engineering sciences, earth and atmospheric sciences, and mathematical sciences (formerly, extramural research). This scheme encourages emerging and eminent scientists for individual centric, competitive mode of research funding. Although renamed, it is still extramural research, meaning the research undertaken by a scientist but outside the normal programme of a college, university, or research institution. There are no upper or lower limits for funding, but the budget is decided based on the actual requirement for its successful completion. The principal investigator should propose a realistic budget considering the infrastructure and resources available at the implementing institution.

Scheme for Funding High Risk-High Reward Research

The scheme for funding 'High Risk-High Reward Research' supports conceptually new and risky proposals, which may have a lasting influence on science and technology. The proposal must have bold research ideas, which may have wide ranging influence and the potential for new scientific and technological innovations. Formulation of new hypothesis or scientific breakthroughs is encouraged.

Scheme for Funding Industry Relevant R&D

The scheme is to utilize the expertise available in academic institutions and national laboratories in solving industry specific problems. All industries are eligible. More than one industry and more than one investigator from one Industry can be associated in a project.

Empowerment and Equity Opportunities for Excellence in Science

The scheme, Empowerment and Equity Opportunities for Excellence in Science (EMEQ), is aimed at providing research support to scientists belonging to the Scheduled Caste and Scheduled Tribe in undertaking research in newly emerging and frontier areas of science and engineering and thus to involve them in the National Science and Technology development process.

Intensification of Research in High Priority Area (IRHPA)

The IRHPA programme supports proposals in high priority areas where multidisciplinary and multiinstitutional expertise is required. The board identifies priority areas in consultation with the stakeholders. Required facilities for implementing the project will be supported through this scheme.

In addition to the above schemes, SERB implements the following programmes as well:

- Ayurvedic Biology Programme
- Mathematical Research Impact-Centric Support (MATRICS) Scheme
- International travel support (ITS)
- Assistance to professional bodies and seminars/symposia
- Start-Up Research Grant (Young Scientists)
- Awards and Fellowships
- Partnership programmes.

For more information on the above schemes and updates, visit the home page of SERB (http://www.serb.gov.in/). Formats and guidelines can also be had from the website.

State Councils for Science and Technology

The decision to form state councils for science and technology was taken in 1971. Karnataka, Kerala, Uttar Pradesh, and West Bengal were the four pioneering states, who had established their State S&T Councils by the end of Fifth Five Year Plan. Subsequently, state science and technology councils have been established in all the States and Union Territories. Several states have also formed a separate department of science and technology. The chairpersons of the state councils are usually chief ministers of respective states or a renowned scientist. The Department of Science and Technology (DST), Government of India played a catalytic role by facilitating state governments in establishing and developing the State S&T Councils and by providing support for their technical secretariats.

In Kerala, the state council is called 'Kerala State Council for Science, Technology, and Environment (KSCSTE)'. Major programmes are under the heads: engineering and technology programmes, ecology and environment, science research schemes, rural technology programmes, biotechnology programmes, fellowships, student projects; seminars/symposia/workshop, travel grant, etc. Similar to Kerala, state councils for science and technology have been created in most other states. Search the websites of state councils for an update (see Annexure).

Indian Council of Agricultural Research (ICAR)

In agricultural research, the main government agency responsible for funding is the Indian Council of Agricultural Research (ICAR). The ICAR sponsors and supports different kinds of research schemes, which aim at filling critical gaps in the specific scientific fields or in the resolution of problems limiting production in agriculture and allied fields. Ad hoc schemes of ICAR were very popular among agricultural scientists until 2005. These were sanctioned for a period of three years utilizing fund from Agricultural Produce Cess funds. The government scraped the levy of cess on export of agricultural produce from the year 2005, necessitating the closure of ad hoc research schemes.

After the closure of ad hoc research schemes, Government of India established a national fund for supporting basic and strategic research through the ICAR system. As outlined by ICAR, the main objective of the scheme is to build capacity for basic, strategic, and cutting edge application research in agriculture, and address issues, which can be solved by intensive basic and strategic research jointly by a team of organizations/institutions. The fund was originally called 'National Fund for Basic, Strategic, and Frontier Application Research in Agriculture (NFBSFARA)', but renamed to 'National Agricultural Science Fund' (NASF) during the Twelfth Plan.

The funding programme promotes collaborative and multiinstitutional research based on innovative ideas of scientists for solving scientific and technological problems in agriculture. Scientists from all research institutions, universities, and private sector in India are eligible to participate. Selection of projects under NASF is done in two stages; in the first stage, a concept note indicating the objectives, technical programme, critical gaps being addressed, details of the organization, project investigators, and estimated budget are to be submitted. Once the concept note is approved, the principal investigator has to submit full proposals for consideration. For more information, visit the site http://www.icar.org.in/nfbsfara/index.html.

The ICAR has launched another ambitious project, 'National Initiative on Climate Resilient Agriculture (NICRA)' in February 2011. A major objective of the project is to increase the resilience of Indian agriculture to climate change and climate variability through strategic research and technology demonstration. Just like NASF, selection of projects is done in two stages, first, the concept note, and later, full proposal, if the concept paper is approved. More details can be accessed from the site http://www.nicra-icar.in/.

From 2013 onwards, The Indian Council of Agricultural Research (ICAR) sponsors and supports short-term result-oriented Extramural Research Projects which aim at filling critical gaps in the scientific field or in resolution of problems limiting production and value addition in agriculture, animal husbandry, and fisheries. These projects can be submitted by ICAR Institutes, State Governments, Agricultural and other Universities, and public, quasi-public, and private institutions capable of undertaking research in the above areas.

Revolving Fund Scheme, Emeritus Scientist Scheme, National Professorship, National Fellowship, Junior and Senior Research Fellowships, etc., are some of the other schemes of ICAR providing grant for research.

Department of Biotechnology

The Department of Biotechnology (DBT) was set up under the Ministry of Science and Technology in 1986 to give a new momentum to the growth and development of modern biology and biotechnology in India. The department provides the following categories of support to scientists working in the fields of biology- and biotechnology-related fields.

- Research and development projects, creation of infrastructural facilities, centres of excellence in the identified areas and demonstration projects
- Biotechnology-based programmes for SC/ST population
- Biotechnology-based programmes for women and rural development
- Bioinformatics
- Human resource development in biotechnology
- Financial assistance for holding seminar/symposium/conference/workshop on thrust areas of biotechnology
- Travel support.

DBT calls for proposals in specified areas. DBT has facilitated online submission at http://dbtepromis.nic.in or http://dbtepromis.gov.in. Sometimes, hard copies of the project document in the prescribed proforma should also be submitted. DBT will receive only online submission for proposals for conference, travel, exhibition, and popular lectures. The URL is http://dbtctep.gov.in for online submission. For an update on call for papers and other details of DBT funded projects, visit http://dbt india.nic.in or http://dbtindia.gov.in.

Indian Council of Medical Research

The Indian Council of Medical Research (ICMR), with its headquarters at New Delhi, is the apex organization in India for the formulation, coordination, and promotion of research in biomedical sciences. The ICMR gets funds from the Government of India through the Department of Health Research under the Ministry of Health and Family Welfare. For strengthening biomedical research in India, the first attempt was made in 1911 by the formation of the Indian Research Fund Association (IRFA). Later on, Government of India made several changes in the organization and activities of the IRFA. In 1949, it was renamed as the Indian Council of Medical Research (ICMR) intensifying its scope and functions.

The ICMR undertakes several activities related to biomedical research. It promotes both intramural research and extramural research. The intramural research is through 32 research centres in different parts of the country. The ICMR supports extramural research activities through (1) creating Centres for Advanced Research in different research areas in selected departments of medical colleges, universities, and other non-ICMR research institutes; (2) setting up task force groups for various studies; and (3) open-ended research, sanctioned based on applications for grant-in-aid from scientists of medical colleges, universities, and other non-ICMR research institutes.

The ICMR also promotes human resource development in biomedical research by awarding research fellowships, short-term visiting fellowships, short-term research studentships, training programmes, seminars, symposium, and workshops. The Council has a scheme to offer 'Emeritus Scientist' positions for retired medical scientists and teachers so that they can continue research on specific biomedical topics. The Council also gives awards and prizes to Indian scientists considering their contributions to biomedical research. For more information, visit the site http://www.icmr.nic.in.

Other Indian Funding Agencies

There are many other agencies giving grants in specified areas. For instance, the University Grants Commission (UGC) provides grant for many research-oriented projects under various heads. In India, the Council for Advancement of People's Action and Rural Technology (CAPART) acts as the major funding agency for voluntary organizations in rural areas. A list of major agencies providing grant support to research in science and technology in India is given in Annexure along with their web addresses. The list is not exhaustive.

Occasionally, certain funding agencies may declare special schemes for research funding, for example, NATP (National Agricultural Technology Project), NAIP (National Agricultural Innovation Project), National Horticultural Mission, and Technology mission on various commodities. These will be operational for a short period only. Scientists should be alert to grab the opportunity. While application forms and processes may be different for different funding agencies, the concepts are similar in that scientists must compete for funds before peer review panels.

Annexure

Agencies providing grant support to research in science and technology.

Funding agency with website address	Mailing address
Department of Science and Technology (DST) (www.dst.gov.in)	Department of Science and Technology, Technology Bhavan, New Mehrauli Road, New Delhi-110016
Science and Engineering Research Board (SERB) (www.serb.gov.in)	The Secretary, Science and Engineering Research Board, 5 and 5A, Lower Ground Floor, Vasant Square Mall, Sector-B, Pocket-5 Vasant Kunj, New Delhi-110070
Indian Council of Agricultural Research (ICAR) (www.icar.org.in)	Indian Council of Agricultural Research, Krishi Bhavan, New Delhi 110114
Department of Biotechnology (DBT) (www.dbtindia.nic.in)	Department of Biotechnology, CGO Complex, Lodhi Road, Block No. 2, Floor 7, New Delhi-110003
University Grants Commission (UGC) (www.ugc.ac.in)	University Grant Commission (UGC), Bahadur Shah Zafar Marg, New Delhi-110002
Indian Council of Medical Research (ICMR) (www.icmr.nic.in)	P.O. Box No. 4911, Ansari Nagar, New Delhi-110029
Ministry of Health and Family Welfare (www.mohfw.nic.in)	Ministry of Health & Family Welfare, Nirman Bhavan, Maulana Azad Road, New Delhi-110011
Department of Health Research (www.dhr.gov.in)	Department of Health Research, Ministry of Health & Family Welfare, Nirman Bhavan, Maulana Azad Road, New Delhi-110011
Indian Council of Forestry Research and Education (ICFRE) (www.icfre.org)	Indian Council of Forestry Research and Education, P.O., New Forest, Dehradun-248006 (Uttarakhand)
Indian Council of Social Science Research (ICSSR) (www.icssr.org)	Director (Research Projects), Indian Council of Social Science Research, (P.O. Box No. 10528), Aruna Asaf ali Marg, JNU Institutional Area, New Delhi-110067
Indian Council of History Research (ICHR) (www.ichrindia.org)	Indian Council of Historical Research, 35 Ferozeshah Road, New Delhi-110001
Department of Atomic Energy (www.dae.gov.in)	Anushakti Bhavan, C.S.M. Marg, Mumbai-400001
Department of Higher Education (https://mhrd.gov.in/higher_ education)	Department of Higher Education, Ministry of Human Resource and Development, Shastri Bhavan, New Delhi-110001
Ministry of Environment, Forests, and Climate Change (www.moef.nic.in)	Ministry of Environment, Forests, and Climate Change Indira Paryavaran Bhavan, Jorbagh Road, New Delhi-110003

(continued)

(continued)

Funding agency with website address	Mailing address
Ministry of Earth Sciences (www.moes.gov.in)	Ministry of Earth Sciences Prithvi Bhavan, Opp. India Habitat Centre, Lodhi Road, New Delhi-110003
Department of Chemicals and Petrochemicals (www.chemicals.nic.in)	Department of Chemicals and Petrochemicals, Ministry of Chemicals & Fertilizers, Janpath Bhawan, (3rd Floor, B-Wing), Janpath, New Delhi-110001
Department of Scientific and Industrial Research (www.dsir.nic.in)	Department of Scientific and Industrial Research, Ministry of Science & Technology, Technology Bhavan, New Mehrauli Road, New Delhi-110016
Ministry of Non-conventional Energy Sources (www.vigyanprasar.gov.in)	The Secretary, Ministry of Non-Conventional Energy Sources, Block No. 14, CGO Complex, Lodhi Road, New Delhi-110003
Council for Advancement of People's Action and Rural Technology (CAPART) (www.capart.nic.in)	Council for Advancement of People's Action and Rural Technology (CAPART), India Habitat Centre, Zone-V-A, 2nd Floor, Lodhi Road, New Delhi-110003
Council of Scientific and Industrial Research (CSIR) (www.csir.res.in)	Head, HRD Division, CSIR, CSIR Complex, NPL Campus, Pusa, New Delhi-110012
Defence Research and Development Organization (DRDO) (www.drdo.gov.in)	Defence Research & Development Organization, Ministry of Defence, DRDO Bhavan, DHQ PO, New Delhi-110011
Oil India Limited (OIL) (www.oil-india.com)	The General Manager, (R&D), Oil India Limited, Duliajan-786602
National Medicinal Plant Board (www.nmpb.nic.in)	The Chief Executive Officer, National Medicinal Plant Board, Chandralok Building, 36 Janpath, New Delhi-110001
Ministry of Food Processing Industries (www.mofpi.nic.in)	Ministry of Food processing Industries, Panchaseel Bhavan, August Kranti Marg, New Delhi-110049
National Horticulture Board (www.nhb.gov.in)	National Horticultural Board, Ministry of Agriculture, Govt. of India, 85, Institutional Area, Sector-18, Gurgaon-122015
Website addresses of some state councils on science and technology	Andhra Pradesh (www.apcost.gov.in); Gujarat (http://dst.gujarat.gov.in/gcst.htm); Haryana (www.dstharyana.org); Karnataka (www.kscst.org.in); Kerala (www.kscste.kerala.gov.in); Punjab (www.pscst.com); Tamil Nadu (www.tanscst.org); UP (www.cstup.gov.in); West Bengal (www.dstwb-counsil.gov.in)

Bibliography

Calnan, J. 1976. One Way to Do Research: The A-Z for Those Who Must. Heinemann Medical Books, London, 250p.

Challa, J., Rama Rao, D., and Virmani, S.M. 2006. Developing Winning Research Proposals in Agricultural Research (2nd Ed.). National Academy of Agricultural Research Management, Hyderabad, 474p.

Day, R.A. and Gastel, B. 2006. How to Write and Publish a Scientific Paper (6th Ed.). Cambridge University Press, Cambridge, 320p.

Gray, D.E. 2014. Doing Research in the Real World (3rd Ed.). Sage Publications, London, 729p.

Hall, M.S. and Howlett, S. 2003. Getting Funded: A Complete Guide to Writing Proposals (4th Ed.). Continuing Education Press, Portland, USA, 180p.

Hamper, R.J. and Sue Baugh, L. 2010. Handbook for Writing Proposals (2nd Ed.), 256p.

Kalla, N.R. 1999. How to get research grant. Curr. Sci. 77(11): 1431–1434.

Kendall, M.J., and Hawkins, C. 1985. Planning and Protocol. In: Research: How to Plan, Speak and Write about It (4th Indian Reprint, 1993). Narosa Publishing House, New Delhi, pp. 12–28.

Kumar, R. 2011. Research Methodology: A Step-by step Guide for Beginners (3rd Ed.). Sage Publications India, New Delhi, 415p.

Leedy, P.D. and Ormrod, J.L. 2010. Practical Research: Planning and Design (9th Ed.), Pearson Education, New Jersey, 360p.

Lester, J.D. and Lester, J.D. Jr. 2009. Writing Research Papers: A Complete Guide (13th Ed.), Longman, London, 416p.

Malmfors, B., Garnsworthy, P., and Grossman, M. 2004. Writing and Presenting Scientific Papers (2nd Ed.). Viva Books, New Delhi, 153p.

Miner, J.T. and Miner, L.E. 2008. Proposal Planning and Writing (4th Ed.). Greenwood Publishing Group, USA, 201p.

NSF [National Science Foundation] 2004. A Guide for Proposal Writing [On line]. Available: http://www.nsf.gov/pubs/2004/nsf04016/nsf04016.pdf. [Accessed 5 Jan. 2019].

Punch, K.F. 2000. Developing Effective Research Proposals. Sage Publications, London. 125p.

Rubenstein, K.D., Heisey, P.W., Klotz-Ingram, C., and Frisvold, G.B. 2003. Competitive grants and the funding of agricultural research in the United States. Rev. Agric. Econ. 25(2): 352–368.

UN [United Nations] 2017. World Population Prospects: The 2017 Revision, Key Findings and Advance Tables. Working Paper No. ESA/P/WP/248. Department of Economic and Social Affairs, Population Division, 46p. Available: https://esa.un.org/unpd/wpp/Publications/Files/WPP2017_KeyFindings.pdf [Accessed 10 Oct. 2018].

Walliman, N. 2011. Your Research Project: Designing and Planning Your Work (3rd Ed.). Sage Publications, New Delhi, 378p.

Ward, D. 2011. Writing Grant Proposals That Win (4th Ed.). Jones & Bartlett Publishers, Burlington, USA, 258p.

Weblinks

ACIAR - http://aciar.gov.au.

Belgian Development Cooperation - https://diplomatie.belgium.be/en/policy/development_cooperation.

Bill and Melinda Gates foundation - www.gatesfoundation.org.

CAPART - http://www.capart.nic.in.

CIDA - http://www.acdi-cida.gc.ca.

CSIR - http://www.csir.res.in.

DANIDA - http://um.dk/en/danida-en/.

DBT - http://dbtindia.nic.in.
Department of Health Research - http://www.dhr.gov.in.
Department of Higher Education - https://mhrd.gov.in/higher_education.
Department of Scientific & Industrial Research - http://www.dsir.nic.in.
Dept. Atomic Energy - http://www.dae.gov.in.
Dept. Chemicals & Petrochemicals - http://www.chemicals.nic.in.
DFID - http://www.dfid.gov.uk/.
DRDO - www.drdo.gov.in.
DST - http://www.dst.gov.in.
Ford Foundation - http://www.fordfoundation.org.
ICAR - https://icar.org.in/.
ICFRE - http://www.icfre.org.
ICHR - http://www.ichrindia.org.
ICMR - http://www.icmr.nic.in.
ICSSR - http://www.icssr.org.
McArthur Foundation - http://www.macfound.org.
Ministry of Earth Sciences - http://www.moes.gov.in.
Ministry of Environment, Forest, and Climate Change - http://www.moef.nic.in.
Ministry of Food Processing Industries - http://www.mofpi.nic.in.
Ministry of Health and Family Welfare - http://www.mohfw.nic.in.
Ministry of Non-conventional Energy Sources - http://www.vigyanprasar.gov.in.
NASF - http://www.icar.org.in/nfbsfara/index.html.
National Horticulture Board - http://www.nhb.gov.in.
National Medicinal Plant Board - http://www.nmpb.nic.in.
NICRA - http://www.nicra-icar.in/.
Oil India Limited - http://www.oil-india.com.
Packard Foundation - https://www.packard.org.
Rockfeller Foundation - http://www.rockfellerfoundation.org.
SERB - http://www.serb.gov.in/.
SIDA - http://www.sida.se/english.
State Council on S&T Gujarat - http://dst.gujarat.gov.in/gcst.htm.
State Council on S&T Haryana - http://www.dstharyana.org.
State Council on S&T Karnataka - http://www.kscst.org.in.
State Council on S&T Kerala - http://www.kscste.kerala.gov.in.
State Council on S&T Punjab - http://www.pscst.com.
State Council on S&T Tamil Nadu - http://www.tanscst.org.
State Council on S&T UP - http://www.cstup.gov.in.
State Council on S&T West Bengal - http://www.dstwb-counsil.gov.in.
State Council on S&T, Andhra Predesh - http://www.apcost.gov.in.
Toyota Foundation - http://www.toyotafound.or.jp/english.
UGC - https://www.ugc.ac.in.
USAID - https://www.usaid.gov.

Chapter 8
Publications and the Library

Books are the carriers of civilization. Without books, history is silent, literature dumb, science crippled, thought and speculation at a standstill.
Henry David Thoreau (1817–1862), American philosopher and author

Information is knowledge communicated through any media, which includes sensible statements, facts, principles, concepts, opinions, interpretations, and ideas. Information becomes knowledge only when it is conceived and understood. We get information from various sources such as documents, organizations, teachers, experts, colleagues, and even lay people. Scientists create knowledge by implementing various research projects or through observation and experience. In most universities, often the information being given to the students is at least 8–10 years old. Chances are that by the time students learn this 'new' information, still new information may have already replaced it! In olden days, whatever we learnt remained fresh for a considerable time—at least the lifetime of an individual. At present, within a few years, the information becomes old and needs updating. Even great professors or scientists, if they are not in touch with the current developments in their respective fields, become outdated and would be of no use to the society. A research worker should remain always a student trying to learn more and more throughout his/her life. She/he has to keep abreast with the growth of knowledge. Like the habit of reading newspapers, scientists should also inculcate the habit of reading journal articles regularly. Reading journal articles and reviews should form a regular feature of every scientist's life.

8.1 Role of Libraries in Information Retrieval

We are living in an era of information explosion. Information is growing at a faster rate, and one has to keep pace with it. For a researcher, the rapid growth of scientific literature leads to difficulties in the search for specific pieces of information. Research workers have to spend much time in the library in search of information.

C. G. Thomas, *Research Methodology and Scientific Writing*,
https://doi.org/10.1007/978-3-030-64865-7_8

They have to update their knowledge base commensurate with the advancement in science and technology. When we plan for a research project, when we start a new project, when it is nearing completion, while the final report is being prepared, and again, when we try to write research papers out of it, we may spend considerable time in the library. A rough estimate indicates that annually more than one lakh scientific journals are published from different parts of the world. All these churn out at least five million research papers. Who can read all these papers? Who will get time? It is impossible, and it is a waste of time to attempt searching or trying to read all this literature. Instead, it would be a better strategy to wean out irrelevant literature, and pick only those that are useful to us. Abstracting journals, indexes, catalogues, bibliographies, and search engines are some tools to help us locate relevant literature.

Any material that conveys information is a *document*, which includes various kinds of books, research papers, theses, popular articles, maps, diagrams, and photographs. The whole body of writings in a particular subject is broadly called its *literature*. *Retrieval* means finding a document or the information contained in a document from a library or database, and selectively recording information. Several tools are employed for information retrieval. In a library, many facilities are available to retrieve documents either as a whole like textbooks, theses, and journal articles or as micro-documents like abstracts and indexes.

What Is a Library?

Library is a place where reference materials such as books, manuscripts, periodicals, e-books, CDs, and many other resources are kept for use, but not for sale. It is an integral part of any academic or research institution (Fig. 8.1). The main function of a

Fig. 8.1 Researchers have to spend considerable time in the library, the repository of information. In the photograph, the lending book section of a library is shown

library is to assemble, organize, preserve, and serve all expressed thoughts available in the form of old manuscripts, books, periodicals, and other documents produced as a means for communication. The libraries help in the transmission of knowledge of the earlier generations to the present ones. Researchers are supposed to spend considerable time in search of relevant literature in the library. The onslaught of e-books, e-journals, and web-based documents may not be able to replace printed books and journals. Just like printed newspapers, printed books, and journals will continue to stay and attract researchers and students. A general awareness on libraries is given in this section.

The library helps research students in the production of excellent literature reviews using all the available literature sources. Most academic libraries also offer training to students on the use of electronic databases, bibliographic management softwares such as EndNote, Internet searching, and library catalogue searching.

Often, libraries are classified into four general types, viz. academic library, public library, national library, and special library based on the clientele and nature of activities.

Academic Library

Libraries attached to an academic institution such as a school, college, or university are called academic libraries. The most important function of an academic library is to provide facilities for study and research for its members. University library, college library, and school library are different academic libraries.

University Library

University libraries, also known as research libraries, stock an extensive collection of print materials including books, reference books, and journals. They also normally have good collection of video and audio materials and other forms such as palm leaf manuscripts, microfiche, and microfilm. Normally, a university library may have a good collection of archival resources and special collections on various subjects. The prime function of a university library is to provide facilities for study and research for its students and members of faculties. University libraries should act as the pivot around which teaching, research, and extension revolve.

Most university libraries have websites containing general information, access to online catalogues, online databases, newspaper databases, electronic journals, e-books, and links to other useful Internet resources. Most university libraries allow the public also to use the library with due permission from authorities and sometimes by charging a fee.

College Library

Every college should have a library of its own. Wherever university library is situated in the vicinity, the college library can scale down some of the costly subscriptions. However, the college library should subscribe to common and routine journals and stock most commonly used textbooks and reference books. A college library should act as the hub, which is to accelerate every learning activity going on in the campus.

School Library
School libraries provide access to all kinds of print and non-print resources for teachers and young students. A school library should stimulate and foster reading habits among children and enlighten immature minds. It should act as the cradle on which the child is intellectually conditioned.

Public Library
Public libraries are established for common people. These should provide every person the education attainable through reading. Public libraries are established under a clear mandate of law, and they are maintained wholly from public grants. Public libraries usually do not charge any direct fees from its users for any of its services. These are free and equally accessible by all citizens regardless of race, caste, creed, religion, age, gender, nationality, language, or educational status.

National Library
A national library is one established by the government of a country with the specific objective to serve as the highest storehouse of information for that country. It has the duty of collecting and preserving for posterity the literary products of that country, which normally include rare, valuable, and significant works. Many national libraries collaborate within the National Libraries Section of the 'International Federation of Library Associations and Institutions (IFLA)'.

In the USA, the Library of Congress performs the functions of a national library. It functions as the research library of the United States Congress. Located in Washington, D.C., it is the largest library in the world in terms of space and number of books. The National Library of Medicine (NLM) and the National Agricultural Library (NAL) are national libraries for the USA within their fields. NAL stores probably the largest collection of agricultural literature in the world. The library attached to Indian Agricultural Research Institute, New Delhi is considered as the de facto 'National Agricultural Library of India' with a collection of more than 0.6 million publications in agro-biological literature.

The 'National Library of India' is at Belvedere, Kolkata (formerly, Calcutta). This has been declared as an 'Institution of National Importance', which comes under the Ministry of Tourism and Culture, Government of India. The library is expected to collect, disseminate, and preserve various printed materials produced in India. With a collection of more than 2.2 million books, it is the largest repository of information in India.

'National Science Library' (NSL) of India is at New Delhi. Started in 1964 to function as a resource centre for all kinds of scientific and technical publications, it comprises of a wide collection of books including reference books and reports. The library also stocks foreign periodicals on CD-ROM and other electronic forms.

Special Library
A library concerned almost exclusively with the literature of a particular subject or a group of subjects can be called a special library. It serves a special clientele located within a single establishment or group engaged in working towards one common

purpose. A special library has three basic elements—special readers, special collections, and special locations. A typical case is a library maintained by a manufacturing company. It exists to serve the personnel of the parent body and caters to their information needs. Stocks in most special libraries consist of non-book materials such as periodicals, serial publications, pamphlets, patents, and theses. Therefore, the layout and arrangement of such libraries is difficult and different. Libraries attached to information centres, industrial organizations, laboratories, libraries of ministries, government departments, public authorities, and commercial banks come under the category of special libraries.

8.2 Laws of Library Science

Dr. S. R. Ranganathan (1892–1972), regarded as the 'father' of library science in India, formulated five 'laws' for application in library science. He was a mathematician and renowned library scientist. The library classification scheme known by the name 'Colon Classification (CC)' is one of his major contributions. His five laws of library science, first published in 1931 (Ranganathan 1931), continue to be the most influential ideas as far as libraries are concerned. Researchers and students who make use of the library should also be aware about these laws. These laws are:

- Books are for use
- Every reader his (her) book
- Every book its reader
- Save the time of the reader
- Library is a growing organism.

Books Are for Use

The first law, 'books are for use', implies that the main task of a library is to circulate books to its users. Open access to the library must be ensured, and there should be no obstruction between readers and their books. To achieve this purpose, obstacles, if any, should be removed making it easy for a user to obtain a book. Obviously, losing a book for whatever reason makes it impossible for readers to use it. Thus, protection and preservation of books are mandatory requirements for satisfying this law. The following are some essential requirements needed to maximize the use of books.

- Academic libraries should be located at the centre of the academic institution or at a distance of few minutes walk from the teaching departments or hostels.
- Libraries should remain open for 18 h a day. In fact, there are certain libraries, for example, the library of University College, London, which is kept open round the clock.
- The library should have enough shelf space for display of books. The library should have sufficient furniture, equipments for specific services, and buildings.

Every Reader His (Her) Book

The second law states that any person has a right to use the publications stocked in the library. All the books stocked in the library must be available to every consumer of the institution or the region in which the library is situated. The library should not have any books or other publications that some section of the population cannot access. Collections should be developed in such a way that every part of the population is interested.

Every Book Its Reader

The third law states that every book must find its reader. This law is concerned with the collection of publications in the library and the users. A book that is not borrowed for reading is a waste. Many books, even the latest editions, sometimes do not have many takers. In such cases, special efforts to expose the book to users are necessary. The law also implies the responsibility of the librarian to make the book reachable to readers. This task includes several techniques like appropriate classification and cataloguing to help the reader to identify and locate the needed book in a quick and efficient manner. In short, the law is all about the need to keep the library in a well-organized condition.

Save the Time of the Reader

The fourth law, 'save the time of the reader', stresses the point that if readers find what they are looking for in the shortest time, they will be more satisfied and likely to feel that their needs have been met. For this, a thorough scrutiny on the techniques of library administration and the methods to find sources of information rapidly are needed. Classification and grouping of books are to be done for locating them easily. This makes library service more efficient. The law is also concerned with the time the library professionals spend to serve the user.

The Library Is a Growing Organism

It is an accepted fact that a growing organism alone will survive. Unless necessary organizational changes take place, the library neither functions properly nor meets its objectives. As the library is being used, new books should be added on routine basis. The library should also get fresh readers. All these necessitate the need for adequate space, equipment, buildings, and staff; therefore, continuous updating of library should also be a regular activity. The library is something that grows forever in accordance with contemporary technologies.

8.3 Classification and Arrangement of Publications

Publications in a library are arranged under specific subjects. The arrangement of books under specific subjects is helpful to the reader, although it is a tedious work to the librarian. Fortunately, most of the libraries use one of the several published classification schemes for arranging publications in shelves. Among them, the *Dewey*

Decimal Classification (DDC) is the most popular classification system throughout the world except in some countries like the USA and Canada. In the USA and Canada, the popular classification system is the *Library of Congress Classification* (LCC) developed by the Library of Congress. The *Universal Decimal Classification* (UDC), developed based on the DDC, is popular in some countries. Another notable classification is that of Dr. S. R. Ranganatham, the *Colon Classification* (CC). As the DDC is the widely used classification system followed in academic libraries, some useful details on this system are given in this section.

The Dewey Decimal Classification, named after **Melvil Dewey** (who developed it in 1876), has been greatly modified and expanded through 23 major revisions. The most recent revision was in 2011. Melvil Dewey introduced the notions of relative location and relative index in a library allowing new acquisitions of books to be added in their appropriate location based on subject. The number given to a book helps the user to find it easily on the library shelves and allows librarians to return it to its proper place.

The DDC employs a numerical notation for classicisation. The whole field of knowledge is organized into 10 main classes numbered from 000 to 900 with the first digit representing its main class. Each main class is then divided into 10 separate divisions with numbers running from 0 to 9, and each division is further divided into 10 sections or subdivisions, also using from 0 to 9 numerical sequences. The classification scheme thus gives 10 main classes, 100 divisions, and 1000 sections. The 10 main classes under the DDC with their class numbers are:

000 General (Computer science, Information, and General works)
100 Philosophy and Psychology
200 Religion
300 Social sciences
400 Language
500 Natural sciences and Mathematics
600 Technology (Applied Sciences)
700 Arts and Recreation
800 Literature and Rhetoric
900 History, Geography, and Biography.

A classification scheme like the above means that books of the same subject will have the same class mark. For example, the number 600 is allotted to 'Technology or Applied Sciences' and number 610 is for medical sciences, 620 for engineering and allied sciences, and 630 for agriculture and related technologies. In the case of agriculture, all agricultural books will be found within the range from 630 to 639. Individual class marks for specific subjects are built up from these class marks by the addition of extra numbers. To illustrate the system, an example from agriculture can be used.

630 Agriculture and related technologies
631 Techniques, equipment, and materials
632 Plant injuries, diseases, and pests
633 Field and plantation crops
634 Orchards, fruits, and forestry

635 Garden crops (Horticulture)
636 Animal husbandry
637 Processing dairy and related products
638 Insect culture
639 Hunting, fishing, and conservation.

According to the DDC scheme, individual class marks for specific subjects are created by the addition of extra numbers. The example of 'soil chemistry' with class mark 631.41 is illustrated here. The number 631, on the left side of the decimal marker indicates the main class, division, and section. In the example, 6 represents the main class (Technology or Applied Sciences), 3 represents the separate division (Agriculture and related technologies), and 1, the distinct section (Techniques, equipment, and materials). The notation requires a decimal point to be placed after the third figure for further classification following the same type of numerical sequence from 0 to 9. In the example, the first decimal point (631.4) represents 'soil science' and second decimal point (631.41), 'soil chemistry'. This can continue with a third or more decimal points, if required. In the instant case, all books on 'soil chemistry' would carry the class mark 631.41. This indicates that books on related subjects are stocked near to each other. See the sequence of notations in the present case.

600-Technology (Applied Sciences)
630-Agriculture and related technologies
631-Techniques, equipment, and materials
631.4-Soil science
631.41-Soil chemistry

A broad outline of Dewey Decimal Classification System with all the main classes is shown in Table 8.1.

How to Make Use of a Library

Every good library will have a printed guide. When you first visit the library, ask for the printed guide. In the library, a diagram showing different rooms and floors for separate sections is often provided in the guide or hung on the wall. The sections of a library usually include: periodical room, reference room, new arrivals (past 15 days), room for microfilm/microfiche (important in documentation, which helps to save space), pamphlet section, theses, text books, rooms for web database search, and so on. The library guide contains the list of periodicals subscribed by that library. It will also have the information about back volumes available in the library, the regulation or procedure of borrowing the book, and facilities for interlibrary loan (exchanges). In a library, facilities for photocopying shall also be available.

Researchers and students should visit the library frequently and use the facilities effectively. They should also learn to make use of various tools to identify relevant literature. For this purpose, several techniques are available at present such as library catalogues, indexing and abstracting services, current contents, web searches, and many others. Every user must be aware of different parts of the library and the services rendered by the library.

In most libraries in India, books are arranged following the 'Dewey Decimal Classification System'. If the user has a preliminary idea about this system, it will

Table 8.1 Dewey decimal classification system

000	**General (computer science, information, and general works)**
000	Computer science, knowledge, and systems
010	Bibliographies
020	Library and information sciences
030	Encyclopaedias and books of facts
040	Unassigned
050	Magazines, journals, and serials
060	General organizations, associations, and museums
070	News media, journalism, and publishing
080	General collections
090	Manuscripts and rare books
100	**Philosophy and psychology**
100	Philosophy
110	Metaphysics
120	Epistemology, causation, humankind
130	Paranormal phenomena
140	Specific philosophical schools
150	Psychology
160	Logic
170	Ethics (moral philosophy)
180	Ancient, medieval, and oriental philosophy
190	Modern western philosophy
200	**Religion**
200	Religion
210	Natural theology
220	The Bible
230	Christian theology
240	Christian moral and devotional theology
250	Christian orders and local churches
260	Christian social theology
270	Christian church history
280	Christian denominations and sects
290	Other and comparative religions
300	**Social sciences**
300	Social sciences, sociology, and anthropology
310	General statistics
320	Political science

(continued)

Table 8.1 (continued)

330	Economics
340	Law
350	Public administration
360	Social services and associations
370	Education
380	Commerce, communications, and transport
390	Customs, etiquette, and folklore
400	**Language**
400	Language
410	Linguistics
420	English and old English
430	Germanic languages (German)
440	Romance languages (French)
450	Italian, Romanian, and Rhaeto-Romanic languages
460	Spanish and Portuguese languages
470	Italic languages (Latin)
480	Hellenic languages (Classical Greek)
490	Other languages
500	**Natural sciences and mathematics**
500	Sciences
510	Mathematics
520	Astronomy and allied sciences
530	Physics
540	Chemistry and allied sciences
550	Earth sciences
560	Palaeontology and palaeozoology
570	Life sciences
580	Botanical sciences/plants
590	Zoological sciences/animals
600	**Technology (applied sciences)**
600	Technology
610	Medical sciences
620	Engineering and allied sciences
630	Agriculture and related technologies
640	Home economics and family living
650	Management and auxiliary services
660	Chemical engineering

(continued)

Table 8.1 (continued)

670	Manufacturing
680	Manufacturing specific products
690	Buildings and construction
700	**Arts and recreation**
700	Arts
710	Landscaping and area planning
720	Architecture
730	Sculpture, ceramics, and metalwork
740	Drawing and decorative arts
750	Painting and paintings (museums)
760	Graphic arts
770	Photography and computer art
780	Music
790	Sports, games, and performing arts
800	**Literature and rhetoric**
800	Literature, rhetoric, and criticism
810	American literature in English
820	English and old English literature
830	Literature of Germanic languages
840	Literature of Romance languages
850	Italian, Romanian, and Rhaeto-Romanic literature
860	Spanish and Portuguese literature
870	Italic literature and Latin
880	Hellenic literature and Classical Greek
890	Literature of other languages
900	**History, geography, and biography**
900	History
910	Geography and travel
920	Biography, genealogy, and insignia
930	History of the ancient world
940	History of Europe
950	History of Asia and Far East
960	History of Africa
970	History of North America
980	History of South America
990	History of other areas

prove helpful to locate the books. When a book is taken from the shelf by a reader, she/he is not supposed to replace it back on the shelf. This stipulation is to avoid chances of misplacing the book. Once misplaced, it would be a formidable task to retrieve it. Let that job be done by the trained library staff. The user should bear in mind that the library is for everyone and should not try to remove pages or make any markings on books and journals.

In a typical library, there would be one main sequence of books for lending and for reference. Most libraries would also have a small separate sequence of quick reference books. These consist of dictionaries, encyclopaedias, yearbooks, and similar works, which enable the readers to find answers speedily on a wide range of subjects. Often, costly books that are in high demand are earmarked as reference and kept separately. These 'reference' books are also issued to the needy users for overnight reading without disturbing the right of other readers.

8.4 Tools for Identifying Literature

For library users, many tools are available for identifying and locating relevant literature without wasting much time. These include (1) library catalogues, (2) abstracting and indexing services, (3) current awareness, (4) reference resources, and (5) online databases and search engines.

Library Catalogues

In a library, the first tool for information retrieval is to search the library catalogue. Library catalogues are generally used to locate book materials of various types. These book materials may be 'primary' publications such as journals, conference proceedings, technical reports, theses, and dissertations; or 'secondary' publications such as textbooks, handbooks, monographs, and other types of compacted publications derived from the 'primary' literature. Separate catalogues will also be maintained for primary journals that publish results of original research. Cataloguing provides a record of all the books in the library. A catalogue provides class number of the books, author/s, title of the book, number of pages, details of illustration, details of edition, publisher, etc. Only when the characteristic class number is given, the reader can look under the name of the subject in the catalogue and find each class number. Catalogue entries are made according to the classification system used in the library.

To a librarian, the classification and cataloguing of a book go together. Both processes are usually done by the same person at the same time. If a book is properly classified, it will always be in the right position on the shelf. At any time in the library, the fate of a book must be known. The book may be on loan to its reader, it may be under repair (e.g. binding), may be shelved in reserve, shelved for special collection, or kept elsewhere.

You should learn to use the library catalogue because it is the only complete record of a library's stock, and the library cannot be used efficiently without it. Library catalogues come in many shapes and sizes, usually on cards or in sheaves.

At present, in most of the libraries, it is in digital form. Libraries may have one or more than one type of catalogues among the following:

Card Catalogue
These are kept in a cabinet with details of books on small cards that fit into the drawers. This is the index of the entire library, listing the publications found in the library with the exception of serially published periodicals. However, card catalogues have become outdated, and most libraries now use Online Public Access Catalogue (OPAC). Yet, card catalogues continue to be used in small libraries. Three types of cards are prepared as a part of the card catalogue system. *Author cards* are arranged alphabetically by the name of an author or authors; *subject cards* are arranged based on the subject; and *analytical cards* provide reference to important sections of a book.

Printed Catalogue
This is in the form of a printed document.

Shelf Catalogue
Shelf catalogue consists of loose leaf binders containing slips.

Visible Index
It consists of a series of metal frames fixed to a wall and holding slips of paper on which details of books are given.

Computer Catalogue
The catalogue information is stored in a computer and printed out as required.

Online Catalogue
Radical developments in information technology (IT) have prompted the librarians to adopt the much useful *Online Public Access Catalogue* (OPAC) discarding all the earlier physical catalogue systems. Simply speaking, an Online Public Access Catalogue is an online bibliography of a library collection available to the public. Users can search a library catalogue chiefly to locate books, periodicals, special collections, and other documents available to them. Online catalogues are easy to search. The information can be searched by author, title, subject and is cross-referenced for easy use.

Abstracting and Indexing Services
Abstracting and indexing services provide access to primary research literature such as journals, monographs, treatises, technical reports, conference papers, and government documents based on subjects, classification, author, and region or country. *CABI Abstracts*, *Biological Abstracts*, and *Chemical Abstracts* are examples. At present, there are several web-based resources too to retrieve documents. More information on these services is given in Chap. 9.

Current Awareness
Periodicals such as *Current Contents* or online databases such as *Current Content Connect* help us to keep abreast of what is being currently published. These kinds of services help us to keep abreast with current literature (Sect. 9.1)

Reference Resources
Certain types of publications are considered as reference resources. These serve a variety of purposes, and one can use them effectively as a part of literature search strategy. Many of these resources can be used as starting points of literature search, which may lead to original research appearing in primary literature. Reference resources can be used to:

- Define terminology (dictionary and glossary)
- Locate accepted knowledge and topical overviews (encyclopaedia, handbook, and review)
- Find factual information (directory and statistical compilations)
- Identify research on specific topics (review and bibliography)
- Locate standard methods and procedures (handbook and manual).

Online Databases and Search Engines
The Internet now offers umpteen opportunities in literature collection. Most of the abstracting and indexing services, current contents, citation indexes, etc., are now available online. A user must learn to know how to use a computer for information retrieval, as it offers unlimited possibilities. Accessing information from the Internet through search engines is the most easy and versatile form of literature collection. There are several online databases; certain databases require a fee on subscriber basis, but many are offered free to the public (Chap. 9).

When Internet resources are used for retrieving data, one must be careful and vigilant about spurious and unreliable websites. Rely only on websites sponsored by established groups, universities, reputed companies, and organizations. Avoid those that appear biased such as from lobbyists and advocacy groups.

8.5 Types of Publications

There are several kinds of publications, which can be classified into different groups based on durability of information and nature of contents. If we consider durability of information, publications or the literature available to us can be divided into two broad categories—*ephemeral* and *enduring*.

Ephemeral Publications
As the name implies, the publications classified as 'ephemeral' are ephemeral in nature, that is, relevant for a brief period only. They are just for informing you the present status of something. Because the information is constantly updated, durability

of such information is lost. Some publications, which come under the category of ephemeral publications, are listed below.

- Local and national newspapers
- Newsletters
- Press releases
- Market reports
- Weather reports and forecasts
- Disease intelligence reports
- House magazines
- Many online publications.

Enduring Publications

Information from enduring publications is durable and has relevance for a long time. Most of the publications we handle are enduring publications. Enduring publications can be further classified into *non-scientific* and *scientific*.

Enduring: Non-scientific Publications

There are several non-scientific publications, which are durable. Although these are classified as non-scientific, occasionally, scientists use them for specific information. See the following list for common non-scientific publications.

- Statutory and official publications (gazette, parliamentary reports, official statistics)
- Publications from agricultural boards, product manufacturers, trade and services
- Popular articles in magazines, books, reviews, reports, and surveys
- Advisory publications, advisory leaflets, farming notes, and extension bulletins
- Diaries
- Interviews (legal proceedings, personal, telephone, and e-mail)
- Letters
- Original documents such as birth certificates, title deeds, or a trial transcript
- Photographs
- Works of literature.

Enduring: Scientific Publications

The enduring scientific publications can be further categorized as *primary, secondary,* and *tertiary* publications based on their position in the publication cycle. A brief description of major information sources coming under primary, secondary, and tertiary is attempted here. The utility and reliability of these information sources differ according to the nature of the source.

Primary Sources Primary sources are reports of original works. These are publicly accessible documents wherein the results of original scientific investigations first appear. These sources contain original, uninterrupted, and unevaluated information based on actual studies. Often, the document containing the results of original scientific investigations is published after subjecting it to stringent review by one or more peers (peer review). Primary sources tend to come first in the publication cycle. Most common examples of primary sources are mentioned below.

Thesis: Thesis is a certificate, which shows that the individual has been trained for doing independent research. A thesis is in the form of a report prepared in a specified style incorporating the findings of a research work done by the student for the award of a higher degree. The internationally accepted maximum length of a Master's thesis is 50,000 words and Ph.D. thesis 100,000 words. Ideally, the size shall be within the range from 40,000 to 60,000 words. For all practical purposes, a thesis is a 'primary' source of information; however, it also functions as a 'secondary' source, as it contains a comprehensive literature review on the topic under study.

Dissertation: A dissertation is also a certificate for education and conduct of research similar to thesis, but it does not indicate that the researcher is capable of doing independent research. A dissertation is comparatively shorter in length than a thesis, and usually the size is 20,000–25,000 words.

Theses and dissertations are classified as 'grey literature' as they are not commonly published, and therefore, not available for reference through regular channels. However, the abstracts of theses from most institutions from the USA are published in *Dissertation Abstracts International.*

Primary Journals: Primary journals are the principal means of communication of original research findings. As they are published frequently, they could provide more recent information compared to textbooks or reviews. Primary journals carry the latest and often very specific account of current research work, new techniques, and unusual and interesting cases. Primary journals are published by government institutions, academies, professional scientific societies, associations, or private publishing houses. See some examples of journals from these sources.

- **Council of Scientific and Industrial Research (CSIR)**: The CSIR through the National Institute of Science Communication and Information Resources (NISCAIR) brings out 17 primary journals. Important journals are *Journal of Scientific and Industrial Research, Indian Journal of Biochemistry and Biophysics,* and *Indian Journal of Biotechnology* (Sect. 9.10).
- **Indian Academy of Sciences**: Currently, the Academy publishes 11 journals. You can access all the journals free (http://www.ias.ac.in). *Current Science* (published by Current Science Association in collaboration with Indian Academy of Sciences) and *Journal of Biosciences* are two prominent journals (Sect. 9.9).
- **Indian Council of Agricultural Research (ICAR)**: *Indian Journal of Agricultural Sciences* and *Indian Journal of Animal Sciences* (Sect. 9.11)
- **Indian Council of Medical Research (ICMR)**: *Indian Journal of Medical Research* (IJMR)
- **Universities**: *Journal of Tropical Agriculture* (Kerala Agricultural University)
- **Professional societies**: *Agronomy Journal* (American Society of Agronomy); and *Crop Science* (Crop Science Society of America)

- **Associations**: *Science* (American Association for the Advancement of Science); and *BMJ* (British Medical Association)
- **Private publishers**: *Nature* (Springer Nature); *The Lancet* (Elsevier); and *Plant and Soil* (Kluwer Academic Publishers, Netherlands).

People usually classify the journals as regional journals, national journals, international journals, and foreign journals. A journal published by a national society or publishing group is a national journal, for example, the *Indian Journal of Agronomy* published by the Indian Society of Agronomy. International journals are published by international societies or international publishing groups with wider readership in several countries. A paper carried in international journals such as *Nature* or *Science* is supposedly of high quality.

Technical bulletin: In a technical bulletin, complete details of an experiment are included. Often, technical bulletins are prepared after the completion of a project including all the details of experiments conducted with analysed and interpreted data.

Technical reports: Technical reports are meant for limited circulation only; therefore, these are printed in limited copies. The progress of ongoing research projects of an institution is usually published for public scrutiny or for evaluation in the form of half yearly reports, annual reports, summary reports, or final technical reports. For funded research, progress reports are mandatory, and they prescribe specific formats for submitting reports. Technical reports often have information about the research conducted before they appear in journal form. Technical reports, however, are not peer-reviewed. Sometimes, the reports may contain data, which appears nowhere else. Because technical reports are for limited circulation, they come under 'grey literature'.

Patent: It is a legal protection for the invention of a new process or method. Consequently, the patent holders get monopoly on the manufacture and sale of inventions granted by a government. Patents are considered as the most important form of publication for applied research.

Specification standards: These are specifications for quality. Accredited national and international organizations specify standards to encourage uniformity in several aspects. International Standards Organization (ISO) takes care of international standards for various products. Bureau of Indian Standards (BIS) publishes standards (ISI specifications) for various products in India.

Working papers: These are reports of works in progress, usually of funded research, on a specific topic ahead of more formal peer-reviewed publications.

Proceedings: It is a record of original papers along with other matters presented or done in a symposium, seminar, or conference.

Secondary Sources

In secondary publications, information from primary sources is synthesized. These sources digest, analyse, evaluate, and interpret the information contained within primary sources. Secondary sources cannot be taken as direct evidences, but rather as a commentary or discussion of evidences. These sources tend to be argumentative, and come second in the publication cycle. Because there is substantial additional work involved in producing secondary literature, especially for surveying the primary literature and collecting information from them, secondary sources are always less contemporary than the primary sources. There are several categories of secondary sources. Some are in book form, and some others are articles in periodicals. Prominent secondary sources are noted below.

Textbook: A textbook is an instruction book for a branch of study, which combines the work of many individuals from both primary and secondary sources after simplifying or eliminating much of the technical material that is of no interest to the general reader. The reader gets a quick and relatively good overall understanding of the topic by going through the textbook.

Handbook: A handbook is a type of reference book intended for providing ready reference. It usually includes the principles, methods, and techniques on a topic in a readymade form, for example, *Handbook on Fertilizer Usage.*

Monograph: It is a book containing comprehensive information on a subject or commodity. Full description, that is, complete information on a subject would be included.

Treatise: It is a multivolume monograph on a subject.

Yearbook: It is an annual volume of current information in descriptive as well as in statistical form.

Review: A review is a survey of primary literature over a given period indicating the developments and trends during the concerned period. Reviews are usually written by experts who survey and critically analyse the literature of a subject over a specific period. It will also carry personal evaluation from the author. An exhaustive list of references at the end of review serves as a good bibliography on the topic. It will be an excellent starting point for anyone unfamiliar with the literature in the field.

Most scientific societies of repute publish reviews, and these are usually identifiable from their titles such as '*Advances in …*', '*Yearbook of…*', and '*Annual Review of …*'. Examples include *Advances in Agronomy, Advances in Soil Science,* and *Annual Review of Plant Physiology.* Occasionally, one or two review articles may appear in primary journals or even abstracting journals. A thesis will also contain a comprehensive review on the researchable topics indicating the past and present status of the problem.

Abstracting journals: A periodical that specializes in providing abstracts of articles and other documents published within the scope of a specific field of study, for example, CABI Abstracts in various disciplines.

Indexing journals: The purpose of an indexing journal is to identify the source material by the use of an index along with name of author, title of article, title of journal or book, publisher, date of publication, and other details.

The arrival of computers has revolutionized both abstracting and indexing services. More details on academic databases including major abstracting and indexing services have been given in Chap. 9.

Tables: It is similar to a handbook, but information is provided in tabular form drawn from different sources, for example, Clark's Tables.

Popular articles: These are extension articles on specific topics written for the public especially, farmers. Normally, one should not quote popular articles in their research papers.

Gazette: Gazette is an official newspaper, which contains announcements of government appointments, notifications, government orders, and similar items.

Translations: Although most of the primary journals are in English, there are several journals in other foreign languages such as French, Spanish, Russian, Japanese, and Chinese. Researchers are supposed to review them also. Even though most of the abstracting journals include abstracts from other languages too, getting the originals may be a problem. In India, National Institute of Science Communication and Information Resources (NISCAIR) have facilities to arrange for translations (see Sect. 9.10). They are maintaining a database and translators for doing the work.

Bulletins: These are known by various names such as circulars, research bulletins, and technical bulletins. Bulletin is the field report from various agricultural experiment stations and does not include many results. They usually give only information of experiments.

Research newsletter: Many professional societies publish newsletters for their members, for example, *Indian Agronomy News* by the Indian Society of Agronomy. News update from the research field and news on the society are often included. This is unique in the sense that it occasionally publishes personal biographies of scientists. Normally, its editorship changes yearly and goes on changing from country to country, if it is an international society.

Pamphlets/leaflets/folders: These are made for common people showing some extension titbits. These are not of much research interest.

Map: A map is a two-dimensional model, usually on paper, of the natural features or topography of an area surveyed (a full map of the earth represented spherically is called ***globe***). Several kinds of maps are prepared. Examples include cadastral maps, topographic maps, and administrative maps.

Atlas: Atlas is a bound volume consisting of a collection of maps, for example, Watershed Atlas of Kerala.

Blog: A blog is a website maintained by an individual with regular entries of comments, descriptions, or materials such as graphics. The term is actually a contraction of the term 'web log'. Presently, blogs are increasingly used by the scientific community as means of disseminating research findings and exchange of ideas.

Tertiary Sources

Tertiary sources consist of information synthesized from primary and secondary sources. They tend to be factual and come last in the publication cycle. In a tertiary material, the information from other sources has been processed, reformatted, and condensed to put into a convenient, comprehensible form. See some examples:

Manual: It is an instruction book describing how to proceed with an experiment, how to operate an equipment, and similar details. Practical manuals and laboratory manuals are examples. When we purchase equipments, the manufactures usually supply operating manuals.

Almanac: An almanac is a publication containing astronomical and meteorological data arranged based on days, weeks, and months of a given year. Often, it also includes a miscellany of other information.

Guides to literature: These guides assist the user to use literature of a specific subject. It directs to various books and publications with an introduction to authors and texts.

Bibliography of bibliographies: It leads to other bibliographies and then goes to primary sources. It lists bibliographies or directs the readers to useful bibliographies.

Some tertiary resources, which have disjointed entries, are called *reference books*. These are designed for consultation from time to time for specific pieces of information. Reference books include dictionary, glossary, thesaurus, encyclopaedia, directory, lexicon, and gazetteer. Some people consider reference books as secondary sources.

Dictionary: A dictionary gives the meaning of words arranged in alphabetical order along with brief details about their pronunciations, meanings, functions, etymologies, and sometimes syntax and idiomatic uses. In general, there are two categories of dictionaries, general dictionaries and technical dictionaries.

Presently, you may find many online dictionaries for free browsing. The *Free Dictionary* is a comprehensive online dictionary by Farlex. For using it, visit the site http://www.thefreedictionary.com/. Another upcoming one is *Wiktionary* run by the Wikimedia Foundation (http://www.wiktionary.org/). Other prominent traditional dictionaries include *Oxford Dictionaries Online* (http://oxforddictionaries.com/), *Cambridge Dictionaries Online* (http://dictionary.cambridge.org/), and *Merriam-Webster* (http://www.merriam-webster.com/).

Glossary: If the dictionary gives only word meaning, then it is called a glossary.

Thesaurus: It is a book of words with synonyms and antonyms.

Encyclopaedia: An encyclopaedia (in print or in electronic form) is a good starting point for anyone newly approaching a subject. It is a compendium of information on diverse topics arranged alphabetically, written by an acknowledged author or group of authors. It contains information in brief about many terms or topics. Encyclopaedia should be updated every 5–10 years. A popular example is *Encyclopaedia Britannica*. At present, several online encyclopaedias are also available. An example of a popular and free online encyclopaedia is Wikipedia (http://www.wikipedia.org/). In *Wikipedia*, readers can act as editors and improve the contents.

Directory: A directory is a classified and alphabetically arranged list of names and addresses of people, institutions, places, or other such details. Directories cover many subjects. A typical example is a telephone directory in which the names and addresses of telephone subscribers with other information are included. Many directories are available in printed and electronic forms.

Wikispecies: It is from Wikimedia Foundation (https://species.wikimedia.org/wiki/Main_Page). It is a free online directory of species useful for biologists for identifying animals, plants, fungi, bacteria, and other life forms.

Lexicon: The term 'lexicon' is often used to represent a dictionary in languages other than English. Similar to a dictionary, it also gives meaning of words arranged alphabetically along with definitions and some additional details.

Gazetteer: It is a geographical dictionary of places arranged alphabetically. It is a finding list of places, oceans, rivers, lakes, mountains, etc., along with pertinent information, for example, Gazetteer of India published by Government of India.

Other tertiary sources include fact books, publisher's catalogues, bookshop catalogues, periodical indexes, company and service catalogues, and reference databases.

8.6 Grey Literature

Grey literature is a term used to refer to non-conventional, non-traditional, or non-book literature, which is not available through traditional book or trade channels. Grey matter is the thinking part of brain; therefore, the term is often used for scientific research. Identification and acquisition of grey literature are important tasks for a librarian. There are some problems with grey literature in its proper identification and stacking. Most grey literature does not bear usual bibliographic details such as names of authors, publication date, or publisher. Another issue is their non-professional layouts, formats, and low print runs, making the organized acquisition of such publications a daunting task in comparison with the dealing of traditional published literature such as journals and books. Most common grey literature consists of:

- Theses and dissertations
- Working papers
- Research reports
- Annual reports
- Translations
- Pamphlets
- Bulletins
- Research newsletter.

8.7 Microfiche

A microfiche is a flat photographic film of size 10.5 mm × 14.8 mm containing printed information in a miniature format. Microfiche, like microfilm, is a stable archival format for storing information. This device allowed many libraries around the world to preserve invaluable printed resources, and they would otherwise never have stored due to paucity of shelf space. A microfiche carries a matrix of micro-images. Frames may be landscape or portrait. Usually, magazine pages require a reduction of 24 or 25. As this cannot be read by the naked eye, a microfiche reader, which magnifies the print to readable size, is used to access information.

There are several advantages for storing information in a microfiche. Compared to the original print materials, this needs very less space for storage. Stability of the format is another benefit. Because of its small size, microfiche can be preserved in storage cabinets, saving much floor space. For this reason, thousands of government documents are archived on microfiche. It is estimated to last a minimum period of 500 years when kept in a temperature-controlled environment, a critical advantage in preserving cultural documents. Reading a microfiche is also easy using a microfiche reader.

When the computers were introduced, everybody believed that in course of time, they would make microfiche redundant. However, a complete replacement did not take place because of certain reasons. CD-ROMs are not as stable for archives as previously believed. They are much more sensitive to temperature changes, and prone to scratches and information losses. Constant use tends to degrade the media, which is not the case for microfiche. However, a CD-ROM can be copied for an infinite number of times unlike microfiche. A major disadvantage of microfiche is the need of a special reader to enlarge print size.

8.8 Digital Libraries

A *digital library* is an online database of texts, still images, audio, video, and other documents in digital media formats. In addition to storage, digital libraries allow for organizing, searching, and retrieving the contents in the collection. An *online library* is a digital library, which can be accessed through Internet. The digital content may be stored and accessed locally or distantly through computer networks. A digital library can be considered as a modern type of information retrieval system. It is considered virtual because the collections are stored in digital formats and accessible only by computers. Earlier, the term ‘virtual library’ was also used to refer to a digital library, but it is now mainly used for libraries, which are virtual in other senses, for example, the libraries that aggregate distributed content. Some major digital libraries, which would be useful for students and researchers, are mentioned below.

Questia.com (http://www.questia.com)

Questia.com is an online digital library containing many academic books and articles. This online repository stores a wide variety of resources including research articles and has links to numerous topics. Because the majority of information sources are under copyright regulations, Questia.com acquired their rights for rental by paying fees to the copyright owners. The library has a good collection of books and journal articles in the social sciences and humanities. As the original pagination has been maintained in the virtual format too, correct citation is easy.

The books and articles listed in Questia.com are accessible on subscription for a particular period. Presently, it provides online reading access to more than 94,000 books, and about 14 million articles from journals, magazines, and newspapers. Please note that Questia does not sell books or e-books for a price for permanent ownership, but sells only monthly or annual subscriptions.

World Digital Library (https://www.wdl.org/en/)

The World Digital Library (WDL) is a popular international free digital library operated by UNESCO and the United States Library of Congress. The mission of WDL is (1) to promote international and intercultural understanding, (2) to expand the volume and variety of cultural content on the Internet, (3) to provide resources for the educationalists, researchers, and public, and (4) capacity building to narrow the digital divide within and among countries. It proposes to make available on the Internet substantial primary materials from around the world including manuscripts, maps, books, musical scores, recordings, films, prints, photographs, architectural drawings, and other cultural materials. The library lists more than 18,000 items from nearly 200 countries dating back to 8000 BC.

The Universal Digital Library (http://www.ulib.org/)
The Universal Digital Library has been created with the Million Book Project of Carnegie Mellon University School of Computer Science and University Libraries from 2007 to 2008. It is a mammoth book digitization project. They plan to include 10 million books within 10 years. Collaborating with government and research partners in India (Digital Library of India) and China, the project scanned books in many languages, using OCR to enable full-text searching and providing free-to-read access to the books on the web. They have already completed scanning of 1 million books and have made the entire catalogue accessible online.

Wikisource (http://en.wikisource.org/)
Wikisource is a free digital library of publications started in 2003 as a project of Wikimedia Foundation to create and collect all forms of public domain texts, films, and similar materials. The library mainly includes source texts previously published by any author, translations of original texts, historical documents of national or international interest, and bibliographies of authors whose works are in Wikisource. Wikisource collects and maintains publications which are in the public domain or available for free redistribution without any legal binds.

Wikisource is also active in many languages apart from English.

Wikibooks (https://www.wikibooks.org/)
A Wikimedia project hosted by the Wikimedia Foundation, Wikibooks serves for the creation of free content e-textbooks and annotated texts. Some books are original, but some others may be texts copied from other sources of free content textbooks found on the Internet. Contributions remain copyrighted to their creators while the licensing ensures that it can be freely distributed and reused subject to certain conditions.

Wikibooks differs from Wikisource in that Wikisource collects exact copies and original translations of existing free content works such as the original text of Shakespearean plays, while Wikibooks is dedicated either to original works, significantly altered versions of existing works, or annotations to original works.

The Internet Archive (https://archive.org/)
The Internet Archive is a non-profit digital library providing free access to digitized materials including websites, software applications, music, movies/videos, and

about 3.0 million public domain books. Besides its archiving function, the Internet Archive works for a free and open Internet. The Internet Archive permits the public to upload and download digital material to its data cluster. However, the bulk of its collected data is drawn automatically by its web crawlers, which work to preserve as much of the public web as possible. The Archive also supervises one of the world's largest book digitization projects.

Open Library (https://openlibrary.org/)

Open Library is a project of the Internet Archive, which has been funded in part by a grant from the California State Library and the Kahle/Austin Foundation. The stated goal of the Open Library is to create 'one web page for every book ever published'. It provides access to many public domain and out-of-print books. It collects information on books from the Library of Congress, other libraries, and Amazon.com, as well as from user contributions. If books are available in digital form, a button labelled 'Read' appears next to its catalogue listing. It also gives links to where books can be purchased or borrowed. Open Library claims to have 6 million authors and 20 million books. About one million public domain books available are digitized books.

Project Gutenberg (https://www.gutenberg.org/)

The digital library offers over 57,000 free e-Books. You can select from among free books, free kindle books, and download them or read them online. It has good collection of world classics with focus on older works for which copyright has expired and now in the public domain. Numerous volunteers digitized and precisely proofread the e-Books. For using the library, no fee or registration is required.

Bartleby.com (https://www.bartleby.com/)

Bartleby.com is another online library containing good and authentic sources, especially for students. The website contains good number of classics from literature and non-fiction including the works of political and social history. The website is named after Herman Melville's story 'Bartleby, the Scrivener'. It was founded in 1993 by Steven H. van Leeuwen as a personal, non-profit collection of classic literature on the website of Columbia University. In 1999, Bartleby.com started to focus on reference works, and by 2013, on its 20th anniversary, more than one billion pages had been accessed.

The Free Library (https://www.thefreelibrary.com/)

Started in 2003, the Free Library owned by Farlex has been offering free, full-text versions of classic literary works. It also stores a massive collection of periodicals from leading publishers covering pure and applied sciences, technology, social sciences, health, humanities, law, politics, government, business, communications, entertainment, recreation, and leisure. The collected works includes numerous articles from 1984 to present that are added to the site on a continual basis.

Virtual library of FAO (http://www.fao.org/library/en/)

The virtual library of FAO (Food and Agriculture Organization of the United Nations) provides access to many FAO publications and rich data on agriculture, forestry, fisheries, and rural development in a variety of forms including the Internet and CD-ROMs. You can read or download many FAO publications free of cost. The online site allows you to search FAO's extensive catalogue and databases, besides providing links to both FAO and other institutional electronic journals and other useful sites. FAO's David Lubin Memorial Library Online, being a member of Agricultural Libraries Network (AGLINET), a voluntary network of global agricultural libraries, provides several other services too such as low-cost interlibrary loan and photocopying services to other member libraries.

The FAO Corporate Document Repository is a digital library that keeps FAO's enormous collection of documents and other publications in full-text digital format. The initiative was started in 1998, and more than 6000 documents have been converted into digital documents allowing users to download them freely. The publications can be viewed online, downloaded as a PDF file, or purchased in hard copy, if available for sale. You can directly access the document repository at http://www.fao.org/documents/en/.

Google Books (http://books.google.com/)

A growing digital database of books, presently, it has information about 25 million books scanned and converted to text. When the user clicks on a result from Google Books, it opens an interface in which much useful information about the books is displayed. The user can view full pages or in part from the book, if the copyright owner has given permission or if the book is out of copyright regulations. For most books, the pages for free view are limited to a preview only. For books where permission for a 'preview' has been denied, two to three lines of text (snippets) may be permitted. However, in most cases, the user can access information about the authors, publishers, number of pages, etc. You can also buy the book through the site.

National Science Digital Library (NSDL), USA (https://nsdl.oercommons.org/)

National Science Digital Library (NSDL) of USA is a free online library for education and research in science, technology, engineering, and mathematics (STEM). NSDL's collections are refined by a network of STEM educational and disciplinary professionals. NSDL offers an organized point of access to these contents, collected and aggregated from different digital libraries and other projects, publishers, and NSDL-reviewed websites. Anyone can access NSDL website and search the library free, but some content providers demand a fee to access their specific resources.

National Digital Library of India (https://ndl.iitkgp.ac.in/)

The National Digital library of India (NDLI) is a free digital library under the Ministry of Human Resource Development, India. Officially launched in 2018, its objective is to integrate several national and international digital libraries in one single web portal. NDLI integrates contents from different institutional repositories such as NCERT, Krishikosh, INFLIBNET (Shodhganga), LIBRIVOX, and NPTEL. It provides a single window search facility for all digital resources. The NDLI provides access to many books in English and the Indian languages free of cost. Currently, it has access to over 18 million publications covering textbooks, theses, videos, audios, educational materials, fiction, etc. It has been designed to benefit many kinds of users such as students, researchers, librarians, and professionals.

National Science Digital Library, India (http://nsdl.niscair.res.in/jspui/)
National Science Digital Library (NSDL) aims at providing comprehensive science and technology information to students of science, engineering, and technology in India. It was started as a 10th Five Year Plan network project of the Council of Scientific and Industrial Research (CSIR). NSDL provides syllabus-based literature to address the information needs of the undergraduate students of science. A discussion forum has also been provided for interaction among the users.

Bibliography

Axelrod, R.B. and Cooper, C.R. 1997. The St. Martin's Guide to Writing (5th Ed.). St. Martin's Press, New York, 791p.
Barrass, R. 2002. Scientists Must Write: A Guide to Better Writing for Scientists, Engineers and Students (2nd Ed.). Routledge, London, 204p.
Best, J.W. and Kahn, J.V. 1993. Research in Education (7th Ed., Indian Reprint, 2004). Prentice–Hall of India, New Delhi, 435p.
Bloom, D.F., Karp, J.D. and Cohen, N. 1998. The Ph.D. Process. A Students Guide to Graduate School in the Sciences. Oxford University Press, Oxford, 209p.
Bond, A. 2007. Your Masters Thesis: How to Plan, Draft, Write and Revise. Viva Books, New Delhi, 135p.
CSE [Council of Science Editors]. 2014. Scientific Style and Format: The CSE Manual for Authors, Editors, and Publishers (8th Ed.). The University of Chicago Press, 840p.
Gray, D.E. 2014. Doing Research in the Real World (3rd Ed.). Sage Publications, London, 729p.
Horowitz, L.1986. A Writer's Guide to Research. Writer's Digest Books, Cincinnati, Ohio, 135p.
Hunt, A. 2005. Your Research Project: How to Manage It. Routledge, London, 147p.
Leggett, G., Mead, C.D., Kramer, M.G., and Beal, R.S. 1985. Handbook for Writers. Prentice- Hall, New Jersey, 558p.
Lenburg, J. 2007. Guide to Research. Viva Books, New Delhi, 560p.
Lester, J.D. and Lester, J.D. Jr. 2009. Writing Research Papers: A Complete Guide (13th Ed.), Longman, 416p.
Ranganathan, S.R. 1931. The Five Laws of Library Science. Madras Library Association (Madras, India) and Edward Goldston (London, UK).
Sharma, C.K. 2006. Practical Handbook of Dewey Decimal Classification. Atlantic Publishers & Distributors, New Delhi, 228p.
Walliman, N. 2011. Your Research Project: Designing and Planning Your Work (3rd Ed.). Sage Publications, New Delhi, 378p.

Weblinks

Bartleby.com - https://www.bartleby.com/.
Cambridge Dictionaries Online - http://dictionary.cambridge.org/.
FAO document repository - http://www.fao.org/documents/en/.

Free Dictionary - http://www.thefreedictionary.com/.
Google Books - http://books.google.com/.
Merriam-Webster - http://www.merriam-webster.com/.
National Digital Library of India - https://ndl.iitkgp.ac.in/.
National Science Digital Library (NSDL), USA - https://nsdl.oercommons.org/.
National Science Digital Library, India - http://nsdl.niscair.res.in/jspui/.
Open Library - https://openlibrary.org/.
Oxford Dictionaries Online - http://oxforddictionaries.com/.
Project Gutenberg - https://www.gutenberg.org/.
Questia.com - http://www.questia.com.
The Free Library - https://www.thefreelibrary.com/.
The Internet archive - https://archive.org/.
The Universal Digital Library - http://www.ulib.org/.
Virtual library of FAO - http://www.fao.org/library/en/.
Wikibooks - https://www.wikibooks.org/.
Wikipedia - http://www.wikipedia.org/.
Wikisource - http://en.wikisource.org/.
Wikispecies - https://species.wikimedia.org/wiki/Main_Page.
Wiktionary - http://www.wiktionary.org/.
World Digital Library - https://www.wdl.org/en/.

Chapter 9
Academic Databases

The problems are solved, not by giving new information, but by arranging what we have known since long.
Ludwig Wittgenstein (1889–1951), Austrian-British philosopher

Several avenues are now available for researchers and students to retrieve the latest information for their academic and research works. In the past, the choices were few, and they had to depend heavily on the meagre library repositories available as printed books, journals, abstracting journals, and indexes. Because of rapid developments in information and communication technology, printed information sources are being replaced by electronic sources such as online databases, e-journals, and e-books. In the recent years, most abstracting and indexing service providers have switched over to online storage discontinuing print format altogether. These online resources are available to you as soon as they are published. By sitting in front of the personal computer, you can retrieve a good chunk of information resources.

The use of Internet for information retrieval is rapidly increasing, thus changing the traditional functions and services of the libraries. The Internet can be considered as a big library for all kinds of information. Most information sources are available online free of charges although some are subscription based. Internet also provides worldwide access to online databases, electronic journals, e-books, online library catalogues, publishers' catalogues, encyclopaedias, online dictionaries, e-conference reports, digital libraries, website of organizations, and many others. With the rise in the use of Internet, most educational and research organizations have developed their own websites for providing the latest information on their services, products, publications, and programmes.

With the advent of computers, networking, and Internet, it is now possible to access information sources from anywhere at any time. Many digital libraries, publishers, and organizations provide access to their online information sources 24 h on Internet. Authors and publishers are also happy to make available their articles online free or for a fee because online accessibility of articles dramatically increase their citation frequencies and impact factors of their journals. Users can access, download, and store the required information in their computers and can use them for academic and research purposes.

C. G. Thomas, *Research Methodology and Scientific Writing*,
https://doi.org/10.1007/978-3-030-64865-7_9

A *database* is a systematically prepared collection of information covering different subject disciplines, which can be accessed in various ways. A *digital database* is a computer programme that organizes, describes, and indexes information. It permits the user to search for specific types of information, depending upon the selected search parameters. A database is made up of records, which consist of smaller units of information called *fields*. Common bibliographic database fields are: author, publication title, article title, subject or keywords, date of publication, volume, issue, and page numbers. A database that is accessible from a local network or the Internet is an *online database*, as opposed to one that is stored on an individual computer. Online databases are hosted on websites, made accessible through a web browser. They may be free or subscription based.

9.1 Bibliographic Databases

Indexes are common in books for easy location of subject terms. An indexing journal publishes indexes to books and articles similar to the above. It gives a citation, which is helpful to identify the source material along with name of author/s, title of article, title of journal or book, publisher, date of publication, and subject indexes. An abstracting journal performs the same function as that of an indexing journal, but in addition, provides an abstract or brief summary of the paper being indexed giving extra information.

Abstracting and indexing databases provide access to primary research literature such as journals, monographs, treatises, technical reports, conference papers, and government documents based on subjects, classification, author, and region or country. Computer technology has brought in great changes in indexing and abstracting services (Fig. 9.1). Presently, print form of abstracting and indexing journals is almost vanishing. Most of the big publishers have stopped publishing print versions of journals and completely switched over to web-based databases. Presently, there are several bibliographic databases, which also provides full texts of articles in collaboration with publishers. Brief notes on important bibliographic services are provided here.

CABI Abstracts

CABI Abstracts, published by the Centre for Agriculture and Biosciences International (CABI, formerly CAB, Commonwealth Agricultural Bureaux, see Sect. 9.8), are very popular in Commonwealth countries. CABI makes abstracts of most primary and secondary sources such as journal articles, conference proceedings, theses, reports, patents, and new books. A record in CABI Abstracts database contains informative abstracts with bibliographic description needed to locate the original article.

CABI Abstracts maintain over 9.0 million records collected during the period from 1973 to the present. The database includes both national and international journals and other sources in agriculture, forestry, environment, veterinary sciences, applied

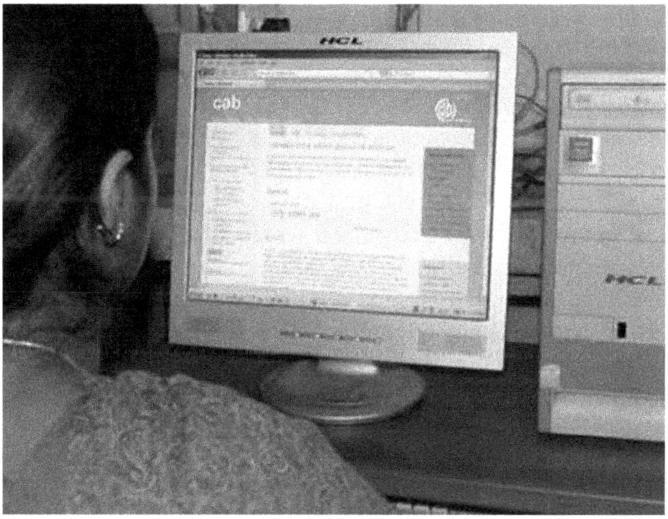

Fig. 9.1 Advances in computer technology enable you to access most of the online databases such as CAB Direct or PubMed by sitting in your room

economics, food science, nutrition, and allied disciplines in life sciences. Journals, proceedings, books, and agricultural serials are covered. Excellent subject index and author index assist the search for relevant abstracts. Abstracts of literature from 1913 to 1972 are archived and available as CAB Abstracts Archive, which has more than 1.8 million records.

Global Health
Global Health is a bibliographic database from CABI, which focuses on literature in public health and medical health science. The database contains 3.0 million scientific records from 1983 to the present derived from over 7300 journals, reports, books, and conferences. Global Health claims to add about 160,000 records each year. The abstracts of global health literature from 1910 to 1983 are archived and available as Global Health Archive, which contains about 0.8 million records.

CAB Direct (http://www.cabdirect.org/)
A subscription-based database, CAB Direct is CABI's online platform, which provides single point of access to all CABI database subscriptions. This online database is a source of references for agriculture and other applied life sciences including medicine with more than 11 million bibliographic records across the applied life sciences and over 340,000 full-text articles incorporating the leading bibliographic databases in CABI Abstracts and Global Health. CAB Direct has Google-like search functionality to enable users to find what they need quickly and easily. CAB Direct uses CAB Thesaurus, an online index from CABI for searching.

Publications from 158 countries in 50 languages are abstracted. Relevant non-English-language papers are translated into English. It also contains good amount of grey literature such as proceedings, patents, theses, electronic publications, and relevant but difficult-to-find literature sources.

CAB Thesaurus (https://www.cabi.org/cabthesaurus/)

CAB Thesaurus provides a controlled vocabulary resource of descriptive terms for the applied life sciences. It acts as the search tool for all users of CAB Direct. Regularly updated, it has broad coverage of agriculture and life sciences, technology, and social sciences with over 2.7 million terms, which includes about 166,000 distinct concepts (preferred terms) and 132,000 synonyms, along with translations of these from English into 10 European languages. It also includes broad, narrow, and related terms to help users find relevant terminology. The thesaurus lists variations on terminology, broader terms are larger in scope and narrower terms are limited in scope.

Chemical Abstracts (http://www.cas.org)

Chemical Abstracts and related products are produced by Chemical Abstracts Service (CAS), a part of the American Chemical Society. It was started in 1907. CAS discontinued print form of Chemical Abstracts from 2010. However, Chemical Abstracts can continue to be accessed by subscription through CAS products. Chemical Abstract Service maintains two principal databases, *CA Plus* and *CAS Registry*. CA Plus database stores bibliographic details and abstracts of all articles from chemistry-related journals; in addition, it also stores articles on chemistry from other scientific journals and publications.

The CAS claims to monitor and cover every year more than 10,000 journals and other products such as technical reports, dissertations, conference proceedings, and new books, in about 50 languages. Patent documents from 63 patent authorities are also covered. CAS Registry stores information covering more than 130 million organic and inorganic substances, and over 64 million protein and DNA sequences. CAS offers access to its databases through search tools such as *SciFinder* and *STN* (Scientific and Technical Information Network).

Biological Abstracts

Biological Abstracts is a database now owned by Clarivate Analytics. Started in 1926, it covers more than 4300 peer-reviewed journals. It includes abstracts from peer-reviewed academic journal articles in biology-related areas such as botany, zoology, biochemistry, biotechnology, pre-clinical and experimental medicine, pharmacology, agriculture, and veterinary medicine. It can be accessed through a number of services, including EBSCO, Ovid, and Web of Science. A similar publication is *Biological Abstracts/RRM* (Reports, Reviews, Meetings), which is meant to cover meetings and conferences, literature reviews, US patents, books, software, and other such sources. The combination of the two is marketed as *BIOSIS Previews*.

AGRIS (http://agris.fao.org/)

AGRIS (International Information System for the Agricultural Sciences and Technology) is a popular database devoted to agriculture, which started functioning in

1975 under the aegis of Food and Agriculture Organization of the United Nations (FAO). AGRIS corresponds to the earlier print publication of FAO AgrIndex, which stopped its publication in print form from 1995.

AGRIS with over 8.0 million bibliographical records covers many aspects of agriculture, including forestry, animal husbandry, fisheries, human nutrition, and extension literature from about 150 participating institutions representing 65 countries. AGRIS also covers good amount of grey literature including theses, conference papers, unpublished scientific and technical reports, and government publications. The database is maintained by Coherence in Information for Agricultural Research for Development (CIARD) movement, a joint initiative co-led by the Consultative Group on International agricultural Research (CGIAR), Global Forum on Agricultural Research and Innovation (GFAR), and FAO, who are working together to make agricultural research information and knowledge publicly accessible. Presently, AGRIS is an international cooperative system to serve both developed and developing countries.

The AGRIS search system permits researchers, students, and others to have advanced searches using keywords from the AGROVOC thesaurus, journal titles, authors, institutions, and countries. In addition to abstracts, a good percentage (around 20%) of bibliographical records have full-text documents, which are retrievable by Google.

AGROVOC

AGROVOC, a multilingual thesaurus, is a collaborative effort coordinated by FAO and maintained by an international community of experts and institutions. Developed in the 1980s, the main objective of AGROVOC is to standardize the indexing process for the AGRIS database to make searching efficient and simpler, thus directing the user to the most appropriate resources. In the present form, AGROVOC is expressed using a concept scheme. It now contains close to 35,000 plus concepts available in 29 languages covering subjects like agriculture, forestry, fisheries, food and nutrition, and environment. AGROVOC is widely used by librarians, information technology managers, and many others for indexing, retrieving, and organizing data in information management systems. AGROVOC is available at: http://aims.fao.org/standards/agrovoc.

AGRICOLA

AGRICOLA (Agricultural On Line Access), started in 1970, now with over 5.2 million records is a database created and maintained by the United States Department of Agriculture. AGRICOLA database serves as the online catalogue and index for the collections of the United States National Agricultural Library (NAL), which covers all aspects of agriculture and allied fields including plant sciences, forestry, farming and farming systems, agricultural economics, extension and education, food and human nutrition, earth and environmental sciences, animal and veterinary sciences, aquaculture, and fisheries.

AGRICOLA includes journal articles, monographs, theses, patents, computer software, audio-visual materials, and technical reports. It consists of two subsets of records. The first contains citations for journal articles that include abstracts. The second consists of bibliographic records describing monographs, serials, audio-visual materials, and online content from around the world.

Other forms of publications such as books, pamphlets, conference proceedings, translations, book chapters, research reports, and government documents are also included in the database. AGRICOLA records also contain links to full-text resources on the web. You can search AGRICOLA free of charge on the NAL website (http://agricola.nal.usda.gov/).

NAL Agricultural Thesaurus (NALT)
The thesaurus used by National Agricultural Library (NAL) is a hierarchical vocabulary of terms related to agriculture, biology, and related disciplines. NALT provides extensive coverage of agriculture, biology, and allied disciplines, which is searchable by 17 subject categories. The thesaurus is updated annually, and presently, it contains over 135,000 terms including 63,000 cross-references. It also gives a glossary of definitions for technical terms. NALT is now used extensively for retrieval in agricultural information systems within and outside USDA. You can access NALT through https://agclass.nal.usda.gov/.

MEDLINE (https://www.nlm.nih.gov/bsd/medline.html)

Published by the National Library of Medicine, USA, once *Index Medicus* in print was the most comprehensive index of the world's biomedical literature. From 2005, the printed publication was discontinued because of the popularity of online resources like MEDLINE and PubMed, which could be used in its place. Medical Literature Analysis and Retrieval System Online (MEDLINE) functions as an important resource for biomedical researchers all over the world. Compiled by the US National Library of Medicine (NLM), Maryland, USA, MEDLINE is freely available on the Internet and searchable through the search engines, PubMed or Entrez. In fact, MEDLINE is the online equivalent of the print resources, Index Medicus, Index to Dental literature, and the International Nursing Index.

MEDLINE provides bibliographic citations and abstracts of over 25 million records from about 5200 professional journals in about 40 languages covering all aspects of life sciences, medicine, nursing, pharmacy, dentistry, veterinary medicine, and health care. It also covers literature in molecular biology, biotechnology, and biochemistry from 1950 to the present. The records are indexed with NLM Medical Subject Headings (MeSH).

MedlinePlus (http://www.nlm.nih.gov/medlineplus/)

MedlinePlus, another product of the NLM (National Library of Medicine), is meant mainly for patients and their families. It offers a wealth of information for health professionals too. MedlinePlus brings together information from the NLM, the National Institutes of Health (NIH), other U.S. Government agencies, and health organizations. This free website provides information in almost 1000 health topics, especially diseases and wellness issues in simple language avoiding many of the medical jargons. MedlinePlus can be accessed for various purposes such as learning about recent trends in treatments, collecting details about a drug or supplement, getting the meanings of technical words, or for watching medical videos.

PubMed (http://www.ncbi.nlm.nih.gov/pubmed/)

PubMed is a free search engine or database maintained by the National Centre for Biotechnology Information (NCBI) at the US National Library of Medicine (NLM) for accessing more than 28 million citations from MEDLINE, journals in life sciences, and online books. Normally, citations may also include links to full-text articles from *PubMed Central* and some other publisher websites. PubMed covers almost all the biomedical fields such as medicine, nursing, dentistry, veterinary medicine, health care, and preclinical sciences.

PubMed Central (http://www.ncbi.nlm.nih.gov/pmc/)

PubMed Central (PMC) is a digital archive to collect and preserve the full text of biomedical and life sciences literature managed by the National Centre for Biotechnology Information (NCBI) at the U.S. National Library of Medicine (NLM). PMC offers free access to all the articles from journals deposited in the archive. Currently, the number of articles archived in PMC is 4.1 million. It is linked to PubMed and is fully searchable. Access to the full text of all PubMed Central articles is free.

MeSH (Medical Subject Headings) (http://www.ncbi.nlm.nih.gov/mesh)

Medical Subject Headings (MeSH) provides the U.S. National Library of Medicine (NLM) controlled vocabulary for indexing articles from about 5200 of the world's leading biomedical journals for the MEDLINE and PubMed database. MeSH can be browsed and downloaded free of charge through PubMed. MeSH terminology is helpful as it provides a consistent way to retrieve information for the same topic. All MeSH terms are arranged alphabetically as well as in subject groups. The subject headings or descriptors are arranged in hierarchical levels. All the terms are arranged from the most general to the most specific, followed by a short explanation, links to related subject headings, and a list of synonyms. Because of the presence of synonym lists, MeSH can also be considered as a thesaurus.

PubChem

PubChem is a database providing information on the biological activities of small chemical molecules. The database system is maintained by the National Centre for Biotechnology Information (NCBI), a component of the National Library of Medicine, USA. PubChem includes information on substances (236 million entries), compound structures (93.9 million entries), and bioassay bioactivity results (1.25 million) through three linked databases—*Substance*, *Compound*, and *BioAssay*. For more information about using each component database, visit the homepage of PubChem (http://pubchem.ncbi.nlm.nih.gov/) and use the links provided.

TOXLINE (http://toxnet.nlm.nih.gov/)

Toxicology literature Online (TOXLINE) provides abstracts of articles on chemicals, pharmaceuticals, pesticides, and pollutants through TOXNET.

TOXLINE is produced by the National Library of Medicine (NLM) of USA.

ScienceDirect (http://www.sciencedirect.com/)

ScienceDirect is an online repository of full-text journal articles and book chapters operated by the Dutch publisher Elsevier. Presently, it has more than 12 million contents from 3800 journals and more than 35,000 book titles and over 11,000 e-books, reference works, book series, and handbooks. The articles are arranged under four main sections and several sub-disciplines. The main sections are Physical Sciences and Engineering, Life Sciences, Health Sciences, and Social Sciences and Humanities. For most articles on the repository, abstracts are freely available; however, access to full text of the articles requires subscription.

EMBASE (https://www.elsevier.com/)

Excerpta Medica Database or EMBASE is a biomedical and pharmacological database produced by Elsevier. The database indexes more than 8500 journals, including over 2900 not available in MEDLINE, and has over 31 million records. It provides clinical and experimental information with extensive coverage of drug research, pharmacology, pharmacy, and toxicology, public health, and mental health topics with abstracts back to 1974. This database is available through subscription.

CINAHL

Cumulative Index to Nursing and Allied Health Literature (CINAHL) provides online access to citations and abstracts in nursing, health, biomedicine, and healthcare. Started in 1961 as the *Cumulative Index to Nursing Literature* (CINL) in print, its title was changed to *Cumulative Index to Nursing and Allied Health Literature* in 1977 when they expanded it to include allied health journals. The index became online from 1984. EBSCO Publishing acquired CINAHL Information Systems, and presently, they bring forth CINAHL. Visit the site:

https://health.ebsco.com/products/the-cinahl-database.

FSTA Abstracts

Food Science and Technology Abstracts (FSTA) is produced by International Food Information Service (IFIS). FSTA is a bibliographic abstracting and indexing database of scientific and technological research, which provides thorough coverage of pure and applied research in food science, food technology, and nutrition. The database is useful for researchers, industry practitioners, and students.

FSTA covers food science, food technology, and nutrition including food additives, biotechnology, food safety, functional and novel foods, and packaging. In addition to about 5500 active and historical journals, FSTA indexes books, trade publications, reviews, conference proceedings, reports, patents, and standards. With records dating back to 1969, FSTA contains information sources in 29 languages, sourced from publishers in over 60 countries. It is available on subscription through

EBSCOhost, Ovid, Proquest Dialog, STN, and Web of Science. Visit the site https://www.ifis.org/fsta for more information.

Inspec

Inspec is a prominent database of scientific and technical literature published by the Institution of Engineering and Technology (IET, formerly IEE, the Institution of Electrical Engineers). The IET is one of the largest engineering institutions with over 168,000 members in 150 countries. It is multidisciplinary to reflect the increasingly diverse nature of engineering. Inspec was started in 1967 as an outgrowth of the *Science Abstracts* service.

The subject coverage of Inspec includes astronomy, communications, computers and computing, computer science, control engineering, electronics, electrical engineering, mechanical engineering, information technology, physics, manufacturing, and production. Inspec also includes information technology for business. You can access Inspec through InspecDirect and avenues such as Web of Science on subscription. Visit the site: https://www.theiet.org/.

Open Access Theses and Dissertations (OATD)

OATD is an index of over 4.6 million electronic theses and dissertations. It is a good resource to find open access theses and dissertations published around the world. They receive information about the theses (metadata) from over 1100 colleges, universities, and research institutions. Many of the records come from their own repositories, but others are from regional or national electronic theses and dissertations (ETD) consortia. The full text of all papers is on the original hosting site, usually the repository of the university that granted the degree. Please note that OATD indexes about the first 30 pages of some theses to show search hits only, but in no case indexes or stores the full text. You can access OATD at https://oatd.org/

Shodhganga (http://shodhganga.inflibnet.ac.in/)

Shodhganga is a digital repository of Indian theses and dissertations. It is maintained by the Information and Library Network (INFLIBNET) Centre, Gandhinagar, an autonomous Inter-University Centre (IUC) of University Grants Commission (UGC) of India. The repository has the ability to capture, index, store, disseminate, and preserve electronic theses and dissertations (ETDs) submitted by the researchers. Note that 'Shodhganga' is a coined term, the word 'Shodh' is from Sanskrit, which stands for research and discovery. The 'Ganga' is the holy river of India symbolizing the long held culture and civilization of India.

Shodhganga provides a platform for research students to deposit their Ph.D. theses and make it available to the research scholars and scientists in open access. The Shodhganga repository was created after UGC made mandatory through regulations for all universities to submit soft copies of Ph.D. theses and M.Phil. dissertations to the UGC for hosting in the INFLIBNET. Currently, it stores over 2 lakhs full-text theses from 355 universities.

Krishikosh (http://krishikosh.egranth.ac.in/)

Krishikosh is a repository of National Agricultural Research System (NARS) of India. It is a digital store house of accumulated knowledge such as theses, books, research articles, monographs, conference proceedings, case studies, and other grey literatures spread all over the country in different ICAR Research Institutions and State Agricultural Universities (SAUs). The Indian Council of Agricultural Research (ICAR) has made it mandatory to upload all Institutional publications such as theses, research articles, monographs, catalogues, conference proceedings, case studies, annual reports, newsletters, brochures, bulletins, summary of the completed projects, and other grey literatures available with the various SAUs in the Krishikosh. MSc and Ph.D. theses (full text) and summary of completed research projects are to be deposited in the Krishikosh repository after completion of the work. The metadata such as the title, abstract, authors, publisher, etc. will be made freely accessible from the time of deposition of the content; however, their free unrestricted use is permitted through open access after an embargo period of not more than 12 months.

BioRxiv (https://www.biorxiv.org/)

BioRxiv (pronounced *bioarchive*) is a free online archive and distribution service for unpublished preprints in the life sciences. It is operated by Cold Spring Harbor Laboratory. By posting preprints on bioRxiv, authors are able to make their findings immediately available to the scientific community and receive feedback on draft manuscripts before they are submitted to journals. Articles are not peer-reviewed, edited, or typeset before being posted online. Authors may submit a revised version of an article to bioRxiv at any time. Once posted on bioRxiv, articles are citable and therefore cannot be removed. BioRxiv accepts preprints of articles covering all aspects of research in life sciences.

ArXiv.org (https://arxiv.org/)

ArXiv (pronounced *archive*) is a repository of electronic preprints or e-prints consisting of scientific papers in the fields of physics, astronomy, mathematics, statistics, computer science, electrical engineering, quantitative biology, and economics, which can be accessed online. ArXiv is owned and operated by Cornell University. Users can retrieve papers from ArXiv via the web interface.

Current Contents

Although abstracting and indexing services are useful, they are mainly for a retrospective search of literature. Current Contents is a publication by Clarivate Analytics providing currant awareness on a variety of subjects. This is published as a weekly in seven editions (Clinical practise; Agriculture, biology, and environmental sciences; Engineering, technology, and applied sciences; Social and behavioural sciences; Life sciences; Physical, chemical, and earth sciences; and Arts and humanities). It reproduces the content pages of the most important journals in their respective fields. Each issue also contains a subject index of the papers in each journal and an index of authors with their addresses to facilitate acquisition of reprints.

Current Contents Connect is the online version of Current Contents series. It is available bundled with *Web of Science.* It is a multidisciplinary current awareness database providing easy web access to the most recently published issues of leading scholarly journals.

9.2 General Search Engines

A *search engine* is a software system specially designed to search for information on the World Wide Web. The search results are generally presented in a line of results, often referred to as *'search engine results pages'* (SERPs). The information may be a blend of web pages, images, and other types of files. They help to mine data available in databases and play an important role in retrieving required information. Search engines provide us with links to other websites relevant to the subject we are searching.

To search the topic of interest effectively, you need to define it first. Decide in unambiguous terms what information you want to get. After doing this, pick out the key words that best describe the concepts. Using these key words, you can search the topics on various search engines or online databases. From among the multitude of web pages and sources listed, you have to pick the most suitable and relevant ones. Many types of search engines are presently available for the users. A few examples, which are helpful to the research community, are included below.

By far, Google, Yahoo!, and Bing are the most popular general search engines. However, there are several other useful search engines too. If one search on a particular search engine fails, try another until you get the required results.

Google (http://www.google.com/)

Google is considered as the world's largest search engine, and in the beginning of 2018, its market share was 90 percent, much ahead of others. Bing occupies about 3.0 percent, Baidu (Chinese) 2.0 percent, and Yahoo! 2.0 percent. Google was founded in 1998 by Larry Page and Sergey Brin while they were Ph.D. students at Stanford University in California. The search engine is immensely popular among users because of its high relevancy of results. By indexing billions of web pages, Google helps the users to search and locate the information they desire. Google has also diversified their web search technology into other search services including images, news, maps, and video.

Google's fast growth since incorporation has initiated a chain of products, acquisitions, and partnerships apart from their core search engine, the *Google Search.* Presently, Google offers several services such as e-mail (*Gmail*), academic work (*Google Docs, Google Sheets,* and *Google Slides*), time management (*Google Calendar*), cloud storage (*Google Drive*), social networking (*Google+*), instant messaging and video chat (*Google Allo, Google Duo, Hangouts*), language translation (*Google Translate*), mapping and navigation (*Google Maps, Waze, Google*

Earth, Google Street View), video sharing (*YouTube*), note-taking (*Google Keep*), and photo organizing and editing (*Google Photos*).

Google Books and *Google Scholar* are two products often used by academic community (described elsewhere). *Google Earth* is a popular interactive mapping tool powered by satellite and aerial imagery, which covers major part of the earth. You can also form discussion groups and participate in the discussions by sending group messages.

Yahoo! (http://www.yahoo.com/)

Yahoo! is also a popular search engine offering many services such as e-mail, besides acting as a search engine. Yahoo! provides a lot of information including the latest news and entertainment programmers and helps users instant access to other Yahoo! services such as e-mail, maps, finance, and group messaging facilities.

Bing (http://www.bing.com)

Bing is Microsoft's search engine. Formerly, it was known in several names as Live Search, Windows Live Search, and MSN Search. Bing provides a variety of search services similar to other popular search engines including news, videos, images, maps, and shopping.

Wiki.com (http://wiki.com/)

This search engine pulls its results from numerous Wikis on the Web. Wiki.com is a specific search engine for those who want to search information from various Wiki products.

DuckDuckGo (https://duckduckgo.com/)

DuckDuckGo (DDG) stresses on protecting visitors' privacy and, therefore, does not collect or store any of visitors' personal information. DuckDuckGo distinguishes itself from other search engines by not profiling its users.

Gibiru (http://www.gibiru.com/)

Gibiru does not install any personalization and tracking cookies on your system. Gibiru provides uncensored and non-personalized web and news results. You can browse the Internet safely.

Yandex (https://yandex.com/)

Yandex is the most popular search engine in Russia. It is also popular in Belarus, Kazakhstan, Turkey, and Ukraine. It also provides some other uses like cloud storage service.

Swisscows (https://swisscows.com/)

Swisscows came up as an alternative for anyone who attaches importance to data integrity and the protection of privacy. Users at Swisscows do not leave any tracks. The owners claim that topics, IP addresses, and personal information of visitors are not stored or used for any additional business. As all the servers are located in Switzerland, they claim that neither the USA nor other data snoopers can access users' information. It uses artificial intelligence to determine the context of a question.

SlideShare (https://www.slideshare.net/)

SlideShare is an exclusive search engine, which allows you to search for documented slideshow presentations. You can also search for e-books and documents. SlideShare also allows you to save slides and even download the entire slideshow for use on your local computer. Users can upload files privately or publicly in PowerPoint, Word, PDF, or Open Document Format (ODF).

Ask.com (http://www.ask.com/)

Ask.com is a question answering-focused search engine based in the USA. Ask.com was formerly known as Ask Jeeves. It allows users to get answers to any types of questions. When you ask a question or type a key word, it will take you to relevant websites, which provides information. It also shows you related searches.

9.3 Metasearch Engines

Metasearch engines work differently from single search engines. They enquire into multiple search engines or databases, and aggregate the results into a single list or display them according to their source. Users need to enter search items only once, and they can access information from several search engines at once. They function on the premise that as the web is too vast for any single search engine to access all the data, more meaningful search is possible using several search engines by a combined operation. This may also save the user from relying on multiple search engines. Actually, metasearch engines create a 'virtual database', instead of compiling a physical database as others do. They take a user's request, pass it to several other search engines and databases, and then compile the results homogeneously based on a specific algorithm.

Certain metasearch engines attempt to search only the most popular search engines, but some others search less popular search engines too including newsgroups and other information resources. Metasearch engines also differ in the method of presenting search results. Some of them just list results according to search engine or database, while some others return results according to relevance. However, as metasearch engines have no direct access to the database of search engines, the results may not be relevant always. The following are some prominent metasearch engines.

Dogpile	http://www.dogpile.com/
Excite	http://www.excite.com/
HotBot	https://www.hotbot.com/
Info.com	https://www.info.com/
Metacrawler	https://www.metacrawler.com/
Searx	https://searx.me/
Startpage.com	https://www.startpage.com/

(continued)

(continued)

WebCrawler	https://www.webcrawler.com/
Yippy	https://yippy.com/

9.4 Academic Search Engines

Some specialist search engines help us to mine academic and scientific contents. Note that most bibliographic databases such as CAB Direct and PubMed also have search functionality.

Google Scholar (http://scholar.google.com/)

Google Scholar is a freely accessible search engine that indexes the full text or metadata of academic literature across an array of publishing formats and disciplines. Started in 2004, Google *Scholar* searches a variety of information sources on the web, which includes journal articles, conference papers, abstracts, theses, books, and other such academic publications cutting across many disciplines and sources. It has a 'cited by' feature, through which *Google Scholar* provides access to abstracts of articles that have cited the article you are screening. This 'cited by' feature is similar to the citation indexing found in Clarivate Analytics' *Web of Science, CiteSeerX*, and Elsevier's *Scopus. Google Scholar* citations help the authors to keep track of citations to their articles. One can check who is citing your publications and compute citation indexes like *g*-index and *h*-index. Google Scholar automatically calculates and displays the individual's total citation count, *h*-index, and i10-index.

It is also possible for you to make your profile public through *Google Scholar*; when someone searches for your name, it will appear along with the results. *Google Scholar* has another useful feature, 'Related articles' through which it gives a list of closely related articles, ranked based on similarities with the original result.

Entrez (http://www.ncbi.nlm.nih.gov/gquery)

Entrez, a search engine for life sciences, is a powerful federated search engine that is designed for searching several linked databases. Federated search allows simultaneous search of multiple searchable resources. Entrez allows users to search many databases related to biomedical sciences at the National Centre for Biotechnology Information (NCBI), a part of the US National Library of Medicine (NLM). To access information from various databases, the user put up a single query supported by Boolean operators, which is then circulated to all the search engines participating in the federation. The search engine then accumulates the results received from various databases. Some textbooks are also available online through the Entrez system.

PubAg

PubAg is the search system of National Agricultural Library (NAL), USDA for agricultural information. It is free on the Internet at: https://pubag.nal.usda.gov/.

PubAg is part of the AGRICOLA family of products. PubAg contains full-text articles relevant to agriculture along with citations to peer-reviewed journal articles and links for full-text access. PubAg searching is accomplished by entering your terms in the search box and clicking the search button. They also provide search suggestions to assist searching. When multiple terms are entered with no connector, they will be combined in the search with an implicit 'AND'. Using the drop-down menu, you can narrow your search of PubAg to terms in the following fields: Title, Author, Subject, or Journal. The default setting is to search 'All Fields'.

PubAg covers most topics in agricultural sciences including nutrition, food safety, food quality, animal and crop production and protection, natural resources, sustainable agricultural systems, rural development, agricultural economics and policy issues, agricultural imports and exports, agricultural statistics, and extension education.

You may wonder how PubAg is different from AGRICOLA. AGRICOLA serves as the public catalogue of the National Agricultural Library. It contains records for all of the holdings of the Library. It also contains citations to articles. AGRICOLA also contains citations to many items that, while valuable and relevant to the agricultural sciences, are not peer-reviewed journal articles. In addition, AGRICOLA has a different interface. Therefore, while there is some overlap between the two resources, they are different in significant ways.

Microsoft Academic (https://academic.microsoft.com/)

Microsoft Academic is a free academic search engine from the Microsoft group. Apart from search results and access to sources, it also gives citation information such as number of citations, *g*-index, and *h*-index. Microsoft Academic has been posited as a competitor to Google Scholar, Web of Science, and Scopus for academic research purposes as well as citation analysis. Using Microsoft Academic, you can search within any of the domains listed in their homepage. The list of domains include agriculture science, arts and humanities, biology, chemistry, computer science, economics and business, engineering, environmental science, geoscience, material science, mathematics, medicine, physics, social science, and multidisciplinary. Some domains may have sub-categories.

BASE (http://www.base-search.net/)

Bielefeld Academic Search Engine (BASE), created and developed by Bielefeld University Library, is a multidisciplinary search engine for locating academic resources on the web. It is one of the world's big search engines for academic web resources. Resources are selected based on academic relevance. Wherever available, the search results are shown with accurate bibliographic details. There are several options for sorting the result list, and search results can be refined by author, resource, document type, language, etc. You can freely access the full texts of about 60 percent of the indexed documents.

CiteSeerX (http://citeseerx.ist.psu.edu/)

CiteSeerX functions as a search engine and digital library for scientific and academic papers with a concentration on computer and information science. It was developed by Dr Isaac Councill and Dr C. Lee Giles of Pennsylvania State University. It can actively search and retrieve both academic and scientific documents on the web. It also uses a citation index, which allow query by citations, and ranking of documents by the impact of citations.

CiteULike (http://www.citeulike.org/)

CiteULike is a free web service for managing and discovering academic references. It allows users to save and share citations to academic papers. Based on the principle of social bookmarking, the site works to promote and to develop the sharing of scientific references among researchers. Using CiteULike, scientists can also share citation information. CiteULike gives access to personal or shared bibliographies directly from the web. New entries are added as public by default, which makes them accessible to everyone. Entries can be added as private, and such entries are available to the specific user only.

PubPsych (https://www.pubpsych.eu/)

PubPsych is a free information retrieval system for resources related to psychology. It provides a comprehensive selection of resources from a growing number of international databases, which cater to the needs of both academic and professional psychologists.

WorldWideScience.org (https://worldwidescience.org/)

WorldWideScience.org is a global search engine for science, specially planned to accelerate scientific discovery and progress. It is operated by the Office of Scientific and Technical Information, a branch of the Office of Science within the US Department of Energy. It enables anyone with Internet access to launch a single-query search of national scientific databases in more than 70 countries. When somebody put a query, it hits databases from all over the world and displays both English and translated results. WorldWideScience.org uses federated searching to offer its coverage of global science and research results. It provides access to 'deep web' scientific databases, which are typically not searchable by commercial search engines.

Science.gov (https://www.science.gov/)

Science.gov is a specialized search engine, which serves as a gateway to United States government scientific and technical information and research. It uses federated search technology. Science.gov searches over 60 databases and over 2200 scientific websites to provide users with access to more than 200 million pages of authoritative federal science information including research and development results. Search results can be filtered by author, date, topic, and format (text or multimedia).

Virtual Learning Resources Centre (VLRC) (http://www.virtuallrc.com/)

The Virtual Learning Resources Centre (VLRC) or the Virtual LRC is a specialized search engine, which hosts an index to the best academic websites selected by teachers and librarians from around the globe. The site provides students and teachers with current and valid information for school and university academic projects. The Virtual LRC acts as a dedicated index of over 10,000 web pages and as a meta-search engine that collects resources from many of the best research portals and university and public library Internet subject guides selected based on the recommendations of teachers and librarians.

Infotopia (http://www.infotopia.info/)

Infotopia is an academic search engine, which pulls from results that have been curated by librarians, teachers, and other educational workers. It allows users to select a category from any of the disciplines as arts, history, literature, science/technology, and so on and then see a list of internal and external resources pertaining to the topic. In addition to the pages of Infotopia, it suggests external sites too.

ScienceOpen (https://www.scienceopen.com/)

ScienceOpen is a search engine for natural and physical sciences, social sciences, and humanities. It provides researchers a wide range of free tools to support their research. It allows relevant research in over 47 million open access articles and article records. It permits multidimensional search in millions of article records for quick orientation.

Educational Resources Information Centre (ERIC) (https://eric.ed.gov)

ERIC is operated by the US Department of Education through a formal review process and provides information on more than 1.3 million bibliographic records of journal articles, books, conference papers, technical reports, policy papers, and many types of online materials. The coverage is focused on education research materials, and the tool is aimed at individuals with an interest in education practices and researchers in education.

SciDev.Net (https://www.scidev.net/global/)

SciDev.Net is one of the leading sources of reliable and authoritative news, views, and analysis about science and technology. Although it is a part of CABI, SciDev.Net is an independent news network. SciDev.Net was founded in 2001 as a way of bridging the sizeable gap in scientific knowledge between rich and poor countries. It engages with development professionals, policymakers, researchers, the media, and the public to use science and technology for global development. The network through independent journalism helps individuals and organizations to apply science for decision-making that would result in sustainable development along with poverty reduction.

SciDev.Net has a global edition based in London, and there are five regional editions too for Latin America and Caribbean (Spanish), Middle East and North Africa (Arabic), Asia and Pacific, South Asia, and Sub-Saharan Africa (English and French).

ResearchGate (https://www.researchgate.net/)

ResearchGate is a massively popular social networking site for academicians and researchers. ResearchGate was founded in 2008 by virologist and computer scientist

Dr. Ijad Madisch with physician Dr. Sören Hofmayer and computer scientist Horst Fickenscher. It started in Boston, Massachusetts, but later moved to Berlin, Germany. As of 2018, it has more than 15 million users. Most of ResearchGate's users are from medicine, biology, agricultural sciences, engineering, computer science, and psychology. Numerous researchers submit their publications to the site for anyone to access. Although it is not a search engine per se that pulls information from external sources, RG's own collection provides a good selection of documents. You can search by publication, subject, and author, or you can even ask questions directly to researchers.

For browsing articles, registration is not required, but people who wish to become members should have an email address at a recognized institution or to be manually confirmed as a published researcher. Each member has a user profile and can upload research outputs including papers, preprints, data, chapters, patents, research proposals, methods, and presentations. Users can also ask questions, follow the activities of other users, and engage in discussions with them.

ResearchGate publishes a citation measurement in the form of an 'RG Score'. Note that RG score is not a citation impact measure. It is based on number of publications, citations, recommendations, questions asked, answers given, and followers. Exact methodology of computing RG score, however, is not known.

Directory of Open Access Journals (https://doaj.org/)

The Directory of Open Access Journals (DOAJ) is an online directory that indexes and provides access to high-quality, open access, peer-reviewed journals. All DOAJ services are free of charge, and all data are freely available. Presently, it has over 12,300 journals from 129 countries.

9.5 Citation Indexes

A citation index is an interdisciplinary index of citations between publications, allowing the reader to trace articles cited by or citing other articles. Citation indexes are helpful to track down articles and documents on similar topics by looking at the citations they have in common. These indexes are also useful to see all the works by a particular author and note the author's impact on the literature. The idea of a citation index actually began in 1955 with the publishing of an innovative paper in the much popular journal *Science* by Eugene Garfield (Garfield 1955).

Garfield was instrumental in establishing the Institute for Scientific Information (ISI) in 1960 (Thomson Reuters acquired ISI, and it was so until 2016, when they sold the Intellectual Property & Science Business to Clarivate Analytics. In 2018, Clarivate Analytics re-established ISI as part of their Scientific and Academic Research group). The *Science Citation Index* (SCI) for natural sciences was the first product of ISI. In SCI, the bibliographic information is indexed in such a way that one can search for specific articles by subject, author, source title, or geographical location or organization of the authors. Later, ISI introduced citation indexes in social sciences (*Social Sciences Citation Index*, started in 1966); and arts and humanities (*Arts and Humanities Citation Index*, started in 1977).

A larger web-based version of Science Citation Index is now available with many easy to use features, the *Science Citation Index Expanded*. Presently, all the above three indexes are accessible through the *Web of Science* collection of indexes, owned and maintained by Institute for Scientific Information (ISI), now acquired by Clarivate Analytics.

Once, citation indexing was almost the monopoly of ISI or Clarivate Analytics. Presently, there are many citation indexes; a notable one is *CiteScore* based on Scopus by Elsevier. Citation indexes allow researchers to identify articles which have cited any particular earlier article or which have cited the articles of any particular author. It is also useful to find out the articles, which have been cited most frequently.

Web of Science

Web of Science is an online mammoth subscription-based scientific citation indexing service originally produced by the Institute for Scientific Information (ISI), now maintained by Clarivate Analytics. The database has current and retrospective coverage from 1900 to the present. WoS platform consisting of several literature search databases is designed to support scientific and scholarly research. It includes 34,200 journals along with numerous books, proceedings, patents, and data sets.

WoS is now available with its core collection bundled with specialist databases. Specialist databases are with a subject focus like Medline, BIOSIS Citation Index, CAB Direct, and Zoological Record. Databases with a document type focus like Derwent Innovations Index (patents) and Data Citation Index (data sets and data studies) and databases highlighting content from regions around the world are also included. If you want to access Web of Science, you require subscription and IP authentication. Visit: https://clarivate.com/products/web-of-science/.

Web of Science Core Collection

Web of Science core collection is the premier resource on the platform, and includes over 20,000 peer-reviewed, high-quality, scholarly journals published worldwide. It also includes open access journals in more than 250 science, social sciences, and humanities disciplines. In addition, the collection contains over 190,000 conference proceedings and over 90,000 editorially selected books. It has about 1.5 billion cited reference connections.

All the journals selected for inclusion in WoS are indexed. For each paper, names of all the authors, author affiliations, abstract, keywords, funding acknowledgements including agency and grant numbers, and all the cited references are included. The Web of Science Core Collection consists of six online databases:

Science Citation Index Expanded: This is the larger web-based version of *Science Citation Index,* which covers more than 8850 major journals across 150 disciplines from 1900 to present.

Social Sciences Citation Index: This citation database covers more than 3200 journals of social sciences across 55 disciplines as well as selected items from 3500 of the world's leading scientific journals and the range is from 1900 to present.

Arts and Humanities Citation Index: More than 1700 arts and humanities journals and selected items from over 250 scientific and social sciences journals from 1975 to present are covered.

Emerging Sources Citation Index: The Emerging Sources Citation Index (ESCI) was launched as a new database within Web of Science platform. It covers more than 5000 'emerging' journals in science, social science, and humanities. All journals submitted to the Web of Science will be evaluated for the ESCI first, and if found suitable, indexed in the ESCI. Please note that journals indexed in the ESCI will not have Impact Factors, but the citations from the ESCI will be included in the citation counts for the Journal Citation Reports. Journals indexed in Emerging Sources Citation Index, based on their acceptability and citations, will be considered for impact factors later.

Conference Proceedings Citation Index: It includes more than 160,000 conference titles in Sciences starting from 1990 to the present day. It contains published reports from important conferences, symposia, seminars, colloquia, and workshops covering different disciplines.

Book Citation Index: This is a database of books, which indexes more than 80,000 editorially selected books from 2005 to present.

Two subject-specific chemistry indexes are also part of the Web of Science Core Collection. These chemistry databases are helpful for creating structure drawings, enabling users to locate chemical compounds and reactions.

Index Chemicus: This database, useful for chemical scientists, students, and industry, includes the structures and related supporting data for new organic compounds reported in leading journals. It indexes 120 organic chemistry journals to search over 2.6 m compounds from 1993 to present.

Current Chemical Reactions: This database indexes more than 350 chemistry journals and contains new synthetic methods reported in leading journals and patents from 39 issuing authorities. It covers more than one million chemical reactions from 1986 to present and INPI archives from 1840 to 1985.

All the above citation indexes are interdisciplinary, and the bibliographic information is indexed in such a way that one can search for specific articles by subject, author, source title, or geographical location or organization of the authors.

Specialist collection

Specialist databases bundled with WoS are with a subject focus. The collection includes:

BIOSIS Citation Index
BIOSIS Previews
Biological Abstracts
Zoological Record
MEDLINE
CAB Abstracts

CABI Global Health
Inspec
FSTA

Regional Collection
From 2008, the Web of Science has been hosting a number of regional citation indices. The following are regional collections.
Chinese Science Citation Database
Russian Science Citation Index
KCI Korean Journal Database
SciELO Citation Index

Data Collection
Data Citation Index

Patent Collection
Derwent Innovations Index (DII): This provides patent literature from more than 40 patent-issuing authorities.

Current Awareness
Current Contents Connect

Scopus
Scopus is an interdisciplinary bibliographic online database launched in 2004 containing abstracts and citation database as a competitor to Web of Science. Scopus is owned by Elsevier, an international publication group. Scopus covers diverse subjects coming under physical and environmental sciences, life sciences, health sciences, social sciences, engineering, business, and management. Access to Scopus is through subscription only. It covers about 36,000 journal titles from more than 11,000 international publishers in scientific, medical, technical, and social science fields, of which nearly 34,000 are peer-reviewed journals. Just like Web of Science and Google Scholar, Scopus also provides citation information.

Scopus provides you with the option of searching for publications based on search terms relating to specific parts of a document such as title, author, keywords, and ISSN. It provides some free features to non-subscribed users too, which is available through Scopus Preview. Scopus Preview provides access to Scopus Sources, but not the source comparison tool. Scopus gives four types of quality measure for each title, which are:

- *h*-Index
- CiteScore
- SCImago Journal Rank (SJR))
- Source Normalized Impact per Paper (SNIP).

Scopus also offers author profiles which cover affiliations, number of publications and their bibliographic data, references, and details on the number of citations each published document has received. For more details, visit http://www.scopus.com/.

RePEc (http://www.repec.org/)

Research Papers in Economics (RePEc)) is a collaborative effort of numerous volunteers from 99 countries to augment the dissemination of research information in economics and related disciplines. It maintains a decentralized bibliographic database of journal articles, working papers, books, book chapters, and others with the help of volunteers. RePEc provides links to about 2.6 million documents from 3000 journals and 4600 working paper series.

Indian Citation Index (http://www.indiancitationindex.com)

Indian Citation Index (ICI) started in 2009 covers peer-reviewed journals published from India. Presently, it covers more than 700 journals in major subject areas such as agricultural, medical, scientific, and technical subjects; social sciences; and arts and humanities. First of its kind in India, the citation database includes both source titles and reference information.

In addition to the above, the online resources such as Microsoft Academic, Scopus, CiteSeerX, and Google Scholar (discussed elsewhere) also provide citation information in various ways.

9.6 Citation Analysis

Most citation indexes were originally designed for information retrieval, but later they began to be used for obtaining citation details. Presently, they are also used for *bibliometrics* and other studies involving research evaluation. Citation analysis is usually done by counting how many times an article or a researcher is cited. It presumes that serious research workers and important publications are cited more frequently than non-serious ones. *Impact factor*, *CiteScore*, *Cited references*, and *Journal rating* are the principal means generally followed for this purpose.

Citation analysis is gaining more importance because of the wide dissemination of the Web of Science subscription databases in many universities and research organizations, and after the arrival of universally available free citation tools such as Scopus, Microsoft Academic, CiteSeerX, and Google Scholar.

Impact Factor

The impact factor (IF) is used as an index to signify the importance of published research papers. Impact factors are calculated for journals (*journal impact factor*) and for authors (*author impact factor*).

Young researchers must try to publish their articles in reputed journals with high rating scores such as *Journal Impact Factor* (Clarivate Analytics) or *CiteScore* (Scopus). Remember that highly acclaimed papers in reputed journals bring attention to scientists and their institutions facilitating easy funding and progress of individual scientists.

Journal Impact Factor

The Journal Impact Factor (JIF) of a journal shows the frequency with which the articles published in that journal are cited in other research journals. Originally conceived in the 1960s by Eugene Garfield, who started the Institute for Scientific Information (ISI) (now part of Clarivate Analytics), IF has gained acceptance as a numerical index of journal quality. For those journals that are indexed by Clarivate Analytics' Journal Citation Reports (JCR), annual impact factors are calculated and revised on a yearly basis.

Journal Citation Reports (JCR) is an annual publication by Clarivate Analytics. Previously, the JCR was published as a part of Science Citation Index. Presently, it is based on citations compiled from the Science Citation Index Expanded and the Social Science Citation Index. It provides information about scholarly journals in the natural sciences and social sciences, selected to have impact factors. It has been integrated with the Web of Science and is accessed from the Web of Science-Core Collections. Impact factors are calculated yearly starting from 1975 for journals listed in the Journal Citation Reports. You can access the latest Journal Citation Reports by searching 'JCR on the web'.

A complete list of journal titles covered in the 2018 edition of JCR is available at: https://clarivate.com/wp-content/uploads/2018/06/Crv_JCR_Full-MarketingList_A4_2018_v4.pdf.

An additional list of journals for the first time listing in the JCR is also available at: https://clarivate.com/wp-content/uploads/2018/06/Crv_JCR_FirstImpact-Factor-List_2018_A4_v3.pdf.

Started as an offshoot of the Science Citation Index, JIF provided a distinctive means of rating journals based on their citations and rapidly became the mantra of journal quality. JIF for a journal is computed based on a three-year period by dividing the total number of current year citations with the source articles published in that journal during the preceding two years. Let us take an example. We want to calculate IF for 2018. Let 'x' be the number of times articles published in 2016 and 2017 were cited in indexed journals during 2018, and 'y' the number of articles and others published in 2016 and 2017. Then, the IF for 2018 would be x/y. Please note that the impact factor of a journal for 2018 can be published only in 2019, as it is impossible to assess the journal until all the 2018 issues come out. You can access the latest list of journals with total citations, journal impact factors, and Eigen factor scores from the web.

JIF is often employed to show the relative importance of a journal in a particular field. Many scientists believe that impact factor is a reasonable indicator of journal quality. For example, journals with higher impact factors are regarded as more important and prestigious than others having lower value. It is also helpful for librarians in selecting journals for library collections. The use of IF as an index of journal quality relies on the assumption that citation frequency precisely measures a journal's importance to its end users. By referring and citing articles from a particular journal in their own papers, research workers are, in fact, approving and supporting that journal.

Some shortcomings of 'journal impact factor' must also be noted. The journal impact factor applies only to a journal, but not to individual journal articles or individual scientists. If you want to assess the author's impact, you have to use 'author impact factor' specifically. Journal impact factor applies to those journals that are indexed by *Journal Citation Reports* (JCR). Naturally, journals not covered by JCR will not have an impact factor. Compared to general scientific journals, specialist journals get lesser citations only. Similarly, among specialist journals, some attract several research workers and citations, because of the nature of that discipline, for example, biotechnology. Certain disciplines may have fewer research workers, and the impact factor for journals in such disciplines would be low. To solve these kinds of problems, some organizations have come up with journal rating based on some specific criteria, for example, the NAAS rating followed in India for agriculture and allied disciplines (discussed elsewhere).

CiteScore

CiteScore is a new journal metric launched by Elsevier in 2016 as an alternative to the monopolistic 'Journal Impact Factor' of Clarivate Analytics. The two companies already have competing bibliographical citation databases in Scopus (Elsevier) and the Web of Science (Clarivate Analytics). For a long time, there was no competition for the Journal Impact Factor. As the Impact Factor is derived from journals indexed in the Web of Science, other journals cannot have an Impact Factor. Elsevier launched CiteScore to counter the JIF of Clarivate Analytics. You can freely access CiteScore on the Scopus Journal Metrics website. At the same time, JCR is a paid subscription. Elsevier also claims that CiteScore is calculated from the list of Scopus journals, which is much larger than the Web of Science list.

CiteScore is computed for a publication as an average of the sum of citations it received in a particular year to documents published in the immediate three previous years divided by the total publications in the same previous three years. Use the following link to find out whether your preferred journal is Scopus indexed. You can also note down 'CiteScore' and other bibliometric values from the site, https://www.scopus.com/sources.

SCImago Journal Rank (SJR)

SCImago Journal Rank (SJR) is a measure of the influence of academic journals that accounts for both the number of citations received by a journal and the importance or prestige of the journals where such citations come from. A journal's SJR is a numeric value indicating the average number of weighted citations received during a selected year per document published in that journal during the previous three years. Higher SJR values are meant to indicate greater journal prestige. SJR accounts for journal size by averaging across recent publications, which is calculated annually. SCImago Journal Rank is powered by inputs from Scopus data and is freely available at https://www.scopus.com/sources.

Source Normalized Impact Per Paper (SNIP)

This is also a Scopus initiated metric. Source Normalized Impact per Paper (SNIP) is an index that accounts for field-specific differences in citation practices. This task is done by relating each journal's citations per publication with the citation potential of its field (the set of publications citing that journal). In other words, SNIP tries to measure contextual citation impact, thus enabling direct comparison of journals in different subject fields. This is necessitated because of the fact that the value of a single citation is greater for journals in fields where citations are less likely, and vice versa. SNIP is calculated annually from Scopus data and is freely available along with CiteScore and SJR at www.scopus.com/sources.

Eigenfactor

Eigenfactor is intended to reflect the influence and prestige of journals. It is a score value based on the total importance of a scholarly journal. Journals are rated according to the number of incoming citations, with citations from highly ranked journals poised to make a larger contribution to the Eigenfactor than those from poorly ranked journals. Nevertheless, the Eigenfactor score is influenced by the size of the journal; therefore, the score doubles when the journal doubles in size, often measured as number of published articles per year. Eigenfactor scores are calculated by eigenfactor.org (http://www.eigenfactor.org/).

NAAS Rating of Journals

Many journals in India do not come under the impact factor scheme of Clarivate Analytics. The National Academy of Agricultural Sciences (NAAS) has recognized this fact and came up with a scheme for agriculture and related disciplines. According to the scheme, each journal is assessed, and its relative merit and standing among researchers is indicated by giving points out of a maximum of 20. Before assigning points, NAAS first ascertain whether the journal is already having an 'Impact Factor' of Clarivate Analytics. For those journals where Journal Impact Factor is available, the scores are assigned as 6.00 + Impact Factor up to 20.00 (category I), and for those journals where Clarivate Analytics Impact Factor is not available, the rating is from 1 to 6.00 only (category II). The academy periodically revises its rating. The rating is available from http://www.naasindia.org/rating.html.

Author Impact Factor

Journal impact factor considers only the impact of particular journals using the mean number of times published articles are cited during the two years after their publication. It is likely that one article in the journal might have been highly cited and another article hardly at all, but both the authors are judged equally based on the high impact factor of that journal. To overcome this problem, 'author impact factors' have been suggested, which measure the impact of individual authors. The author impact factor is based on the scientific value of a given researcher or author. You can try the *h*-index, *g*-index or compile cited references by using Web of Science, Google Scholar, Scopus, or Microsoft Academic.

***h*-index**

We know that total number of papers published by an individual may not reflect the quality of his/her scientific activity, while total number of citations can be disproportionately affected because of a single article of major influence. The *h*-index has been proposed to measure concurrently the quality and quantity of scientific output of researchers, by computing an index based on the highly cited papers of the author and the number of citations these papers have received from other publications.

The *h*-index was proposed by Jorge E. Hirsch, a physicist at the University of Carolina, San Diego, to measure concurrently the quality and quantity of scientific output of a scientist based on the highly cited papers of the author and the number of citations these papers have received from other articles. It was originally meant as a tool for determining theoretical physicists' relative quality (Hirsch 2005), which is sometimes also called the Hirsch index or Hirsch number. The *h* in *h*-index stands for Hirsch and 'highly cited'.

The *h*-index is computed based on the highly cited papers of the author and the number of citations these papers have received from other articles. The *h*-index gives a measure of the number of articles (*h*) of an author that have received at least (*h*) citations up to that time. For instance, an *h*-index of 6 would mean that an author has 6 articles that have received at least 6 citations each. As the *h*-index attempts to measure a researcher's impact over time, the longer a researcher has been publishing, the higher the *h*-index would be! Naturally, established and veteran researchers have better chance to achieve higher *h*-indices!

For calculating *h*-index, several information resources such as *Web of Science, Scopus, Google Scholar, Microsoft Academic,* and *ResearchGate* are used, which include the necessary citation data. Please note that because of differential coverage, each database is likely to produce a different *h*-index for the same scholar. In addition, you can also use 'Publish or Perish' database to determine your own *h*-index (https://harzing.com/resources/publish-or-perish).

i10-Index

i10-index has been created by Google Scholar as an alternate author level metric to *h*-index. It is a whole number indicating the number of publications of a particular scientist with at least 10 citations. This simple index is used only by Google Scholar as a measure to gauge the productivity of a scholar.

***g*-index**

Similar to the *h*-index, the *g*-index is also used for quantifying scientific productivity of a scientist, taking into consideration his/her publication record. It was suggested by Leo Egghe in 2006 (Egghe 2006). It is also a simple metric similar to *h*-index in its functionality, but *g*-index also takes into consideration the performance of the highly rated articles by the author.

The *g*-index is determined depending on the distribution of citations received by the publications of a particular author. A particular author's *g*-index would be *g*, if his/her number of highly cited articles is *g*, and that each of them has brought a mean of *g* citations. For example, a *g*-index of 20 would mean that an author who published 20 articles had an average of 20 citations for each article. Although this is similar to the *h*-index, the number of citations for a single article is not critical unlike *h*-index. The *g*-index attempts to address limitations of the *h*-index, and *g* is expected to be highly correlated with the total number of citations an author has received. At the same time, *h*-index is in good correlation with the highest number of citations. Consequently, *g*-index is found to be always a greater number than *h*-index.

Cited References

'Cited references' is a feature in certain online databases such as Google Scholar. This feature helps us to locate other publications that are related by topic or subject to the original publication. Cited references or the references that cite an individual article can be used to gauge the utility and impact of a cited publication. A frequently pointed out drawback is that sometimes author self-citing or by publishing in an open access journal can defeat the usefulness of 'cited references'.

'Cited references' search is one of the features in the Web of Science database. The number indicated in the citing articles column of Web of Science shows the number of times the reference has been cited in Web of Science until then. Remember that 'cited references' in WoS may not include all the known citations of the paper, but only those appeared in journals covered by WoS.

As discussed earlier, Google Scholar is a freely accessible search engine for academic literature that searches a variety of information sources on the web. Each Google Scholar search result, in addition to showing bibliographic details, indicates a 'cited by' link, displaying a list of documents that have cited the viewing document, originally retrieved in the search. Note that this includes resources indexed by Google Scholar only. Microsoft Academic and Scopus also feature 'cited references'.

9.7 Online Searching Methods

When searching for information from databases such as CABDirect, PubMed, or other online databases, problems may arise in matching the key words we used with what is stored in the databases or search engines. Use any of the approved techniques to narrow your search and find out relevant literature quickly. Two widely used methods are phrase searching and Boolean searching.

Phrase Searching

In this case, we use a proper name or distinct phrase enclosed in double quotations, for example: 'watershed management in India' or 'web search engines'. Before you start searching, think critically the key phrases, proper names or organization, which you want to search. When the search query is put in double quotations, you are asking the search engine to use all of these words, in this specific order.

Boolean Searching

Boolean searching allows the user to have a meaningful search by combining words and phrases using the specific words, AND, OR, and NOT (also known as 'Boolean operators' or 'search connectors'). Use of these words helps us to limit, widen, or define the search. Most online databases, Internet search engines, and web directories support Boolean searches. For a good web search, you should know how to use basic Boolean operators. Boolean search operates based on a method of symbolic logic, developed by **George Boole**, an English mathematician. Boolean search techniques allow the user to conduct effective searches culling out many unconnected or irrelevant documents.

There are two options, you can either use the standard Boolean operators (AND, OR, or NOT) or their equivalents using math symbols. The user can decide the method. For example, the Boolean search operator AND is equal to the '+' symbol; NOT is equal to the '—' symbol; and OR is the default setting of any search engine, returning automatically all the words you type in.

Suppose that we want to find documents about 'resistance' and 'tolerance'. To get a general idea of what is available on the web, we perform three keyword searches before focusing on some aspect of resistance or tolerance as shown below.

- Resistance AND tolerance: This search finds articles that compare resistance and tolerance. By using AND, you can 'narrow' a search by combining terms, and it will retrieve documents that use both the search terms you specify.
- Resistance OR tolerance: This search finds articles that discuss either resistance or tolerance. When OR is used, it 'broadens' a search to include results that contain either of the words you type in.
- Resistance NOT tolerance: In this case, the search finds articles that are exclusively on resistance. That is, using NOT will narrow a search by excluding certain related search terms, allowing you to retrieve documents that contain one, but not the other of the search terms you enter.

Implied AND

Many databases and search engines such as Google and many others put AND between words by default. Implied AND means that a search for 'water harvesting' will retrieve 'water AND harvesting' and not the phrase water harvesting. Some web search engines have an implied OR rather than AND.

Nested Searches

Nested searches combine the use of AND and OR in the same search statement. In such cases, we use parenthesis instead of search connectors. In certain cases, a focused search is useful, which combines several related words for one idea with another word or words, as in the search: (rehabilitation or conservation) and watershed. In this search, we used Boolean OR to combine the related words: rehabilitation or conservation. Then we used Boolean AND to combine these words with watershed. Parenthesis indicates relationships between search terms, and it ensures that we get the expected results from our searches. It forces the computer to process our search terms in the order we intended and to combine them in the way we want.

Truncated Searches

In certain situations, we may search for part of a word or a whole word that may have similar beginnings or endings. In such cases, by using a 'wildcard', commonly an asterisk (*) (some search engines use other symbols), we can improve search results and reduce searching time. For example, to find engine, engines, engineer, and engineering, use engine*.

9.8 Centre for Agriculture and Biosciences International (CABI)

The 'Centre for Agriculture and Biosciences International' (CABI), previously known by the name 'Commonwealth Agricultural Bureaux' (CAB) is an international, not-for-profit inter-Governmental development and information organization established by a United Nations treaty level agreement between 49 member countries including India. CABI's stated mission is 'to improve people's lives worldwide by providing information and applying scientific expertise to solve problems in agriculture and the environment'. The head office of CABI is in Wallingford, England.

The history of CABI can be traced to 1910, when an Entomological Research Committee has been formed. Later, in 1930, it developed into a Commonwealth organization called the Imperial Agricultural Bureaux. The bureaux was renamed in 1947 as the Commonwealth Agricultural Bureaux (CAB) by combining various activities. In 1986, CAB became CAB International. CABI has three main areas of operation—scientific publishing, development projects and research, and microbial services.

A main product of CABI is CAB Abstracts, the world-leading bibliographic database covering agriculture and the environment. CABI Abstracts are popular and used by scientists from agriculture and life sciences in a big way. CABI abstracts covers a vast area of research literature in agriculture and allied fields such as forestry, veterinary and animals sciences, biotechnology, environmental sciences, and nutrition. A similar database, which focuses on public health and health science, is Global Health. Presently, CABI Abstracts and Global Health are available online through CAB Direct (Sect. 9.1). CABI also publishes many other products such as books, e-books, multimedia compendia, and full-text electronic resources with the objective of further promoting science and its practical application. For more information, visit the site https://www.cabi.org/.

9.9 The Indian Academy of Sciences

The Indian Academy of Sciences was formed in 1934 as a society with the main aim of promoting and upholding the cause of science both in pure and applied branches. The major activities include publication of scientific journals, organizing conferences, seminars, symposia, and discussions on important topics. The Academy also promotes scientific talent by recognising their merits through various means such as awards. It also strives for improvement of science education in the country besides taking up other issues of concern to the scientific community. Publication of scientific journals and other literature is an important activity of the Academy.

The Academy's journals are 'open access', and full text is available as PDF files on each journal's website. The Academy currently publishes 12 journals covering all major disciplines in science and technology. The journals are:

Bulletin of Materials Science

Current Science (published by the Current Science Association in collaboration with the Academy)

Dialogue: Science, Scientist and Society

Journal of Astrophysics, & Astronomy

Journal of Biosciences

Journal of Chemical Sciences

Journal of Earth System Science

Journal of Genetics

Pramana (Journal of Physics)

Proceedings-Mathematical Sciences

Resonance (Journal of Science Education)

Sadhana (Academy Proceedings in Engineering Sciences)

In addition to regular journals, the Academy also brings out special publications of topical interest frequently. All these journals are available online in the open access mode from http://www.ias.ac.in/.

9.10 National Institute of Science Communication and Information Resources (NISCAIR)

National Institute of Science Communication and Information Resources (NISCAIR) is an important information provider in India functioning under the Council of Scientific and Industrial Research (CSIR).

The National Institute of Science Communication (NISCOM), a leading institute of the CSIR had been in existence for the last six decades. During the period, NISCOM diversified its activities through a multitude of its information products such as research journals, popular science magazines, books, monographs, encyclopaedia, and other information services. Another prestigious institute of the CSIR, the Indian National Scientific Documentation Centre (INSDOC), was engaged in providing information and documentation services in science and technology through various means such as abstracting and indexing literature, design and development of databases, and translation services. It was also providing access to international information sources, in addition to other activities like human resource development. INSDOC was also functioning as the host to the National Science Library (NSL) and the South Asian Association for Regional Cooperation (SAARC) Documentation Centre. In 2002, by merging NISCOM and INSDOC, the National Institute of Science Communication and Information Resources (NISCAIR) was created.

With the formation of NISCAIR, all the activities of NISCOM and INSDOC have been combined, elevating NISCAIR as an institute capable of undertaking new missions in the field of communication, dissemination, and information management systems and services in science and technology. The main functions of NISCAIR include collection, storage, publishing, and dissemination of information related to science and technology in India through a mix of traditional and modern means.

Currently, NISCAIR brings out 18 primary and 2 secondary scientific journals. The primary journals are:

Annals of Library and Information Studies
Bharatiya Vaigyanik Evam Audyogik Anusandhan Patrika.
Indian Journal of Biochemistry and Biophysics
Indian Journal of Biotechnology
Indian Journal of Chemical Technology
Indian Journal of Chemistry-Sec A
Indian Journal of Chemistry-Sec B
Indian Journal of Engineering & Materials Sciences
Indian Journal of Experimental Biology
Indian Journal of Fibre & Textile Research
Indian Journal of Geo-Marine Sciences
Indian Journal of Natural Products and Resources
Indian Journal of Pure and Applied Physics
Indian Journal of Radio and Space Physics
Indian Journal of Traditional Knowledge
Journal of Intellectual Property Rights
Journal of Scientific Temper
Journal of Scientific and Industrial Research

The institute brings out two abstracting journals, Indian Science Abstracts (ISA) and Medicinal and Aromatic Plants Abstracts (MAPA). Science Reporter is a popular magazine in English brought out by NISCAIR.

The Indian Science Abstracts (ISA) published fortnightly, which has been in existence from 1961, covers scientific works done in India. The ISA is available online from 2000 onwards, which can be accessed by visiting the website http://isa.niscair.res.in/. It is searchable by document type, title of article, journal name, subject, author name, keyword, and geographic term.

The Medicinal and Aromatic Plants Abstracts (MAPA), being published as a bimonthly, covers research works from around the world on agronomy, botany, pharmacognosy, and biological activities of medicinal and aromatic plants. Advanced search and full-text search facilities are presently available online for MAPA (www.niscair.res.in).

Another major activity of NISCAIR is translation of science and technology documents from foreign languages into English. Currently, the institute undertakes translation from 20 major foreign languages. You can utilize the facilities available at NISCAIR by visiting the website http://www.niscair.res.in.

9.11 ICAR Initiatives for Knowledge Management

The Indian Council of Agricultural Research (ICAR) has established a 'Directorate of Knowledge Management in Agriculture' for dissemination and sharing of agricultural knowledge and information through information products in print, electronic, and web mode. The books published by ICAR, especially those in the handbook series, are very popular among researchers and students. The ICAR brings out three research journals,

Indian Journal of Agricultural Sciences

Indian Journal of Animal Sciences (both monthlies)

The Indian Journal of Fisheries (quarterly)

The ICAR also brings out two popular semi-technical magazines, 'Indian Farming' and 'Indian Horticulture'.

All the above five journals are available in open-access mode. In addition to these journals, the ICAR supports many other open access journals. You can access all these journals from the site, http://epubs.icar.org.in/ejournal/.

Krishikosh (http://krishikosh.egranth.ac.in/)

Krishikosh is a digital database of accrued knowledge in agriculture and allied sciences with a huge collection of valuable books, old journals, thesis, research articles, popular articles, monographs, catalogues, conference proceedings, success stories, case studies, annual reports, newsletters, pamphlets, brochures, bulletins and other grey literatures spread all over the country in different ICAR Research Institutions and State Agricultural Universities (SAUs).

Under the ICAR's open access policy, it is mandatory to upload all Institutional publications available with the SAUs in the Krishikosh. Along with this, full texts of M.Sc. and Ph.D. theses and summary of completed research projects are submitted in the Krishikosh repository when the work is completed. The metadata, for example, title, abstract, authors, publisher, etc., are freely accessible from the time of deposition, but their free and unrestricted use through open access can be made after an embargo period not more than 12 months.

Currently Krishikosh has more than 83,000 articles items, which include more than 35,000 Ph.D. and M.Sc. theses submitted by various SAUs and ICAR deemed universities with full-text searchable facilities.

Consortium for e-Resources in Agriculture (CeRA)

The ICAR has established the *Consortium for e-Resources in Agriculture* (CeRA) in 2007 under the National Agricultural Innovation Project (NAIP) of ICAR for providing online access to e-journals and resources in libraries attached to ICAR institutes and State Agricultural Universities (SAU's). The Indian Agricultural Research Institute (IARI), New Delhi, functions as the host for the entire programme. The network connectivity already created across various ICAR institutes and SAUs made CeRA a reality, and through this facility, selected journals are made available for researchers and students avoiding duplication and wastage of money.

At present, more than 2900 journals related to agriculture and allied sciences from selected publishers are accessible through CeRA. Full text of articles from journals marked full text (FT) can be accessed. Provision has also been made to access an article from a journal, which is not subscribed under the consortium but available in the library of a particular institution. If any researcher wants articles from such journals not covered under CeRA, the 'Document Delivery Request' (DDR) system can be used. Those journals marked 'Request Article' (RA)comes under this group. When you click on a journal marked RA, you will be shown a link 'Request the Article', which is meant for this purpose. For obtaining such articles through DDR, the user has to submit a DDR form. Please note that individuals cannot join CeRA as a member and access the information, but CeRA is for an institution in NARS. If your institution is a member of CeRA, you can access information through the institutional route. Internet Protocol (IP) authentication is required for the institution. A web-based application was developed for facilitating online access through CeRA, using the URL: www.cera.jccc.in. For more information on CeRA, visit the site http://cera.iari.res.in/index.php/en/.

Bibliography

Barrass, R. 2002. Scientists Must Write: A Guide to Better Writing for Scientists, Engineers and Students (2nd Ed.). Routledge, London, 204p.

Bloom, D.F., Karp, J.D. and Cohen, N. 1998. The Ph.D. Process. A Students Guide to Graduate School in the Sciences. Oxford University Press, Oxford, 209p.

Bond, A. 2007. Your Masters Thesis: How to Plan, Draft, Write and Revise. Viva Books, New Delhi, 135p.

Egghe, L. 2006. Theory and practice of the g-index. Scientometrics 69(1):131–152.

Garfield, E. 1955. Citation indexes for science. Science 122 (3159): 108–111. https://doi.org/10.1126/science.122.3159.108.pmid14385826.

Garfield, E. 1999. Journal impact factor: a brief review. Can. Med. Assoc. J. 161(8):979–980.

Hirsch, J.E. 2005. An index to quantify an individual's scientific research output. Proc. Natl. Acad. Sci. 102(46):16569–16572.

Horowitz, L.1986. A Writer's Guide to Research. Writer's Digest Books, Cincinnati, Ohio, 135p.

Hunt, A. 2005. Your Research Project: How to Manage It. Routledge, London, 147p.

Leedy, P.D. and Ormrod, J.L. 2010. Practical Research: Planning and Design (9th Ed.), Pearson Education, New Jersey, 360p.

Lenburg, J. 2007. Guide to Research. Viva Books, New Delhi, 560p.

Lester, J.D. and Lester, J.D. Jr. 2009. Writing Research Papers: A Complete Guide (13th Ed.), Longman, 416p.

Saha, S., Saint, S. and Christakis, A.D. 2003. Impact factor: a valid measure of journal quality? J. Med. Libr. Assoc. 91(1): 42–46.

Walliman, N. 2011. Your Research Project: Designing and Planning Your Work (3rd Ed.). Sage Publications, New Delhi, 378p.

Weblinks

AGRIS - http://agris.fao.org/.
AGROVOC - http://aims.fao.org/standards/agrovoc/.
ArXiv.org - https://arxiv.org/.
Ask.com - http://www.ask.com/.
BASE - http://www.base-search.net/.
Bing - http://www.bing.com/.
BioRxiv - https://www.biorxiv.org/.
CAB Direct - http://www.cabdirect.org/.
CAB Thesaurus - https://www.cabi.org/cabthesaurus/.
CABI - https://www.cabi.org/.
CeRA - http://cera.iari.res.in/index.php/en/.
CeRA web-based application - www.cera.jccc.in/.
Chemical Abstracts - http://www.cas.org/.
CINAHL - https://health.ebsco.com/products/the-cinahl-database.
CiteSeerX - http://citeseerx.ist.psu.edu/.
CiteULike - http://www.citeulike.org/.
Clarivate Analytics - https://clarivate.com/.
DOAJ - https://doaj.org/.
Dogpile - http://www.dogpile.com/.
DuckDuckGo - https://duckduckgo.com/.
eigenfactor.org - http://www.eigenfactor.org/.
Elsevier - https://www.elsevier.com/en-in/.
EMBASE - https://www.elsevier.com/.
Entrez - http://www.ncbi.nlm.nih.gov/gquery.
ERIC - https://eric.ed.gov/.
Excite - http://www.excite.com/.
FSTA - https://www.ifis.org/fsta.
Gibiru - http://www.gibiru.com/.
Google - http://www.google.com/.
Google Scholar - http://scholar.google.com/.
HotBot - https://www.hotbot.com/.
Indian Academy of Sciences - http://www.ias.ac.in/.
ICAR - https://icar.org.in/.
ICAR e-journals - http://epubs.icar.org.in/ejournal/.
Indian Citation Index - http://www.indiancitationindex.com/.
Info.com - https://www.info.com/.
Infotopia - http://www.infotopia.info/.
Inspec - https://www.theiet.org/.
JCR (regular) - https://clarivate.com/wp-content/uploads/2018/06/Crv_JCR_FullMarketing-List_A4_2018_v4.pdf.
JCR(first IF) - https://clarivate.com/wp-content/uploads/2018/06/CrvJCR_First-Impact-Factor-List_2018_A4_v3.pdf.
Krishikosh - http://krishikosh.egranth.ac.in/.
Master journal list - http://mjl.clarivate.com/.
MEDLINE - https://www.nlm.nih.gov/bsd/medline.html.

MedlinePlus - http://www.nlm.nih.gov/medlineplus/.
MeSH - http://www.ncbi.nlm.nih.gov/mesh.
Metacrawler - https://www.metacrawler.com/.
Microsoft Academic - https://academic.microsoft.com/.
NAAS rating - http://www.naasindia.org/rating.html.
NAL website - http://agricola.nal.usda.gov/.
NALT - https://agclass.nal.usda.gov/.
NISCAIR - www.niscair.res.in/.
PubAg - https://pubag.nal.usda.gov/.
PubChem - http://pubchem.ncbi.nlm.nih.gov/.
Publish or perish - https://harzing.com/resources/publish-or-perish.
PubMed - http://www.ncbi.nlm.nih.gov/pubmed/.
PubMed Central - http://www.ncbi.nlm.nih.gov/pmc/.
PubPsych - https://www.pubpsych.eu/.
RePEc - http://www.repec.org/.
ResearchGate - https://www.researchgate.net/.
SciDev.Net - https://www.scidev.net/global/.
Science.gov - https://www.science.gov/.
ScienceDirect - http://www.sciencedirect.com/.
ScienceOpen - https://www.scienceopen.com/.
Scopus - http://www.scopus.com/.
Searx - https://searx.me/.
Slideshare - https://www.slideshare.net/.
Startpage.com - https://www.startpage.com/.
Swisscows - https://swisscows.com/.
TOXLINE - http://toxnet.nlm.nih.gov/.
Virtual LRC - http://www.virtuallrc.com/.
Web of Science - https://clarivate.com/products/web-of-science/.
WebCrawler - https://www.webcrawler.com/.
Wiki.com - http://wiki.com/.
Wikipedia - https://www.wikipedia.org/.
WorldWideScience.org - https://worldwidescience.org/.
Yahoo! - http://www.yahoo.com/.
Yandex - https://yandex.com/.
Yippy - https://yippy.com/.

Chapter 10
The Literature Review

One can measure the importance of a scientific work by the number of earlier publications rendered superfluous by it.
David Hilbert (1862–1943), German mathematician

The whole body of writings on a particular subject produced by scholars is broadly called its *literature*. A 'review of literature' is a critical and evaluative description of publications on a chosen topic. Literature reviews are integral parts of research proposals, theses, and research papers. Exclusive literature reviews are also published in review series of publications such as annual reviews of reputed publishers. Sometimes, one or two review articles, mostly invited, may find a place in primary journals too. In a thesis, literature review is an important chapter, providing the background and justification for the research undertaken. A good review should show the knowledge that exists and the gaps in knowledge related to the research work undertaken. Research students should master the art of reviewing the literature critically and objectively.

A literature review is different from an *annotated bibliography*. An annotated bibliography is a bibliography giving the summary of different sources. An *annotation* is a short explanatory note about the contents of a source, and it gives the reader a summary along with an evaluation of the source. At the same time, a literature review provides a narrative using the references in the annotated bibliography. A literature review is organized around ideas and not the sources themselves in contrast to an annotated bibliography.

10.1 Importance of the Literature Review

The literature review can be time consuming and demanding, but it is an essential part of the research process. The primary purpose of literature review is to demonstrate that the author has extensively surveyed the literature of a particular subject and critically evaluated its quality by classifying and summarizing the findings in comparison with previous research works and literature reviews.

C. G. Thomas, *Research Methodology and Scientific Writing*,
https://doi.org/10.1007/978-3-030-64865-7_10

A critical review of literature is essential before embarking on a research project because the process of reviewing the literature helps you to understand the subject area better. The literature survey plays an important role in shaping your research problem. It helps you to find out whether it has already been done by somebody; and if so, to what extent it was successful. Sometimes, only a part of a broader subject area may have been investigated, and there may still have several unexplored areas. The review also helps you to acquaint with the methodologies used by peers in the same field and the problems or pitfalls, if any, while using those methodologies. Thus, reviewing the literature brings clarity and focus to your research problem, improves your methodology, and broadens your knowledge base. This is the main reason of compulsory inclusion of a chapter titled 'Review of Literature' in theses and dissertations. You have to continue the literature survey started before finalizing the research problem until the report is finished. This is important to properly 'discuss' your results.

While writing a literature review, it is not enough to narrate simply what others have discovered, invented, or explained. The works should be reviewed with a critical mind, and it must show the facts and ideas so far generated in a subject along with their strengths and weaknesses. A good review must have a clear focus, and be presented in a logical and well-organized style. The objectives of conducting a literature survey can be summarized as follows (Hartley 2008; Randolph 2009; Kumar 2011):

1. to understand the current level of knowledge on the chosen subject
2. to identify seminal works in the proposed subject area
3. to identify shortcomings in the existing knowledge base on a subject
4. to see whether the work has already been investigated
5. to identify opposing views
6. to carry on from where others have already reached
7. to identify other scientists working in the same field
8. to find out how a research project is related to the works already done by others
9. to identify the methodologies and research techniques that have been used
10. to relate new findings to previous findings in the discussion section of a thesis or research paper.

These objectives can be fulfilled only by reading as much as you can! Find and read on all the information sources related to the proposed research topics. The preliminary review of literature concentrates more on general textbooks and existing reviews of previous works that summarize the state of knowledge on a particular topic. Textbooks and reviews are useful as they combine information from many primary sources into a single book. If the literature review is part of a thesis, the review must be comprehensive covering all the research works done on the topic. When it forms part of a research paper, it must cover all the recent major works that has been done on the topic, but it is not necessary to include all the research works on the subject.

10.2 Planning a Literature Search

Identifying and locating relevant literature for writing a review is a formidable task. You may use indexes, abstracts, databases, online sources, and many other devices to identify suitable and appropriate literature. Researchers should develop a search strategy to locate relevant literature in their subject of interest. Consider the following aspects before finalizing a search strategy.

Determine the Scope of Review

Before you start searching for the literature, define your research question appropriately. It is important to assess what you already know about the area of search. The search is influenced by the purpose. Sometimes, the information may be a central element of the topic or background information, for example, for writing the 'introduction' part of a research paper or thesis or just additional information to carry on the research. If the search is for writing a review paper, you have to undertake the search on a wider platform. You must have clarity on what you want to study and review.

Decide the Period to be Covered

Most researchers and students try to access a wide range of literature in their area of concern. It is also important that the references should reflect recent findings and developments in one's field of interest. Therefore, before embarking on the search, decide whether you need everything ever written on the topic or in the last 20 years, or just the last 5 years, or a similar period. If your intention is model building or forecast, you may probably need to look back over a long period.

Settle on the Range

You may be trying to gather information pertaining to a local practice or with an international appeal. Appraise yourself whether the information needed is of regional, national, or international importance. You have to fine-tune your search strategy based on the range. For national or internationally important information, the search has to be broadened, especially concentrating on sources from international publishers.

Identify the Key Concepts and Key Words

Identifying the key concepts involved in the study is an important step. Similarly, keywords are the basic units of any literature search. Using a list of keywords or key phrases, you will be able to search the literature easily and effectively (Hartley 2008). If you are clear about the concepts, you can easily find out the major key words. Check for other words such as synonyms and variations in spelling, which could be used as alternate key words. You may also try alternative phrases for concepts and ideas.

Use the Library Effectively

For a good literature review, you should know how to use the library effectively (Chap. 8). Your efforts as a reviewer should be to develop the skill needed to conduct exhaustive bibliographic searches, especially by using various reference resources, indexes, and abstracts. You should also have the ability to organize the collected information meaningfully to describe, critique and relate each source to the subject of the inquiry; to present the organized review logically; and to cite correctly all sources mentioned using the recommended style of reference citation (Chap. 15).

Use Various Types of Information

You can access several kinds of information sources in both print form and electronic form. Determine the type of publications (e.g., journals, books, government documents, directories, magazines, and websites) required for information retrieval. You may find it helpful to consult a subject dictionary, encyclopaedia, or glossary for the common terms coming under the subject area. The use of a thesaurus is also advisable to define useful terms. For general theoretical information, you may consult textbooks; for current analysis and comments, journals; and for names of contacts, directories.

In the category of *books*, several secondary sources of information are included. Textbooks, handbooks, subject review books, manuals, monographs, treatises, and yearbook are important examples of books. *Reference books* such as dictionary, encyclopaedia, directory, and gazetteer are also books. Often, books are more accessible to the reader than journals and magazines. However, books may not include recent developments, because of the long preparation time involved in its making. Reliability of books is another issue. This is especially so with the books written by lobbyists and activists who are actually not real experts in the fields. For certain post-modern movements, peer-reviewed research reports may be hardly available. However, you may find umpteen number of books based on anecdotes and invalid proofs. Sometimes, the book could be sensationalized to enhance the sale volume.

The category of sources of information called *journals* is subject-specific periodicals brought out frequently. Journals that publish the results of original research are often called *primary journals*. Since primary journals are published periodically, they could provide information that is more recent compared to textbooks or reviews. They carry the latest and often very specific account of current research work, new techniques, and unusual and interesting cases. Before a paper is accepted for publication in a journal, it should go through a rigorous process of quality evaluation, called *peer reviewing*. Peer reviewing ensures that the paper is read, analyzed, commented, and judged by more than one expert in the field of study for appropriateness in the subject and quality. Research papers published through journals are well respected because of the reviewing process. Since the papers contain extensive cross-referencing, they serve as an excellent introduction to other references that you may not have come across.

The journals that concentrate on publishing review articles or teaching articles are usually called *secondary journals*. Annual review series being published by many organizations are very helpful in this regard. They usually cover up-to-date work in a new perspective. Reading good reviews strengthens one's broad areas of interest. They also serve as an excellent introduction to other references in the field.

Use Online Databases Efficiently

There are several sources designed to make your search for articles, and these can save valuable time. These include indexing journals, abstracting journals, citation indexes, and several web-based sources including search engines. All these can be accessed online now. For a comprehensive account on information sources and website addresses, refer Chap. 9.

10.3 Locating Relevant Literature

Information available in a number of formats can be accessed from different sources. However, it is important to understand the significance of various formats so that researchers know what is more suited to their information requirements.

When you plan, you should aim for a comprehensive literature search.

Select Appropriate Tools

There are several methods for locating relevant literature. These include searching through library catalogues, abstracts, indexes, general subject guides, CD-ROM databases, search engines, online databases such as CAB Direct and PubMed, checking library links on the web, and consulting many other Internet resources (more details in Chaps. 8 and 9).

Trace Relevant Publications

Using the selected tools, search through the subject indexes to trace relevant articles. You may use appropriate key words for an effective search. Use synonyms and alternative terms of key words also. Try to find out as many reference sources as possible on the topic. However, if you find too many articles, use keywords that are more specific. If you are not able to find appreciable number of references, you have to change your search strategies. Broaden your search using more general terms. Still, if you are not able to trace out much information, the reason may be that you are using an inappropriate index or there are not many previous works on it. Sometimes, the topic selected may be so new that there has been no time for published research to appear in print.

Note Making

When you go through the sources, it is imperative that you record notes for future use. Note making is an active and focused activity where you understand the information from a source, be it a research paper or book, and note down what is relevant and important for you. In other words, note taking is almost verbatim recording from a book or lecture, and note making is a summary representation or comment in your own words on what you read, heard, or thought about something. Make good notes including the details of all the relevant sources you found and read (Best and Kahn 1993; Walliman 2011). Refer Chap. 22 in this chapter for more details on note making with examples.

Keep Detailed Bibliographic Records of All the Sources

While you write notes, it is important to ensure that bibliographic details such as name of author/s, title of publication, name of publisher, and page numbers are accurately written for later use (Barrass 2002). There are slight variations in bibliographic details needed for correct citation depending on the type of sources. Refer Chap. 15 for a detailed treatment on referencing. Record all the bibliographic information on the note card itself. This is important to cite and list the sources accurately avoiding plagiarism (Chap. 24). Include the following bibliographic details in the case of books and research papers:

- Name of author/s, editor/s, or organization
- Title of article, book, or chapter (if it is an edited book)
- Volume and issue numbers (for journals)
- Edition and reprint (for books)
- Name of publisher (for books)
- Place of publication (for books)
- Year of publication
- Inclusive page numbers (for articles and chapter articles)
- Total pages(for books)

In the case of electronic publications, especially online publications, the details required are different. Note down the following details:

- Name of author/s, editor/s, or organization
- Title of the article
- Title of the site (i.e. homepage)
- Date of creation of the site
- Date, last modified or updated
- Date of access
- Digital Object Identifier (DOI) or Full address of URL.

10.4 Managing Information

Collected information must be handled carefully. Follow an organized system to keep track of references. When you locate a reference source, you can put the bibliographic details of the source directly into your referencing system. The following methods are commonly used to keep track of references:

Index cards: Using index cards is an old but still popular filing and management system of bibliographic information. Bibliographic information is collected on index cards (also called reference cards) along with pertinent details.

Reference management software: The easiest method of managing information at present is to use any of the popular reference management software. You can use either priced or free bibliographic software such as Mendeley, CiteULike, Refbase, and Zotero (Sect. 15.10).

Electronic card files: You can also use a computer application such as an electronic card file or database programme for managing information.

10.5 Reliability of a Source

All the sources you come across cannot be taken for granted. You have to assess the trustworthiness of the sources. A thorough evaluation of identified literature materials is needed before using them. Read critically and actively, and it is important to keep your purpose for reading in mind while you read. To present a critical review of literature, ask some critical questions yourselves and sift through information. You may consider the following questions and reflect on all possible aspects that may affect your assessment. Having done this assessment, you can use the source for your work.

Who Are the Authors?

When you assess the credibility of a source, it is important to determine how believable it is. For this, the first thing you must do is to check the credentials and background of the author. In general, authors who have advanced degrees in their fields; those who have specialized training; those who work for reputed organizations such as renowned universities, government agencies, and international organizations such the UNO; or those who have published books or articles previously are likely to be more reliable and credible than authors who lack any of these credentials. In other words, qualifications provide some justification that the source is correct and trustworthy. Still, experts are liable to make mistakes, and you should never accept at face value an author's words without documented evidence. Books and propaganda materials of experts from the Non-Governmental Organization (NGO) and private sector must be subjected to greater scrutiny. Check whether they are real experts and determine whether the material is a fact, interpretation, imagination, or simply opinion. Verify the facts with other sources to see how accurate or how trustworthy the source is. If you are unable to verify such a dubious source, never use it. If available, author impact factors such as the *h*-index or the *g*-index are good indices to assess the scientific value of a given researcher or author (Sect. 9.6).

Who Are the Publishers?

When assessing trustworthiness of a publication, assessing who is the publisher is also important. When you consider who wrote the material, you might also check where and when this work was published. Well-known national and international publishers, especially those affiliated to reputed universities and those who have a long tradition of publishing, are likely to be more reliable because they have reputations to uphold.

Most established publishers have expert editors and have a peer review system to upkeep the quality of their products.

Consider the Impact Factor or Rating

In the case of journals, the impact factor gives you some idea about the credibility of journals. The first thing you have to confirm is whether the journal is following a peer reviewing system. Peer-reviewed journals are assessed based on their impact factor or a rating system, for example, NAAS rating followed in India (Sect. 9.6). A journal with high impact factor is supposed to be of good quality and standard. The articles published in such journals also get that esteem.

Identify the Purpose of the Publication

Before you start taking notes, think for a while, and identify why and for whom the material was written. What was the author trying to find out and why is this work important? You will be able to understand easily, if it is a propaganda material.

The significance of the article or book must be clearly established.

The Subject Matter Coverage

Assess whether the literature has been presented and evaluated clearly and objectively. Read the material critically and find out the focal point of the article. The supporting information must be well researched and accurate. In the case of an article or thesis, have an objective assessment about the aspects included or omitted both from literature review and from the results and discussion sections. The findings, discussions, and conclusions must be logically arranged. It is important to understand how the perspective presented in the publication differs from other sources and how the information given in the source is relevant to your own work.

The Treatment Given to the Content

Try to find out the theoretical perspective of the author and check whether this differs from the current belief prevalent among the research workers about the topic. The concern and the philosophy that influenced the writer's perspective and how the author developed ideas are important. You should be able to observe any biases or inconsistencies in the information presented. Observe whether the author evaluated the literature relevant to the problem or issue. Partiality of the author is another problem. Check whether the author has included any literature taking positions s/he does not agree with.

The Structure Followed

The structure of the source must be evaluated. If the source is a research paper, it should be presented in the IMRAD format with the sections such as Introduction, Materials and Methods, Results, and Discussion (Sect. 11.1). A logical relation must be there between the sections that affects the overall discussion. The article should evaluate the designs and methodological issues sufficiently well as discussed in Chap. 11. You may also assess whether the author uses appeals to emotion, one-sided examples, or rhetorically charged language and tone, especially in materials written for a popular readership. Conflicting and complementary ideas should be presented and discussed. There must be an objective basis to the reasoning. Beware of authors who merely reiterate or try to prove what s/he already believes.

Avoid Biased Sources

Avoid sources that appear biased such as those from lobbyists and advocacy groups, which tend to present only evidences that support one interpretation. You may also evaluate the reasoning in a source's argument. Beware of sources that depend upon logical fallacies or false arguments that tend to divert or mislead readers instead of material evidence or logical reasoning (Chap. 14). Consider timeliness of publications too. Current literature, including those published recently, is likely to be more reliable and convincing than older sources. As new knowledge gradually becomes available to all, researchers will be able to ascertain their reliability. Therefore, prospective authors may not like to take a risk by presenting unreliable information. You also want to look for publications or other sources whose authors may have considered various explanations and interpretations.

Maintain Extra Vigil on Electronic Sources

Presently, you may come across innumerable web-based resources, which are relatively easy to access. Most often, this information can be accessed free; therefore, researchers, students, and the public are increasingly searching these sites to obtain information. Everybody agrees that it is easier to publish materials on the Internet than to publish in a reputed print journal. Remember that most journals accept articles for publication based on the preview comments from peer reviewers. Therefore, print sources tend to be more credible than sources you find on the Internet. When you use Internet resources for research, you must be extremely careful about spurious and unreliable websites. Rely only on websites sponsored by universities and reputed organizations. When you are using an electronic source, you have to be more careful and critical about the validity of the information, which it conveys. Check the authors, their credentials, and perspective, and avoid articles that have no author's name or date of creation or revision attached to them.

The above critical questions for evaluating a source material are not comprehensive. An assessment of the above aspects will help you to form a good basis for your written review. You may probably think that considering these aspects as you read will impede your reading process, but starting this analysis early will make the process of writing a review much easier. If you are able to take notes in your own words as you read and think, you may have already completed most of your work before you start to write the review. If that were the case, the final writing part would not be a problem at all.

10.6 Writing a Review

For writing a literature review, it is imperative that you located and read a wide range of reference materials on the topic. When you exhausted all the sources and made notes on selected sources, you have to compile the information in an appropriate manner. Develop a broad plan for the review outlining headings and subheadings. One of the difficult aspects of writing a literature review is to decide when to stop searching and reading and start writing the review. It is incorrect and meaningless to simply listing sources by describing or summarizing one source after another. Do not try to list all the materials published, but synthesize and evaluate them according to the guiding concept of the thesis or research question.

Some authors start every paragraph with the name of an author giving the review a monotonous and boring look. A better approach is to organize the literature review into various sections, with headings and sub-headings, presenting themes or trends including relevant theory. It is a good idea to group items into sections, which helps to indicate comparisons and relationships. If it is a lengthy literature review, it is always recommended that you arrange the review under headings and subheadings.

A literature review is usually written in a narrative form or as a discussion. The review can have three parts—an introduction, the body of review and a conclusion. Use an overall introduction to state the scope of the literature search and to formulate the problems or concepts related to the chosen topic. In the conclusion part, you can provide a summary of your review along with your final comments. As writing a review is part of the whole process of your study, you should begin writing as soon as you have a basic understanding of your topic area. By doing so, you can assess the need for further reading, if any. Regularly review your drafts, and add comments, questions, and ideas. If your sources are from a long period, it is meaningful to compare and show connections between different articles.

Review Formats

There are many ways to organize a literature review. An *essay-based* review is readable and comprehensible, as it tells a story referring to other works during the process. This requires some expertise to do it in a meaningful way. A *bibliography-based* review lists the literature found on the subject and says something about each one. It starts with the earliest work on the topic and moves through to the latest. The review is organized chronologically. A *theme-* or *topic*-based review divides the literature into certain broad topic areas related to the work. For theses, theme-based review is often followed. In thematic format, a broad range of loosely associated literature is grouped into a set of common subject areas or themes. Arranging the material around issues and findings relevant to your work not only provide a framework but also reinforces its importance.

A good literature review will have a good flow of arguments based on materials of others with appropriate citations. Your opinion on the literature also matters. Simply listing them without saying any thing is not correct. You are not expected to write everything related to your topic but you are supposed to create a story that leads to your specific study. You should also maintain a sense of balance between your ideas with the idea of others. Therefore, make use of all those critical comments made while reading notes to express an academic opinion. Make sure that examples, citations, and quotations are used wherever appropriate connecting each section of the review. Correct documentation of all bibliographic details is also essential. Remember that both in-text citation and accurate listing of references are important for a good review (see Chap. 15).

Review Papers

Sometimes, review papers are published in secondary journals specifically intended for that purpose. A *review paper* is an exhaustive review of literature in which the reviewer attempts to trace out all the possible information on a certain topic. It is usually written as a survey of primary literature over a given period indicating the developments and trends in the concerned period. Before attempting an exhaustive review, try to define the scope of the review in such a way that it is within bounds, and the number of articles to review is manageable. The main objective of writing a review paper is to appraise previously published literature and to put them into some kind of perspective. An exhaustive list of references at the end of review serves as a good bibliography on the topic.

While writing reviews, remember that the intended audience for review papers and primary papers is different. The primary paper is highly specialized, and its audience consists of peers of the authors. Ordinarily, a review paper is long, ranging from 10 to 50 printed pages (Day and Gastel 2006). The subject would be general in scope unlike original research papers. A good review paper offers critical evaluation of the published literature, and often provides important conclusions based on the literature. The organization of review is also different. There will not be sections as "Materials and methods", "Results", and "Discussion". Prepare an outline first, similar to a term paper or assignment before you start writing. While writing, slight modifications can be made in the outline. Many review journals give an outline of review at the beginning of the paper where it serves as a convenient "Table of contents" for prospective readers.

Some review journals demand critical evaluation of the literature, while some others are more concerned with bibliographic completeness. Based on the requirement of the journal, you decide the strategy. Most annual review journals publish reviews designed to compile and annotate the papers published on a particular subject during a defined period. Most professional societies of repute publish exclusive review publications, and these are usually identifiable from their titles such as '*Advances in ...*', '*Yearbook of...*', and '*Annual Review of...*' Examples include '*Advances in Agronomy*', '*Advances in Soil Science*', and '*Annual Review of Plant Physiology*'. Occasionally, primary journals and abstracting journals also carry review papers on certain chosen subjects. A thesis will also contain a comprehensive review on the researchable topic indicating the past and present status of the problem.

Bibliography

Allen, M. 2004. Smart Thinking: Skills for Critical Understanding and Writing (2nd Ed.). Oxford University Press, Oxford, 193p.

Bailey, S. 2003. Academic Writing. A practical Guide for Students (Indian Reprint, 2004). Routledge, Taylor and Francis Group, London, 191p.

Barrass, R. 2002. Scientists Must Write: A Guide to Better Writing for Scientists, Engineers and Students (2nd Ed.). Routledge, London, 204p.

Best, J.W. and Kahn, J.V. 1993. Research in Education (7th Ed., Indian Reprint, 2004). Prentice–Hall of India, New Delhi, 435p.

Bryman, A. 2008. Social Research Methods (3rd Ed.). Oxford University Press, Oxford, 748p.

Davis, M. 2005. Scientific Papers and Presentations. Academic Press, Massachusetts, 356p.

Day, R.A. and Gastel, B. 2006. How to Write and Publish a Scientific Paper (6th Ed.). Cambridge University Press, Cambridge, 320p.

Hartley, J. 2008. Academic Writing and Publishing: A Practical Guide. Routledge, New York, 196p.

Jenkins, S.R. 1985. Searching the literature. In: Hawkins, C. and Sorgi, M. (ed.), Research: How to Plan, Speak and Write about It (4th Indian Reprint, 1993). Narosa Publishing House, New Delhi, pp. 29–59.

Kumar, R. 2011. Research Methodology: A Step-by step Guide for Beginners (3rd Ed.). Sage Publications India, New Delhi, 415p.

Leedy, P.D. and Ormrod, J.L. 2010. Practical Research: Planning and Design (9th Ed.), Pearson Education, New Jersey, 360p.

Lester, J.D. and Lester, J.D. Jr. 2009. Writing Research Papers: A Complete Guide (13th Ed.), Longman, 416p.

Malmfors, B., Garnsworthy, P. and Grossman, M. 2004. Writing and Presenting Scientific Papers (2nd Ed.). Viva Books, New Delhi, 153p.

Marder, M.P. 2011. Research Methods for Science. Cambridge University Press, 227p.

Randolph, J.J. 2009. A guide to writing the dissertation literature review. Pr. Assess. Res. Eval. [e-journal] 14(13). Available: https://pareonline.net/getvn.asp?v=14&n=13 [Accessed 10 Aug. 2011].

Walliman, N. 2011. Your Research Project: Designing and Planning Your Work (3rd Ed.). Sage Publications, New Delhi, 378p.

Chapter 11
Preparation of Research Papers and Other Articles

Twice and thrice over, as they say, good is it to repeat and review what is good.
Plato (427 BC–347 BC)

Research workers engaged in scientific pursuits bring out new ideas and information continually on various aspects of science. However, such research results, if not made available to the widest possible audience within a reasonable time, are worthless. New research is often based on the work of others. One may start where another scientist stopped. Research workers, therefore, after the completion of research projects, must compile the results and publish them in scientific journals with possible interpretations and clear conclusions. Only then, one can say that the research project is completed. Apart from research papers, there are some other established means of research reporting. These include theses, dissertations, conference proceedings, electronic publications, progress reports, and completion reports. For obtaining Ph.D., M.Phil., or Master's degree in professional subjects such as agriculture, medicine, and engineering one has to submit a thesis or dissertation—another form of research report.

As a scientist or science student, you are expected to write essays, articles, and reports effectively. The objective of scientific writing is to accurately and clearly communicate with students, scientists, funding agencies, and lay people. The following are major types of publications one scientist may write in course of his/her career.

1. Thesis/dissertation
2. Scientific papers
3. Short communications/notes/letters
4. Conference papers/abstracts/ extended abstracts/posters
5. Research/project proposal
6. Project reports/ progress reports
7. Review papers
8. Critical analysis of popular concepts
9. Semi-scientific articles/popular articles

C. G. Thomas, *Research Methodology and Scientific Writing*,
https://doi.org/10.1007/978-3-030-64865-7_11

10. Textbooks, handbooks, monographs, etc.
11. Book chapters
12. Book reviews in journals
13. E-publications
14. Creating Web pages
15. Writing in social media groups such as ResearchGate/Facebook.

11.1 The IMRAD Structure for Research Reporting

Theses and research papers in any part of the world follow almost a uniform style. Researchers and research students have to read many original research articles to keep abreast of the developments in their fields. They may also be writing and publishing original articles frequently. Therefore, a familiar format with clear component parts is adopted for research reports. This format, universally followed for theses and research papers, came to be called the *IMRAD*, an acronym for Introduction, Materials and Methods, Results, and Discussion. Most scholarly journals and conference organizers have adopted the IMRAD structure for their papers. Both authors and readers, wherever they are, recognize this format. The IMRAD has become very convenient to researchers, as it allows research results to be presented logically by summarizing the research process in a perfect sequence without having to go for unnecessary details.

It is interesting to examine the method of science communication before the IMRAD structure came to stay. According to Day (1989), research journals were started for the first time in AD 1665. At that time, research papers were 'letters' written by scientists. The style of writing was descriptive, adopting a chronological sequence. By the latter half of the nineteenth century, the style of writing scientific papers began to change. Louis Pasteur (1822–1895) is credited to be the first scientist who wrote his findings in IMRAD form on 'germ theory' demolishing the 'spontaneous generation theory' myth. Although Pasteur did not use the terms, Introduction, Materials and Methods, Results, and Discussion, he argued his case sufficiently well following this structure. The prime intention of Pasteur in adopting this structure was that any person could repeat them and get the same results. In fact, as everybody could understand now, he was recognizing the power of 'reproducibility of results' as a part of research process. We should note that Pasteur was arguing for his case in the midst of strong supporters of the then widely believed theory of 'spontaneous generation.' Using the IMRAD structure, he could argue effectively and persuasively in support of his ideas.

After the 1950s, especially after the World War II, many countries began to invest heavily in scientific research. When funding for research increased, the only way to assess effective utilization of money is to see the outcome of research through articles and patents. Naturally, professional societies and journals grew in numbers, so also research papers. Universities began to churn out PhDs in large numbers. To put some sense of quality in these papers, journal editors began to insist compactly written articles following IMRAD structure. Gradually, almost all universities and journal establishments adopted the IMRAD structure for theses and research papers. Austin Bradford Hill (1897–1991), a renowned medical statistician and editor, in a 1965 conference of medical editors gave further momentum to the spread of IMRAD by the now famous '*Bradford Hill questions*'; why did they start?, what did they do?, what did they find?, and what do the results mean? (BMA 1965).

In 1972, with the publication of the 'American national standard for the preparation of scientific papers for written or oral presentation,' the IMRAD format was accepted as standard in the USA. At present, IMRAD is accepted in not only natural sciences but also social sciences, arts, and humanities. When the research paper is prepared, a brief abstract should also go with it. Ever since the abstract has become a mandatory component of a paper, the 'A' of abstract is often added to 'IMRAD' so that we get 'AIMRAD.' However, bear in mind that the abstract should function as an independent text.

Theses, dissertations, and scientific papers must be written in a formal and structured style with IMRAD, and it is essential that the information they carry must be authentic. Therefore, before finalizing these reports for submission, students and researchers have to make several drafts and amendments. Write the first draft by answering the following:

1. Why did you start the experiment? (The introduction)
2. What did you do? How was it done? (The materials and methods)
3. What did you find? What was learned? (The results)
4. What do the results mean? How do they relate to what is already known? (The discussion).

Use these questions to analyze your completed research work before you start writing. If you have good answers to these questions, a major part of your writing job is over. Having completed these sections, you can complete other parts of the report such as references, table and figure legends, abstract, running title, key words, and acknowledgements. Researchers should note that many of these points are equally applicable to theses and research papers.

As thesis preparation is elaborate than a research paper, it is considered separately in the next chapter (Chap. 12).

11.2 Research Papers

The research reports that are published in scientific journals are commonly called *research articles, research papers, scientific papers*, or simply *papers*. For a scientist, publication of research papers is the major accepted means to get recognition from the public and fellow scientists. The pressure on the academia to publish papers at any cost shaped the popular phrase, '*publish or perish*.' Because both quality and quantity of publications are important, try to publish your articles in reputed journals with good impact factor or rating. Most often, the funding agency who has supported the research project with grant would like to see the results published as soon as the project is completed. In this chapter, useful guidelines for the preparation of common scientific publications with emphasis on research paper are given. A research paper and a thesis are similar in structure and arrangement of information, but major differences are with respect to size and treatment of various sections.

The standard method of transmitting original scientific information is to publish it as a formal scientific paper or research paper in a valid 'primary publication'. In this connection, it is important to define a valid 'primary publication'. Day (1983) paraphrased the original definition of the Council of Biology Editors (1968) and defined a primary publication as '(1) the first publication of original research results, (2) in a form whereby peers of the author can repeat the experiment and test the conclusions, and (3) in a journal or other source documents readily available within the scientific community' (Note that the 'Council of Biology Editors' is now the 'Council of Science Editors').

A research paper is prepared following the IMRAD format, and the major intended audience is fellow workers and research students. There is near unanimity among scientists on the following aspects of a scientific paper.

- The paper must be prepared following the IMRAD format.
- The peers of the author must be able to repeat the reported study, if required.
- The readers must be able to evaluate the result, discussion, and conclusion.
- The paper must be readily available to the scientific community without restrictions.
- The form of the paper must be permanent.
- The paper must undergo a peer reviewing process.
- The paper must be made available to the information retrieval services such as CAB abstracts, Biological Abstracts, and Medline.

Apart from publishing in academic journals, research results can also be presented in conferences and published in conference proceedings. In the past, it was customary to bring out conference proceedings immediately after the conference, compiling the text of papers actually presented along with questions and answers. As the presentation of papers by the authors and critical evaluation through question and answer sessions were serving as a mechanism of 'peer reviewing,' these were considered as valid primary publications. However, most conference organizers have stopped this practice now, and they bring out 'proceedings' in advance! In most of such conferences, you do not have a mechanism to ascertain whether the 'paper' listed in the 'advance proceedings' has actually been presented! This indicates that most of the present day 'proceedings' of conferences, which contain reviews, abstracts, or extended abstracts, are not valid primary publications. In other words, all the criteria mentioned above apply only to primary academic journals.

The following definition of scientific paper seems appropriate in the present context: A scientific paper is a report of original research prepared by the researchers concerned in the IMRAD format, and published in a valid primary publication after peer reviewing.

The main purpose of an *academic journal*, sometimes also called *research journal, scientific journal,* or *scholarly journal*, is to publish peer-reviewed original research articles or papers relating to a particular academic field. A journal may be owned by a professional society, association, educational institution, research organization, government department, or a private publishing group. Although the main intention of such academic journals is introduction and presentation of new research through original research papers, it may also serve as a forum for the evaluation of existing research through occasional review articles and book reviews.

Research articles received for publication are normally peer reviewed for ensuring quality before accepting it for publication (Sect. 11.6). This mechanism of publishing papers in journals is part of the research process, and the information becomes part of the literature of the subject. Note that papers published in peer-reviewed journals typically enjoy higher status as they undergo stringent scrutiny. Journals are usually rated based on their *impact factor* or *rating* (Sect. 9.6), and you may try to publish you papers in journals with high impact factor or rating. The high status of a journal is usually reflective of its impact factor.

11.3 Selecting a Journal

There are many kinds of academic journals. Professional societies vie with each other in bringing a journal for their group. However, not all are alike in credibility, promptness, punctuality, prestige, or popularity. Therefore, selecting a journal for submitting your much-valued papers deserves careful attention. Do not allow good research papers to be buried in inappropriate journals. Decide the journal and the targeted readers before the writing stage itself. The factors influencing the selection of a journal include impact factor or rating, periodicity, promptness, punctuality, cost involved, the area of circulation (international, national, or regional), prospective audience, and the language of the journal. The preferences of the researchers also play a part in selection, for example, they may prefer to submit their article to a journal published by the professional group to which they are life members.

Impact Factor/Rating

A major factor that governs the researcher's choice of journal is the impact factor or rating of a journal, which is a reflection of its prestige and reputation (Sect. 9.6). As the status of a research paper is considered directly proportional to the rating of the journal in which the paper is published, try to publish the papers in journals with higher impact factors. Most scientific establishments have evolved a mechanism to rate the articles of their scientists based on impact factors.

Periodicity

Periodicity of journals is an important issue to be considered. Certain journals are published weekly (*Nature, Science*), fortnightly (*Current Science*), monthly (*Indian Journal of Agricultural Sciences*), bimonthly (*Journal of Biosciences*), tri-monthly

(*Indian Journal of Agronomy*), or bi-annual (*Range Management and Agroforestry*). Bear in mind that the 'publication lag' of a weekly or monthly journal is always shorter than that of a quarterly or bi-annual journal. Whatever is the periodicity, possible delays due to peer reviewing and editing is almost similar. The publication lag of a monthly including the time for editorial review usually ranges from 4 to 7 months. You can add an extra 3 or 6 months, if it is a quarterly or bi-annual as the case may be. Most reputed journals publish date of receipt of publications. If that is the case, you can easily find out the average time they take for getting a paper in print form.

Area of Circulation
The area of circulation is an important issue of primary journals. You may find international, national, and regional journals. Try to publish your articles in reputed international journals to reach a wider audience.

Language
Journals are published in many languages. Language is a personal choice and convenience. Select a language so that a good number of people go through your paper enhancing its impact. A good number of journals are published in English. You may also find sizable number of journals in French, Spanish, German, Russian, Chinese, Japanese, and others. Most abstracting and indexing services include these journals too employing translators.

Promptness and Punctuality
Promptness and punctuality of a journal are two major factors, which decide the rating of a journal. Certain journals are notorious in delaying the issues with many issues of 'backlogs'. Some may lag for years and regularly hoodwink the readers by combining issues altogether. For example, if the journal is a bi-annual, it should bring out two issues in a year regularly and not one single issue by combining two issues. Weed out such notoriously irregular and backlog journals. A primary journal must be up to date. Usually, journals accept research papers not more than five years old. If it is again delayed by some more years, what good it is going to do to the researcher and the society? What good will it serves, if the information is stale and old? Remember, regularity of journals adds prestige and reputation to the journal. In any case, if your paper is going to be delayed by more than one year, it is a serious matter, and you should avoid such an eventuality.

Prospective Audience
Another major factor is the prospective audience. The article may be interesting for a small group of fellow experts or a wider scientific community. Accordingly, the researcher can select a general scientific journal or a specialist journal for publication of articles. While preparing for general scientific journals, you have to change your style to meet the requirements of a general audience, as specialist journals have their own jargons and abbreviations, which may not be understood by general readers.

Cost Involved

This is also a major factor, which you should take into consideration. Most professional societies publish articles from their members only, which means the prospective author has to become a member (annual or life member) by paying a fee. Some journals, especially by private publishers, demand page charges. Still, another group of journals demands only subscription from you. For paying the charges, most often, authors can utilize their research grant or employer's fund specially set apart for the purpose. In the case of exceptionally good research papers from poor counties, many international journals waive all or a part of the fee. You have to weigh all these options. If your institution or funding agency bears the cost of publication, you need not worry much about the cost factor; all you have to worry is other factors such as promptness and journal rating.

Type of Journals

Researchers can select any of the three categories of journals mentioned below that publish original articles.

General Scientific Journals

These include journals such as *'Nature'*, *'Science'*, *'Current science'*, and *American Journal of Science*. Most general scientific journals carry not only original research articles but also editorials, review articles, news, letters, and other materials to keep varied groups of readers well informed about the current events in the scientific world. Most of these journals are weeklies or fortnightlies, and their rejection rates of articles are usually 80–85 percent. These journals often employ full time editors and staff. In India, *'Current Science'* is an important general scientific journal published by the Current Science Association in collaboration with the National Academy of Sciences.

General Discipline Journals

Science and technology fields such as agriculture, animal science, and modern medicine are composed of numerous sub-disciplines and specialties. Some journals cater to all the sub-disciplines and specialties of a particular broad area. For example, in agriculture, the journals such as *Experimental Agriculture, Tropical Agriculture* (Trinidad), *Canadian Journal of Plant Sciences*, and *Journal of Agricultural Sciences* (Cambridge) and *Indian Journal of Agricultural Sciences* (ICAR) can be categorized as *general agricultural journals*. The journals published by most State Agricultural Universities (SAU's) also fall under this category. These journals accept research papers on all aspects of agriculture.

Similar to general agricultural journals, there are *general animal science journals*, and *general medical journals*. Similarly, general academic journals are available in other disciplines such as fisheries, biology, botany, and zoology.

Some general animal science journals are Animal Science Journal, Journal of Animal Science, Indian Journal of Animal Sciences, and Indian Journal of Animal Research. Some general medical journals include British Medical Journal, the Lancet, the New England Journal of Medicine, Journal of the American Medical Association, and Indian Journal of Medical Research.

Specialist Journals

These groups of journals cover major specialties, subspecialties, and sometimes selected areas of research. These are monthly, bimonthly, quarterly, or half-yearly, and usually published by professional groups or societies. Often, the editors of such journals are scientists who edit the journal in their spare time. There are umpteen number of specialist journals suitable for each disciplines and sub disciplines. Their rejection rates are in the order of 50–60 percent.

Open-Access Journals

Open-access journals are research journals that are freely available on the Internet. However, only peer-reviewed open-access journals qualify as valid publications. Many journals, which come under the three categories mentioned above, are now available as 'open access'.

'Open-access journals' are the results of widespread use of Internet and online publishing. Most of these journals are available in print form too. They have several plus points such as enabling free access to research papers, lowering the costs for research in both public funded institutions and private concerns, and improving access to information for the public. Probably, a major advantage to the authors is higher citation rates, which will be reflected in both journal and author impact factors.

Open access is achieved in three ways. (1) Many open access journals are made free utilizing the subsidy they get through grants from funding agencies. Subsidized journals are financed by an academic institution, professional society, or a funding agency. In such cases, neither the authors nor the readers pay any money for publishing and accessing the papers. (2) Some open access journals do so by accepting page charges or charges per article from authors. For example, PLOS (Public Library of Science) journals demand specified fees for an article. (3) A third method is to publish the articles in a traditional subscription based journal like Nature, and after a certain period (six months or so), make their paper freely available on the Internet by self-archiving it in their institutional repository (IR).

You can access directories of open-access journals to see a list of them. An important one is *Directory of Open Access Journals* (DOAJ). For more information, visit their site: https://www.doaj.org.

11.4 The Structure of a Research Paper

Research papers published in journals follow a set pattern, some minor differences notwithstanding. Some journals restrict the size of articles, for example, 4000 words (about 20 typed pages; double-spaced) in the case of a full-length article and 1000 words (about 5 typed pages) in the case of short notes or short communications. Usually, half page table or illustration is taken as equivalent to 200 words. Currently, most journals demand manuscripts in the form of soft copy in a compact disk or through e-mail as an attachment. Normally, they may also ask for two hard copies of the manuscript along with electronic copies. The organization of a full-length paper is in the following order:

- Title
- Running title
- Name of author/s
- Institution and address
- Abstract
- Key words
- Introduction
- Materials and methods
- Results and discussion
- Conclusion
- Acknowledgements (if any)
- References
- Tables and Figures (if any).

Title: The title is the first part of an article that attracts the readers. While deciding a title, try to balance between brevity and preciseness so that the title gives a clear indication of the subject and scope of the work in a few words. Ensure that the title is direct and informative. Including key words and phrases in the title is recommended to indicate the contents of the paper. A good title must attract the attention of all those who might benefit from reading the whole paper or selected parts. Do not use abbreviations in titles. However, if the abbreviations are more familiar than the words they stand for, for example: DNA and pH, such abbreviations can be used in titles. If you are not sure about this aspect, use the term in full and its abbreviation in parenthesis.

Titles can be shortened by deleting familiar phrases, which give no additional information. A title may become lengthy and awkward, if you add several prepositional phrases to qualify it. Similarly, avoid the use of unnecessary articles in titles. Before finalizing the title, try several attempts and decide the most appealing one.

As narrated in the case of thesis, avoid the use of low impact expressions and phrases in titles of papers such as 'Effect of…', 'Study of…', 'Studies on…', 'Preliminary studies on…', 'A study of…', 'Investigations on…','Response of…', 'Observations on…', 'Aspects of…', 'Notes on…', 'A note on…', 'An enquiry in to…', and 'An analysis of…'.

Running title: It is a shortened title of the article, which is given at the top of every page of an article in the journal while printing. Often, a running title is between 30 and 50 characters. Running title gives an idea about the contents of an article, especially when you open a journal in the middle of an article. While making a running title, focus on the keywords.

Author's name and address: In a published research paper, the names and addresses of authors are given immediately after the title. However, most journals insist to write the title and the names of the authors separately on a sheet of paper while submitting the manuscript.

The addresses of authors both at the time of experiment and the present must be indicated as transfers or changes in positions are routine affairs in research organizations. For example, a Ph.D. student might have joined a different institution after taking the degree. The usually followed custom in such cases is to give the address of the institution where the experiment was originally done as byline, and indicate the present address of the author as a footnote. In the case of multiauthored papers, make it clear who would deal with the proofs (*the author for correspondence*).

Key words: Key words are important words or short phrases, normally 3–10, which will reflect the content of the article. These are intended to assist indexers in cross-indexing the article in various indexing and abstracting journals and for easy retrieval of relevant information. In an article, key words are given below the abstract.

Introduction

Unlike a thesis, only a brief introduction is given in a research paper. Start the paper with a brief introduction by answering the question, 'Why did you start the research?' The purpose and scope of the work must be stated clearly for justifying the work. Briefly describe the state of knowledge before the research was started, define the gaps in knowledge, and state your objectives. Unlike a thesis, there is no separate section as 'Review of literature' in a research paper. As a part of 'Introduction', gaps in knowledge are highlighted briefly through a few references.

Materials and Methods

As described under thesis, the reader should get the answer to the question, 'What did the author do in the study?' after reading this part. The answer must include the details of the materials and methods used by the authors. The statistical techniques used to analyze the results should also be mentioned. The computer programme used by you for analysis should also be mentioned with reference details. It is unethical to publish results based on inadequate statistical analysis. Remember that presently most journal editors seek a statistical opinion before accepting an article, if it contains substantial statistical works. Many scientists read the 'Materials and methods' section to understand the techniques the authors employed and to check whether there is sufficient details for them to repeat the work themselves. Therefore, journal editors usually insist to include sufficient details on materials used and methods employed. Sometimes, the journal office may return the article for want of sufficient details on materials and methods.

Results and Discussion

In most research papers, the results and discussion are combined and given as a single section 'Results and discussion'. This saves space, because if these are presented separately, some aspects might be repeated. After going through this section, the readers should be able to get answers to the questions, 'What did the authors find?', 'What do the results mean', and 'how do they relate to what is already known?' The findings supported by data in tables, diagrams, graphs, and photographs are presented and discussed with reference to previous articles related to the subject as discussed under thesis. However, the discussion must be brief in an original article. Conclude the discussion with the message of the article as a final paragraph. The author may also indicate some future lines of work based on the findings discussed. This is helpful for fellow research workers, who are interested to plan experiments on similar lines.

Acknowledgements
Although not mandatory, some research papers carry 'Acknowledgements'. This is the place to mention the names of all those who deserve recognition for their contributions in one way or other for completing the study and research paper. Put acknowledgements in a brief format. You may express thanks to the head of an institute/department who has allowed the facilities to be used and to a sponsoring organization for research grants. Use courteous words to write this section. Normally, 'Acknowledgements' is given at the end of a paper just before 'References'.

References
Citation of works by other researchers in the text and listing the sources at the end of the article must be done carefully. A complete list of works cited in the text should be given as per the citation and listing system followed by the journal. Most journals in India follow the Name-Year system. However, there may be variations between journals in the actual style. Consult the latest issue of the journal and go through the 'Instructions to Authors' section before finalizing 'References'. Refer to Chap. 15 for more details on citing and listing references.

Tables and Illustrations
Most journals insist that the tables and illustrations should be prepared on separate sheets and attached at the end of the article. This is to make printing works easy, as the setting of text, tables, and illustrations can be done separately. When you put tables and illustrations at the end of a research paper, make sure that these can be easily identified from any attached appendix materials. While preparing tables, ensure that you have numbered consecutively each table and figure, and that they are complete in all respects including captions. Remember that the caption must go under a figure, but in the case of a table, the title with description must go above it.

Tables should be carefully numbered in Arabic numerals, and a reference to each table must be given in the text. Tables and figures should have self-explanatory captions and complete enough to be meaningful. Photographs, however, should have their legends typed on a separate sheet. Identify the figures (including the photographs) on the back by the figure number, the authors, and the title of the paper. This should be written lightly with a pencil as the firm pressure of a pen may dent the surface of the photograph and ruin it. Photographs should not be attached to the paper with paper clips, as these may also damage them. The tops of photographs should be indicated, if it creates confusion. It is better to send electronic copies of the photographs and figures so that these can be directly used for printing. If you are sending electronic copies of the photographs, you need not worry much about these aspects. Before sending the paper, make sure that all numbers given in tables and figures are correct, and that they agree with the figures in the text (see also Chap. 13).

Abstract

As mentioned earlier, an abstract should summarize the research concisely, and must include the objectives, study design, important results, and conclusion. In a research paper, an abstract always appears at the beginning just below the title and name and address of the authors. However, we write the abstract only last after completing all the other parts of the article. Earlier, it was a custom to give a summary at the end of scientific articles. Some journals, however, began to put it first to give the readers an overall view of the article. Later on, it began to be used by abstracting journals. Now, most journals include 'abstract' instead of 'summary'. Keep in mind that it is not intended to be a part of the article, and should not include any material that is not part of the article. The title is a part of the abstract; therefore, it should not be repeated. As mentioned under thesis abstract, avoid ambiguous statements such as '...are described' or '...will be presented.' Include a gist of the rationale and objectives, methods, main findings, and main conclusion. The abstract will help the reader to grasp things at a glance. It should also enable the reader to assess whether the article is worth reading. For a journal article, normally, the abstract is limited to 150 words.

Please note the differences between an abstract, summary, executive summary, and synopsis clearly. A *summary* is the summing up of the information, which is given at the end of an article. An *executive summary* is a short write up that summarizes a document in such a way that readers can quickly become acquainted with the long document without having to read it fully. It is rare to see executive summaries with research articles; however, most funding agencies insist to have an executive summary along with a project proposal. An *abstract* is usually shorter than a summary, which is intended to provide an overview rather than being a miniature version of the full document. An abstract can be considered as a brief summary. Another related term is *synopsis*, which is a brief or condensed note providing a general view of an article, usually prepared by the journal editor rather than the author. Synopses are also not given along with research papers.

11.5 Other Considerations in the Preparation of Research Papers

The features mentioned in previous sections mostly cover how to prepare a research paper for publication. As authors, you should bestow attention to the following aspects too while preparing the paper.

Avoid Duplicate Publications

Avoid redundant and duplicate publications. Writing two different versions of an article using the same data and sending them to two separate journals is improper and unethical. You may have some doubts about the status of papers presented in conferences. Many consider publication of research papers in *Proceedings* (full-length papers incorporating discussion on the topic by the participants) as valid publication. However, if the outcome of the conference or symposium is in the form of 'Abstracts' or 'Extended Abstract,' you need not have any doubt about their status; these are not valid publications. The authors must prepare such original articles in the IMRAD format and publish them in peer-reviewed primary research journals to make them valid.

Avoid Unnecessary Splitting of Papers

In a situation of fierce competition where number of papers matters (*publish or perish!*), authors have a tendency to write as many papers as possible out of the research work done. Unnecessary splitting of papers to increase the number of articles is not good for science; and having recognized this problem, many recruiting bodies are now insisting on impact factors for journal papers. It is always desirable to have a few robust and meaningful papers in journals with good impact factors rather than many splinter papers in journals with low impact factors.

Follow Instructions to Authors

Most journals insist to have a specific style for their journal, and they follow certain norms in accepting articles. Instructions on preparation of research papers and other matters are usually given as 'Instruction to authors' and included in each issue of a journal or in some selected issues. You can also refer to the 'Instruction to authors' by visiting their official websites. Researchers must read the section on 'Instructions to authors' carefully before starting to write. Some journals may levy a charge on authors. These charges are designed to offset some of the costs of assessing and publishing manuscripts. Certain journals owned by scientific societies insist that the prospective authors should be members of the society.

Authorship of Articles

Often, the authorship of research papers is a difficult and sensitive issue. Normally, the individual who actually writes the paper may have played a major part in doing the work, and therefore, the convention is to put his/her name first in the order of authors. Similarly, it is also customary that the Head of the Department, if he/she has had some part in the work, should go as the last author. The order of remaining authors, if any, can be made depending upon their relative contributions. It is a usual convention to refer to the first author as the ***senior author*** as he/she is the primary progenitor of the work being reported. Even when the first author is a PG student, and the second is the Head of the Department, it is now accepted to refer the first author as the 'senior author' and to assume that he/she did most or all of the research.

In a scientific paper, only those who contributed substantially to the work should be listed as authors. Note that multi-authorship weakens the whole effort and adversely affects the prospects of real investigators. A simple help in taking observation, analysis, or interpretation do not entail anybody for authorship. Instead, acknowledge their help in a proper way. A casual attitude to authorship is not good for the progress of science.

Avoid fraudulent authorship such as *gift authorship, honorary authorship, ghost authorship,* and *hierarchical authorship.* Sometimes, authorship is given as a gift based on many extraneous considerations. Giving authorship to someone in return for a routine technical help, to a colleague in return for including his/her name in a published research paper, and to a relative (often, author's spouse who works in the same department or institution) are some typical examples of gift. Giving authorship to the head of the department or institution, simply because he/she is the head is an example of *hierarchical authorship.* The habit occasionally seen to put somebody's name as author for the prestige it confers to the paper (*honorary authorship*) must also be condemned. *Ghost authorship* is the most despicable form of fraudulent authorship, wherein, somebody else writes for a person. It is unethical to name somebody who has not made a substantial contribution to the work as an author.

All authors should know that their names have been included as authors! Including names of persons as authors without their consent is unethical and may create problems. For this reason, certain journals require the corresponding authors to state when they submit papers that all authors have seen and approved the final version. Some journals insist that all authors should sign the covering letter. Similarly, if an author is later added or removed from the paper, his/her signature must also be obtained.

Editing by the Author

There is not much difference between the editing style of a thesis and research paper before submission. In the case of a research paper, a minimum of two drafts have to be made and edited. If more than one individual author has been listed as authors, all of them must edit the paper. If needed, it can be given to a language expert to check the language and style. You should be vigilant about the types of errors you make quite often.

Unlike a thesis, in a scientific paper, it is unusual to number either headings or paragraphs. Instead, follow a hierarchy of headings; main headings could be capitals and centred, second order headings in capitals but not centred, and third headings in bold face flushed with left margin. For most purposes, three grades of headings suffice. However, if fourth grade of headings are needed, they can be made in italics and the text run on after a period on the same line.

It is a good idea to involve one of your colleagues in the editing process. This is easy, if the paper is a jointly written one. Otherwise, find a helpful colleague.

Submitting the Manuscript

If you have made the manuscript in a presentable form, you need not delay it further for dispatch. Check the instructions for authors carefully. The text should be neatly typed according to the instructions. The correct number of copies must be sent. Usually, journals give specific instructions on these aspects. Most journals now encourage sending a copy through e-mail or CD-ROM. You may send the paper with a covering letter addressed to the editor in a strong envelope. If you are sending the soft copy of the material in a compact disc or including photographs, some additional protective covering is needed. Most journals usually acknowledge the receipt of the manuscript, which can be used for future correspondence. Always keep a copy of all the original materials sent, including illustrations, tables, and photographs.

11.6 Review and Peer Review

For most journals, there will be a managing editor, editor, associate editors, and editorial board or committee. Often, the editor is the deciding authority to accept or reject manuscripts. The editor of a journal may be a scientist of repute. The managing editor is normally a full time paid professional, who often functions as a copyeditor too; whereas, editors are unpaid volunteer scientists. As soon as the paper arrives at the office of the journal, the process of vetting begins. Normally, a paper will be read first by the editor or an editorial committee member, who will then decide whether it should go to a referee and whom. Some journals, usually specialist ones, send nearly all papers to referees, whereas others initially weed out those that are clearly unsuitable. Most journals send the paper to two reviewers or referees. The reviewing of research papers, popularly called *peer review*, is done by peers. A peer is one belonging to the same group of the author based on some criteria such as age, education, or profession; in this case, a scientist working in the same field. Peer review is the evaluation of research articles prior to publication by people from the same field in order to uphold or enhance the quality of the work. The reviewers read, judge, and make comments on the article for its appropriateness and quality. The reviewer must be constructive by locating and fixing technical inaccuracies.

Scientists who write research papers should also be prepared to act as peer reviewers. You must view peer reviewing seriously, as it gives many opportunities to you as a writer. When you peer-review another author's work, you are supposed to appraise it based on both technical and stylistic considerations. You may also criticize the logic and reasoning and suggest improvements, if any. Your criticisms and suggestions are promptly communicated to the author. This process of peer reviewing is beneficial to you as you learn from many of the deficiencies of articles sent to you for review. Referees usually remain anonymous. If you are undertaking peer reviewing for the first time, you may be a bit perturbed about criticizing someone else's work. Take it as a challenge, go through the article critically, and comment on its shortcomings for improvement. If it is poorly written and not appropriate for the journal, do not hesitate to state your opinion or even recommend for rejection!

Referees are expected to comment on the methods used, reliability of the work, and its originality. Often, journals provide a checklist for proper reviewing. Some examples of the type of questions put to the referees are:

1. Does the article come under the subject area of this particular journal?
2. Does the article contain original work or is it a repetition?
3. Is the title clear, concise, and effective?

4. Is it scientifically reliable—including the ethical and statistical aspects?
5. Are the experimental methods used are adequate?
6. Are the interpretations of results and discussion in conformity with relevant facts?
7. Is the style of writing and use of English appropriate for the journal?
8. Is the abstract comprehensive and concise?
9. Comments on the quality of illustrations/photographs
10. Does the manuscript conform to the journal's requirements as indicated in the 'Instruction to authors'.

Make sure that your article fulfils these criteria before submitting it for publication. Referees or reviewers should not hold the articles with them for a long time, and it is not proper to take more than a month to return the paper to the editor. If the first two referees disagree and the paper has to go to a third one, it will take more time. In certain cases, it may also be sent to a statistician, if the paper includes substantial statistical analysis. Usually, referees will also be asked to comment on its acceptance as shown below:

- Accept the paper without any modification
- Reject the paper without any review
- Acceptance, subject to modification as suggested by the reviewers
- Rejection, but with an offer to reconsider, if it is drastically revised as suggested.

Accordingly, authors can expect a rejection letter, acceptance letter, or a letter asking for modification or revision with referee's comments. Acceptance of papers without modification is rare, but you can expect acceptance subject to modifications as suggested by the reviewers. You may also be provided with a copy of reviewers' comments. Go through the comments and modify your article suitably, and return the modified article promptly without much delay. Normally, they may inform you acceptance details including the issue number of the journal in which it would be published. Once an article has been accepted for publication, you may not be able to make any changes to it. Nevertheless, if an author recognizes that something crucial has been left out or got some figures wrong, he/she should inform the editor immediately of the mistake.

The reasons for rejection of an article can be many. Normally, the reasons may also be mentioned in the rejection letter. If rejected due to the reasons mentioned by the referees, try to revise your article and send it to a new journal. In case, the work is not original and merely repeats the work already published or unsuitable for the journal by way of its mandate, the article will not be accepted. Sometimes, the article may be too long or the article is written badly and is difficult to read. If the journal editors have doubts about the validity of results, the article will be rejected. If the authors feel that the referee's criticisms are unfair and invalid or that the referee has misunderstood your interpretations, you may write to the editor detailing your points. The author can explain why he/she disagrees with the referees and suggest some modifications to make the paper more meaningful. If the points raised by the authors are reasonable, most editors send the paper to a third referee.

11.7 Review Papers

A review paper is a secondary publication, and its main purpose is to review critically the previously published literature on a topic. It is a survey of primary literature over a given period indicating the developments and trends during the concerned period. An exhaustive list of references at the end of review serves as a good bibliography on the topic. As a research student, you may write two or three mandatory 'review papers'. One is for your credit seminar (for Ph.D., generally two) and the other one for your thesis. The purpose of seminars is to train the student to prepare and present review papers. This is also helpful to get hands-on experience in preparing the 'review of literature' for your thesis (see also Chap. 10 on literature review). Remember that presenting a review paper in a seminar is different from simply delivering a lecture (Chaps. 20 and 21).

Certain secondary journals concentrate on publishing review articles mostly covering up to date information in a new perspective. Most professional societies of repute publish reviews, and these are usually identifiable from their titles such as 'Advances in ....', 'Yearbook of...', and 'Annual Review of ...' Examples include '*Advances in Agronomy*'; '*Advances in Soil Science*'; and '*Annual Review of Plant Physiology*'. Occasionally, one or two review articles may appear in primary journals or even abstracting journals too. A thesis will also contain a comprehensive review on the researchable topic indicating the past and present status of the problem.

While writing reviews, remember that the intended audience for review papers and primary papers is different. The primary paper is highly specialized, and its audience is peers of the authors. A review paper is often long, ranging from 10 to 50 printed pages. The subject would be general in scope unlike original research papers. A good review paper offers critical evaluation of the published literature, and often, provides important conclusions based on the literature. The organization of review is also different. There will not be any sections on Materials and Methods, Results, and Discussion. Prepare an outline first similar to a term paper or assignment before you start writing. While writing, slight modifications can be made in the outline. Many review journals print the outline at the beginning of the journal where it serves as a convenient 'Table of contents' for prospective readers.

Most review journals demand critical evaluation of the literature along with bibliographic completeness. The strategy is to compile and annotate the papers published on a particular subject during a defined period. Most prefer reviews that provide a new understanding of a rapidly moving field, and the recent literature on the subject is evaluated and catalogued. Historical analyses are seldom given now, and whether to include historical analysis is a matter of choice. However, if you are reviewing a subject that has not been previously reviewed or one in which many misunderstandings has developed, you must cover the relevant historical aspects adequately.

11.8 Conference Papers

Conference papers have become a very substantial portion of the total literature in many areas of science. It may take many names—conference, congress, symposium, seminar, or workshop (Sect. 20.1); and many professional societies organize them frequently as an easy fund raising method. Scientists and research students are also attracted to such events because of the lure of easy acceptance of their abstracts or papers. Before sending papers to such conferences, you should check some important aspects. Conference reports are normally not subjected to peer review, and because of any real quality control, most publishers and institutions give 'Proceedings' paper a low status only. Therefore, think twice before sending papers to such 'conferences'.

If the conference reports are not in full proceedings form (papers with the IMRAD structure, peer reviewed, and with the questions and answers at the conference), it is not considered as validly published primary data. If the report is in 'abstract' or 'extended abstract' form, it is clear that it is not validly published; therefore, you should publish the data in a regular primary journal after redesigning it in the form of a full paper. An abstract is typically a one-paragraph write-up summarizing the results, which will not carry separate sections as Introduction, Materials and methods, Results and Discussion. No tables or figures are allowed in such abstracts, and the results are presented in a brief format.

Presently, many reputed conference organizers have switched over to 'extended abstracts' rather than 'abstracts' or 'full papers'. Even international conference organizers, who were publishing proceedings with full papers along with questions and answers in the past, have also shifted towards extended abstracts. There are some reasons behind this trend. A major reason is the status of a proceedings paper as discussed above. If it is included as a full paper in the proceedings, primary research journals may not accept it for publication as it would be a duplicate publication. Naturally, those authors having good research results may not like to send their papers to such conferences. Another problem is the cost and time involved in bringing out conference proceedings. Publishing full papers along with questions and answers after the conference and making them available to all the participants is cumbersome and expensive. Some organizers, who are also publishers of academic journals, circumvented this problem by publishing a compilation of abstracts before the programme, and later publishing full research papers as a special issue of the concerned academic journal of the organizers.

Normally, an extended abstract should not exceed 1000 words in double space (equivalent to five A4 size pages with double spacing) inclusive of text, tables, and illustrations. In print form, an extended abstract may have a size ranging from 1 to 2 pages. The organization of a typical extended abstract of 1000 words is as follows: (1) the title; (2) Names and addresses of the authors; (3) Introduction (about 100 words); (4) Materials and Methods (about 250 words); (5) Results and Conclusions (about 600 words including tables and illustrations); and (6) References (about 50 words). Both organizers and authors of abstracts work on the assumption that the original paper would later be sent for publication in primary journals. There is one advantage for the authors who present abstracts or extended abstracts. They can utilize all the criticisms and suggestions made while presenting the abstract at the conference for improving the quality of paper.

You can approach the 'conference' in a different way. Scientific conferences or meetings can be exciting because they do serve as the forum for presentation of new ideas. Conference presentations (both oral and poster) are usually used to present ongoing works. This gives the researcher an opportunity to disseminate new findings and to get valuable feedback from the audience. This kind of presentation at a meeting is useful for the preparation of the paper for later publication because the audience's questions may probe weaknesses in the work or the researcher's account of it, and shows where he/she needs to add or explain. Moreover, attending conferences provide you a chance to interact with eminent workers in your field.

In a conference, the 'visual and verbal' mechanism in the case of a presentation or 'visual' only mechanism in the case of a poster assures credit for discovery, and establishes that a particular worker is studying a particular topic. It is also a way out to make sure the researcher's right to continue an already started line of research. Other scientists normally will accept the signal and not interfere with the original scientist's right to investigate further a topic. Occasionally, certain topics are so significant and economically attractive that several researchers will compete for solution to the problem in order to gain scientific prominence or economic gain.

Speaking at conferences demands a different style from writing with more repetition and emphasis, because listeners, unlike readers, cannot go back over a point to absorb it more fully. You may use computer aided visual display of slides with points of emphasis along with tables and diagrams. However, make the slides carefully. For example, a table with large amounts of data may be acceptable for a printed article, but it makes an unreadable slide during a presentation. Such points on preparation of slides and presentation are discussed at length in Chap. 20.

11.9 Electronic Publications

Electronic publications or 'E-publications' are now widely accepted as a mode of information transfer through Internet and CDs. The Internet is a global system of interconnected computer networks, which is now offering umpteen opportunities as a source of research information. It offers instantaneous access to millions of computer files related to almost any subject. The World Wide Web (WWW) or simply Web or web is one of the applications that run on the Internet, which can be accessed by web browsers such as Google Chrome, Safari, Internet Explorer, Microsoft Edge, or Mozilla Firefox. You can present the information as HTML (Hypertext Markup Language) files, PDF (Portable Document Format) documents, Blogs, or electronic mails. These can be rapidly distributed and downloaded at will by Internet operations. Many journals are available free on the Internet (Open Access Journals), while some others can be subscribed 'online' by paying fees.

Electronic journals or E-journals have appeared in many scientific fields as an alternative to paper based journals. The electronic mode of publication appears to be the major medium in which we will distribute scientific information in the future. However, permanence of the publications is not yet assured because most websites are transient and the owners change the content frequently.

You can write for electronic publications in various ways. You can start your own World Wide Web site or blog, and use it for disseminating your ideas. Another opportunity is to participate in the editing processes of online free encyclopaedias like Wikipedia. You can also submit research papers to E-journals, a rapid way of research communication indeed! However, wean out spurious E-journals and those without impact factors or rating.

11.10 Short Notes and Other Forms of Reports

Regular research papers published in research journals are sometimes called full research papers, as they follow IMRAD structure and cover a particular aspect of study fully. However, most journals also admit research reports in short note form and some in letter format too. Some other forms of research reports are project reports, progress reports, semi-scientific articles, and popular articles.

Short Notes
Most journals have a section for *'short notes'* or *'short communications'*. The maximum length of short notes is normally 1000 words. It is one of the most rapid means of publishing current information and is suitable for reporting isolated observations, which do not require lengthy comments or excessive documentation. It can be considered as a concise, but complete, description of an investigation of limited scope. These are mainly the reports of ongoing projects or very short reports such as the report of the discovery of a new insect or a plant. The text is usually given in a narrative form following IMRAD but without the usual headings and sub-headings. If tables are included, often it is confined to one or two only. Limit the references to a maximum of two to three or without references.

Letters
Letters are regular features in weekly or fortnightly journals such as *Nature*, *Science*, and *Current Science*. Unlike short communication, only a few journals accept and publish letters from researchers. A letter can be considered as a useful and rapid form of communication of findings. Normally, a letter should not be more than 250 words. Similar to short notes, limit the number of references to two or maximum three. A letter can also be without any references. Like short communication, the letter format is one of the most rapid means of publishing current information and is suitable for reporting isolated observations. Letters also allow researchers to exchange opinions with other workers while the subject is still new. The above kind of journals also

encourages fellow scientists to write letters related to something that has already appeared in the journal.

Project Reports

Project work, which is part of a degree programme, involves the preparation of a report and its assessment. If the project involves an investigation and observations, the project report should be written as a scientific report with the usual headings of a research report. However, the size of a project report will be rather small compared to a thesis.

Progress Reports

These are meant for limited circulation only and are printed in limited copies. The progress of ongoing research projects of an institution is usually published for public scrutiny or for evaluation in the form of quarterly reports, half yearly reports, annual reports, summary reports, final reports, etc. For funded research, progress reports are mandatory, and they prescribe specific formats for submitting reports.

Semi-scientific and Popular Articles

Semi scientific papers and popular articles are two breeds of popular writing meant for non-scientist readers. Semi-scientific papers are written for the professionals not engaged in research. Popular articles are meant for non-professionals. For these articles, the style of writing is entirely different. A popular non-technical style devoid of any technical jargons, which stimulates the interest of the readers, must be used for these kinds of writing. These are useful to disseminate scientific principles in a simple manner to the wider public.

Semi-scientific articles are usually published in scholarly magazines (e.g. *Science Reporter*), which are intended for a broader expert audience. However, popular magazines are designed for common people. Usually, popular magazines are published weekly and the time taken for handling articles is too short. This tells upon the quality of reviewing process. The articles in popular magazines may not be technically deep. The emphasis is more on entertainment, training, stimulating, and advertisement. More often, you may find advertisements or lobbying pieces disguised as articles. Therefore, consult popular articles only for general reading and updating, but do not quote them for theses or research articles without proper follow up through other sources.

11.11 Copyediting and Proofreading

Each journal will have an editorial policy. The amount of editing or editing style varies between journals. The quantum of editing depends mainly on the quality of the paper submitted, for example, a technically sound but a poorly written paper requires heavy copyediting. The size and resources of the journal establishment also affects the editing process. Big journal establishments may have full time editors.

As expected, editing by a full time editor and part time editor will be different. Nevertheless, all journals normally check spelling, punctuation, grammar, style, lay out, and position and arrangement of tables and illustrations.

Editors are also responsible for ensuring internal consistency, for example, that the numerals given in tables match those in the text and that the cited references are correctly arranged according to the accepted style of the journal. Most journals have their own policy on nomenclature, abbreviations, citation and listing of references, units of measurement, spellings, etc. For example, journals from the UK and the USA usually follow different styles, most prominent is differences in spellings of many words. It is now easy for the editorial staff to contact the authors through e-mail for clarifications, if any. Try to address such queries immediately avoiding further delays in getting your paper published.

Once accepted for publication, the next job is to get the article ready for printing. The advent of sophisticated word processing facilities and offset printing has revolutionized printing industry. As most journals now demand electronic copies of the manuscript in addition to hard copies (certain journals demand only electronic copies!), typing, lay out, page setting, etc. became much easier. All these have made proof and proofreading relatively easier tasks. Printing errors were rampant during the era of letterpresses, and at that time, proofreaders were important part of the printing industry. In spite of all these developments, we cannot simply do away with proofreaders; you can still expect some problems such as spelling mistakes, grammatical errors, and lay out errors. Most journals send final proofs to authors for correction and approval. Computer typeset copies are usually sent to the author for proofreading. The author should read the proof carefully, make corrections, and return it promptly.

Copyediting

The phrase *copyediting* is used to denote the work that an editor performs to improve the grammar and style, accuracy, and formatting of a manuscript; and the term *copy* in this context means a written or typed text for final typesetting and layout for publication. Unlike general editing, copyediting might not involve changing the text substantially. Copyediting is done before the final typesetting and layout. An editor who undertakes this kind of editing works is often called a *copyeditor*. In general, the phrase 'copyeditor' is used in book publishing works, but the term *sub-editor* is used in newspaper and magazine publishing industry.

A copyeditor ensures that an author's manuscript or copy is correct in terms of spelling, grammar, punctuation, and usage so that readers can grasp his/her ideas easily. A scientist may be skilled at explaining a procedure or discussing the results properly, but the copyeditor is the one who makes sure that the manuscript's syntax is smooth, and that the writing adheres to the conventions of grammar and style. The copyeditor may also suggest some reorganizing and probably some changes to chapter titles and subheadings. If the work is a lengthy one such as a book, the copyeditor usually prepares a style sheet, a statement of overall editorial policy, for example, use of serial comma. A copyeditor is also responsible to avoid awkward errors of facts and point out any possible legal problems. Copyediting is usually done using the same symbols used for proofreading as given in Table 11.1. Not all the symbols may be relevant for copyediting, but the symbols for insertion, delete, and punctuation are frequently used. With the universal adoption of computer technology and Internet, copyediting is also entering a phase of great changes. Copyediting is now generally achieved by entering comments in a wordprocessing programme such as 'Word' instead of putting symbols or marks and making comments on a typewritten manuscript.

Proofreading

For proofreading, use standard proof correction marks only for making corrections as given in Table 11.1. Proofreaders' marks are symbols or short notations that indicate the changes to be made to the final copy of a text before printing. In a book-publishing firm, proofreading is generally done by editors, proofreaders, and typesetters. Authors should also know how to proofread using proofreading marks. They are responsible for correcting the page proofs of their work—be it a research paper or a book. Learn and use common proofreading symbols and techniques to proofread your work effectively.

There is a major difference between copyediting (described above) and proofreading. Unlike copyediting, proofreading is done on a true copy of the finished page, a ***proof***. Proof reading is usually done on a hard copy with a pen or pencil. However, copyediting is usually done on a double-spaced manuscript or on-screen; and when copyediting is done, editing symbols or marks are usually positioned directly in or above lines of text, as sufficient space is available between lines to indicate additions or revisions (this is the reason for using double-spaced manuscript). On the other hand, proofreading is done on a print ready material, and enough space may not be available between the lines. To take care of this limitation, proofreading marks are placed in two places on the page, in the margin and corresponding line of text. The correction, the word, or letters to be inserted or changed are indicated in the margin, line-by-line, and appropriate marks are made in the corresponding line of text to show their exact location. For example, the symbol to show the place where something is to be inserted (the caret symbol) is marked in the line itself, but the word or letter to be inserted is written in the margin only. If there is more than one correction in the same line, they should be written in the margin from left to right as they appear in the line separated by a virgule or slash (/). Start from the left margin and after exhausting the available space in that line, go to the right margin, and complete corrections.

Probably, insertion mark (caret) may be the most commonly used copyediting and proofreading symbol. The insertion mark indicates where a letter, word, phrase, or punctuation mark is to be inserted in the line. If something new is to be inserted after correction, it must also be indicated in the margins after putting a horizontal line through the word to be corrected and caret symbol over it. Another symbol in common use is the delete sign. If you want simply to delete something, strike through the word to be deleted, and show the symbol for deletion in the margin. A horizontal line delete mark is used for deleting a phrase, sentence, or paragraph; while a vertical or slanting line delete mark is used for deleting a single letter or mark of punctuation. If you want to ignore the changes you have already made and

Table 11.1 Standard copy editing and proof reading symbols

Instruction	Marginal marks	Example with in-text markings
Core symbols		
delete as indicated	ꝗ /	What is known known is knowledge.
delete and close-up (use only when deleting letters within a word)	ꝗ /	What is knoþwn is knowledge.
insert as margin indicates	is/art/	Speaking/an/. However….
replace	proper/	Use appropriate words.
insert a space	# /	What is knownis knowledge.
reduce space	less # /	What is known is knowledge.
close up	⊂ /	What is kno wn is knowledge.
Insert space between lines	# /	Anyone can become a speaker by hard work.
reduce space between lines	less # /	Anyone can become a speaker by hard work.
stet (let it stand)	stet /	What is known is knowledge.
transpose	trs /	See the on section results.
query to author: Is this right?	OK?	
omission here, see the original copy	out, see copy	A degree. The generally accepted ..
Formatting symbols		
wrong type face ype)	wf /	Speaking is an art However….
spell out figure or abbreviation	sp /	I purchased 2 water melons.
abbreviate	m/	The height of the plant is 3 metres
change to uppercase (capitals)	caps /	Go through agricola database.
change to small capitals	sc /	Speaking is an art. However….
Change to lower case	lc /	This is a RARE specimen.
change to italics	ital /	The practice is called Koottumundakan
change to roman	rom /	This is a rare specimen.
change to bold face	bf /	See the section on results.
change to bold face italics	bf ital /	Research Methodology
change to light face	lf /	See the section on **results**
underscore	us	Answer any five.
remove underscore	‡‡‡	Do not forget to reply.
superscript	2 /	The surface area measures 5000m2
subscript	2 /	MOP contains 60 percent K2O.

(continued)

Table 11.1 (continued)

Instruction	Marginal marks	Example with in-text markings
Punctuation symbols		
insert period	⊙ /	Speaking is an art⁁However…
insert comma	⁁, /	Similarly⁁weeding is also…
insert semicolon	; /	This is true⁁however…
insert colon	(:) /	It includes the following ⁁
insert hyphen (-)	=̂ /	He is a well⁁known teacher.
insert en dash (–)	1/n /	During 2010⁁2012
insert em dash (—)	1/m /	The test is over⁁disastrous.
insert interrogation mark	? /	What are the benefits⁁
insert exclamation mark	! /	It is simply enthralling⁁
insert apostrophe	' /	The students⁁efforts are laudable.
insert quotation marks	" / " /	He said⁁We must start it now⁁
insert parenthesis	(/) /	Competitive Grant Fund⁁CGF⁁
Insert ellipsis	--- /	I already told you that⁁
insert brackets	[/] /	⁁15-5-2013⁁
Insert slash	(slash) /	Preparation of thesis⁁dissertation
insert plus sign	(plus) /	a–b⁁c = x
insert minus sign	(minus) /	a+b⁁c = x
insert equals sign	(equals) /	a+b–c⁁x
Lay out symbols		
equalize space	eq # /	What is ⁁ known is ⁁ knowledge.
single space	(ss)	Anyone can become a speaker by hard work.
double space	(ds)	Anyone can become a speaker by hard work.
break and start next line	⌐ /	He searched.⌐He found.
take over the last word to the beginning of next line	(to) /	Anyone can become a speaker by [hard work.
take back the first word of a line to the previous line	(tb) /	Anyone can become a speaker by hard work]
new paragraph	¶ /	Speaking is an art. (However…
run in (no paragraph, continue)	(run in) /	Anyone can become a speaker. However, …
move to left	[/	[Speaking is an art. However…
move to right	] /	]Speaking is an art. However
flush left	⌐ /	⌐What is known is knowledge.
flush right	¬ /	¬What is known is knowledge.
centralize	][/	]Chapter I[

need to retain the original as it was before the correction, put continuous dots under a word, phrase, or sentence and note the word 'stet' (let it stand) in the margin. In the final draft copy, proofreaders may also consider artistic issues, for example, a small word such as 'it' alone in the last line of a paragraph (*orphan!*) or a very short final line of a paragraph at the top of a page (*widow!*) is awkward and has to be corrected. To minimize the impact of resetting, try to replace a word or phrase with a word or phrase of the same number of letters and spaces. This aspect of resetting impact must be considered when introducing other changes as well.

The proofreader is assigned with the important task of ascertaining the appearance of the finished product, the printed matter. Remember that a proofreader's job is not revision, but correction; to eliminate all the typographical errors crept into the manuscript after copyediting or inadvertently introduced in the production stage. Proofreaders also verify page numbers and other recurring elements at the top or bottom of a page. They ensure that the typeface and font size for one text element match another element of that class. They must also check the captions of illustrations and tables to ascertain whether they match the contents of illustrations or tables.

Remember that proofs are not for rewriting, adding some new ideas, or to change the style of writing. The prime intention is to verify that nothing has been missed out, and that the intended meaning has not been changed inadvertently in the process of converting the manuscript into print. Therefore, authors of research papers themselves have to examine the final proof. It is better to read the paper twice, first with the original manuscript to ensure that nothing has been missed out, and the second to ascertain that there are no typographical errors. Check the tables and illustrations carefully; sometimes, these may have been retyped or redrawn. The author should also answer any queries marked on the proof by the editors. Sometimes, spellings, capitals, punctuation, etc. have been changed by the editor. They might have been altered to conform to the journal's house style, and a journal's house style is not open to negotiation. Journals often indicate a date for the return of proof, and if delayed, the paper cannot be included in the early issue of the journal.

The time the journal publishers take from acceptance to printing a paper is dependent mainly on the periodicity and reputation of the journal. If the journal is a weekly, it may take only a few weeks to see the paper in print. If the journal office has intimated the acceptance of a paper, the author can cite this paper in his/ her forthcoming papers as 'in press'.

Some journals give authors some free reprints or offprints. An *offprint* is an additional copy of the article made when the journal issue is printed. Therefore, it will be an exact replica of the article as it did in the journal. At the same time, a *reprint* is a copy of the article reprinted separately once the regular printing of journal is completed. It may have a different layout.

Bibliography

Barrass, R. 2002. Scientists Must Write: A Guide to Better Writing for Scientists, Engineers and Students (2nd Ed.). Routledge, London, 204p.

BMA [British Medical Association] 1965. The reasons for writing (Report of Editors Conference). Br. Med. J. 2: 870–872.

Bond, A. 2007. Your Master's Thesis: How to Plan, Draft, Write and Revise. Viva Books, New Delhi, 135p.

Booth, V. 1996. Communicating in Science: Writing a Scientific Paper and Speaking at Scientific Meetings (2nd Ed.), Cambridge University Press, 96p.

Cargill, M. and O'Connor, P. 2009. Writing Scientific Research Articles: Strategy and Steps. John Wiley & Sons, 173p.

Davis, M. 2005. Scientific Papers and Presentations. Academic Press, Massachusetts, 356p.

Day, R.A. 1983. How to Write and Publish a Scientific Paper (2nd Ed., Reprint, 1997). Vikas Publishing house, New Delhi, 160p.

Day, R.A. 1989. The origins of the scientific paper: The IMRAD format. Am. Med. Writers Assoc. J. 4(2): 16–18.

Day, R.A. and Gastel, B. 2006. How to Write and Publish a Scientific Paper (6th Ed.). Cambridge University Press, Cambridge, 320p.

Gustavii, B. 2008. How to Write and Illustrate a Scientific Paper (2nd Ed.). Cambridge University press, Cambridge, 168p.

Hartley, J. 2008. Academic Writing and Publishing: A Practical Guide. Routledge, New York, 196p.

Lester, J.D. and Lester, J.D. Jr. 2009. Writing Research Papers: A Complete Guide (13th Ed.), Longman, 416p.

Malmfors, B., Garnsworthy, P. and Grossman, M. 2004. Writing and Presenting Scientific Papers (2nd Ed.). Viva Books, New Delhi, 153p.

Matthews, J.R. and Matthews, R.W. 2007. Successful Scientific Writing: A Step-by-Step Guide for the Biological and Medical Sciences (3rd Ed.). Cambridge University Press, 256p.

Rubens, P. 2004. Science and Technical writing: A Manual of Style (2nd Ed.). Routledge, New York, 427p.

Smith J. 1985. Publication. In: Hawkins, C. and Sorgi, M. (eds), Research: How to Plan, Speak and Write about It (4th Indian Reprint, 1993). Narosa Publishing House, New Delhi, pp. 153–169.

Sollaci, L.B. and Pereira, M.G. 2004. The introduction, methods, results, and discussion (IMRAD) structure: A fifty-year survey. J. Med. Libr. Assoc. 92(3): 364–371.

Whimster, W.F. 1985. What the critical reader looks for in an original article: A guide for writers. In: Hawkins, C. and Sorgi, M. (eds), Research: How to Plan, Speak and Write about It (4th Indian Reprint, 1993). Narosa Publishing House, New Delhi, pp. 85–109.

Chapter 12
The Structure of a Thesis

But man has still another powerful resource: natural science with its strictly objective methods.
Ivan Pavlov (1849–1936), Russian Nobel laureate

Students enrolled for Master's degree in agriculture and allied subjects have to conduct a short-term original research work, and present the outcome in the form of a thesis. A thesis is also mandatory for higher degrees in medicine and engineering. In basic sciences such as physics and botany, although thesis is not mandatory for Masters programme, the present trend is to have a dissertation or at least a project report. All the disciplines, whether basic or applied, demand the submission of a thesis for the award of Ph.D.

12.1 Thesis or Dissertation?

Research students often hear the terms thesis and dissertation, and may wonder about the exact differences between them. Although many of us treat thesis and dissertation as interchangeable terms, please note that there are some differences, especially with respect to size and purpose. *Thesis* is a certificate that the author has been trained for doing independent research. A thesis is in the form of a report along with some mandatory certificates, which includes the findings of the research work done by the student in partial fulfilment of his/her academic requirement. The internationally accepted maximum length of a Masters thesis is 50,000 words and Ph.D. thesis 100,000 words. Ideally, the size shall be within the range of 40,000–60,000 words.

A *dissertation* is also a certificate for education and conduct of research, but it does not indicate that the researcher is capable of doing independent research. Often, the research reports submitted as a part of Master's degree and M.Phil. programme of traditional universities are called dissertations. A dissertation is comparatively shorter in length than a thesis, and the size is usually 20,000–25,000 words. Please note that in the USA, thesis and dissertation are just the opposite of what we have

C. G. Thomas, *Research Methodology and Scientific Writing*,
https://doi.org/10.1007/978-3-030-64865-7_12

already seen; for them, the research report submitted by a student for a Ph.D. degree is a dissertation and the one submitted for a Masters degree is a thesis!

The research work for students must be designed in such a way as to give sufficient exposure to them in the intricacies of literature collection, project preparation, conducting experiments, taking observations, analyzing data, and finally, presenting results and interpretation in the form of a thesis and later as research papers. Normally, the research work for a Masters programme in agriculture and allied subjects is of one-year duration with sufficient time to prepare the thesis. Please note that in disciplines such as agriculture where Masters programme is offered with research work followed by the submission of a thesis, there will not be any separate M.Phil. programmes. Research works for Ph.D. programmes are normally of 2–3 year duration.

Do not think that the job of a research student is completed by the submission of a thesis and the award of degree! They must prepare concise research articles from the thesis, and publish the papers to reach a wider audience. Probably to make the students and guides comply, some universities have made it mandatory to prepare at least one article for M.Sc. and two for Ph.D. for awarding the degree!

Choosing a topic and developing it into an outline of research work have already been discussed in Chap. 7. In this chapter, guidelines for developing various parts of a thesis and typing instructions are discussed. For more details, refer to exclusive sources on thesis writing such as Murray (2006), Fisher and Thompson (2014), and Eco (2015).

12.2 Parts of a Thesis

A thesis has three main parts—*preliminary pages, subject proper,* and *references.* Sometimes, an additional part, *appendixes* (although not mandatory) may also be attached. Individual items within these parts can be adjusted in the following order. *Abstract,* although not a part of the thesis, is also considered along with it.

Preliminary Pages

Title page

- Signature page
- Declaration by the student
- Certificate from the guide (Major Advisor)
- Acknowledgements
- Table of contents
- List of figures
- List of tables
- *List of symbols, abbreviations,* or *nomenclature* (optional).

Subject Proper

Introduction

- Review of literature

Materials and methods
Results
Discussion
Summary and conclusion.

References

Appendixes

Abstract

12.3 Preliminary Pages of a Thesis

Title Pages

The title of a thesis and scientific paper made out of it may be different. In the case of a thesis, the title already approved by the academic bodies must be used. Therefore, sufficient care must be taken to select attractive titles as mentioned in Chap. 7 when you formulate your research problem and submit the ORW (Outline of Research Work) for approval. However, scientific papers made from the results of the thesis can have titles according to the contents of the articles. For a thesis, two title pages are usually given; the first is the cover page with title of the thesis, author's name, and department and college name. In the second title page, title name, author's name, mandatory name of the thesis (whether the degree is Master of Science or Doctor of Philosophy), faculty name, name of the department, and name of the institute or college are given. Write the title of the thesis in capital letters (but scientific name, if any, in italics).

Certificate Pages

After the title pages, the *'signature page'* having the signatures of all the members of the Advisory Committee is given. This is to serve as an official endorsement. The signature page is immediately followed by the *'declaration by the student'*, and then, *'certificate from the guide'* (Major Advisor) is given. There may be slight variations in style between universities. These two certificates declare that the thesis is a bonafide wok, and unethical practices were not employed in its production. For the above certificates, the language and style and the formats should be as prescribed by the institution.

Acknowledgements

The 'Acknowledgements' section has almost attained the status of an obligatory part of a thesis, and a separate page is earmarked for it. It is usually given at the beginning, just before the 'Table of contents'. Students have to acknowledge the help from their research supervisor or guide, all the members of the advisory committee, the department, the financing authority, and others, if any. However, avoid effusive and out of context acknowledgements.

While writing acknowledgements, it is better to use first person. The general feeling among the academia that authors are more scientific, if they write in the third person is obsolete now. Using the third person in acknowledgements, for example, the style occasionally seen, 'The author wishes to acknowledge the help…' is rather old fashioned and not straight forward. Instead, you may start the sentence as, 'I wish to acknowledge the help…'.Try to avoid colloquial usages such as *Bhai, Didi, Chettan, Chechi*, etc. Gratitude may be expressed to your friends also but never go on adding names!

Table of Contents

Except title pages, certificates, and acknowledgements, all other major divisions of the thesis should be listed in the 'Table of contents'. List the entries in the 'Table of contents' as given in the text. If sub-headings of a chapter are given in the 'Table of contents', parallel sub-headings of all other chapters should also be given. If you have followed decimal numbering for headings, include these also in the 'Table of contents'. Further, ensure that the page numbers entered in the 'Table of contents' are correct.

List of Tables

All the tables given in the thesis should be listed and the same captions should be used in the list as given with the tables.

List of Figures

Figures or illustrations are listed separately and should be numbered in Arabic numerals (1, 2, 3, and so on). All types of illustrations (maps, diagrams, photographs, etc.) should be referred to as 'Figure' in the text. However, the term 'Plate' is often used for a full page of photographs. The plates are numbered in capitalized roman numerals, for example: Plate I, Plate II, and so on, but individual photographs on the plates are numbered in Arabic numerals. For example, a third photograph on a plate will be referred to in the text as Plate III Fig. 3. A separate list of such plates has to be provided after the list of tables and figures.

Dedication

There is a tendency among some students to dedicate the thesis to someone they love or remember or even to almighty. This practice should be discouraged. Research guides should advise the students properly in this regard. **Thesis is strictly an academic document—a certificate; therefore, it should not carry any dedication.** If you are very particular, instead of dedication, express your indebtedness in the 'Acknowledgements' section.

12.4 The Subject Proper

The IMRAD structure (Day 1989) already discussed must be followed for your thesis. In a thesis, each part of IMRAD is prepared in separate chapters in a comprehensive manner. The main body of a thesis is divided into six chapters, namely *Introduction, Review of literature, Materials and methods, Results, Discussion,* and *Summary*. In the case of a research paper, a separate section as 'Review of literature' is not given but a few pertinent references are reviewed and given as part of 'Introduction'. However, in a thesis, a comprehensive 'Review of literature' is given as an independent chapter immediately after 'Introduction'. Similarly, the 'Results' and the 'Discussion' are presented in a detailed manner in separate chapters with appropriate tables, diagrams, and other illustrations unlike a research paper where they are usually combined. The 'Abstract' is generally attached at the beginning (before the Introduction) or end of a thesis and bound with it, usually with a facing sheet. The main body of the thesis begins with Chap. 1, page 1 and contains all the subsequent chapters duly paginated. Each chapter should bear a distinct title.

Introduction

The 'Introduction' part in a thesis must be sufficiently comprehensive but brief as this forms the first chapter. The 'Introduction' introduces the reader to the state of knowledge before starting the research, defines the gap in knowledge, and states what the student planned to do. The aim of an 'Introduction' is to tell the reader the justification and rationale in taking up the research work. State the significance of the study as answer to the question: 'Why did you start the experiment?' The hypothesis or research question, its importance, the clientele, and the benefits or practical utility of the study are to be mentioned as a part of introduction.

The purpose and scope of the work must be stated clearly for justifying the work. You may also set out precisely your objectives in taking up the research. Include a statement of the problem, any background information leading to the recognition of the problem, and a clear statement of the hypothesis or research question (remember that for exploratory or descriptive studies, you may not have research hypotheses, but research questions only). As stated in Sect. 3.3, most hypotheses contain an independent variable and a dependent variable. Make sure that your hypothesis includes both variables, and that the hypothesis is reasonable. If you conceive that a particular relationship exists between the independent and the dependent variables, state clearly with previous research results or observed evidences, what really prompted you to believe that your assumption might be correct.

While writing 'Introduction,' references to earlier works shall be indicated only in relevance to the need of the study. If you have published any preliminary research note or research paper on related works previously, you may mention them in the introduction. This should be followed by the importance of the research with practical implications. The objectives of the study should be clearly spelt out. If the answer to the question 'Why did the student select this topic?' could not be understood after reading the 'Introduction', it has failed, and it will be rated as a poor 'Introduction'.

Review of Literature

In a thesis, the 'Introduction' part of the IMRAD structure is generally expanded to give a comprehensive review of literature as a separate chapter. A comprehensive review of the works done on the subject selected for the study and the lacunae in the subjects are presented in this chapter. After going through the review, one should get answer to the question: 'What was the level of knowledge on the chosen topic just before the experiment?' The review must be divided in to sections with sub-headings. Present the findings of various authors interconnecting them with your observations. Some students may simply list the findings in a chronological order, and the quoted references may not have any relationships! This is of no use and the reader will not be in a position to deduce anything out of it. In a thesis, the review is meant to present not only up-to-date knowledge of the field but also the intellectual ability of the student to synthesize and put together information from different sources. Proper guidance from the research guide is necessary to have a good review of literature.

The literature on a subject is to be reviewed in a concise manner. It should include only those works, which are relevant to the problem investigated. Students may be confused about the number of references to be reviewed in a thesis. This is mainly dependent upon the topic you have chosen for the study. For some topics, you may find numerous references, and it is impossible to review all of them. You then chose the most relevant and the latest ones. On the contrary, for certain topics, there may not be many references directly relevant to the study. In such cases, you have to search sources of information on related topics. Remember that it is not the number of references that matters but their relevance. The review should be up to date, indicating that the student was following the literature during and after the study. It is assumed that the candidate has read all the original references.

List all the references cited in your thesis following Name-Year system. In a thesis or research paper, as far as possible, use only primary literature, that is, original research articles authored by original investigators. However, you can use secondary literature, especially reviews, to substantiate your claims in the discussion section. Exercise extra caution when using websites as references. Anyone can put just about anything on a website, and you have no definite way of knowing, if it is a fact, opinion, or simply imagination (for more details on literature review, refer to Chap. 10).

Materials and Methods

After reading the chapter on 'Materials and methods,' one should get answer to the question, 'What did the student do?' The answer must include the details of materials used—land, soil, plants, animals, patients, chemicals, or others; and the methods or techniques employed in both the laboratory and the field. Sampling procedures adopted and the time and method of observation are usually described. Details on experimental design, replication, blocking, and the statistical techniques (with authority) used to analyze the results are also mentioned. If you have used a computer programme for analyzing the data, its name shall also be mentioned with reference details. Sometimes, the research involves only non-experimental or descriptive studies. In such cases too, you have to mention the techniques or methodology followed to obtain information. In the case of field experiments with plants, details on cultural operations are usually provided.

Unlike a research paper, where the 'Materials and methods' are given in a concise manner, these are described rather exhaustively in a thesis. Even then, if a well-known technique has been employed, it is sufficient to give only the relevant reference to the original source. However, modifications, if any, made to the published techniques must be described in detail. For example, if you have used 'Walkley and Black rapid titration method' to determine organic carbon status of soil, just state that you have used Walkley and Black rapid titration method to determine organic carbon status of the soil followed by the source as (Jackson 1958). The method called 'Walkley and Black rapid titration method' and the book '*Soil Chemical Analysis*' by M. L. Jackson are popular among agricultural workers. However, if the technique employed is available in an obscure source only and not known widely, you must describe it with full details.

You must also mention the ethical requirements of your research work, especially if the work is related to human or animal subjects. Details such as the criteria for exclusion or inclusion of subjects, informed consent, and approval of ethical committee are required (see Chap. 23).

Results

The data collected and compiled are described in the chapter 'Results.' The 'Results' is the most important part of a thesis. In this section, you present your findings with supporting data. This chapter must answer the question, 'What did the student find?' The 'Results,' therefore, should communicate the facts, measurements, and observations gathered by the student supported by data in tables, diagrams, graphs, and photographs. In the IMRAD structure of a research report, in fact, all the 'MRAD' part is concerned with the results. The 'Materials and methods' section shows how the researcher obtained the results, and the 'Discussion' section explores the significance of the results. In other words, the 'Results' section is the backbone of a thesis. This section, therefore, should provide the most critical information about the study supported by data allowing the researcher to discuss how the hypothesis is (or not) supported.

Before you start writing 'Results,' go through all the data you gathered and evaluate critically how the data is related to your hypothesis. Tabulation of data after subjecting them to appropriate statistical tests is the first step in presenting results. Narrate the results after putting them in presentable tables, charts, or graphs. If necessary, photographs can also be used. Describe each result pointing out the most relevant observations. If deemed necessary, you can also include observations that were not presented in a table or figure. However, do not attempt to discuss, interpret, or explain anything in the 'Results' section. Keep aside all such interpretations and explanations for the 'Discussion' section.

Tables and illustrations are important parts of the 'Results' section. The data can be arranged in several formats as tables, charts, graphs, line diagrams, and photographs (see Chap. 13). Extreme care is needed in arranging data in tables. Presenting data in horizontal rather than vertical columns is preferred as it facilitates better comparison of data. Watch for the size of tables. Note that through a little rearrangement, tables with even cumbersome dimensions can be put in to a manageable size. The title, column headings, and footnotes—all have value in helping a table to tell its own legend. In the tabular format, only the mean values are presented after statistical analysis (If applicable, LSD (Least Significant Difference) and SEm (standard errors of the mean) must also be given). Normally, raw data are not given in a research paper, but in a thesis, raw data are sometimes given as an appendix. Each table and figure must be numbered consecutively and complete with captions and a reference to each must be given in the text. Do remember that caption goes under the figure, and title with description goes above the table. Try to make each figure and table as complete as possible so that it could have a separate identity.

Discussion

In any research report including thesis, the 'Discussion' is the most difficult section to write. The discussion should answer the question 'What do the results mean and how do they relate to what is already known?' It is in the 'Discussion' that the student contributes something to the existing knowledge.

The primary purpose of the 'Discussion' section is to show the relationships among observed facts. This chapter must show how far the objectives set out or assumptions made at the start of the experiment are proved or disproved. The main principles or conclusions revealed by the study should be clearly brought out by giving evidences. Exceptions and opposing theories, if any, should be explained. Sometimes, the true meaning of the data may be completely masked because of a faulty interpretation. Do not encourage the tendency of some students to conceal negative results. The research guide should advise the students not to conceal negative results, if any.

In the discussion section, we 'discuss' the results and do not repeat the results once again. Students must clearly understand that the function of a discussion section is to interpret the results properly, and not to simply repeat the results in some other ways. Try to interpret the data in appropriate depth. In other words, when you explain a phenomenon, you must describe the likely reasons or mechanisms behind that phenomenon. Some students, due to lack of experience or proper guidance, present a superficial interpretation, sometimes, simply re-stating the results without bothering to explain underlying mechanisms. Any attempt to prove or substantiate a point should take into account the findings of earlier research workers. State clearly how your results and interpretations agree or contrast with the published works of others. Remember that in life sciences, as phenomena are so complex, it is possible to have more than one interpretation. While interpreting your results, include previous results or comments to buttress your arguments. You need not mention all the previous works, but choose the most relevant works and include them based on their relative importance to the present study.

If your results differ from your expectations, explain why that has happened. If your results are in agreement with your expectations, then explain the theory behind the phenomena. It is not acceptable to simply state that your data are in expected lines without further explanations. For example, the statement 'The results are in agreement with…' will not reveal anything to the reader. Similarly, we will not say that the hypothesis is 'proved' or 'disproved' or that it is 'correct' or 'incorrect.' The preferred practice is to use words like 'supported,' 'indicated,' or 'suggested'.

The theoretical implications of the work as well as any practical applications must be discussed as a part of 'Discussion,' taking care to avoid speculative explanations.

You can use correct logic for argumentation; and based on the arguments, draw your conclusions. Due to lack of confidence, novice researchers may write, for example, 'This seems to indicate…' or 'This might be due to …' Avoid such ambiguous statements. Similarly, take extreme care to avoid logical fallacies as explained in Chap. 14. All reasonable explanations for an observation based on one's own or other's work should be given. When you cite a source, do not forget to add the source immediately in the list of references using an appropriate style taking care to avoid plagiarism. You may point out lacuna in the present level of knowledge, but a general statement as 'further studies are needed' is not appropriate. It is advisable to conclude the discussion with an overall message of the study as a final paragraph. Probably, in this paragraph, you may point out some future lines of work as well based on identified gaps in knowledge.

Summary

The gist of the thesis is prepared as a separate chapter and given under the title 'Summary' or 'Summary and conclusion.' In fact, a summary is the summing up of all the messages. This part in a thesis must include an introduction, materials and methods in brief, main findings, and conclusion. Begin the summary with a problem statement briefly describing the context of the thesis work, its importance, and significance. State the original objectives of the study, which you have aimed at the beginning of thesis. You may also state your hypothesis or research question, followed by summarizing the materials and methods adopted to achieve the objectives. Thereafter, you narrate your salient results or contributions. You can list your results in order of decreasing significance, that is, with the most important contribution first. While writing conclusion, advancement to knowledge and positive gains to science must be clearly stated. As a part of summary, you may also suggest possible future line of work that could be done as a future thesis or departmental work.

References

The section 'References' indicates the sources from which the student has obtained information. There are two parts to a reference citation; first, the way in which the item is cited in the text when it is quoted or discussed, and second is the way in which it is listed in the 'References' section at the end of a document. Citation and listing of works of other researchers must be done carefully. Prepare a complete list of works cited in the text according to Name-Year system. These aspects are examined in detail in Chap. 15.

Appendixes

Sometimes, materials such as original raw data, survey schedules, and meteorological data, which may detract the reader from the main theme or make the text unduly large and poorly structured, are given as appendixes after the list of references.

12.5 The Abstract

An abstract is a brief summary of a document. The abstract of a thesis should be a correct précis of the thesis. Remember, a 'summary' is different from an 'abstract', which usually contains more information than an abstract. The abstract is often written as the last part of the thesis after completing all the other sections of a thesis regardless of its position in a thesis—before the introduction section or at the end of thesis depending upon the style adopted by your institution.

A good abstract should be a miniature version of the thesis, and provide a brief summary of each of the main chapters of the thesis. A typical thesis abstract should have information on four aspects—(1) the rationale and main objectives of the study, (2) the materials and methods used, (3) the summary of the results obtained, and finally, (4) the conclusion of the entire thesis. The abstract should be self-contained and able to stand alone without the need to consult the full text. An abstract should not contain citations or references to tables or figures given in the main part of the thesis. It should not contain any information that is not in the thesis. The title is a part of the abstract; therefore, it should not be repeated. Avoid ambiguous statements such as '... are described' or '... will be presented'. The results should be described in the past tense. Avoid the use of unfamiliar terms, acronyms, and abbreviations in the abstract. If they are to be used at all, they should be defined at first mention.

The abstract of a thesis shall not normally exceed 300 words in the case of Masters thesis and 500 words in the case of Ph.D. thesis. It shall be of high quality and properly edited. A copy of the abstract is generally attached with the thesis and bound with it, usually with a facing sheet. Normally, in addition to attaching a copy of the abstract in the thesis, two additional copies of abstract prepared according to the style of thesis abstracts should be submitted to the university. This is insisted as abstracts of theses approved for the award of degree by most universities are published as 'Thesis abstracts.'

12.6 Formatting Requirements of a Thesis

The thesis should be neatly formatted and sufficient copies made. The formatting requirements of a thesis being discussed here are general in nature. There may be minor variations in the styles prescribed by different universities. Consult the latest 'Academic Handbook' or 'Instruction Manual' before finalizing your thesis. If there are no specific instructions, you can definitely follow the instructions given here as such. Establish a formatting sequence, and follow it throughout the thesis. Bear in mind that consistency in thesis formatting is an important aspect, which contributes greatly to the attractiveness and readability of a thesis. This means that the student formulates and follows certain conventions on spacing, typeface, font, page numbering, heading sequence, page setting, and other aspects of appearance.

In normal case, students have to prepare a minimum of five copies of the thesis of which one copy (the original copy returned after *viva voce* examination) is for the student, one copy for the department, one for the college library, one for the major advisor, and one for the university library. Sometimes, the student has to make one extra copy, if he/she is in receipt of a fellowship and the funding agency demand a copy of the thesis.

12.7 Paper and Text Spacing

Use good quality A4 size (21.0 cm × 29.7 cm) pure white bond paper for printing. A bond paper often contains rag fibre (cotton) pulp making it a stronger and lasting paper. Several institutions insist to use paper with at least 25 percent rag content so that it lasts long without appreciable fading, discoloration, or deterioration. Some institutions allow use of A3 size (29.7 cm × 42.0 cm) paper for oversized figures and tables.

The thesis must be word-processed with a letter-quality printer. All the pages should be printed only on one side of the paper with a line spacing of double or one-and-a-half spacing. When manual typesetting was the norm, double spacing was compulsory. After the universal adoption of computer typesetting, most universities now permit one-and-a-half spacing for regular texts, and single space for block quotes, footnotes, headings, tables, table and figure captions, reference entries, and for the 'Table of contents.' When setting paragraphs, you need not give any extra space between paragraphs, if you are following one-and-a-half spacing or above. However, if single spacing is used in the 'Table of contents', leave an extra space between main (chapter) headings. Similarly, if single spacing is used in the 'References' section, leave an extra space between each entry.

Margins

Give sufficient margins on all sides. The most commonly recommended margins on right and bottom sides of the paper are 25 mm (1″) and left and top sides 38 mm (1.5″). On the left side, a higher margin is given to facilitate binding; and on the top side, extra margin is needed to accommodate page numbers. Indent the first line of every paragraph by 12.5 mm (0.5″). Some exceptions on margins are commonly allowed. You may use a larger top margin on main heading pages. Similarly, smaller margins may be used with appendix material, but the page numbers must appear in the same location as on the text pages.

Page Numbering

All the pages in a thesis must be numbered. However, page numbers are not shown on title pages and main heading pages. Put page numbers on top of the pages in the same typeface and size as the text. Preliminary pages shall be numbered with lower case roman numerals starting with 'ii' on the signature page (title page is 'i', but not shown). The body of text is numbered with Arabic numerals starting with

'1.' Do not display a page number on a full-page figure, but include these pages in the page numbering. All the pages must display a page number, numbering the pages consecutively throughout the thesis. Provide at least 2 cm margin between page number and top of page, and put page numbers preferably in the centre above text. Placement of page numbers must be consistent throughout the thesis; the numbers must be in same location in all pages. Do not put any punctuation before or after page numbers.

Design and Formatting of Chapters

The design specifications given below are on the assumption that you are following one-and-a half spacing and font size of 12 points. In general, theses shall not be more than 5 cm thick (greater than about 300 pages), as thickness above this renders it unwieldy. If there are more than 250 pages, usually two volumes are made, dividing the manuscript at the end of the last chapter that comes immediately before the size limitation. Provide a title page for both volumes with 'Volume 1' or 'Volume 2' under the title on each title page. Number the text in the second volume continuously from the end of the first volume. Make one 'Table of Contents' page and place it in the first volume with entries from all volumes included. As discussed already, a thesis normally has six chapters—Introduction, Review of literature, Materials and methods, Results, Discussion, and Summary. Authors should arrange the thesis in such a way that it is easy for an examiner to read. The charts, tables, and diagrams must be put close to the relevant text.

Always begin a chapter on a new page. A chapter of a thesis will have a chapter number and chapter title. In most cases, it has to be divided into sections and sub-sections depending upon the nature of its contents. The placement of headings of sections and sub-sections should follow the same order in all the chapters. The headings should be short captions rather than sentences. The word 'Chapter' is typed in capital letters and its number in capital roman numerals with no punctuation. It is centred horizontally 5 cm from top of the page. After chapter number, chapter title is given, which is centred one-and-a-half spaces below the chapter number and written in bold capital letters. You may use a larger font (16 points) for chapter titles. You may begin the first line of text or the heading three line spaces below the chapter title.

Indicate the major divisions of a chapter by centre headings. There must be at least two major divisions of a chapter, if centre headings are to be used. For centre headings, capitalize all the words in boldface. Give one-and-a-half spacing between the heading and the immediate text or the next heading. Along with centre headings, side headings can be used to indicate chapter sub-divisions. If no centre headings are possible as noted above to indicate major chapter divisions, major side headings can be used.

Instead of centre headings and side headings as discussed above, the thesis can be arranged with major side headings, sub-headings, and sub-subheadings. The major side heading is typed flush with the left margin and could be in capitals without boldface separated by one-and-a-half spaces from the text above and the text below. The major heading may be followed by side sub-heading in bold title case (initial letters of all key words except articles, prepositions, and conjunctions in capitals). Use bold lower case italics for the sub-subheadings. Paragraph headings represent the last level divisions within side headings or side-subheadings. These are indented like paragraphs and can be set in lower case italics, and the text run on after a period or colon on the same line.

Numbering the Sections

When preparing a thesis, especially to facilitate cross-referencing, various sections can be identified by decimal numbering or point numbering. In using this method, except the chapter title, no other headings are centred. The chapter title is numbered 1 (arabic numeral), and the first major heading in this chapter is numbered 1.1, the next 1.2 and so on. The sub-headings below the major heading 1.1 are numbered 1.1.1, 1.1.2, 1.1.3, and so on. Although it is possible to continue this decimal numbering (e.g., 1.1.1.1 or 1.1.1.1.1), soon this becomes cumbersome and confusing to both the authors and readers. Therefore, if you are using decimal numbering, it should not go beyond two points (e.g., up to 1.1.1).

Lay-Out of Tables

Tables are used to present extensive numerical data within a short space and with greater clarity. A table should be self-contained and self-explanatory (for more details on tables, see Chap. 13). In no case, however, should a table be used for sharing information, if it has already been explained adequately in the text. Each table should have a number and caption given at its top. The title should be in sentence case with only the first letter capitalized. Use a smaller font (11 points) and single spacing in order to make the table compact. If a table cannot be accommodated in 'portrait' orientation, you may put it in 'landscape' format. If needed, the table can be continued to the succeeding page with a repeat of appropriate column headings and captions indicating the continuation (e.g. Table 12 contd.).

If the data are simple, it is better to present them in textual form rather than tabular form (Day and Gastel 2006). For example, the following data (fictitious) need not be presented in tabular form (Table 12.1).

Instead of the above table, you may write, 'A survey of 150 villages in 3 districts of Kerala showed that 120 men and 160 women were affected by vector borne diseases during 2011–12.'

Language and Style

Language and style commonly used for technical writing must be followed for writing thesis also. If the language is apt and expressive, the results of research

Table 12.1 Number of persons affected by vector borne diseases during 2011–12 (fictitious data given as example only)

No. of districts	Villages surveyed	Persons affected	
		Men	Women
3	150	120	160

become comprehensible and appreciable. Therefore, apart from caring for the technical aspects of the thesis work, research students should enthusiastically take care of language aspect too. Use simple and familiar language. Words and sentence structure should be chosen carefully deploying your linguistic skills to bring out the focal point. Avoid long words and phrases, clichés, redundant words, and repetitive usages. Bestow special attention to use precise words and grammatically correct sentences. Refer Chaps. 16–18 for a brief overview of some hints for improved technical writing.

Typeface and Fonts
The old manual typewriter had very few typefaces, and the choice of typefaces was very limited. However, after the advent of computer aided printing, publishers can select a variety of typefaces. In this connection, please note that *typeface* and *font* are not synonymous. The overall visual appearance or style of lettering is called typeface, and the term font is often used to mean a specified appearance such as roman, bold face, or italic fonts. For example, a given typeface such as Times New Roman may include roman, italic, or bold face fonts. Typefaces belong to two broad categories, *serif* and *sans serif*. A serif is a short line or finishing stroke at the top or bottom of letters as in 'T' of Times New Roman. Some of the most common serif typefaces are Times New Roman, Georgia, Palatino, Minion, Berkeley, Bookman, Garamond, and Courier. A typeface without these flairs is called sans serif. Ariel, Helvetica, Myriad, Verdana, Optima, and Trebuchet are some of the most common sans serif typefaces. Serif typefaces give a visual illusion of connecting the letters in a word. Therefore, it is generally agreed that a serif typeface is easier to read than a sans serif typeface in print form. However, for online reading and for presentations, sans serif typefaces are more appealing, and therefore, recommended.

Do not use too many typefaces in a single document. Sometimes, your institution prescribes a specific typeface as Times New Roman. If there is no specification, select any appealing typefaces as you like such as Times New Roman, Ariel, Helvetica, Optima, Myriad, Verdana, Palatino, Trebuchet, Berkeley, Bookman, Garamond, or a similar typeface. Use the selected typeface throughout the document. However, a different typeface may be used within figures and tables and for equations. Select a font size of 12 point for main texts (type sizes are measured in points; 72 points = 1 inch). Use a larger font (up to 16 points) for main chapter headings; but use a smaller font for footnotes or superscripts and subscripts. Make sure that all text, page numbers, and captions are in the same font. Certain institutions specify the typeface and font size to be used. If that is the case, you have no other choice but to obey their specifications.

12.8 Thesis Editing

The thesis draft should undergo several rounds of editing. The research guide and committee members will act as 'editors', and help the student by giving their suggestions. Thesis editing involves two aspects—*technical editing* and *grammar and style editing* (similar to *copyediting*). The guide and committee members should identify lapses in clarity, problems in logic, precision, and consistency. The writing should be scientifically clear and precise. Revise the scientific language for usage, flow, and clarity by checking for awkward phrasing, biased language, clichés, and jargons that is inappropriate for a thesis. The guide must make sure that facts, data, and scientific units have been used precisely and consistently. Proper instructions must also be given for formatting the text to comply with the guidelines or style manual prescribed by the university.

Examine each paragraph, sentence, and word in the text to locate ambiguities and distractions both technically and grammatically. See that the thesis is free from errors in spelling, grammar, and punctuation. Some typical cases of misunderstandings on grammar and usage related to technical writing have been discussed in Chaps. 16–18. You should know the basics of grammar and usage to edit well. Certain individuals frequently make errors of the same type, and you should be vigilant about the types of errors you make most often. If you have problems with subject–verb agreement, singular–plural forms, correct prepositions, or others, look for such errors and rectify them. Sometimes, there may be a common pattern in your errors; identity them, and try to eliminate the more serious and frequent errors. Once you are able to locate and solve such major errors, you can continue to check for less serious errors. Your research guide may be focusing more on technical aspects of the thesis; however, if the student is weak in writing, the research guide or committee members have the responsibility to intervene and help, especially by correcting sentences and checking overall style. If required, you may also consult a language expert.

Editing can be done in several ways. If you are writing at the computer, always check your work promptly on the screen and run a spell-check occasionally. Facilities are also available on word processing softwares such as 'Word' for checking grammar and writing styles. You can enable grammar and style options by checking relevant boxes. Normally, you will be shown the likely grammar and style errors by green lines and possible alternatives. In addition to on-screen editing, conduct an editing after taking a printout too. Go through the draft meticulously touching every word with a pencil looking for anything you may have missed out. Ask yourselves questions of the following type and correct them. Is the tense form used is correct? Do subjects and verbs agree? Does each pronoun have a clear antecedent? Do the use of 'which' and 'that' are correct? Are the prepositions correct and in order? Did you drop any *s* or *ed* at word endings where it was actually required? Are there any clichés or worn out words in the text? Whether there are any gender specific or discriminatory terms? Can the sentences be shortened? Continue questions like these and rectify them. While editing, check for the consistency in formatting also.

While editing, your research guide and others may be using copyediting and proofreading symbols to correct mistakes. You should be familiar with various copy-editing and proofreading symbols and their use. When editing is done, editing marks or symbols are usually placed directly in and above lines of text because an editor generally works on a double-spaced manuscript, where space is available between lines to indicate additions or revisions. A frequently used symbol in editing is the caret or insert symbol, which shows where a substituted or additional letter, word, phrase, or punctuation mark is to be inserted. Another commonly used mark is a delete mark, which is made through a phrase, sentence, or paragraph. If you want to abandon the changes you have already made, and on second thoughts, need to retain the original as it was before the correction, put continuous dots under a word, phrase, or sentence and show the word *stet* ('let it stand') in the margin. For more information on the use of copyediting symbols, see Sect. 11.11 and Table 11.1.

Bibliography

Bond, A. 2007. Your Masters Thesis: How to Plan, Draft, Write and Revise. Viva Books, New Delhi, 135p.

Day, R.A. 1989. The origins of the scientific paper: The IMRAD format. Am. Med. Writers Assoc. J. 4(2): 16–18.

Day, R.A. and Gastel, B. 2006. How to Write and Publish a Scientific Paper (6th Ed.). Cambridge University Press, Cambridge, 320p.

Eco, U. 2015. How to Write a Thesis (transl. Italian, Farina, C.M. and Farina, G.). The MIT Press, Cambridge, 229p.

Fisher, E. and Thompson, R. 2014. Enjoy Writing Your Science Thesis or Dissertation! (2nd Ed.). World Scientific Publishing Company, 337 p.

Gustavii, B. 2008. How to Write and Illustrate a Scientific Paper (2nd Ed.). Cambridge University press, Cambridge, 168p.

Hartley, J. 2008. Academic Writing and Publishing: A Practical Guide. Routledge, New York, 196p.

Jackson, M. L. 1958. Soil Chemical Analysis (Indian Reprint, 1967). Prentice Hall of India, New Delhi, 498p.

Lester, J.D. and Lester, J.D. Jr. 2009. Writing Research Papers: A Complete Guide (13th Ed.), Longman, 416p.

Matthews, J.R. and Matthews, R.W. 2007. Successful Scientific Writing: A Stepby-Step Guide for the Biological and Medical Sciences (3rd Ed.). Cambridge University Press, 256p.

Murray, R. 2006. How to Write a Thesis (2nd Ed.) Open University Press, UK, 301p. Available: https://manaraa.com/upload/55b319e2-e557-4fdd-b3620d3268dd602b.pdf.

Oliver, P. 2008. Writing Your Thesis (2nd Ed). Sage Publications, New Delhi, 174p.

Rubens, P. 2004. Science and Technical writing: A Manual of Style (2nd Ed.). Routledge, New York, 427p.

Chapter 13
Tables and Illustrations

Given a large mass of data, we can by judicious selection construct perfectly plausible unassailable theories—all of which, some of which, or none of which may be right.
Paul Arnold Srere (1925–1999), American biochemist

Facts or information measured and used as a basis for reasoning, calculation, or decision-making are called data. The collected data have to be presented in research reports in a meaningful manner, so that readers grasp the contents easily. Condensing data in tables is the most common method of presenting data followed by charts, graphs, line diagrams, and other illustrations. Tables facilitate easy comparison and offer a useful form of condensation of data as the requirement of detailed explanation is reduced. Often, instead of tables, charts, and graphs can be conveniently used. Charts and graphs free you from long winded clarifications and save much space for the text. The choice to present the data in table form or graph form depends primarily on the type of data (Rubens 2004; Day and Gastel 2006; Mathews and Mathews 2007). When the data show a trend, the graph form is ideal. A mix of tables and charts is the most ideal for the presentation of data. You can also use drawings and photographs judiciously. As tables, graphs, charts, and other illustrations are considered as indispensable parts of theses and research papers, you should learn to prepare and use them appropriately.

13.1 Making of a Table

You cannot imagine a research report without tables. The data, which forms the heart of the research report, are presented through tables. It is important that the data presented in the tables must be easily comprehensible. The tables should also be attractive and be able to catch the attention of the readers. Therefore, never take a casual attitude to the formation of tables, but try to make legible, neat, and attractive tables. Presently, you can use computer applications such as 'Word', 'Excel', or similar ones to prepare any kind of tables easily and elegantly in various forms. It is

C. G. Thomas, *Research Methodology and Scientific Writing*,
https://doi.org/10.1007/978-3-030-64865-7_13

Title

Table 13.1. Mean number of tillers at 60 days after sowing and number of ear bearing tillers of wheat

Box heading

Seed rate (kg/ha)	Number of tillers/m²		Number of ear bearing tillers /m²	
	1993-94	**1994-95**	**1993-94**	**1994-95**
50	353.3	310.0	257.4	271.4
100	443.8	492.6	311.1	290.5
150	499.4	674.9	349.6	350.6
200	531.5	827.9	377.9	449.4
SEm +	27.2	33.4	11.2	20.3
LSD (0.05)	80.9	99.3	33.2	60.2

Stub

Field

Fig. 13.1 Parts of a typical table

also easy for you to format the size of tables, columns, rows, and lettering by using these applications.

A table has five main parts—the title, the stub, the box headings, the field, and the supplements (Fig. 13.1).

Title: The title gives an indication about the type of information included in it. An appropriate title should be given, and it must be written at the top of the table. The title normally includes the table number, heading (caption), and sub-heading (sub-caption) of the table. Tables included in a document must be numbered sequentially, and these numbers only be used along with titles. The title should be short and descriptive, but must be attractive too. Brevity is important; the title itself should explain the contents. The caption is written in 'sentence case', with only the first word and proper nouns in capital initials. As the titles are not given as complete sentences, do not put periods after them.

Provision for head note below the caption: If more explanations are needed, for example, when units of quantity are represented, give enough space just below the title and present this information in parentheses.

Number the tables serially: Each table should be given its own number as we need to refer to the tables when interpreting the data in 'results' and 'discussion' parts of a research paper or thesis. Number the tables serially in Arabic numerals as they appear in the text for easy reference. For example, in the first chapter, table numbers might be Tables 1.1, 1.2, and 1.3; and for the second chapter, these might be Tables 2.1, 2.2, and 2.3. Numbering is given along with the title or above the title.

Because the table has its own number, it is a proper noun; therefore, write 'table' in title case (the first letter in capital as Table). However, when it is not followed by a number, it should not be capitalized. Similarly, do not write 'Table No. 1'; just 'Table 1' is enough.

Box headings and stubs: The data are presented in vertical columns and horizontal rows. The headings given to columns are called 'box headings' or 'column headings' and row headings, 'stubs'. The stub can be considered as the first column of the table, which lists the items about which information is provided in the table in horizontal rows to the right. Often, columns are made wider and rows narrower. Represent the unit of measurement of each character below the caption in respective columns, for example: yield in kg, height in cm, and so on. The unit of measurement should be abbreviated in the table.

Field of table: The field or body of table means the cells containing the data. For ordinary cases, it is better to go for approximation and use natural numbers. However, when the differences between the data are very narrow, include decimals too. The lettering should be sufficiently big enough to be readable. Ensure uniformity in the presentation of data, for example, instead of writing 3.0, 1.54, 6.5, and 2; write 3.00, 1.54, 6.50, and 2.00. Write the number or figure, even if it is a repeat, and avoid putting ditto marks ('do'/,,). Similarly, avoid putting '–' marks for zero values. Instead, put '0'. If a value is not available, mark it as NA and write a footnote as NA-Not available. Never use abbreviation alone.

Supplements: The supplements, mostly in footnote form, are given below the table. The supplement usually includes footnotes about the sources of the data given in the table, special features of the table, and symbols used in the table. Four types of supplements are generally used—source notes, general notes, notes on specific parts of the table, and notes on the general level of probability. If the data are from a different source, indicate it at the bottom of the table. Notes on specific parts of the table may be shown by putting an asterisk.

13.2 Arrangement of Data in Tables

Several conditions have to be satisfied while arranging data in tables. The following points must be considered critically while preparing tables.

Size of tables: Plan the tables in such a way that they do not exceed the page size of the thesis. Large tables may be typed in 'landscape' orientation (horizontally, along the long way of the paper). However, a better comparison of data could be made, if they are presented in a 'portrait' or vertical format. If a table does not fit on to a page, it may be carried over to the next page or pages. In such a case, the table number and the word 'continued' (e.g., Table 7 contd.) should be given on all succeeding pages to which the table is carried over, and the box headings should be repeated. There is no need, however, to repeat the caption of the table. If this method too does not work, reduce large tables photographically to desired sizes.

Avoid overcrowding: Do not overcrowd tables incorporating many ideas and large numerals. A maximum of five or six related characters can be condensed and represented in a table. For example, when biometric characters like plant height, leaf number, and leaf area are to be presented, try to put them in a single table.

Arrange logically: While presenting the data, follow some logical way of arrangement. For example, certain type of data can be arranged in chronological order. In the case of a growth analysis study, the data obtained from different stages (tillering, flowering, and harvesting) must be presented only in their order of occurrence.

Condense the data through approximation: In tables, approximate the numbers to the nearest significant figures by following rounding up or down the figures to the nearest integers. Avoid surplus zeros from very large or small numbers by choosing the factors or units in the column headings carefully to keep numbers between 0 and 1000; for example, 60,900 micromoles can easily be converted to 609 millimoles, and written as 609 in a column headed 'mmol'. Similarly, avoid numbers with exponents (15^5, 15^{-5}) as factors in headings. Instead, convert the data into appropriate units (mega, milli, micro, nano, and so on) as factors in headings and write the units such as Mg (one Megagram is equal to one tonne), mg (milligram), mm (micrometre), and nL (nano litre) at the top of the column.

Data with common letters: When you represent data after conducting analysis of variance and applying Duncan's multiple range tests by using common letters, you can make comparisons easily and say which treatments are homogeneous and which are not. See the example given as Table 13.1. In the example, T2, T3, and T5 have the common letter 'c'. In drafting tables after statistical analysis, wherever required, the test of significance used should be made clear, giving the probability (p) values and the standard errors of the means.

Some General Tips for Using Tables

When a table presented in the 'Results' section of a research article or thesis is mentioned in 'Discussion' section, the table number must be indicated for reference. You need to mention only the number and no need to include caption, for example, Table 1 (*not* table 1 or Table No. 1). Always make a mention of the table in the text,

Table 13.1 Mean height of wheat at 30 days after sowing

Treatments	Mean height (cm)
T1	19_{bc}*
T2	20_{bc}
T3	17^{c}
T4	16_{cd}
T5	21^{b}
T6	25^{a}
T7	15^{d}
T8	22^{b}

*Treatments having common letters do not differ significantly

that is, table numbers should find a place in the text. If possible, position the table immediately after it is mentioned in the text. Explanation of certain values may be required; for example, if the yield from a particular treatment is found to be too low, it is better to explain the reason as footnote to avoid doubt and confusion in the minds of the readers. If there is some text above the table, place the title of the table three spaces below the text.

The table and the arrangement of its content should facilitate easy comparison of data. To facilitate relative comparison, it is better to give percentages along with values, or give row totals and column totals. Similarly, if two sets of data are related and needed comparison between them, try to arrange the data one after the other as neighbouring tables.

If you have used secondary data for inclusion in the table, mention the source as a footnote. Inconsistency between tables and text is perhaps the most serious fault in research papers and theses; therefore, every care should be taken to avoid its occurrence.

13.3 Illustrations

An *illustration* is any material that cannot be set in type, which includes photographs, line drawings, charts, graphs, and tracings. Generally, four kinds of illustrations—*charts*, *graphs*, *line drawings*, and *photographs*—are used in scientific documents.

The function of an illustration is the same as that of a table—to make the subject readily comprehensible. All the illustrations included in a report should be clear, artistic, and meticulously planned, so that they convey the information effectively to the reader. Similar to tables, illustrations should also be self-explanatory. All the illustrations are referred to in the text as 'figures' and denoted by 'Fig.' spelled with capital 'F', when combined with numbers as Fig.1. The term 'figure' can be applied to a wide variety of graphs, charts, maps, sketches, diagrams, and drawings. As in the case of tables, the figures should preferably follow immediately after a reference in the text. The figures for a thesis should be planned keeping in view the fact that their ultimate place would be in journals. Therefore, they should be fit for reproduction.

Charts and graphs are the most important illustrations used in theses and research papers. These allow the researchers to present numerical information in a pictorial or illustrative form, facilitating better understanding of the data. Although many people consider the terms charts and graphs synonymously, they are not the same. Understand the differences clearly and use the correct term only depending on the figure. For example, you may use bar chart or bar diagram but not *bar graph*! A chart (sometimes also called diagram) is merely a visual representation of variable data, in which the data are represented by representative figures such as bars in a bar chart (bar diagram) or sectors of a circle in pie chart (pie diagram). Graphs are, however, different from charts as they are built by plotting the values of a function along an axis that represents some or all the possible values taken by a variable. A graph is considered more scientific, as it usually contains X, Y, or Z-axis. You may specifically note that in a graph, no data is represented without using axes.

General Instructions for the Preparation of Charts and Graphs

The advent of computers and user-friendly application softwares revolutionized chart and graph making. Earlier, all these were made manually, and researchers had to waste much precious time in their making. At present, you can use spreadsheet programmes such as Excel or chart and graph-making programme such as Delta-Graph for the creation of charts or graphs easily, which can subsequently be imported into your document. However, for advanced graphs such as scatter diagrams or for 'curve fitting', statistical packages such as SPSS and CropStat are the preferred tools (CropStat is a freely downloadable statistical package from IRRI; see Sect. 7.11). Computer applications allow you to put titles and values of x-axis and y-axis smartly in the chart. You can also put gridlines, legends, and data labels.

Remember that computer applications are to help you in creating the charts or graphs; but you are the designer and must give proper and correct instructions to the computer to get fine and beautiful charts and graphs. As in the case of tables, take note of some points especially when designing them.

Create Simple Figures

Figures should be simple and convey a clear idea. Most importantly, it should be comprehensible without giving much description in the text proper. Present the data carefully and accurately to avoid oversimplification, misrepresentation, or distortion of facts. Ensure that numerical data upon which the figure is based are presented in the text or in an accompanying table unless they are included in the figure itself.

Use Figures Sparingly

Avoid including too many figures in the document. Note that figures are used to highlight certain points. However, if too many figures are used, they may only detract the readers from the paper instead of elucidating it.

Titles to Figures

To explain what a figure is about, a title or caption is placed at its bottom, one double space below the figure. The title is preceded by the figure number (e.g., Fig. 6. A rice plant). Titles of figures, similar to caption of tables, must be concise, self-contained statements of what is presented in the figure. Note the differences between table captions and figure titles carefully. For a table, the caption must be given above the table, but for a figure, it must be below the figure only. Similarly, you should not put periods after table captions unless they are full sentences; but for figure titles, periods should be added even if they are not full sentences.

Numbering of Figures and Pagination

All figures in a thesis should be numbered in consecutive series. The pages containing figures should also be numbered in the same series as those of the text.

The page number should appear at the same position as on the text pages.

Footnotes to Figures
Symbols used in illustrations, source of illustration (if it is not author's own), and special features of illustration should be explained in footnotes. The first line of the footnote begins one double space below the title of the figure. Put the footnote matter in single spaces. However, between footnotes leave one double space.

Select the Most Appropriate Type of Charts
If you have decided to use a chart, choose the most appropriate one for your data. A wide variety of charts and graphs can be prepared. Charts are generally classified as shown below:

- **Bar charts**: These are generally made for showing simple comparison of data at a given point of time. These are appropriate when the treatments are discrete and relatively few. For example, performances of a set of cultivars of rice or sales figures from various offices in an organization can be easily presented through bar charts.
- **Pie charts**: Pie charts are useful for illustrating proportions such as the percentage contributions of various components to the cost of cultivation of a crop.
- **Line charts/area charts**: These are generally used for representing changes over time, for example, mean rainfall data over a 10-year period.

Remember that for categorical or discrete variables, you can construct only bar charts or pie charts, but for continuous variables, you can also draw line charts or trend lines. Sometimes, two-dimensional (area) or three-dimensional (volume) charts are also made. When variability is larger than that can be accommodated by common one-dimensional diagrams, two-dimensional area diagrams can be made. Comparison is often made between the area of figures such as squares, rectangles, and circles instead of length of lines. Among these figures, comparison is much easier when squares are used. When variation is too large to be accommodated by two-dimensional diagrams, three-dimensional diagrams can be used. In these diagrams, usually, cubes are made instead of squares. The side of a cube is equal to the cube root of quantity to it represents. Detection of simple differences, however, is easy with one-dimensional diagrams.

13.4 Bar Diagrams

When the variable is discrete or categorical, which do not require continuity, and when they are relatively few, bar diagrams or charts can be ideally used. Bar charts are easy to construct, simple, and easy to understand. These are generally used when a striking difference is to be stressed, and when it is not important to maintain high degree of precision of the individual mean values. The following are the common types of bar diagrams:

- Simple bar diagram
- Multiple bar diagram
- Component/Subdivided bar diagram
- Percentage bar diagram
- Deviation bar diagram
- Pictorial bar diagram (pictogram)
- Cartogram.

Simple Bar Diagram
In simple bar diagram, rectangles (sometimes, lines too) of equal width drawn to scale length, arranged either horizontally or vertically, are used to compare data (Fig. 13.2). This type of bar diagrams is generally used to represent discrete variables such as people, crops, crop varieties, years, or months. Simple bar diagrams may be vertical or horizontal in orientation.

Vertical bar charts are the common form of bar charts used to make comparisons between two or more groups at different times. In this type, the width of rectangles is kept constant, but the height of each rectangle varies according to its value. The height of rectangles is drawn in proportion to the quantity. When the items to be compared are less, say 5–7 or less, we generally use rectangular type bar diagrams (Fig. 13.2a). However, if too many items are to be shown, it can be represented as slender rectangles or single lines (Fig. 13.2b). However, avoid overcrowding with too many slim rectangles or lines. Sometimes, *horizontal bar charts* are also made especially to compare components at a particular time.

In bar charts, the spacing between two bars must be equal. Sometimes, the values are given at the top of rectangle bars to provide more accuracy. If colour charts are permitted, select attractive colours for the bars. Markings are made on *Y*-axis only, as only heights of bars are compared. When we represent various homogenous treatments, treatment means can be represented as heights of bars. When the values are large for certain items, sometimes, comparison may not be possible through ordinary charts. In such exceptional cases, broken bars can be used for the extraordinary item.

Some researchers have a tendency to join the top points of bars as we do in the case of histogram. This practice is not acceptable as the data involved are discreet data. Such type of graphical representation is possible with continuous variables only.

Multiple Bar Diagram
If the objective is to represent more than one character for an item, then multiple bar diagram can be used. Suppose, we want to represent grain yield and straw yield of rice obtained from two treatments. The arithmetical means of these characters can be represented as shown in Fig. 13.3. Follow a particular order in all the bars, for example, in the above figure, give the first bar to grain yield, second bar to straw yield and follow this order in all the bars. Avoid overcrowding; a maximum of 3–6 treatments can only be accommodated in this type of charts.

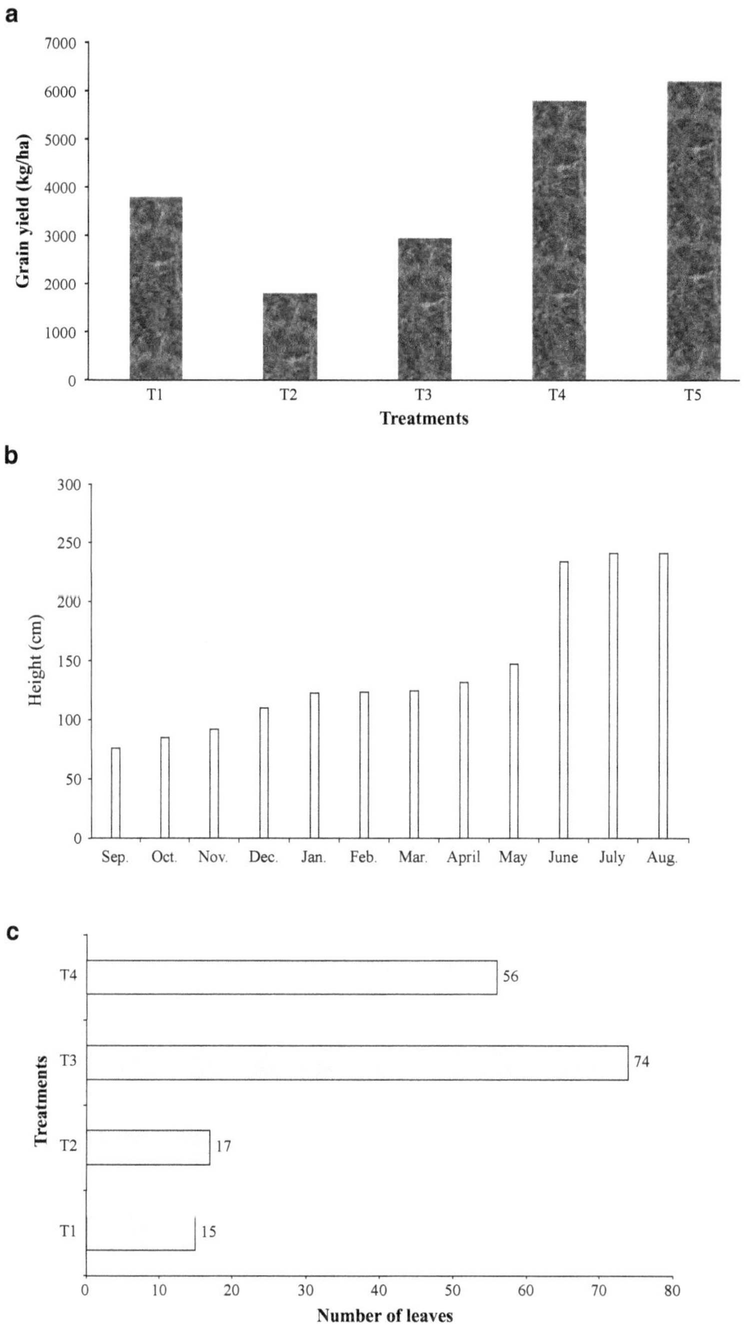

Fig. 13.2 **a** Rice yield from different treatments (vertical bar diagram). **b** Plant height at monthly intervals (vertical bar diagram with slim rectangles). **c** Number of leaves at 45 days after sowing (horizontal bar diagram)

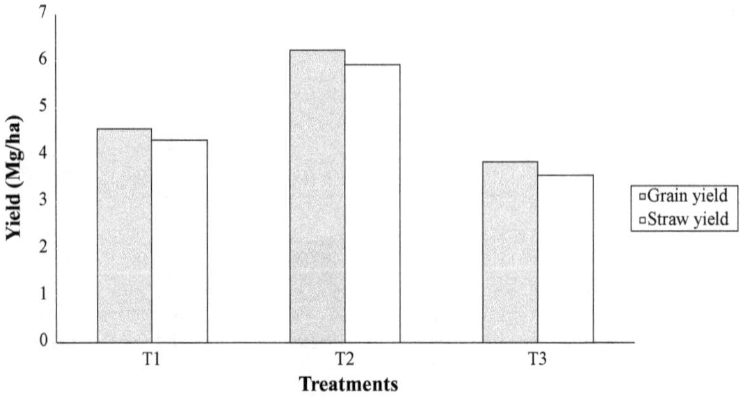

Fig. 13.3 Grain and straw yield of rice (multiple bar diagram)

Subdivided Bar Diagram

A subdivided bar chart, component bar chart, or stacked bar chart represents the components of a character in one bar. The whole character can be split or divided into small components. Different individual components are combined together to form the whole, for example, in Fig. 13.4, total biomass of a plant is split into roots, stem, and leaves. Arrange the bars in a logical order. Each section of a bar shows the proportion of the whole it represents in relation to others.

Percentage Bar Diagram

In percentage bar diagram, only relative comparison is done on a variable and its components. It is similar to a component bar chart, but the sub-categories for a particular bar add up to 100. All the bars may have the same height as the total is made up of 100. Each bar is sliced into portions in relation to their proportion out of 100. For example, if we want to compare the cost of production of milk through

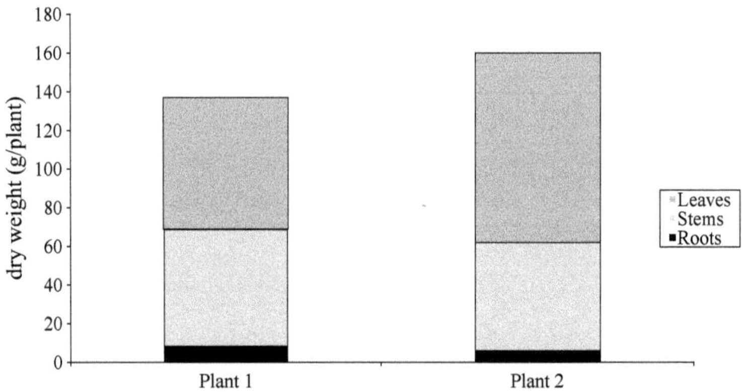

Fig. 13.4 Distribution of biomass in two plants (subdivided bar diagram)

two different management systems, it can be represented to include the cost of feed, labour, housing, and maintenance (Fig. 13.5).

Deviation Bar Diagram

Deviation bar diagrams are used when positive and negative fluctuations are to be represented. The bars are directed upward or downward depending upon the trends. Upward bars indicate positive values, while the downward bars indicate negative values. For example, when we want to depict profit and losses, especially when economic analysis is done, deviation bar diagram is ideal. In Fig. 13.6, profit and losses from different rice cultivation systems are shown as a deviation bar diagram.

Pictograms

Pictograms or pictographs are similar to bar charts but these are made attractive using human figures, animals, tractors, cycles, plants, flowers, panicles, and the like. In

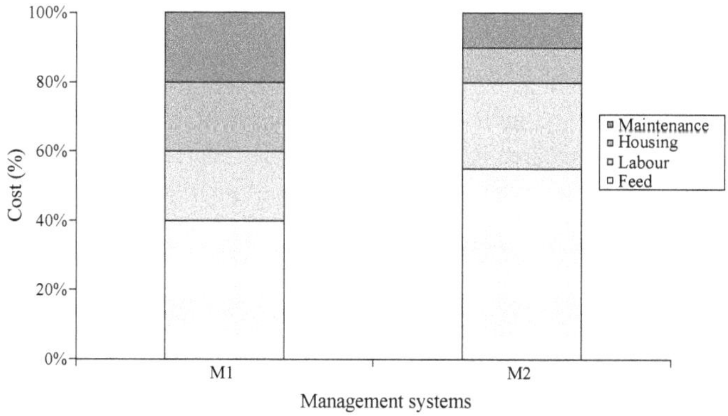

Fig. 13.5 Cost of production of milk in two management systems (percentage bar diagram)

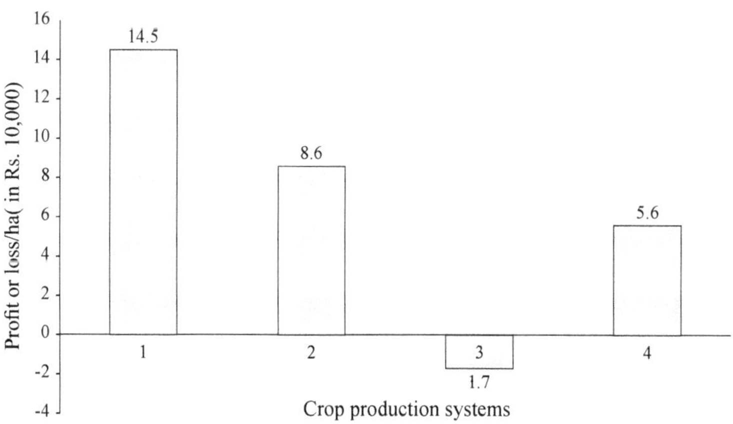

Fig. 13.6 Profit–loss analysis in four crop production systems (deviation bar diagram)

pictograms, instead of the usual lines or rectangles, pictures of the objects are used. They are useful for display purposes and convey the message easily. The picture size should be the same in all items, but their numbers differ according to the data. Sometimes, instead of numbers, heights of pictures are adjusted according to their values.

Cartograms

When we want to depict statistical data according to geographical regions, maps of these regions can be used to depict data using symbols. For example, area of crops, production of crops, livestock population, human population, etc., can be shown using cartograms. Either the number of symbols or the size of symbols is made proportional to the quantity being represented.

13.5 Pie Diagrams

Pie diagrams, also known as pie charts, sector chart, circle chart, or circular graph, are useful to represent the division of a whole into different parts. These are two-dimensional diagrams as area is represented. In pie diagrams, various components of a single item are represented as various sectors of a circle. As there are 360° in a circle, the full circle can be used to represent the whole population or 100 percent. First, a circle is drawn. Then, divide the circle into sectors one by one starting from 12 O' clock position. Represent the major item in the first sector. While dividing sectors, the angle of the sector is made proportional to the magnitude of the item being represented. Miscellaneous items are represented only at the end. An example of a pie diagram representing family budget is shown in Fig. 13.7.

In a pie diagram, the whole circle is divided into component parts based on their proportional size; but there should be limit for the number of segments for clarity. A useful thumb rule is to limit the number of slices to five or maximum seven and to place the data (percentage or number) inside the pie segment. If the space is insufficient to put the data in the segment, place them outside the segment and draw a small arrow from the identification term to the segment. You can also use colour for pie diagrams, if it is allowed.

13.6 The Line Graph

Line graphs are usually used when the variable plotted along the x-axis is continuous, for example: time, temperature, or distance. You can use line graphs only with continuous data. Comparison of trends of different items is possible through graphs. For depicting models and for 'curve fitting', graphs are used. Graphs need more accuracy while charts do not need much accuracy as we are making only relative comparisons. Accuracy is important in the case of graphs because we have to make

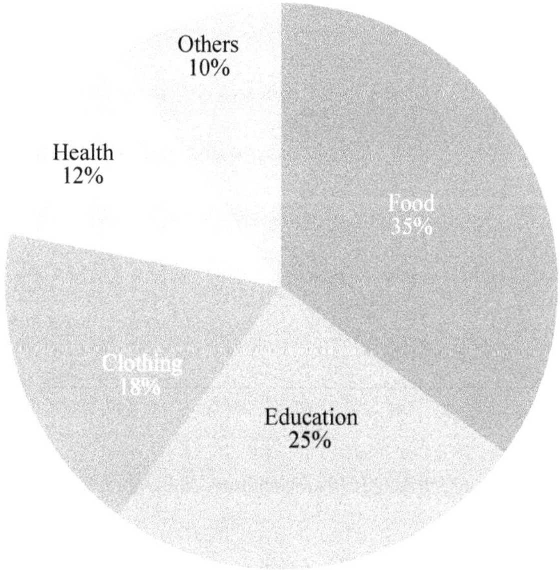

Fig. 13.7 Family budget of a middle class family (pie diagram)

scientific interpretations from them. In addition, we can have interpretation of hidden facts, which are revealed through graphs.

Line graphs are the most widely used and versatile type of diagrams for showing a change in data relationships over a period. These are most useful for showing trends, and sometimes, for identifying whether two variables are related to each another. For example, most parameters of growth analysis such as plant height, leaf area, leaf weight, and dry weight can be represented through line graphs (Fig. 13.8). Similarly, periodic changes of meteorological parameters such as temperature, evaporation, and wind speed can also be shown through line graphs.

Line graphs simply use a line to connect the data points that are plotted. Usually, the horizontal axis measures the independent variable, and the vertical axis, the measured characteristics of the dependant variable. Use the horizontal axis to represent response and the vertical axis to represent treatment levels. Choose the scales for the vertical axis and the horizontal axis in such a way as to highlight important points, avoiding distortion. The lines may be straight line or curved.

Do not include more number of lines on a single graph paper, as we cannot distinguish these lines easily, if overcrowded. Limit the lines to 3–4 per graph. Different colours or marks can be given to different graph lines. Clear distinction between colours should be ensured. If colours are not used, use thick line (—), broken lines (---), dotted lines (…..), or alternate dot and broken lines. The arrangement of lines should proceed from left to right on the horizontal axis and from top to bottom on the vertical. Usually, the line starts from 'zero' origin. However, we can represent

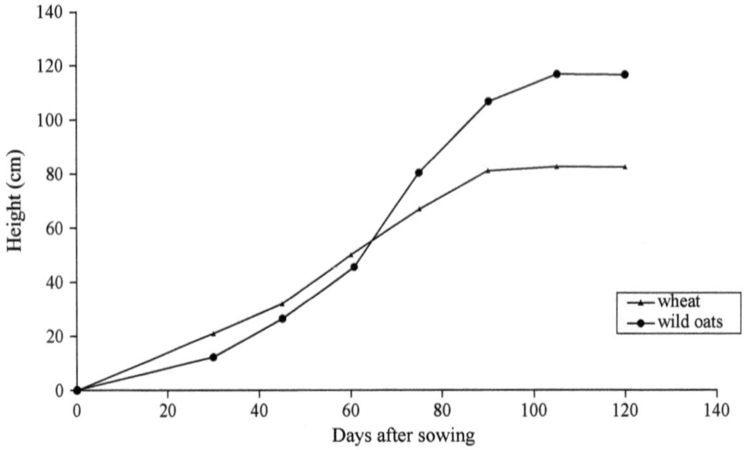

Fig. 13.8 Height of wheat and wild oats in monocultures (multiple line graph)

a 'false base' and start from 100, 50, or any number, especially when the last value represented is too large.

In a line graph, for showing direct relationships, an arithmetic scale is commonly used. Sometimes, a logarithmic scale is used to show relationships that vary as a power of a number or for comparing widely differing data.

Regression equations can be fitted, if the number of treatments is adequate, and the resultant line will be a smooth curve (Fig. 13.9). Otherwise, simply draw a line to connect observed points and incorporate appropriate test of significance. We can also use a false base, when we need to compress values on Y-axis. When it is necessary to accommodate two scales of measurement in the same graph, include one unit at

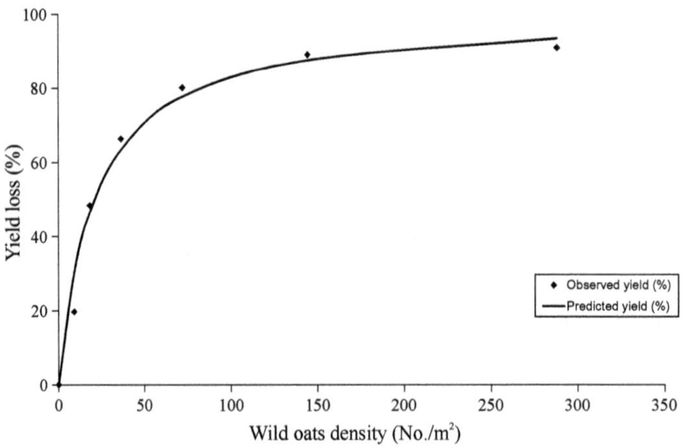

Fig. 13.9 Wheat yield loss as a function of wild oats density (a graph obtained after 'curve fitting')

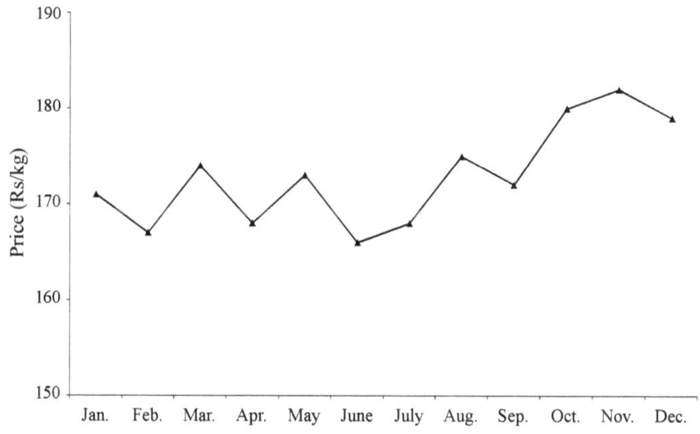

Fig. 13.10 Market fluctuations of commodity (time series graph)

the left and another at the right. Markings should be done such that the mean values used coincide.

When the y-axis indicates a quantity or percent and the x-axis represents units of time, the line graph is often referred to as a *time series graph* (Fig. 13.10). For example, by plotting the market price of a commodity, it is easy to see the fluctuations during the course of a year. Line graphs can also be used to depict multiple series data.

13.7 Line Drawings

Line drawings and photographs can create lasting impressions compared to other methods of data presentation. These are not normally intended as proof to the idea presented but to illustrate the point clearly. Line drawings help us to avoid confusion by directing emphasis to those things we consider essential to our argument. For example, the habit of a plant, the parts of its flower, and the floral diagram help us to establish the identity of a plant specimen. A line drawing can be used as an aid to observation, and each line or stroke in it should be an accurate record of an observation. In other words, the completed drawing should be a summary of observations! A line drawing may be a better option than a photograph for illustrating very small objects, as photographs may not be effective in revealing minute details.

Normally, only lines are used to represent the things observed in line drawing. Shading is avoided in such line drawings. Accurate and effective labelling is essential to buttress your arguments and direct attention to parts named in the text. Try to draw the diagrams according to a scale, and the scale must be marked on the drawing. Inaccuracies in drawing are usually the result of faulty observation and insufficient knowledge of the subject. Consider the art of drawing as a part of the learning and training process of a scientist or engineer. You can use the help of computer programmes such as DrawPlus, CorelDRAW, OpenCanvas, Artweaver, Skencil, or similar programmes for creating and improving drawings.

13.8 Photographs

With the advent of digital cameras at affordable costs, the use of photographs in research reports and presentations has increased. The advantages of using photographs are many (Barrass 2002; Gustavii 2008).

1. They are self-explanatory. They reduce the number of words required in the description.
2. Normally, they cannot deceive you. Remember the old dictum 'seeing is believing'. However, bear in mind that in the era of advanced digital technologies, photographs can also be 'made'.
3. Photographs show a universal language, and therefore, irrespective of the language barriers, everybody can appreciate them.
4. Multiple copies can be made, and the message may reach many hands quickly.
5. Photographs simplify interpretation.

Every scientist should master the art of photography. With the new generation cameras (both SLR and non-SLR) having many user-friendly features, you can take beautiful pictures even without knowing much about cameras or photography. However, it would be advantageous if you study some basics about the types of cameras, lenses, lens systems, focusing, zooming, aperture, shutter speed, etc. Similarly, try to acquire some skills in using the softwares such as 'Photoshop'. Do not affix photographs simply to give colour to the report as certain students do by including irrelevant photographs in their thesis. Do think twice before using photographs. While including photographs in theses, research reports, or papers, consider their relevance, scientific merit, and sharpness of photographs in terms of focus, lighting, and contrast.

Bibliography

Barrass, R. 2002. Scientists Must Write: A Guide to Better Writing for Scientists, Engineers and Students (2nd Ed.). Routledge, London, 204p.

Best, J.W. and Kahn, J.V. 1993. Research in Education (7th Ed., Indian Reprint, 2004). Prentice-Hall of India, New Delhi, 435p.

BSI [British Standards Institution]. 1992. Guide to Presentation of Tables and Graphs (BS 7581: 1992). BSI Standards, London, 28p.

Cargill, M. and O'Connor, P. 2009. Writing Scientific Research Articles: Strategy and Steps. John Wiley & Sons, 173p.

Davis, M. 2005. Scientific Papers and Presentations. Academic Press, Massachusetts, 356p.

Day, R.A. and Gastel, B. 2006. How to Write and Publish a Scientific Paper (6th Ed.). Cambridge University Press, Cambridge, 320p.

Gustavii, B. 2008. How to Write and Illustrate a Scientific Paper (2nd Ed.). Cambridge University press, Cambridge, 168p.

Hartley, J. 2008. Academic Writing and Publishing: A Practical Guide. Routledge, New York, 196p.

Kumar, R. 2011. Research Methodology: A Step-by step Guide for Beginners (3rd Ed.). Sage Publications India, New Delhi, 415p.

Lester, J.D. and Lester, J.D. Jr. 2009. Writing Research Papers: A Complete Guide (13th Ed.), Longman, 416p.

Malmfors, B., Garnsworthy, P. and Grossman, M. 2004. Writing and Presenting Scientific Papers (2nd Ed.). Viva Books, New Delhi, 153p.

Mathews, J.R. and Mathews, R.W. 2007. Successful Scientific Writing: A Step-by-Step Guide for the Biological and Medical Sciences (3rd Ed.). Cambridge University Press, 256p.

Rubens, P. 2004. Science and Technical writing: A Manual of Style (2nd Ed.). Routledge, New York, 427p.

SSC [Statistical Services Centre] 2000. Informative Presentation of Tables, Graphs and Statistics. The University of Reading Statistical Services Centre, 24p. Available: https://www.reading.ac.uk/ssc/n/resources/Docs/PresentationOfTablesGraphsAndStatistics.pdf [Accessed 11Feb.2009].

Chapter 14
Reasoning in Research

The moment we want to believe something, we suddenly see all the arguments for it, and become blind to the arguments against it.
George Bernard Shaw (1856–1950), Playwright and Nobel Laureate

After doing several experiments and carefully analysing the observations recorded, scientists come to certain conclusions. For arriving at correct conclusions, a systematic approach to *argument* or *reasoning* is essential. The first systematic approach to reasoning, attributed to the Greeks, was the method of deduction. They used appropriate logic to prove their point. *Logic* is the science of argument evaluation and includes methods and criteria for deciding whether arguments are reliable. Before the modern concept of research and experimentation surfaced, the term the old philosophers used to denote 'research' was logical reasoning. Therefore, it is natural that some of the essential characteristics of logic have passed over into present-day research. Thus, the inductive and deductive methods of argument or reasoning became part of modern research. Knowingly or unknowingly, we use many of such logical concepts when we prepare a research report or while presenting a scientific case. Research students should learn to argue their cases satisfactorily using scientific logic.

14.1 Some Logical Terms

Reliability of evidences or premises is a major issue in arguments. We use several methods and criteria for deciding whether our arguments are reliable. We should use appropriate logic to interpret our results; therefore, let us first familiarize with some of the logical terms.

Arguments and Premises
In the context of logic, the term '*argument*' has a meaning quite distinct from its everyday use when referring to a dispute as a 'difference of opinion'. Each of you can have opinions on different subjects, and there can be differences of opinion among

C. G. Thomas, *Research Methodology and Scientific Writing*,
https://doi.org/10.1007/978-3-030-64865-7_14

you. Opinion is simply an unsupported claim while an argument is a claim supported by evidences. In logic, an argument is a series of statements, consisting of *evidences* and a *conclusion* or *main claim.* Evidence statements are called *premises*, and the conclusion is to follow from these premises. Premises directly support the conclusion. A simple argument must have two premises and a conclusion, but a more complex argument may contain many claims. In research, 'argument' is same as 'reasoning'.

Syllogism

Syllogism was a model of thinking existed among early philosophers. Aristotle, who is considered as the 'father of logic', cherished syllogisms, and encouraged them as the main devices for argument evaluation. The fundamental tool of deductive logic is syllogism, consisting of a *major premise, a minor premise*, and *a conclusion.* A major premise is a self-evident statement concerning a relationship, and a minor premise is a particular case related to the major premise. From the logical relationship of these premises, conclusions can be drawn, for example: what is the relationship between the two classes A and C, if we know the relation of both A and C to the third class B? Consider the following arguments.

All A's are B's.
All B's are C's.
Therefore, all A's are C's.

Let us take a true example.

All dogs are mammals.
All mammals have mammary glands.
Therefore, all dogs have mammary glands.

The above example consists of three statements. The first and second are the premises, and the third is a conclusion that follow from the premises. In the third statement, the conclusion contains a subject and a predicate. The first statement is a premise dealing with the subject, and the second is a statement dealing with the predicate.

Assumptions

Normally, premises for arguments are put up as self-evident statements. Because the person who puts forth the argument considers the premise self-evident, asserts that the argument does not require any further support or evidence. These premises, which are self-evident, may be called *assumptions, presumptions,* or *suppositions. Postulates* and *axioms* also serve the same purpose.

Statements

In arguments, all sentences can be divided into three types—true, false, and those that cannot be either true or false. Sentences, which are *commands* ('Do it now'), *exclamations* ('He won!'), and *questions* ('Where is it?'), cannot be true or false. Exclamations and commands are rare in arguments. Nevertheless, certain type of questions, called *rhetorical questions*, can be found more frequently in reasoning.

These questions are asked in such a way that they make a point without requiring an answer. The majority of sentences in reasoning that can be either true or false are called *statements* or *claims*.

Specific and Non-specific Claims

Certain statements or claims can be specific or non-specific. Some statements serve as conclusions, premises, or support in an argument. *Specific claims* contain or imply language or figures of an exact nature, for example, 'Due to recession, half of the investment in stocks has been lost'. In the example, 'half' represent specific information. These statements are authoritative and convincing in an argument. However, these kinds of statements are prone to criticisms because a single different case is sufficient to dispute them. The most common specific statements are universal ones in which the figure involved is either 100 or 0 percent; usually expressed by words such as *always, never, all, none, everyone,* and *no one.*

Non-specific claims are those in which no specific number is cited. Because of this non-specificity, it is very difficult to attack such claims, for example, *Due to recession, about half of the investment in stocks has been lost*. The example qualifies 'half' with the word 'about' making the statement weaker but harder to disprove!

Facts and Opinions

While presenting arguments, certain details such as a date, statistical data, or other details that most people accept as irrefutable come under the category of *facts*. People generally think that 'facts' are much more reliable and convincing than *opinions*. Opinion is simply an unsupported claim of an individual based on several considerations. However, many 'facts' such as statistical data, survey results, scientific measurements, and historical events are based on 'opinions'. In arguments, there is not much difference between facts and opinions. We commonly use both facts and opinions to support arguments; and remember, sometimes, strong expert 'opinions' can outweigh weak or inconsistent 'facts'! Note that much of what we know about the physical sciences is based on hypotheses or theories, that is, opinions, which are in different stages of confirmation. In science, only those assertions called 'laws' are considered as irrefutable facts.

Validity and Soundness of Arguments

In evaluating an argument, if we find nothing wrong with its *form* (i.e., how the argument is made) and nothing wrong with its *content* (i.e., the assumptions on which the argument is based), then we must accept its conclusion. That is, if the form and content are correct, we should agree with the conclusion. We first analyse the form of an argument followed by its content, as evaluating the form of an argument is easier than its content. When the premises and conclusion of an argument are in proper relationship, its 'form' is acceptable, and we say that the argument is *valid*. In other words, a valid argument is one that is in an acceptable form; and invalid argument is one in an unacceptable form. If an argument is found to be *invalid*, all judgment based on it must be rejected. We cannot accept an invalid argument. However, the conclusion of an invalid argument might not necessarily be wrong as could be seen from the examples given under deductive reasoning (Sect. 14.3).

If the form of an argument is found to be valid, then the content of its premises must be evaluated to determine whether they are true or false. A true premise is one that everybody believes, verifiable, or is self-evident. Justification for an argument usually comes in the form of empirical evidence, expert opinion (authoritarian evidence), or other supporting arguments.

If an argument is 'valid' and its premises are 'true', it is a *sound* argument, and its conclusion must be accepted. However, not all valid arguments may be accepted as completely correct. The more complex the argument is the less likely that it will be considered conclusively sound. In such cases, we may consider the 'relative soundness' of an argument by describing it as *strong* or *weak*. A strong argument is valid in form and true in content making a convincing case for its acceptance. A weak argument is also valid in form, but its uncertain premises do not force its acceptance.

14.2 Evidences

In an argument, we try to present different kinds of evidences to assert our point of view. A premise can be made more acceptable when it is supported by different kinds of evidences, for example: statistical data, historical information, physical evidences, observations, experimental results, and eyewitness accounts. However, the genuineness of evidences could be disputed by various means. For example, we could challenge the materials and methodology used for a study, statistical tools used, the interpretation of results, and the methods used for collecting physical evidences. Eyewitness account could be disputed by questioning about eyesight and motivation. Empirical evidence and authoritarian evidence are the two major groups of evidences.

Empirical Evidence

The evidence that anybody can experience by the processes of seeing, hearing, touching, tasting, or smelling is *empirical evidence.* It can also be called *natural evidence,* as it is the evidence found in nature. Among various types of evidences, empirical evidence is the most reliable evidence. It is repeatable, and therefore, empirical evidence can be checked by anybody when knowledge claims are made by an individual. As empirical evidence is the only one possessing these attributes, scientists and critical thinkers use them to arrive at meaningful conclusions. When you are conducting an experiment, you are creating empirical evidences.

Authoritarian Evidence

In olden days, people believed without questioning whatever the authorities like kings, priests, parents, and teachers said or preached, commonly called authoritarian evidence. Acquiring knowledge through authority is widespread among people because of certain reasons. Most people are conditioned from the day they are born with positive and negative suggestions from their parents to believe and obey authorities. In most cases, authoritarian teaching is an efficient method to pass on knowledge. However, authoritarian knowledge should be authenticated through empirical evidences, logical reasoning, and critical thinking. As far as science is considered, if we accept the authority of politicians or public figures like poets, literary critics, sports persons, or activists, it would be the end of science. We should accept only 'experts' in their own fields as authorities in science. Nonetheless, even the postulates of an expert will be modified or even rejected, if another expert put forth contrary facts and evidences.

The old type of authoritarian evidence is no longer acceptable in science. People should not take for granted the information passed on to them based on authority. However, the problem is that the authorities, especially religious and political authorities who have some sway over the people, take them for a ride and try to meddle with even science for parochial ends. Sometimes, it may reach fanatical proportions. Do not consider such kinds of authorities as infallible.

Authoritarian evidences have some uses in both everyday and scientific field, if the authorities are real experts. Sometimes, we may not be able to judge the supporting evidences given in the text for ourselves. For example, when there is excessive quantity of evidences and evidences are highly technical, we have to seek the help of some experts. Sometimes, the evidences may not be directly accessible. In such cases too, we often rely on the judgments of others, the 'authorities'. Similarly, while writing discussion section of a thesis or paper, we rely on authoritarian evidence to back up our claims.

We usually think of experts in science and technology including agriculture, medicine, and other fields as authority in their domains, but authority can also come from religions, scriptures, politics, folk wisdom, and proverbs. The strength of an authority as evidence depends on how agreeable a person is to accept the judgment of that authority. If the authority is reliable, authoritarian evidence can also be reliable. However, as many authorities are not reliable, we must check the reliability of each authority before we accept the evidence. This unequivocally points out that we must be our own authority and rely on our own powers of critical thinking to know what we believe is reliable and true.

Other Types of Evidences

In everyday life, we can see many individuals presenting some other kinds of evidences too to win their cases. Reliability is a major problem with such 'evidences'. It is easy to distinguish empirical evidence with these kinds of evidences to understand its value.

Anecdotal evidence: Anecdotes are some first-hand stories narrated to assert arguments. We may accept a premise, if we are given some specific examples or anecdotes. A major problem with anecdotes is that anybody can concoct one and present it as a reliable example. It might simply be a fiction! In a scientific argument, such anecdotes are not given the status of evidences or authority, and we cannot accept a scientific premise simply based on anecdotes. However, anecdotal evidences may affect our understanding of a premise and may influence our judgment. Therefore, critically evaluate the anecdote and ask for additional evidences, if you find some merit in the anecdote. Religious groups and proponents of pseudo sciences or alternate sciences rely mainly on anecdotes to present their cases.

Hearsay evidence: Hearsay evidence is what someone says about hearing from another person and present it as evidence to bolster a point. Many people start believing what they hear without verification. We should not accept hearsay evidence as reliable or valid, if it is impossible to check its source.

Testimonial evidence: Testimonial evidence, the evidence given by a witness, is more reliable than hearsay evidence. Witness accounts are allowed in courts of law and can be accepted depending upon the reliability of the person who makes the testimony. Remember, however, that even testimonial evidence could be unreliable.

Circumstantial evidence: Circumstantial evidence is also a common type of evidence employed by both scientists and non-scientists. Courts of law also allow circumstantial evidence, the circumstances such as the means and motives, but this may also be unreliable.

Spectral evidence: Another kind of evidence prevalent among common people is spectral evidence, the evidence attributed to ghosts, spirits, and other supernatural or paranormal entities. Again, this cannot be accepted as evidence, as it is simply unreliable.

Revelatory evidence: If somebody says that the evidence was revealed to him/her by a supernatural power or divine being, it is classified as revelatory evidence or simply revelation. However, this is also unacceptable for science, as it cannot be verified. Repeatability is also a problem with such evidences.

Emotional evidence: Emotional evidence is evidence derived from one's subjective feelings. However, as it is applicable to one person only, emotional evidence is unreliable and cannot be accepted as evidence.

14.3 Inductive and Deductive Reasoning

As already discussed, we use two basic kinds of logic, inductive and deductive, in our arguments. When the major premise of an argument is based on observation or experience, it is an *inductive argument*. On the contrary, a *deductive argument* is one in which the major premise is based on a theory, rule, law, principle, or general understanding. Deductive reasoning is the process of reasoning from a general assumption to a specific application, while inductive reasoning is reasoning from a specific application to a general assumption. Look at the following conversation.

Thomas: 'Whenever I spray 2, 4-D for controlling weeds in cowpea, it produces phytotoxic symptoms. I am sure that this herbicide is not suitable for cowpea'.

Meera: '2, 4-D is a selective herbicide that cannot be used for the control of weeds in pulses such as cowpea. If it is used in cowpea, it will produce phtytotoxic symptoms'.

Thomas is using inductive logic arguing from observation, while Meera is using deductive logic arguing from the already known character of 2, 4-D. Meera's argument is from the general (selectivity of 2, 4-D) to the specific (phytotoxic symptoms); while the argument of Thomas is from the specific (whenever he spays 2, 4-D, it is producing phytotoxic symptoms) to the general (the prediction that a similar event will result in a similar situation in the future). As we could see, the difference between inductive and deductive reasoning is mostly in the manner in which the arguments are presented. In several cases, inductive arguments can also be presented deductively and vice versa. It is important that one should be able to distinguish whether the argument is inductive or deductive because of the specific requirement of supporting evidences. Inductive argument of Thomas is supported by his previous observations, while Meera's deductive argument is supported by her understanding of the character of 2, 4-D. Thus, Thomas could provide evidences by detailing those observations, while Meera could provide support by already known character of 2, 4-D.

In an argument, we presume that if the premises are true, then the conclusion is also true. However, for the conclusion to be true, two critical preconditions must be met. Firstly, the premises must be true, and secondly, the 'form' of the argument must be 'valid'. In a valid deductive argument, the conclusion must be true, if the premises are true. However, validity or invalidity of arguments depends on the 'form' of the argument only, and not based on the accuracy of the premises! The following is an example of argument with correct 'form'.

All A's are B's.
All B's are C's.
Therefore, all A's are C's.

The following examples of arguments show the distinction between the roles of premises and the 'form' of arguments in deciding the correctness of a conclusion:

- All dogs are cats. All cats are animals. Therefore, all dogs are animals (All A's are B's. All B's are C's. Therefore, all A's are C's.)

The form of argument is valid, but the major premise is false (*All dogs are cats*); therefore, the argument is incorrect. The conclusion here happened to be true, but might not be true always, if we follow this form of arguments in other cases. See other forms:

- All dogs are mammals. All cats are mammals. Therefore, all dogs are cats (All A's are C's. All B's are C's. Therefore, all A's are B's.)

True premises, but form is invalid; therefore, the argument is invalid, and we cannot have a wrong conclusion like this.

- All dogs are mammals. All mammals are animals. Therefore, all dogs are animals (All A's are B's. All B's are C's. Therefore, all A's are C's.)

Form is valid, true premises; therefore, the argument is correct, and the conclusion must be true.

A critical evaluation of arguments is not necessary for the above examples, as we already know which conclusions are true and which are false based on our knowledge on the subject. However, this may not be so simple always. When employing reasoning as a part of scientific research, one should consider separately the two preconditions; premise correctness and argument form. Correctness of premises can be evaluated based on expertise in the subjects. Nevertheless, evaluation of the form of argument is objective. We could find several instances of invalid arguments in publications. In research articles, authors may commit this error while discussing the results. With some practice and guidance, one can avoid using invalid argument forms and recognize them.

Induction in science usually involves generalization from the behaviour of a sample to that of a population. For example, after observing a number of crows, we could make the inductive statement: 'All crows are black'.

It is possible to depict the confirmation of an inductive hypothesis using symbols as:

$$A \rightarrow B, \quad B, \quad \therefore A$$

(A implies B; B is observed; therefore, A must also be true).

The strength of an argument as above is mainly based on the accuracy and comprehensiveness of observations and the connection with the cause. Although confirmation of an induction is often uncertain, the hypothesis can be disproved by a single experiment using deductive logic. We can write this in symbols:

$$A \rightarrow B, \quad -B, \quad \therefore -A$$

(A implies B; B is not observed; therefore, A must not be true).

The scientist proposes a *hypothesis* through induction and then tries to deduce the probability that it is false through empirical evidences, often collected through experiments. If not done properly, chances of error in such experiments can be very high. The main source of error in an inductive statement is the sample size. When the sample size is too small, the generalized statement may not be reliable. The samples must be representative of the population as a whole; otherwise, chances of occurrence of fallacies are great. Generalization from observations of small or unrepresentative samples to the likely behaviour of the entire population is a *fallacy*. To overcome the problems of sample size, bias, and non-uniformity of experimental materials, we resort to various techniques such as randomisation, local control, replication, and other statistical techniques.

In science, we use both inductive and deductive arguments, but before accepting them, they should be evaluated for genuineness. Inductive arguments are evaluated as 'strong' or 'weak' based on the stringency of evidences, but deductive statements are stated as 'valid' or 'invalid' according to the 'form' of argument and also on the possibility that whether true premises entail a correct conclusion. Of the two, induction is more useful for the advancement of scientific knowledge than deduction. Induction produces new knowledge in the form of a generalization that did not previously exist, whereas deduction produces no new knowledge at all. In contrast to inductive reasoning, the conclusions of deductive reasoning must be as valid as the initial assumption. Deduction is *definite*; but induction is *indefinite* and uncertain. Deduction, however, can be very useful in testing poor hypothesis using the deductive technique as already shown.

Inductive reasoning is more open-ended and exploratory, especially at the beginning of a study. Deductive reasoning is narrower in scope, but is useful for confirming or testing hypotheses. Both inductive and deductive approaches can be considered as different ways of approaching the same objective, and in practice, a combination of induction and deduction is used. The observations made through induction are further verified deductively through applications to new situations. A scientist may propose a hypothesis through induction and then tries to deduce the probability that it is false through empirical evidences. This is what is commonly known as the *hypothetico-deductive* method or *inductive-deductive* method, recognized as equivalent to scientific method or experimental method mentioned already (see Sects. 1.6 and 1.8).

Many scientists and philosophers battled over whether science should be deductive or inductive. After the realization that both deduction and induction are necessary aspects of science, we could observe a phenomenal leap in scientific progress. However, as anticipated, theoreticians value deduction, and empiricists value induction, but most scientific disciplines now use both.

14.4 Fallacious Reasoning

Everybody wants to win an argument and presents evidences accordingly. We usually employ logic to support claims in our argument or articles. A fallacy is a statement that is logically false, but often appear to be true. This causes an argument to be invalid, unsound, or weak. Just as common people do, scientists may also present fallacious arguments. Fallacious arguments are presented because of many reasons such as superstitions, religious beliefs, astrological beliefs, simple naivety, and sometimes deliberately created to deceive others.

It is human nature to indulge in arguments. However, it becomes a nuisance when fallacious arguments are presented. Fallacies affect the researchers and the public in several ways. Some distract the readers from understanding what is really going on. Some make use of vagueness or ambiguity to cause confusion. Some appeal to emotions rather than logic and reason. You can easily make out some of these fallacies if you keenly read the newspapers or watch discussions on television channels. Journalists and politicians are experts in creating fallacies. They create fallacies to 'prove' their point of views. Scientists may also succumb to fallacies. Scientists and the public should be able to understand defective arguments, explanations, and conclusions.

An argument in everyday life usually involves tension, confrontation, anger, and sometimes, personal attacks and counter-attacks. Nonetheless, when used in an academic sense, the meaning of argument is very different. In 'academic argument', a position is taken on a subject, and that position is defended with certain kinds of evidences. Scientific or academic arguments should always be made for or against ideas only and not for or against people who put forth these ideas. Make such arguments complete with data, ideas, opinions, and facts. In any case, do not present fallacious arguments. Therefore, it is important to identify and eliminate fallacies in scientific reports and communications.

14.5 Formal and Informal Fallacies

We usually use syllogistic arguments as explained earlier without knowing the intricacies. Some arguments may be fallacious although these can appear to be structurally valid. In such cases, even if the premises are true, the arguments yield only a false conclusion. In general, fallacies can be broadly classified into two categories, *formal* and *informal*. A formal fallacy is an error, recognizable simply by examining the logical form of an argument rather than any specific statements. Informal fallacies are also faults, but their deficiencies can be identified only through an analysis of the actual content of the argument.

Formal fallacies are found only in deductive arguments with identifiable syllogistic patterns. Although the structure of the argument makes them appear reasonable, the form is invalid. See the example:

- All dogs are mammals (major premise).
- All cats are mammals (minor premise).
- Therefore, all dogs are cats (conclusion).

In this example of syllogistic logic, the middle term (mammals) is in the predicate of both universal affirmative premises, and therefore, is undistributed. This is an example of 'fallacy of undistributed middle', wherein one fails to distribute the middle term properly in the sentences. Although both premises in this argument are true, the conclusion is false. Therefore, the argument is invalid and does not yield this conclusion. The above 'fallacy of undistributed middle' is an example of a formal fallacy. It can be made clear by reducing the argument to its fundamental form of deductive reasoning. The above example of formal fallacy has the logical form:

- All A's are B.
- All C's are B.
- Therefore all A's are C.

B is assumed to cover all items in its category. It is immaterial what do A, B, and C represent. See what happens if we replace them with 'rose', 'jasmine', and 'scented'. For example:

- All roses are scented
- All jasmines are scented

- Therefore, all roses are jasmines.

Informal fallacies are also common in arguments. However, their faults can be identified through an analysis of the content of the argument instead of its basic form. For analysing such arguments, you should have enough knowledge on the subject. See the example:

- All dogs are cats.
- All cats are animals.
- Therefore, all dogs are animals.

Let us break the above argument to its basic form:

- All A's are B.
- All B's are C.
- Therefore, all A's are C.

The form of above argument is valid, but it involves a false premise. The argument, therefore, is incorrect although the conclusion happens to be true. Because of the inclusion of a false premise, the fault cannot be a formal fallacy, it must be categorized as an informal fallacy. This fault is identifiable only by examining the content. When we look at the content, we could find that the first premise is false. The argument may not be as simple as given here, and the public may take the argument as granted. In many complicated arguments, a thorough appraisal of the content is necessary to identify the faults. Informal fallacies affect researchers and public in several ways. Some make use of vagueness or ambiguity to cause confusion. Some appeal to emotions rather than logic and reason. Whether formal or informal, many of these fallacies are genuine pitfalls to scientists. Post modernists and some environmental activists frequently make use of fallacious arguments to win over their opponents. You can identify these kinds of fallacies, if you closely go through their works.

14.6 Common Fallacies

Fallacies are not new, and they were here probably from the time humans started arguments! The great philosopher Aristotle could identify and describe 13 such fallacies. Now, there are several, more than 200 have been named. Many of these have traditional Latin names, revealing their classical origins. If the error is more frequent within public discussion and debates, the chances of having a name for it is more. Labossiere (2010, 2011) described 72 fallacies. Jarrard (2001) gave an account of fallacies relevant to science and technology. A few common fallacies identified and given names have been presented in this section. Only those fallacies that have some relevance to scientific arguments are included. For a comprehensive account on fallacies, refer to sources such as Tindale (2007), Copi and Cohen (2007), and Damer (2008).

1. **Suppressed evidence**

A person commits the fallacy of 'suppressed evidence' when deliberately suppressing relevant and significant information. Intentional omission of adequate evidence weakens one's own conclusion. This fallacy is common among both scientists and ordinary people. Scientists may not deliberately hide evidence, but they may suppress evidences by a conscious decision that the evidence is faulty to deserve mention. For example, while interpreting results, if some of the statistical evidences relevant to the experimental results are purposefully hidden or overlooked, this fallacy is committed.

This fallacy of 'suppressed evidence' is used extensively in advertising. Most marketing campaigns will highlight only the greatness of a product, but will ignore its drawbacks.

You can also notice 'suppressed evidence' in several post-modern movements. Several examples can be cited from agriculture. Take the case of movements such as organic farming, natural farming, homa farming, zero budget natural farming, and system of rice intensification. Supporters of these movements always highlight the virtues and conveniently suppress negative evidences.

2. **False dichotomy**

It is also called 'false dilemma' or 'fallacy of excluded middle'. This fallacy occurs when an arguer presents a few alternatives only for solving a problem when more possibilities exist. By focusing on the choice, the other person is diverted from the fact that there may be other alternatives. If you succumb to select one among those choices only, you accept the premise that those choices are the only possibilities. In most similar cases, only two choices are presented; and therefore, the phrase 'false dilemma'. Often, it occurs as an incorrectly exclusive 'either …or…' statement in one of the premises. When one option is eliminated by a premise, the other alternative is accepted incorrectly as the conclusion.

Consider the example, 'This experiment will either prove or disprove the notion that astrology is a science. The results did not disprove the hypothesis. Therefore, it must have proved it'. However, the inability of disproving the hypothesis is not enough proof of its validity. You have to check the reasons objectively.

Another example where extreme views are presented ignoring a central position: 'If we do not switch over to organic farming, the world is doomed'.

3. **Black-or-White**

The black-or-white fallacy is a type of 'false dilemma' fallacy that unfairly limits you to only two choices. This fallacy tries to eliminate the middle choice. A typical black-or-white is the assertion, 'You are either with me or against me'.

Another example: 'You decide. Will you oppose the construction of this hydro-electric dam, or are you on the side of environmental destruction'? A mere opposition to the construction will not make an individual 'environment friendly'. A major test to identify the black-or-white fallacy is to find out whether the limited choices are fair or unfair.

4. **Slippery slope**

The arguer may base the argument on the assumption that if a particular event occurs, other undesirable events follows. Take the example of a person who claims that a first step in a chain of causes and effects will lead to a second step that in turn will probably lead to a third step and so on until ending in a final step of big trouble. The purpose of the arguer is to prevent that first step to take place because if the person climb onto the first step of a slippery slope, that may cause a chain reaction in that the person to slide all the way down to some disastrous situation. However, the hypothesis that a full chain reaction always follows the initial event is wrong. Each step in the chain of events requires causality.

The following is a typical example of slippery slope. 'If we ban smoking, the tobacco farmers may face difficulties, they cannot sell their produce, they may commit suicides, and the suicide rate among farmers will go up and up. Therefore, don't ban smoking'.

5. **Might makes right**

This is a threatening argument in which the arguer forces the listener to accept the arguer's point of view or suffer the consequences. The intimidation may be physical abuse or some other undesirable action. Although the threat is irrelevant to the validity of the conclusion, it may affect the listener's decision-making. The arguer threatens the audience explicitly or implicitly with dire consequences, if the claim is not believed. For example, 'Everyone knows that organic foods are superior with better taste and quality, and if you try to prove otherwise, you will destroy your credibility as a scientist'.

6. **Appeal to authority**

This is the fallacious claim that an argument should be accepted by everybody as some experts endorse it. Although the arguer quotes authorities to demonstrate that the claim is valid, the authorities may not be true experts in the field, their opinions may be taken out of context, or other experts in the same may differ with the cited authorities. Celebrity endorsements of commercial products are often used as fallacious appeals to authority. A person may be doing well or an expert in one avocation such as acting, music, poetry, novel writing, or cricket, but it does not entail his/her claims or opinions any additional merit in an entirely different area such as farming, environment, health care, diet, or investments.

Scientists should not appeal to authority; they must evaluate evidences personally. In practice, however, we limit such inquiries mainly to our own field, and we cautiously accept the prevailing wisdom of scientists in other fields. The appeal to authority must be considered logically based on how much more experience the 'authority' has than the arguers. Most reasoning of this kind is not fallacious. However, it becomes a fallacy when the authority appealed to is not really an authority in the subject, when the authority cannot be trusted to tell the truth, and when authorities disagree on this subject. For example, when an agronomist considers a physics argument, it is valid to give weight to what a physicist believe. Nonetheless, when a physics expert considers an argument related to agriculture, it is a fallacy to accept it merely because some 'great' physicist believes it.

7. **Appeal to common practice**

Many people observe or do something probably as a common practice. They believe that because of common practice, it is a reasonable thing to follow suite. However, in most cases, the appeal to 'common practice' is used as an excuse. When we are not sure about something, we look at what other people follows and begin to follow them as though they are correct. This is often used as a tactic to influence people to obey the norms of the group to which they belong.

The observation of common practice is mainly based on superstitions. Sometimes, the argument is based on unacceptable reasons usually due to unreasonable fear of the unknown, trust in astrology, black magic, or an obviously false idea of omens or bad luck (for example, avoiding the number 13). Such reasoning deserves to be called 'superstitious' and not 'common practice'. Superstitious beliefs are rampant among lay people and even among scientists. For many effects, they may attribute superstitious reasons. Breaking coconuts before the launch of scientific ventures, lighting lamps, avoiding inauspicious days, etc., are also based on superstitious thinking, but people defend them based on common practice. For example, 'It's common practice to break coconuts before starting a new venture. Therefore, arrange a coconut to break it before the launch'.

8. **Appeal to tradition (Old is better/Traditional wisdom)**

If we imply that a practice must be accepted simply because it has been the apparently wise practice in the past, we commit the fallacy of 'traditional wisdom'. Traditional practices might have a good justification, but merely saying that they are always correct, as they have been practicing them as a part of tradition is not acceptable. For example, it is fallacious to say, 'We've been doing this practice for many years, and we've never had any problems with it'. An established tradition becomes a cultural obsession where people follow it without much thinking and defend it with all might. However, the traditional practice may not hold grounds in the modern world or in the changed environment. Not all the traditional practices may endure, and they have to be evaluated case by case.

9. **Appeal to novelty (New is better)**

Appeal to novelty is a kind of fallacy wherein an arguer claims that new is always better than the old one. The arguer claims that a new finding has better effects simply because of its novelty. However, being new does not make an idea more correct. Simply because something is new, we cannot say it is good and superior and can replace what has gone before. For example, 'This model of car has ten new features, and therefore, it is superior to the old one'. A new model need not be better than the older version.

10. **Appeal to pity**

It is an appeal for accepting the conclusion out of pity. The arguer asserts that approval of the argument will help someone in trouble or that its rejection will cause undue hardship to the arguer or someone else. However, the circumstances are irrelevant to the validity of the basic arguments. Students often use this fallacy on their teachers; subordinates on their superiors; and lawyers use it on judges. For example, if the examiner found serious inconsistencies in a thesis, the student may plead with the examiner not to press for drastic action such as rejection of the thesis as it may affect his/her future. Some scientists also use it for personal gains before their chiefs or directors.

11. **Appeal to ridicule**

This fallacy is also common among both lay people and scientists. Ordinary people may use it daily during arguments and scientists may use it when confronted with an uncomfortable question during a discussion or scientific conferences. They may ridicule the claims and arguments of their opponents in the field. Other participants or onlookers may laugh at such deriding comments. The comments like 'Oh, you are supporting that movement? Incredible, did you lose your senses?' or 'What you are talking is simply nonsense' are made with this objective. This ridiculing is done deliberately to downgrade the victim's image and social position. People usually hesitate to associate themselves with a person who has been relegated to a lower social position, because of the fear that they may also be pulled down to that position.

12. **Appeal to fear**

In this case, some form of intimidation is directed towards the person who comes up with counter arguments. The threat may be physical, emotional, or spiritual. For example, 'Don't go too far with that hypothesis; if the Director comes to know about this, he would not be happy'. The threatened individual may withdraw fearing reprisal such as adverse confidential remarks, act of vengeance in interviews, or other reprisals such as transfers. It can also lead to other uncomfortable emotions such as dislike and hate, which can bounce back upon the arguer.

13. **Bandwagon (Appeal to common belief)**

The fallacy of bandwagon is prevalent among both lay people and scientists. If the arguer suggests that the claim is correct simply because it is what most people believe, the arguer is committing this fallacy of bandwagon. Similarly, if you argue that someone's claim or argument is wrong simply because it is not what most people believe, then again you have committed the fallacy. However, what everyone believes is likely to be true, and if one defends a claim on those grounds, we cannot dismiss it fully as a fallacy. Still, if you believe the worthiness of a new idea or trend solely on the grounds of its new popularity, it is a fallacy. See the following examples.

Example 1: 'You should see that film. It's the most widely seen film this year'. This is fallacious because it is implicitly suggesting the questionable premise that the most widely seen film this year is the best film for you.

Example 2: 'You should attend this seminar on zero budget natural farming. Everyone is accepting this farming method and producing crops without spending any extra cost'.

14. **Mob appeal (Social conformance)**

'Social conformance' or 'mob appeal' is a fallacy of using mob pressure instead of rational argument to get agreement on a particular subject. The arguer seeks a simultaneous group response through the lure of inclusion or threat of exclusion. If you do not agree, then you will be rejected by your peer group. Social acceptance is very important to most people, especially, with those they consider friends and peers. Thus, the danger of rejection is taken very earnestly, and people will change their stand and even their beliefs to maintain their social status. Political parties and organizations make use of this fallacy to keep their members disciplined. For example, 'If you keep on disagreeing with the leader, you will be out of our party'.

15. **Red herring**

In an argument, a 'red herring' is a tactic to distract someone, a reader or listener, instead of building an argument. Often, a distracting smoke screen is created to divert attention from an argument's weakness. This fallacy got the name 'red herring' from a technique used in training hunting dogs. The trainer drags a sack of red herring, a strongly scented fish across the scent track that the dog is supposed to take to divert the attention of the dog. The expertise of the trainer is to train the dog to stick to the main track without being distracted by the red-herring track. This method of distraction is the secret of many magicians' tricks.

The fallacy of red herring is not usual in scientific arguments. However, a distraction is sometimes adopted deliberately. The arguer demolishes a minor criticism of his/her argument, giving the false impression of careful and objective analysis, while obscuring a brief mention of a serious weakness in the argument. For example, during a discussion on the problems of increased use of alcohol in the society, if somebody raises arguments like, 'Some of the world's greatest writers were alcoholics', the statement would be a red herring.

16. **Missing the point (Avoiding the issue)**

If an arguer, instead of addressing an issue with valid evidences, goes off on a deviation is committing the fallacy of missing the point or avoiding the issue. The fallacy can be detected easily by deciding the acceptable conclusion from the premises, and then comparing this expected conclusion to that presented by the arguer. See the examples,

Example1: 'Gender related atrocities are increasing. It must be because of the modern style of dressing among women'.

Example2: 'There has been an increase in crimes in the area. It must be because there are more immigrants in the area'.

The causes cited here have no relationships with the statements of the arguers. Instead of addressing the issue with valid reasons, they are presenting misleading reasons. Many times, people who want to prove something but do not know how, use this fallacy to win over others. They simply begin an argument and then take their desired conclusion on to the end. This is something that the politicians and activists often do.

17. **Begging the question (Circular reasoning)**

This is also known as '*circular reasoning*' because the conclusion appears both at the beginning and at the end of the reasoning. The difference between a valid deductive argument and the fallacy of 'circular reasoning' must be clearly understood. In circular reasoning, we fallaciously assume that what we are supposed to be proving is true. However, in a valid deduction, the conclusion derived from the premises or assumptions is definitely true. The major difference is that in circular reasoning, the conclusion is from a single premise; while in a deductive argument, the conclusion is derived from more than one premises. The author makes one or more statements that are obviously true and then states that as the first statement is true, the second must be true. In fact, there may be no logical connection between the accepted premises and the stated conclusion. See the example: 'A stem when placed in a horizontal position will bend upwards because it is negatively geotropic'.

18. **Wishful thinking**

The name of this fallacy 'wishful thinking' is noteworthy. Lay people and scientists are not bad in wishful thinking. An arguer who suggests that an argument is true or false simply because he or she strongly hopes it so is committing the fallacy of wishful thinking. Although the falsehood of this fallacy is obvious, people often indulge in wishful thinking. It is also startling to know how often people do not realize that they are committing these kind of fallacies. In fact, while doing so, they are expressing their subconscious desires through assumptions of truth. See the example: 'I won't believe she has received foreign money. She is an environmental activist, and an activist would never do such a thing'.

19. **Non-sequitur reasoning (Fallacy of false cause)**

In an argument, if one reason follows another in a sequence, it increases validity of the argument. However, when a conclusion is based on irrelevant or weak reasons, the argument is fallacious and non-sequitur (literal meaning, 'it does not follow'). Sometimes, you do it without taking the troubles to generate enough evidences taking recourse to fallacious solutions. Any deductively invalid inference is a non-sequitur, and you cannot rescue it by saying it is common sense. Examples:

(a) 'Pesticide use is a risk, but everything in life involves a risk. Every time you travel in an aeroplane, you are taking a risk. If you're willing to fly, you should be willing to allow the use of pesticides'.

(b) 'Light is always necessary for photosynthesis; therefore, an increase in light intensity will increase photosynthesis'.

20. **Straw man**

The fallacious strategy for refuting an argument by singling out a weak part of the argument and making a great fuss about it is called 'straw man'. The strategy is to misinterpret it first, refute the misinterpreted version, and then conclude that one has refuted the original argument! The term is a take-off on the concepts of scarecrows and burning in effigy. Deliberately, attention is diverted from the stronger reasons that should have been the main part of the argument. By putting up this fallacy, you dream that you have set up a straw man and easily knocked it down, asserting that you have knocked down the real thing. For example, 'Astrology may be unproven, but it is a science as nobody has proved it to be false'. The argument is nothing but a straw man!

21. **Hasty generalization**

We know that one of the main objectives of scientific research is to establish reliable theories and general laws that can be used to explain and predict events. Inductive reasoning, which leads to generalization, is done based on experience or observation, especially from the behaviour of a few samples to that of a population. The main source of error in an inductive statement is the sample size. When the sample size is too small, the generalized statement may not be reliable. A 'hasty generalization' is a fallacy of inductive extrapolation to an entire population, based on a non-representative sample. Often, it may not be a representative at all. However, bear in mind that smallness alone does not imply the fallacy of hasty generalization; for example, a sample of only two or three is enough when dealing with a uniform and representative property as with many experiments in physics. If we want to have reliable generalization, the samples must be representative of the population as a whole.

Lay people are more prone to hasty generalization than scientists. Common people usually ignore the need for a representative sample. See the following statements:

(a) 'Three coconut palms in front of the school building are giving bumper yield without any cultural operations including irrigation or fertilizers. So, there is no need to give irrigation or fertilizers to coconut'.
(b) 'I have met four farmers in this village and all of them are rearing cows. I am sure that all the farmers in this village are rearing cows'.

22. **Jumping to conclusions**

When we draw a conclusion without taking the trouble of acquiring all the relevant data or evidences, we commit the fallacy of jumping to conclusions. This is all the more condemnable, when there is sufficient time to obtain additional evidences, and that the efforts to gather the evidences are not difficult. For example, 'See this rice plant. It produced 60 tillers. I am sure this particular treatment is superb and fetches bumper yields'. You are jumping into conclusion based on an exception. Other plants may not have these much tillers or the number of plants per unit area may be less compared to other treatments. You can gather data by conducting a replicated 'controlled' experiment. Only if the results agree, you can come out with such a sweeping conclusion.

23. **Converse accident**

If we argue by stressing on exceptions, and generalize based on the exceptions, we succumb to this fallacy of converse accident. In fact, it is a kind of 'hasty generalization' already discussed. See the example:

'I tried organic farming and conventional farming in one acre of paddy each, and in the first year, I got 2000 kg grains from the former and 2400 kg from the latter. However, in the second year, I got 2500 kg from organic plots and 2450 kg from conventional plots. I am sure that yield from organic plots would increase in the coming years and will outyield conventional plots'. The arguer places too much trust in this exception of just 50 kg difference in yield in the second year and generalizes on it to produce the faulty conclusion that organic farming gives more yield than conventional farming. The small differences may be due to experimental error or simply because of random error. If you subject these values to statistical test, most probably you may get the values at par.

24. **Unrepresentative sample (Biased sample)**

This is also a kind of 'hasty generalization' fallacy, which happens when a biased or otherwise statistically invalid sample is used to declare the results significant and conclusions are drawn based on this defective interpretation. Most people believe that they are good at making statistical assessments. In fact, we are generally poor at it, and we may fall easily into many fallacy traps. Drawing conclusions based on unrepresentative samples or biased samples is one of the typical mistakes committed by ordinary people. For example, see the statement: 'I asked 10 people in the street and 7 liked coca cola, indicating that 70 percent of people like coca cola'. As the fallacy of biased sample is committed when the sample is biased or unrepresentative, it is important to have samples that are truly representative for generalizing. The best method to accomplish this is to take samples in a manner that avoid bias, for example, by random sampling.

25. **False extrapolation to the whole**

This fallacy occurs when a wrong conclusion is made overall as one or more parts exhibit some of its characteristics. This conclusion is acceptable in certain cases, but it is not universally valid. To decide whether the statement is correct, careful evaluation of the content is required, because sometimes extrapolation to the whole is unjustified. For example, 'One of the students of that private English medium school got first rank in the examination; it must be a good school and I am going to send my daughter to that school'. This need not be correct as the receipt of first rank might be an exception, and others were just above average in scores.

26. **False extrapolation to parts or individuals**

This fallacy is the reverse of the previous one, 'false extrapolation to the whole'. This fallacy is committed when someone erroneously apply a generalization to an individual case, especially when we make conclusions about parts or individuals based on group data. A general rule may be inappropriate for this kind of specific cases. One cannot conclude that all the parts exhibit the characteristic of the whole unless s/he has taken pains to ascertain facts. Assume that after the final secondary school examinations, a particular school in the state has been declared the highest scorer with maximum percentage of distinctions. If we happen to see one of the students of that school, we may probably think, 'She must be a brilliant kid'. This assumption might go wrong! She could be one among the low scorers in the said school that otherwise consists of many students with distinction. See another fallacious argument of this kind: 'That conference is regularly held every year and worth attending, and therefore, every presentation at the conference is worth attending'. The conference might be popular and worth attending, but there might be a few boring or worthless presentations! Therefore, do not jump into the above conclusion.

27. **Stereotyping**

Stereotypes are general beliefs to categorize people, objects, and events. However, these beliefs are overstatements that should not be taken literally. For example, consider the statement: 'She's a feminist, so she's going to be arrogant'. This conveys a mistaken impression of all feminists. Most people will not understand that even the best stereotypes are correct only when taken probabilistically. This kind of fallacious reasoning is at the core of gender bias and racism and many superstitions. For example, a misogynist, who sees a woman making a driving error, may conclude: 'See how incorrectly she is taking that turn. I am sure women cannot drive properly'.

28. **Ad hoc rescue**

It is quite natural that we try to rescue many of our long held beliefs from being refuted. When faced with contradictory data, we will try to mention how the contradiction will disappear, if some new assumption is taken into account. However, if the argument is just to save our cherished belief from refutation, the effort would simply be a fallacy. This type of effort is usually called 'ad hoc rescue'. For example, those who have strong belief in astrology usually resort to this kind of ad hoc rescue, whenever something contrary to astrological predictions occurs. They will definitely provide you some instant explanations for not happening the way the astrologer predicted or they believed.

29. **Personal attack (ad hominem)**

Some arguers indulge in 'personal attack' instead of presenting evidences, sometimes as a diversionary tactic. The meaning of the Latin phrase 'ad hominem' is 'to the person'. 'Personal attack' is a criticism of the opponent's persona in a debate rather than refutation of the opponent's arguments.

As the name suggests, the arguer weakens the credibility of the opponent by attacking his/her character, personality, or psychological health, although those factors are irrelevant to the argument being 'refuted'. Logical reasoning or argument in no way is connected with the morality and conduct of somebody. If anybody makes an inappropriate attack on the arguer, and claims that this attack challenges the argument, this fallacy is committed. See the statement: 'I am sure that these scientists manipulated with the evidences and came to this wrong conclusion after receiving money from multinationals'. This statement is made to tarnish the image of the scientists concerned and to lower their repute among the public.

30. **You too fallacy (appeal to hypocrisy/tu quoque)**

This is a form of 'personal attack' fallacy, which is called tu quoque, the Latin version of 'you too'—because it often occurs when a person is attacked for doing what s/he is arguing against. This fallacy is committed when someone defend an error by pointing out that the opponent has made the same fault. This kind of argument attempts to discredit the opponent's position by attempting to defend oneself from criticism by turning the criticism back against the accuser. However, whether the accuser is guilty of the same or a similar wrong is irrelevant to the truth of the original accusation. As a diversionary tactic, 'you too' can be very effective, because the accuser is put on the defensive and frequently feels compelled to defend against the accusation. See the following 'you too' argument: 'You ask me to stop copying in the examination, but I know that when you were a student you had been debarred from examination once for copying, so your advice can't be worth listening to'. Politicians quite often use these kinds of arguments to belittle their opponents.

31. **Poisoning the well**

'Poisoning the well' is a type of personal attack. This fallacy is a defensive attack on somebody in order to discredit his/her argument in advance. The arguer 'poisons the well' by reasoning fallaciously and makes others unreceptive to opponent's reasoning. For example, before giving a chance to the opponent, the arguer may declare, 'Everybody knows that endosulphan is causing all the troubles in Kasargod in Kerala. Professor Thomas, do you have something to say on this issue?' You can imagine the plight of Professor Thomas! He may have some other explanations to make, but as the environment is vitiated, he may keep silent!

32. **Affirming the consequent**

The arguer uses an 'If…then …' statement in reasoning. The arguer then confirms the 'then' part of the statement and derives the 'If' part, thus committing a serious logical mistake. For example, 'If it rains, then my car will get wet. My car is wet; therefore, it rained'. It might not be due to rains. Your car could have been splashed by the sprinkler when irrigating the lawns!

33. **Denying the antecedent**

In this case too, the arguer employs an 'If…then…' statement. Here, the arguer denies the 'if' part so that the occurrence of 'then' part can also be denied. However, we cannot simply assert that the 'if' part of the argument does not happen, the 'then' part will also not materialize. The 'then' part could result for some other reason. For example, 'If the use of endosulphan is not banned, many people will get cancer and suffer from other illnesses. Endosulphan has been banned by the Government. Therefore, many people will be freed from the risk of cancer and other illnesses'.

34. **Anecdotal evidence**

If you ignore evidences gathered by experimental studies or observation, and go for some anecdotes narrated by somebody, you are committing the fallacy of anecdotal evidence. Anecdotes are some first-hand stories narrated to assert arguments. A major problem with anecdotes is that anybody can concoct one and present it as a reliable example. Many religious groups smartly make use of anecdotes to prove what they preach. The following is a typical example of using anecdote fallaciously. 'I've read the warnings on alcohol consumption and I know about all that findings on drinking; but my elder brother drinks, and he says he's never been sick a day in his life. Therefore, I am sure drinking can't really hurt you'.

35. **Unfalsifiability**

According to the principle of 'falsifiabilty', a scientific statement must be falsifiable, which means that if the statement is false, this could be shown by observation or experiment (for more information on falsifiability see Sect. 2.6). When the argument or discussion contains claims that are not falsifiable, the fallacy of 'unfalsifiability' occurs. They cannot be tested and inferred from other well-accepted claims. As an example, consider the statement: 'Biodynamic farming produces wonder yield because of the interplay of astral and terrestrial forces' (biodynamic farming is an 'alternate farming' system). There is no way to collect evidences for or against the claim that astral and terrestrial forces are acting, therefore, presenting it as the correct explanation for 'wonder yield' is an error.

36. **Teleological fallacy**

The word 'teleological' is derived from the Greek word *telos*, meaning 'end' or 'purpose'. According to teleology, everything in nature has a purpose or end and design. See a typical teleological statement often found in textbooks: 'Flowers develop colour to attract insects'. The statement implies that the purpose of attracting insects for pollination is directly responsible for the development of colour in flowers. These kinds of explanations suggest that the advantage obtained from a particular structure or arrangement is an acceptable reason, and therefore, no further search for a mechanism is required. In fact, numerous beautifully coloured flowers are there in nature in which pollination is accomplished by wind, water, or other means. Therefore, the above teleological statement can be branded as a fallacy. Another example,

'Plant roots grow deeper in to the soil in search of water'. You may find several teleological statements like the above in biology (Wright 1948).

37. **Tokenism**

If anybody tries to interpret a mere token gesture as an adequate substitute for the real thing, a fallacy called tokenism is committed. An argument against women's reservation in Parliament goes like this: 'You cannot say that in India there is any discrimination for women. They can even become Prime ministers and Presidents. You should not forget that one of our past President was a woman. We have had a woman prime minister too. Therefore, there is no need to give women one-third reservation of seats in Parliament'. If you accept this line of reasoning, without actually assessing the involvement of women in real administration, you have committed the fallacy of tokenism.

See another example: 'How can you call our organization gender biased? After all, our director is a woman'. However, you are ignoring the fact that the director is a woman simply because she is the daughter of the company owner!

38. **False cause (Questionable cause)**

Cause-and-effect reasoning is a valid form of deductive logic, but only if there is a causal relationship. If we say that a particular cause produces effects without sufficient proof that a causal relationship actually exists between these two, we commit the fallacy of 'false cause'. The two things may go together and assume a cause-and-effect relationship. However, both may be effects of a prior common cause. See the statement: 'Eating fresh vegetables will keep you well'. It may contribute to health, but in reality, it may not be the only cause for health, several other causes are also contributing.

39. **Post hoc fallacy**

Post hoc fallacy occurs when an argument presumes a causal relationship without sufficient grounds. It is also a form of 'false cause' fallacy. This fallacy is named after a Latin phrase post hoc ergo propter hoc, meaning 'after this, therefore because of this', which suggests that one event caused another simply because it took place first. It is likely that other factors have caused the second event. See the following statement: 'I won this essay competition by writing with my blue pen; therefore, my blue pen is auspicious and I should use it whenever I go for competitions again'. Here, the arguer tries to establish a causal relationship between the blue pen and the success in essay writing without ascertaining the real causes. This kind of fallacies is common among people. Good omen (*Sakunam*) and linking it to some happening is an example. Most of the superstitions prevalent in the society are post hoc fallacies; and they are often worded so vaguely that they will almost predictably be fulfilled in the normal course of events!

Post hoc fallacy is also a problem with ex post facto experiments such as case control studies used by researchers when true experiments cannot be conducted (see Sect. 5.5). Because two factors go together, you cannot conclude that one must be the cause and the other the effect. For example, linking cancer with pollution, pesticide application, fast foods, or other such causes require caution. As there is a danger of confusing symptoms with causes and the chances of committing post hoc fallacy are greater, you should be very cautious in using case control designs for research. Such study designs should test not just one hypothesis but other logical alternatives or competing hypotheses as well to rule out fallacies of this kind.

40. **Shifting the burden of proof**

'Shifting the burden of proof' is a fallacy committed by many when confronted with strong arguments. In order to win, the person who would normally have the burden of proof makes an effort to change that burden to the other person. The arguer challenges those with an opposing view to defend their arguments. This puts the arguer in a position in which s/he can deny the opposition's assertions. For example, an arguer may defend that biodynamic farming (a form of alternate farming) is superior to conventional farming because of its divine connection. Although the arguer is unable to produce any empirical evidence, s/he pushes forward the agenda with a challenge to the audience: 'If any of you don't agree that biodynamic farming is much superior to conventional farming, then come with evidences to prove it otherwise!'.

Bibliography

Allen, M. 2004. Smart Thinking: Skills for Critical Understanding and Writing (2nd Ed.). Oxford University Press, 193p.

Bird, A. 1998. Philosophy of Science, UCL Press, London, 313p.

Carey, S.S. 2011. A Beginner's Guide to Scientific Method (4th Ed.). Wadsworth, Boston, 160p.

Copi, I.M. and Cohen, C. 2007. Introduction to Logic (13th Ed.). Pearson Education, 688p.

Damer, T.E. 2008. Attacking Faulty Reasoning: A Practical Guide to Fallacy-free Arguments (6th Ed.). Wadsworth Publishing Company, 234p.

Hughes, W. and Lavery, J. 2004. Critical Thinking: An Introduction to the Basic Skills. Broadview Press, Canada, 403p.

Jarrard, R.D. 2001. Scientific Methods [E-book]. Available: https://www.emotionalcompetency.com/sci/sm_all.pdf. 236p. [Accessed 10 June 2018].

Labossiere, M.C. 2010. Forty-Two Fallacies (For Free) [A pdf book]. Available: https://records.viu.ca/www/ipp/pdf/42_fallacies.pdf [Accessed 05 Nov. 2012].

Labossiere, M.C. 2011. Thirty More Fallacies [A pdf book]. Available: https://blog.talkingphilosophy.com/wp-content/uploads/2011/11/30morefallacies.pdf [Accessed 05 Nov. 2012].

Losee, J. 2001. A Historical Introduction to the Philosophy of Science (4th Ed.), Oxford University Press, New York, 328p.

O'Hear, A. 1989. An Introduction to the Philosophy of Science. Oxford University Press, New York, 239p.

Tindale, C.W. 2007. Fallacies and Argument Appraisal. Cambridge University Press, Cambridge, 236p.

Walliman, N. 2011. Your Research Project: Designing and Planning Your Work (3rd Ed.). Sage Publications, New Delhi, 378p.

Wright, K.E. 1948. The teleological approach in botany. Am. Biol. Teach. 10 (7): 187–189.

Chapter 15
References: How to Cite and List Correctly

Everything deep is also simple and can be reproduced simply as long as its reference to the whole truth is maintained. But what matters is not what is witty but what is true.
Albert Schweitzer (1875–1965)

When we write an essay, research paper, thesis, or book, it is normal to include information from the work of others or support our arguments by reference to other published works. All such academic documents draw heavily on the ideas and findings of previous and current researchers available through various sources such as books, journals, theses, newspapers, magazines, government reports, or Internet sources. In all these cases, proper referencing is essential in order to ensure easy retrieval of information. *Referencing* is the name given to the method of showing and acknowledging the sources from which the author has obtained ideas or information.

Different referencing systems have evolved depending upon the nature of disciplines. For example, the way we handle reference sources in linguistics would be different from that required in chemistry. Nevertheless, all the referencing systems depict the sources, which have contributed or supported a specific idea or information in your writing. Similarly, enough details of sources are also given so that the readers could find the sources themselves.

In-Text References and End References

Referencing involves two aspects; the first is the way in which the information is cited or included in the text. We may use ideas of others in a paraphrased form or include a direct quotation; but in both cases, the sources must be acknowledged appropriately. Within the text of a document, these references are presented in an abbreviated format referring to the end references. This process of acknowledging a source of information in the text of a publication is called *in-text reference* or *in-text citation*. The second aspect is the way in which the references are listed at the end of the documents. Once you have cited the works of others appropriately in the text, the next task is to prepare a list of cited references, which forms the '*end references*'. These must show full details of the sources so that any reader can trace out the original sources, if required.

C. G. Thomas, *Research Methodology and Scientific Writing*,
https://doi.org/10.1007/978-3-030-64865-7_15

In most referencing systems, all the references that contributed to the work are presented as a list at the end of the document. The title given to the end references section indicates the nature of the sources. The term '*Bibliography*' means a complete list of published works along with details that were consulted in preparing the report or influenced your thinking. The works listed in a bibliography may not necessarily be cited in the report. For an original article, thesis, or review, use the term '*References*', as the list of sources comprises of complete bibliographic details of every publication. Some prefer to use '*Works cited*' or '*Sources cited*' to include non-traditional sources such as online sources, video recordings, and personal communications (e-mail, interview, etc.). In some textbooks, you will see titles such as '*Selected Bibliography*' or '*Selected References*' suggesting that the author has narrowed the list to important works, but that s/he has not cited all of them in the book. '*Books for further reading*', '*Books for supplementary reading*', and '*Further reading*' are some other ways of listing bibliographic sources in textbooks.

Researchers and research students should master the skill of referencing source materials accurately. However, much confusion is prevalent in the whole referencing process. The confusion of researchers and students is often compounded when they see substantial variation in referencing styles between publishing houses and journals. Even in a particular named referencing system, there may be variations in the style of handling individual references. Therefore, when you write a scientific paper, first consult the 'Instructions to Authors' and the latest copy of the journal. For thesis writing, the concerned university or institute may provide specific 'style manual', which has to be consulted for its preparation.

In this chapter, some broad aspects of referencing and documentation styles are given followed by a detailed account on Name-Year system of referencing prescribed by the Council of Science Editors (CSE 2014) with appropriate examples.

15.1 Importance of Referencing in Documents

Often, researchers initiate new works after reviewing the extent of works others have done on the subject. Fellow researchers get an idea of the importance of completed works only by reading research papers made out of them and the related source documents cited and listed in such papers. Some articles with important findings or information will be cited by numerous researchers. Citations can be considered as the rewarding system of scientific publications. We cite somebody's work in our document mainly to acknowledge that person's impact on subsequent works and on our work. If we do not cite sources properly, we are depriving other researchers of the information contained in those sources, leading to duplication of efforts. Nevertheless, bear in mind that the sources quoted must be relevant, and they must be quoted accurately.

We may use ideas of others in a paraphrased form or borrow a direct quotation of someone else in our works; but in both cases, we have to acknowledge the sources appropriately. Similarly, when we use factual data too, we have to show the source.

This is also required when a table, diagram, chart, picture, photograph, or any other illustration is reprinted with permission. In the discussion part of a thesis or scientific paper, you have to defend your results (Remember the question: 'What do the results mean?'). You need to show the readers that you have evidences for a statement or argument you have made. The information that originally appeared in other reports will help you to discuss and substantiate your claims. Sometimes, the reader of an article may wish to find out additional information about the subject by reading some of the books, research articles, and other sources you have used. Unless you have cited and listed these works in an appropriate way, the reader may not be able to trace out the sources.

Referencing should not be insisted for all the information you have included in the text. It is not usually necessary to provide a reference when you use your own knowledge such as a personal anecdote or your own research findings. However, the reader must be able to understand that you are using your own experiences and findings. Similarly, when you use common knowledge in your own words, a reference need not be cited (see Sect. 15.5).

If you do not cite and list the sources properly in your documents, you may also be blamed for *plagiarism* (Chap. 24). This involves a serious breach of ethics, which is undesirable for the growth of science. In fact, most cases of plagiarism are due to ignorance of rules of documentation rather than a deliberate effort on the part of the student or the researcher to cheat the system.

15.2 Citation and Listing Systems of References

Citation and listing systems of references fall into two broad categories, the *parenthetical systems* and the *notation systems*. In parenthetical systems, when a source is referred to, a brief information about the author is included in the body of the text in parentheses. In most scientific disciplines, these usually include the author and the date of publication, for example (Thomas 2018). Hence, the system is also called '*Author-Date*' or '*Name-Year*' system. In humanities including linguistics and fine arts, however, the author's name is followed by page numbers instead of year of publication in the body of the text, for example (Thomas 112). This is called *Author-Page* system.

Author-Date systems are more common with science disciplines. In this system, sources are briefly cited in the text by the surname or last name of the author closely followed by the year of publication, often in parentheses. All the in-text citations must match with the list of end references, where full bibliographic information is given. *Harvard system* of referencing is an Author-Date system. In the arts and literature, date of publication may not be the most important information about a particular document. This is not the case with a research paper on biology, where particular date has more relevance. As literary works may go through numerous editions or translations long after the original publication, *Author-Page* systems were devised by humanities scholars. Modern Language Association (MLA) recommends Author-Page system for arts, literature, and humanities.

In *notation systems*, when a source is referred to, a number (a superscript number or number in parenthesis) is placed in the text. This number leads the reader to a footnote or endnote that provides full bibliographic details about the source. Footnote method and Author-Number systems come under notation systems. *Vancouver system* uses Author-Number system, a preferred system for health and medical sciences.

Certain publishing houses insist on a particular style of citation in all their publications. Each journal will also have a style for citation and writing references. When the authors have the freedom to choose a style, they should exercise this with great caution. Remember that whatever style you use, it should provide a logical sequence, concisely expressed, enabling another interested scientist or student to retrieve the document to understand what has been done and with what results.

There are five major citation and listing systems for books and journals, which include the *Harvard system*, the *Vancouver system*, the *Citation-Name system*, the *Footnote method*, and the *Author-Page* system.

Harvard System

Harvard system, also called the *Author-Date* system or the *Name-Year* system (N-Y system), is a generic term for any style of citation and referencing having Author-Date references in the text. Probably because of its simplicity, Harvard system is the most widely used style for referencing. In this system, emphasis is placed on the author and the year of the work to identify it distinctly. All the three leading documentation style guides, CSE Style Manual (CSE 2014), APA Style Manual (APA 2013), and Chicago Style Manual (UChicago 2017), recommend Harvard system of referencing with some minor variations.

Name-Year system is a parenthetical system, wherein, in-text citation consists of the surname or last name of the author or authors (without initials) followed by the year of publication in parenthesis. End references are not numbered but arranged in alphabetical order by author and year of publication. For listing of references, in addition to the last name, initials of other parts of the name (first name, second name, etc.) should also be given preceded by a comma after the surname. The year of publication then follows. For example, a work by Norman E. Borlaug will be cited in the text as Borlaug (1992) and listed in the endnote section as Borlaug, N.E. 1992 or Borlaug, N.E. (1992). An example is shown below with in-text reference (common) and end reference with three variant forms.

In-text reference

The most common design of competition experiments well suited to agronomic objectives is the additive series (Cousens 1991).

End reference

CSE style

Cousens, R. 1991. Aspects of the designs and interpretations of competition (interference) experiments. Weed Technol. 5(3):664–673

APA Style

Cousens, R. (1991). Aspects of the designs and interpretations of competition (interference) experiments. *Weed Technology,* 5(3):664–673

Chicago style

Cousens, Roger. 1991. 'Aspects of the designs and interpretations of competition (interference) experiments'. *Weed Technology* 5 (3): 664–73

Although CSE Style Name-Year system is the preferred referencing system for science and social sciences including agriculture and life sciences, what we actually see in many journals is a mix of all the above three styles!

A major advantage of Harvard system is convenience to both authors and readers. Usually, a researcher from a particular specialization may easily identify the author/s in a citation without the need of crosschecking in the references section, thus saving much time. Another advantage is that if the same reference is cited more than once, even a novice reader not familiar with the author may remember the name. Probably, the most favourable advantage for the author is that unlike other systems, no renumbering is required when the order of in-text citations is changed or additional sources are brought in. As the references are not numbered, these can be added or deleted easily at any time before finalizing the manuscript. In other words, without the help of reference management software (Sect. 15.10), you can neatly arrange the references manually with this system.

Some disadvantages of Harvard system must also be noted. It requires more space in the main body of text for in-text citations. The system may be distracting to the readers, as they have to confront the in-text citations too often; this is worrisome, if the document is a long review. The style may become a problem for the authors with old books or papers with no date of publication. A way out is to write [date unknown] at the place of year in such undated documents (Sect. 15.6).

Vancouver System

Vancouver system, also called *Citation-Sequence* system (C-S system) or A*uthor-Number* system, is a broad term for a notation system, wherein, numbers act as note identifiers in the text. It is also called *Endnote referencing* system. Each in-text reference is given a number according to the order in which it first appears in the text, and the end references are listed in the sequence in which they first appear within the text. A reference retains the same number throughout the document, and if the source is referred to again, the *note identifier* is repeated. For example, if a reference by Thomas were the first one cited in the text, then, Thomas would be listed as number 1 in the end references with complete reference details. Vancouver is a preferred system for health and medical sciences.

The citation in the text is indicated by an Arabic numeral, often set as superscript as[1] or a number in parentheses as (1) or in square brackets as [1]. If a single reference points to more than one source, list the source numbers in a series, for example, as [1,3,6]. Use a dash to separate more than two numbers as [1–3], if these form a sequence. However, use a comma to separate two numbers as [1,3] (without space in between), if these do not form a sequence. After finalizing citations in the text, the reference sources will be listed at the end of the document in the order of their position in the text. In this system, the year of publication is often given after the publisher's name in the case of books or after the journal's name for articles. See the example:

In-text reference

The most common design of competition experiments well suited to agronomic objectives is the additive series[1].

Note: The in-text reference marked as[1] is the first entry in the article in sequence.

End reference

CSE style

1. Cousens, R. Aspects of the designs and interpretations of competition (interference) experiments. Weed Technol. 1991; 5(3):664–673

AMA Style

1. Cousens, R. Aspects of the designs and interpretations of competition (interference) experiments. *Weed Technol.* 1991; 5(3):664–673

Citation-Sequence system is considered as a reader friendly referencing system because one can quickly go through the details of references cited in the order of their appearance in the text. It may be a useful system for a journal publishing short articles and short notes with only a few references. Nevertheless, for a long paper with many references, it may not be a good system, if you consider the authors' predicament in arranging entries, especially when addition or deletion of citations demand renumbering. You may use a reference management software to tide over this crisis.

Two more variants of Vancouver system are prevalent where references are identified by notation, namely, the Citation-Name system and the Footnote method.

Citation-Name System

The *Citation-Name system* (C-N system) or the *Alphanumeric* method is an improvement over the Citation-Sequence system. CSE style recommends Citation-Name system too along with Name-Year and Citation-Sequence system.

It can be considered as a hybrid between Citation-Sequence and Name-Year systems. In this system, the end references are listed alphabetically by the surnames of authors and numbered consecutively. For example, if the authors are Bhatt, Chatterjee, Abu, Abraham, Aditya, and Menon, the first entry authored by Abu would get the number 1; Abraham, number 2; Aditya, number 3; Bhatt, number 4; Chatterjee, number 5; Menon, number 6; and so on. If you are doing it manually, you may rearrange file cards in alphabetical order by the surnames of the first authors and numbered. Those numbers must then be inserted in the text.

The numbers assigned to the end references would be used for in-text citations irrespective of the sequence in which they appear in the text. For example, if a work by Sruthi were number 14 in the reference list, each in-text citation to Sruthi would get the number 14. A major drawback with this system (equally applicable to the C-S system already described) is that the readers would not get an idea of the sources when they see the numbers unless they deliberately look for the sources each time in the reference section. See an example:

In-text reference

The most common design of competition experiments well suited to agronomic objectives is the additive series[4].

End reference

4. Cousens, R. Aspects of the designs and interpretations of competition (interference) experiments. Weed Technol. 1991; 5(3):664–673

Note: When the references are arranged in alphabetical order, the entry by Cousens gets the number 4. Therefore, the number[4] is used for in-text reference.

Footnote System

Footnotes are notes to sources of information, which appear at the foot or bottom of a page. Footnote referencing is similar to Endnote referencing (Citation-Sequence), but notes are collected at the bottom of each page, rather than at the end of paper. Footnotes and endnotes serve the same purpose, but vary in their utility.

In a footnote referencing system, you indicate a reference by putting a small number usually as superscript above the line of type directly following the source material. You put the same number, called a note identifier, followed by details of your source, at the bottom of the page. Footnotes should be numerical and chronological; the first reference is indicated by 1, the second is 2, and so on. An advantage of this method is that the reader will be able to locate the references instantly with ease. A major drawback is repetition of references, sometimes on each page, and therefore, expensive. The Chicago Manual of Style recommends the footnote method for fine arts (art, theatre, and music) and humanities (history, philosophy, and religion).

In-text reference

The most common design of competition experiments well suited to agronomic objectives is the additive series[2].

Footnote reference

[2]Robert Cousens, 'Aspects of the designs and interpretations of competition (interference) experiments', *Weed Technology,* vol. 5, no. 3 (1991): 664–673

Note: The reference by Cousens was the second footnote. Therefore, when it was listed as footnote, it received the second position.

Author Page System

In language, literature, linguistics, and fine arts, date of publication may not be the most important information about a particular document. As literary works may go through numerous editions or translations long after the original publication, importance has been given to titles and pages. In this system, page number is given in the body of the text in parenthesis, for example, (Thomas 21). This is called *Author-Page* system.

Modern Language Association (MLA) recommends Author-Page system for arts, literature, and humanities. For in-text citation in MLA, the author's last name and the page number from which the quotation or paraphrase has been taken is indicated. See an example:

In-text reference

The most common design of competition experiments well suited to agronomic objectives is the additive series (Cousens 665).

End reference

Cousens, Robert. 'Aspects of the designs and interpretations of competition (interference) experiments'. *Weed Technology*, Vol. 5, no. 3, 1991, pp. 664–673.

15.3 Common Documentation Styles

As scientific disciplines differ in the use of research methods and literature sources, professional associations have developed or adopted documentation styles appropriate for their subject fields. For each style, style manuals have been brought out. Along with the style of preparation of documents including grammar, style of writing, use of symbols, abbreviations, etc., they also prescribe specific citation and referencing systems. Brief accounts on the Council of Science Editors (CSE) Style, the American Psychological Association (APA) Style, the American Medical Association (AMA) Style, the Modern Language Association (MLA) Style, and the Chicago Manual of Style are given below.

The Council of Science Editors (CSE) Style

The CSE (formerly, the Council of Biology Editors or CBE) recommends ***CSE style*** for all scientific disciplines. Among other things, the CSE Manual (CSE 2014) recommends three methods for citing and documenting sources:

- the Name-Year (N-Y) system (Harvard)
- the Citation-Sequence (C-S) system (Vancouver)
- the Citation-Name (C-N) system (Alphanumeric)

In CSE style, the system of in-text citation that you adopt will decide the order of end references. In all the three systems mentioned above, the end references have essentially the same format, except for the placement of the date of publication in the Name-Year system. However, note that even in CSE style, slight journal-to-journal variations can be noticed.

If you know how to write the sources in N-Y system, it is easy to change them to C-S system or C-N system. For complete information on CSE style of documentation, refer to 'Scientific Style and Format: The CSE Manual for Authors, Editors, and Publishers' (8th Ed.), The University of Chicago Press, 722p (CSE 2014). It is also available online for subscription (https://www.scientificstyleandformat.org/Home.html).

The American Psychological Association (APA) Style

This system is commonly used for referencing in social and behavioural sciences such as psychology, sociology, and education as described in the style guide of the American Psychological Association (APA), the 'Publication Manual of the American Psychological Association' (6th Ed.) (APA 2013). APA is an author-date-based style similar to N-Y system under CSE style (https://www.apastyle.org/manual/).

The American Medical Association (AMA) Style

The American Medical Association prescribes AMA Manual of Style for journals published by the American Medical Association (AMA 2007). The manual (AMA Manual of Style: A Guide for Authors and Editors) was first published in 1962, and the latest is the 10th edition. AMA style is the preferred one for medicine and related health fields. For referencing, AMA adopts Vancouver system (Citation-Sequence system). You can also purchase online edition. Visit: http://www.amamanualofstyle.com/oso/public/index.html.

The Modern Language Association (MLA) Style

The Modern Language Association developed the MLA Style for language and literature. For referencing, MLA uses Author-Page system (MLA 2016).

The Chicago Manual of Style

The Chicago Manual of Style (CMOS) recommends two formats of referencing, *Notes-Bibliography* and *Author-Date* (UChicago 2017). Notes-bibliography format uses a numerical system of notes, footnotes, endnotes, or both, and often with a bibliography. This system is preferred in disciplines such as literature, history, and the arts. It can accommodate a wide variety of sources including unusual ones that cannot be easily fitted into the Author-Date system. For example, in history, footnote citation is widely used, because important references in history are often to archive sources or interviews, which do not readily fit into a standard citation format. The Author-Date format under Chicago is very similar to the N-Y system of CSE. For the science disciplines, the Author-Date is preferred. CMOS is the most popular style guides in the USA, which deals mainly with American English.

15.4 CSE Style: The Name-Year System

We will now consider in detail the steps involved in the in-text reference and listing process according to the Name-Year (N-Y) system as recommended by CSE style. Many of these are equally applicable to other systems but conventions vary. Occasionally researchers have to prepare their works in other styles too such as the Citation-Sequence (C-S) system recommended by CSE or AMA style, especially when the paper is intended for a biomedical journal or a US-based journal. At present, computer programmes (Sect. 15.10) are available to convert easily from one style to another. You can also do it manually, and if you know how to write the sources in one style, it is easy to change them to other styles.

An important aspect to remember while preparing any in-text citation or reference list is to have consistency. Take care to put all in-text citations and reference list entries for the same type of information sources uniformly throughout your document. For doing this smoothly, we must have enough bibliographic details about each information source we used. There are some differences in the bibliographic details to be recorded for books, articles, and online sources. Take care to collect these details at the time of literature collection itself. It would be very difficult to gather missing details, if any, later.

Books: In the case of books, bibliographic details include the names of all the authors or editors, year of publication, title, edition, whether it is a reprint, volume number (if from a multi-volume work), total pages, publisher, and place of publication.

Articles: The term article is generally used to refer to news, features, reports, research papers, reviews, etc., published in newspapers, magazines, and journals. For using an article as a source of information, the details needed are names of authors, year of publication (date in the case of a newspaper or weekly), title of the article, title of the periodical, volume and issue number, and page numbers of the periodical in which the article appears.

Online sources: In the case of an online source, bibliographic details must include details like author of the information source, date or year of publication, title of the work, database provider or sponsor of the webpage, date of access (the first view date of the resource), webpage address or Uniform Resource Locator (URL), and DOI, if available. A digital object identifier (DOI) helps us to identify digital objects on the web. It consists of some numbers, letters, and symbols. The object may change physical locations; however, the DOI assigned to that object will never change. Anybody can easily locate a document from a DOI. Publishers assign DOIs to electronic articles in journals.

15.5 Citation of Sources in the Text

All the sources of information used for developing the written works must be acknowledged properly. As discussed earlier, whenever you refer to someone else's information in your work such as a thesis or article, you must insert an 'in-text citation' at the appropriate place within the text. Sometimes, you may have to do this many times throughout your book or article. Examples of dealing with specific cases are mentioned below.

When Using Common Knowledge

In general, facts and information in common use, which can be verified from a number of sources and are likely to be known by a large number of people, are known as common knowledge. As a thumb rule, information that occurs in five or more sources can be considered as common knowledge. This usually includes facts available from various secondary sources such as textbooks, handbooks, manuals, dictionaries, directories, or encyclopaedias. When a common knowledge is paraphrased for use in a document, a source need not be cited. For example, consider the statements, 'Light is essential for photosynthesis'; 'Oxygen is released during photosynthesis and the source is water'. These are now common knowledge and form the very bases of studies on photosynthesis. Likewise, bulk of the knowledge bases of different scientific disciplines through repeated use has become common knowledge. As a caution, if you were unsure whether the information comes under the purview of common knowledge, it would be better to give a reference to the source. However, when you use such established knowledge for documents such as textbooks, all the sources from which you have drawn the material shall be listed under the title 'bibliography'.

When Quoting a Sentence or Passage

Sometimes, you may copy the exact sentences of an author in your works. When using an author's exact words, use quotation marks around the sentences or words copied. Ensure that you have copied accurately, both words and punctuation marks as used by the author. If the borrowed information is longer than 40 words or 4 lines margin to margin, instead of marking it as a quotation, indent the material from the margins. Further, cite the source in the text of the paper, and include the source in the list of references. You should also take care whether you are infringing on copyright rules, as in some cases, permission is required for reproduction. In general, consent is required for the reproduction of any material unless the principle of 'fair use' or 'public domain' is applied thus allowing the material to be used without permission (Chaps. 23 and 24).

Use of direct quotes is discouraged in science literature. However, when you write a literary review or a paper in history, sociology, current events, and such related fields, direct quotes or partial quotes may become necessary. Sometimes, we use direct quotes in popular articles related to science also, for example, 'Science is 99 percent perspiration and 1 percent inspiration' (Louis Pasteur) or 'Everything else can wait but not agriculture' (Pandit Jawaharlal Nehru). In most cases, direct quotes are used to avoid much explaining! Bear in mind that it is not a good practice to quote large sections of text as such and rely too heavily on the original author's words. In scientific papers, you should express your own thoughts and interpretations, and not those of someone else. This problem can be overcome easily, if you learn the art of paraphrasing (Chap. 24).

When Using Visuals

Visuals include figures, cartoons, photographs, charts, and other illustrations. Tables should also be considered like visuals. Sometimes, you copy the entire table or illustration for your review paper or book. In such cases, you should cite the source in the text of the paper (and usually right below the visual) and include the source in the list of references at the end. Sometimes, copyright rules may not permit you to copy a visual without the written consent of the author or publisher. Sometimes, you have to pay a fee for getting the permission. In that case, the case of permission must be mentioned and the source acknowledged.

Non-retrievable Sources

The sources you cite in your document must be retrievable. Sometimes, authors are forced to include a few non-retrievable sources. However, they should bestow extreme care while citing non-retrievable sources in their written works. Personal communications, articles submitted but not yet accepted, etc., should not be given as full references. When these are used, they are cited in the text in parentheses as personal communications giving the name of the source. See the examples given in this chapter.

When Using Abstracts

Quoting or citing 'abstracts' from abstracting journals without seeing the original is not a good practice. The reader works on the understanding that every reference quoted has been seen by the authors and that it does say what it is quoted as saying. Citing a source of information unless you have referred and read the original is an unhealthy practice. With interlibrary loan systems, translations, and photocopying services, it is now possible, although sometimes expensive, to see every reference. In a thesis, however, if the reference is very important and the original is in some foreign language that cannot be retrieved easily, it is permissible to cite it. List such references with an asterisk and footnote: 'originals not seen'. Still, avoid using too many references, the originals of which have not been seen.

When Paraphrasing

Paraphrasing is the process of expressing someone else's ideas or thoughts obtained from an article or book in our own words, syntax, and style, but preserving the tone of the original and maintaining approximately the same length. It is done mainly to avoid *plagiarism.* We also try to paraphrase sentences of others to make it more comprehensible. Paraphrasing is required when we write a review for our thesis, assignment, term paper, seminar paper, or review for a journal. For a successful paraphrase, it is essential that you understand the original passage and restate it in your own words, and remember that mere substitution of certain words with synonyms, changing tenses, or shifting words are not enough. Do not forget that as

Table 15.1 Common verbs helpful for author-prominent referencing

Add	Disagree	Point out
Affirm	Dispute	Predict
Agree	Find	Question
Assert	Highlight	Reason
Claim	Imply	Remark
Clarify	Justify	Report
Confirm	Maintain	Show
Contend	Note	State
Contest	Offer	Suggest
Describe	Observe	Theorize

we are borrowing someone's thoughts, we must document the source when we use the paraphrase in our paper (see Chap. 24).

While paraphrasing, in-text references can be written at least in three ways. Give prominence to authors, if they are well-recognized persons in their fields. Two styles have been shown as examples for author-prominent citation. The verbs commonly used for author-prominent referencing are listed in Table 15.1. To avoid monotony, vary the verbs as frequently as possible. If you want to highlight the information rather than the author, place the in-text reference at the end of the sentence in parenthesis. See the examples given below.

Prominence to authors

Thomas (2008) stated that shade-tolerant fodder grasses could come up very well in coconut gardens.

Please note that instead of *stated* in the sentence, you could use other suitable verbs such as *remarked, reported, claimed,* or *confirmed* according to your liking. Sometimes, starting the sentence with 'according to', the same meaning with author prominence can be implied as shown below:

According to Thomas (2008), shade-tolerant fodder grasses come up very well in coconut gardens. *No prominence to authors*

Shade-tolerant grasses can come up very well in coconut gardens (Thomas 2008).

How to Include the Names of Authors

The treatment given to the names of authors varies according to the context involved. On the title pages of documents, the names of authors are usually written in full. While citing in the text, only the surname or last name is used, but while listing the sources at the end of the document, the surname or last name is used along with initials. For specific examples for in-text citation and listing using the names of authors, see Table 15.2.

Normally, the name of a person constitutes two parts, the *given name* and the *surname* (family name). The first part of the name is the given name (sometimes, first name, forename, or Christian name), and the second part is the surname. In Western

Table 15.2 Different styles of writing names of authors in technical papers

On the title page	In-text citation	Listing
Type 1		
Philip Rubens	Rubens	Rubens, P.
Clifford Hawkins	Hawkins	Hawkins, C.
Type 2		
Norman Ernest Borlaug	Borlaug	Borlaug, N.E.
Paul Jackson Kramer	Kramer	Kramer, P.J.
Type 3		
John Cyriac	Cyriac	Cyriac, J.
George C. Thomas	Thomas	Thomas, G.C.
C. George Thomas	Thomas	Thomas, C.G.
Type 4		
Savitha Anthony	Anthony	Anthony, S.
Jeethu M. Gopalan	Gopalan	Gopalan, J.M.
Nima Mariam George	George	George, N.M.
Type 5		
Nisha Lakshmy	Lakshmy	Lakshmy, N.
A. Liz Elizabeth	Elizabeth	Elizabeth, A.L.
Type 6		
Priyanka Choudhary	Choudhary	Choudhary, P.
Meera V.Menon	Menon	Menon, M.V.
Gurpreet Singh	Singh	Singh, G.
Type 7		
William Mc Kelm	Mc Kelm	Mc Kelm, W.
Manuel Livera-Munoz	Livera-Munoz	Livera-Munoz, M.
Type 8		
Alphons de Condolle	de Condolle	de Condolle, A.
Olga de Wit	de Wit	de Wit, O.
Type 9		
Yuan Longping	Yuan	Yuan, L.
Zhang Ziyi	Zhang	Zhang, Z.
Type 10		
S. Renu	Renu	Renu, S.
P.V. Sindhu	Sindhu	Sindhu, P.V.
A. Praveen	Praveen	Praveen, A.

cultures, 'surname', 'family name', and 'last name' are synonymously used, as these terms indicate the same thing, family names.

The names with a given name and surname are typical, and there may not be any confusion in using surnames in such cases (type 1). Sometimes, the name may have three parts—given name, middle name, and last name. Americans often write their name with a middle initial, for example, Nyle C. Brady. Middle name of a person is normally chosen by the parents at the same time as the first name, and it may be written in full or as initial. In such cases, surnames are followed by the initials of first and middle names (type 2).

In some countries, the last name may not be the family name alone. Some people attach the first name or sole name of their fathers to their given names, for example, Mary Anthony (Anthony is Mary's father). In this case, we have no other choice but to take the last name of the author, although it may look a little awkward (type 3 and type 4). Sometimes, women authors may attach their mothers' first names (type 5). In India, several people attach caste names to their given names, which acquire the status of surnames (e.g. Menon, Pillai, and Nair). In such cases, the caste names are taken as surnames (type 6). In non-English-speaking Western countries, you may sometimes find compound words as surnames (type 7 and type 8).

In most countries, surnames are attached at the end of the given names of individuals, but in South-East Asian countries, they put their surnames in front of given names. In such cases, take care to put the first word of the name as surname (type 9). In some regions, the custom may be to have the given names followed by just initials only indicating surnames or parents' names (type 10). In that case, you have to use the given name itself for in-text reference

Use of Periods and Commas

In scientific writing, a modern trend is to do away the practice of putting periods after initials and journal abbreviations, for example, 'Thomas CG', instead of 'Thomas, C.G.' and 'J Tropic Agric', instead of 'J. Tropic. Agric'. Some variation in comma use both in the text and the list of references is evident between journals and publications. Popular style is '(Thomas 2008)', but the latest CSE Manual allows the style as '(Thomas 2008)'.

Use of Italics for In-Text Citation

Many Latin terms have been accepted in English, which can be used without translation. For example, the terms such as in vivo, in vitro, e.g., i.e., etc., and et al. are widely used in English, and therefore, these terms need not be italicized. Note that, for example, if more than two authors are involved, we write Thomas et al. (1997) and *not* Thomas *et al.* (1997).

15.6 Examples of in-Text References

Specific recommendations for citing sources after paraphrasing in research documents such as books, theses, and research papers are given here with examples.

One Work by Single Author

For arranging the in-text reference, place the year of publication in parenthesis immediately after author's surname (without initials). If you are giving the source after writing a sentence, it shall be surname followed by a comma and the year, both in parenthesis. See the examples given below. You may adopt one of the three methods after deciding whether to give prominence to authors as discussed earlier.

- Thomas (2008) claimed that shade-tolerant fodder grasses could come up very well in coconut gardens.
- According to Thomas (2008), shade-tolerant fodder grasses come up very well in coconut gardens.
- Shade-tolerant grasses can come up very well in coconut gardens (Thomas 2008).

One Work by Two Authors

The entries are similar to the above. Decide on the prominence issue and adopt the most appropriate one.

- Renu and Thomas (2000) claimed that for preventing the infestation of weeds in dry seeded rice, stale seedbed technique could be practised.
- According to Renu and Thomas (2000), stale seedbed technique is an effective strategy for preventing the infestation of weeds in dry-seeded rice.
- For preventing the infestation of weeds in dry seeded rice, stale seedbed technique can be practised (Renu and Thomas 2000).

One Work by Multiple Authors (more than Two)

Thomas, Abraham, and Sreedevi are the authors. Write the name of the first author followed by et al., meaning, 'and others' as:

- Thomas et al. (1997) confirmed that *Sacciolepis interrupta* (Willd.) Stapf. is a serious weed of semi-dry rice in Kerala.
- According to Thomas et al. (1997), *Sacciolepis interrupta* (Willd.) Stapf. is a serious weed of semi-dry rice in Kerala.
- *Sacciolepis interrupta* (Willd.) Stapf is a serious weed of semi-dry rice in Kerala (Thomas et al. 1997).

  Note: As "et al." is an accepted term in English, it need not be put in italics. You may also note that you should read 'et al' as 'and others' and *not* as 'et al'.

A Chapter from an Edited Book

When you are using information for in-text citation from a compiled book where each chapter is written by different authors with an overall editor or editors, cite only the name of the author/s of the chapter in your work. However, while you list the entries in the 'References' section, the names of editor/s should also be mentioned in addition to the names of authors. See Sect. 15.8 for specific examples for listing.

Multiple Works by the Same Author, the Works are Written in Different Years
While making an in-text entry, list the publication years in chronological order separated by commas after author's name in parenthesis.

> The validity of using grasses for soil conservation works has been described (Thomas 2008; 2010).

Multiple Works by the Same Author, the Works are Written in the Same Year
When there are more than one work by the same author, all the works written in the same year, differentiate the entries using lower-case letters a, b, c, d—along with the year. In the list of references, arrange the entries alphabetically by the above lower-case letters.

> Biswas (1999a) found that… Biswas (1999b) from another study reported that…

Works by Different Authors with the Same Surname but Different Initials Published in the Same Year
You may confront this kind of situation of two or more authors whom have the same surname but different initials but all of them have written their works in the same year. In such situations, first add their initials after the surnames and separate the names from the initials with a comma. Use these surnames with initials for in-text citation. Then arrange the reference list entries alphabetically by initials.

> Biswas, C. (1999) found that… However, according to Biswas, V. (1999)…

More than One Work to Support a Statement
Sometimes, more than one work can be acknowledged in a single in-text reference, if you find that two or more authors arrive at the same finding or conclusion about the same subject. It may be a similar statement, a view, or a finding. Cite all the names of the authors along with the years in chronological order.

> The competitive ability of weeds depends mainly on their time of emergence relative to that of crop (Hakansson 1983; Cousens 1987).

No Author's Name, but There Is a Sponsoring Body (a Corporate Author)
Sometimes, you may have taken information produced by a university or a similar organization instead of specific individuals. As several individuals are involved in generating such sources of information, credits could not be given to a few of them. In fact, all these persons are working for the institution. This is known as *corporate authorship*. In such cases, give an in-text citation as usual, but cite the organization as the author. If the name of the organization is a long one, use its standard abbreviation in the text. For example:

According to the FAO estimates, 821 million people all over the world suffered from chronic hunger during 2017 (FAO 2018).

The standard abbreviation for Food and Agriculture Organization of the United Nations is FAO. When you list the references at the end, provide full expansion of FAO in square brackets as FAO [Food and Agriculture Organization of the United Nations] FAO [Food and Agriculture Organization of the United Nations] 2018. If you do not give full expansion in the list of references, readers have to waste some time to correctly identify what the abbreviation, FAO, stands for.

Citing Unpublished Material (Personal Communications, Oral Presentations, Interviews)

Any information that is pertinent to the article but is not available in literature may be cited as 'personal communication' (when the information is from someone other than the present author/s), or 'unpublished data' (when the information is from one or more authors of the current paper). Note the type, source, and year in the text. However, only works available to the readers shall be included in the reference list. Interviews, if published, can be listed in the reference section as well. You can cite such sources as shown below.

Case 1. When the information is from one or more authors of the current article and all the authors are responsible for the data.

In Wayand, patch budding is highly successful in rose (Devadas, V.S. and Thomas, C.G., unpublished data, 1987; unreferenced)

In the example, Devadas and Thomas are the authors of the present paper. 'Unreferenced' indicates that it is not listed in the 'References' section.

Case 2.When the information is from someone other than the author.

Litchi comes up very well in Wayanad (Devadas, V.S., personal communication, 2001; unreferenced).

In the above example, Devadas is not an author of the present manuscript but responsible for the data referred to in the paper.

Citing References Not Seen Directly (Secondary Citation)

When you read an article, your assumption is that all the references cited in the paper have been seen by the authors. Sometimes, you may have noticed an important source of information quoted in a textbook or a research paper. You should try to trace out the original, if you want to use it for your work. Do not cite a source unless you have read the original. However, if all your attempts to trace out the original fails and still you feel that it is an important information, you can quote the reference in an indirect way as:

Sheng (1989) cited by Thomas (2010) states that treatment oriented land capability classification is ideal for the humid tropics

It is permissible to cite an unobtainable reference as shown above. This is called a *secondary citation* indicating that the author has not seen the material in its original form, but rather has obtained the information from another document that cited the original source. In the example, Thomas is responsible for what Sheng said. You will then list Thomas (2010) in the list of references and not Sheng (1989). This is a better and safe practice instead of pretending to have seen the original reference! You must ensure that the list of references at the end of your document should contain only those works that you have seen and read.

15.7 Listing References in the "References" Section

All the sources cited in the text must be listed in the 'References' section. Various steps involved in listing references according to Name-Year system are outlined below.

Titles of Articles and Books
Write the titles of books, theses, and articles in full without any shortening. The titles of books shall be in 'title case'. Title case means every word, except prepositions and conjunctions, in a title are capitalized. However, the name of an article in a journal or other such works shall be written in 'sentence case' (written as a sentence with the first letter capitalized). The title of a thesis must be written like a journal article (i.e. in 'sentence case') as it is considered as an unpublished work. Please note that although the APA Style Manual recommends title case italics for title of books and journals, in the CSE style for Name-Year system, regular fonts are recommended without italicization or underlining.

Inclusive Pages
Giving inclusive pagination (first and last page number of the article) gives a hint to the potential users to distinguish between one-page notes and 50 page or more long review articles. In the case of textbooks, monographs, technical bulletins, and theses, give the total number of pages. For example, a book of 505 pages shall be listed as 505p. Please note a point here. In the case of a compiled book, if you have referred an article coming between pages 45 to 55, write pp. 45–55 ("pp." is the abbreviated form of pages and 'p.' stands for page). However, if you have referred only one page, for example, 45th page in a compiled book (which may be a short note), then write p.45.

Volume and Issue Number: Journal Article Paginated by Issue
In the case of a journal, periodical, or review series, include volume and issue number, if each issue is paginated separately. Earlier, some publications followed the style of using bold fonts for volume numbers. However, as there is no valid justification or additional advantage for using bold fonts, most journals now write the volume and issue numbers in ordinary fonts only. For example:

Yaduraju, N.T. and Mani, V.S. 1987. The influence of delayed planting and seedbed preparation on the competition of wild oats on wheat. Indian J. Agron. 32(3): 299–301.

Volume and Issue Number: Continuous Pagination for the Volume

In the case of journals that follow continuous page numbering for a particular volume, there is no harm in omitting the issue number. However, it is always advisable to include issue number, if provided, along with volume number to locate the issue of the journal easily. For example:

Dyer, W. E. 1995. Exploiting weed seed dormancy and germination requirements through agronomic practices. Weed Sci. 43: 498–503.

Supplementary Issues of Journals

If a particular volume has a supplement, part, or special number, place it after the volume number in abbreviated form, as ‘Suppl’, ‘Part’, or ‘Spec. No.’ as shown below:

- 23 Suppl.
- 23 Suppl. 1
- 23 (Part 1)
- 23 (Part A)
- 23 Spec. No. 2

Example: Vadiraj, B.A., Thomas, C.G., and Krishnakumar, V. 1993 Comparative efficiency and economics of weed management in cardamom plantation. J. Plant. Crops 21 Suppl.: 16–20

Journal Abbreviations

In CSE style, instead of writing journal titles in full, abbreviated title words are used. In APA style, the custom is to write the journal titles in full without abbreviation. Significant space and printing expenses can be saved by abbreviations. However, some amount of uniformity is required in abbreviations. Do not form your own abbreviations! Note that articles, conjunctions, and prepositions are omitted from abbreviated titles.

Use only the title word abbreviations as approved by International Standard Serial Number (ISSN). ISSN currently maintains a list of more than 56,000 words and their abbreviations in 65 languages. An international standard, ISO 4, is recommended, which advocates a uniform system for the abbreviation of serial titles. Anybody can access the words in the ‘List of Title Word Abbreviations (LTWA)’, by visiting the website http://www.issn.org/services/online-services/access-to-the-ltwa/. You can also obtain abbreviations of journal titles from the following databases.

Web of Science https://images.webofknowledge.com/images/help/WOS/A_abrvjt.html

Journal database at NCBI (National Center for Biotechnology Information) http://www.ncbi.nlm.nih.gov/nlmcatalog/journals

CAS Source index search tool http://cassi.cas.org/search.jsp

Genamics JournalSeek http://journalseek.net/

Genamics JournalSeek is a database of journal details collected from the Internet. The database presently contains more than 39,000 titles. You can also note journal abbreviations from the latest Journal Citation Reports produced by Clarivate Analytics.

Useful Directions for Journal Abbreviations

By knowing some general abbreviation rules, you may be able to abbreviate even unfamiliar journal titles without reference to a source list. For example, if one knows the abbreviations of words commonly used in journal titles, most journals can be abbreviated easily. All the journals whether English or non-English shall be abbreviated, if the intended journal is following that style. You may mention the language of the journal in parentheses immediately after the journal abbreviation, if it is not English.

1. Follow the title word abbreviations as approved by ISSN (see Annexure for the most commonly used abbreviations).
2. One-word journal titles and titles in character-based languages, such as Chinese, Japanese, and Korean, are never abbreviated, for example: Nature, Science, and Biochemistry.
3. Single-syllable words and words of five or fewer letters in singular form are not usually abbreviated. Some such examples are Acta, Blood, Cell, Crop, Crops, Dairy, Drug, Fauna, Fish, Flora, Freshwater, Root, Soil, Tissue, Weed, and Methods. However, there are some exceptions, for example, human is abbreviated Hum.
4. Certain country or city names are abbreviated, for example: Am. for American, Br. for British, Can. for Canadian, Jpn. for Japanese, Calif. for California/n, Lond. for London, and Camb. for Cambridge. However, some country terms such as India/Indian and Sweden/Swedish are not abbreviated.
5. Often, all the '…ology' words are abbreviated at the 'l', for example: physiology is abbreviated as Physiol., mycology is abbreviated as Mycol., and so on.
6. Omit articles, prepositions, and conjunctions from the abbreviated title of journals. For example, Advances in Agronomy will be abbreviated as Adv. Agron. dropping the word 'in'
7. All the abbreviations in the journal title should start with a capital letter.
8. Put a period after the abbreviation, for example: J. (Journal), Sci. (Science), Rev. (Review), and Prot. (Protection). However, if the abbreviation is formed by the first and last letters of the word, then period shall be omitted. Examples include Stn (Station), Wkly (weekly), and Natl (National). Please note that CSE Style encourages dropping of periods after title word abbreviations.
9. The names of conferences or proceedings are not abbreviated and should be written in full. However, there are certain journals, the names of which start with 'Proceedings'. In such cases, the titles of journals shall be abbreviated starting with 'Proc'.

10. Please note that most journals from the UK omit periods after all journal title abbreviations (e.g. J Agrl Sci) retaining only single spaces between them. However, in India, most journals follow the system with periods or full stops.

Using Italics and Underlining

The CSE style manual does not specify the use of italics or underlining in reference entries, leaving such matters to the discretion of writers and editors although it prefers regular upright fonts. With respect to the use of italics, there are only two options; either you follow italics for all the title words of books and journals or go without italics for these words in the list of references. The CSE style wants to simplify rules, and discourage italicization or underlining the titles of books or journals because all these entail extra work at the keyboard. Please note that APA style recommends italicization for titles. Taking a cue from CSE style and popular usages, in the present book, the titles of books or journals have been set in regular upright fonts avoiding italicization.

More than one book or journal article by the same author/s in the same year

This case occurs when you have used more than one source by the same author or groups of authors in identical order with the same publication year. In such cases, the entries up to the year of publication will be the same. List the entries in alphabetical order by looking at the first word of the title of the article or book and indicate with lower-case letters (a, b, c, and so on) immediately after the year without any space, for example: 2009a, 2009b. After the surnames, if there is still a tie, arrange by volume number chronologically. However, if the references are in the same volume, arrange by page numbers. See the examples.

Bhagat, R.L., Prasad, N.K., and Singh, A.P. 1986a. Effect of N and P on the fodder production of dinanath grass. Indian J. Agron. 31 (3): 215–218.

Bhagat, R.L., Prasad, N.K., and Singh, A.P. 1986b. Studies on forage and food based cropping sequences. Indian J. Agron. 31 (4): 384–386.

More than One Book or Journal Article by the Same Author/s in Different Years

Sometimes, you might have obtained reference materials of the same author or group of authors published in different years. The following instructions may be followed to deal with such cases.

- If the entries are by the same author in different years, arrange chronologically by the year of publication, the earliest first.
- When the references are with the same first author but different second and subsequent authors, list them alphabetically by the surname of the second author and then by the surname of the third author and so on.
- In the case of references with the same group of authors in the same identical order, enter them chronologically by the year of publication, the earliest first.

References with Missing Details

Sometimes, we may have to use certain references with missing details. Normally, references with full particulars only are accepted as valid publications. In certain cases, you may feel that certain references are important and must be included because of their historical or strategic importance. You may use such references with missing details as outlined below. However, you should be very cautious about using materials with missing details as supporting evidence.

With certain references, the problem may be the absence of a clear publication date. In case a publication date is not traceable, check the content and references to work out the earliest likely date, for example:

1936?: probable year
ca. 1936: approximately 1936 (ca. stands for *circa*, Latin)

If your efforts in finding a probable date fail, and if you still feel that it is an important document to cite, put [date unknown]. When citing homepages, databases, or websites, use the dates of update/revision.

In some other cases, the problem may be to identify an author, place, or publisher. This applies particularly to 'grey literature' such as government documents, leaflets, and such other materials. If these details are not available readily, the following devices are useful. When no city or place can be located or inferred, use the words 'place unknown' within square brackets. You can trace out the place of publication and publisher from other sources and use them in the document. However, details that are not originally part of the document but traced from other documents should be put in square brackets when using them.

Similarly, find out publisher information too, which would be available from the title page, the back of the title page, or the cover of monographs. If the publisher's name cannot be located like this, look for other areas such as the preface. If the publisher's name is obtained through indirect means, use square brackets around the name. Note that publisher name is not required for references from journal articles but is required for all other types of 'book type' publications. If you fail in deciding the name of publisher, put the words in square brackets as [publisher unknown].

Arrange the List Alphabetically

The final step in listing is arrangement of reference sources in alphabetical order according to the surname of the author (or editor) or that of the first of a number of authors or editors. If the author is an editor (ed.) or translator (trans.), it is indicated by an abbreviation after the name in parenthesis. Then, the year of the publication is given followed by a period or full stop. In N-Y system, do not use numbers, letters, or bullet points along with names to begin each entry. After writing the year, you write the name of the article or the book. In the case of books, name of the publisher, place of publication, and inclusive page numbers are also given. When more than one place of publication is given, write only the first one along with the publisher's name.

In the case of a journal article, the order for entry in the list is—author, year, name of the article, name of the journal, volume number, issue number, and page numbers. Publisher's name is not given in the case of journals.

15.8 Examples of Writing References

Some typical cases of listing references according to Name-Year style following the principles outlined in Sect. 15.7 are included here as examples. Note the differences carefully.

Book by One Author

Thomas, C. G. 2010. Land Husbandry and Watershed Management. Kalyani Publishers, Ludhiana, 716p.

Book by Two Authors

Kohnke, H. and Bertrand, A.R. 1959. Soil Conservation. McGraw-Hill Book Company, New York, 298p.

Book by Three or more Authors

Leggett, G., Mead, C.D., Kramer, M.G., and Beal, R.S. 1985. Handbook for Writers. Prentice-Hall, New Jersey, 558p.

Edition of a Book

After entering the title of the book, note any edition beyond the first in parenthesis. If the book is 10th Edition, then write (10th Ed.) or (10th Edition).

Thomas, C. G. 2008. Forage Crop Production in the Tropics (2nd Ed.). Kalyani Publishers, Ludhiana, 333p.

Salisbury, F.B. and Ross, C.W. 1992. Plant Physiology (4th Ed.). Wadsworth Publishing, California, 682p.

Reprint of a Book

Many young researchers are confused about the year of publication to be used in the case of reprinted books. As there are no changes in the content of the book, use the original year of the first print after the author's name. However, the year of reprint shall also be mentioned in parenthesis after the title of the book to show that you referred the reprint only.

Jackson, M. L. 1958. Soil Chemical Analysis (Indian Reprint, 1967). Prentice Hall of India, New Delhi, 498p.

Book by a Corporate Author

'Corporate author' means an institution such as a university, research institution, or organization. Normally, the names of such institutes are long, and for convenience, abbreviations are used. For in-text citation, usually abbreviations are used. When you list the references at the end, provide full expansion of names in square brackets. If you do not give full expansion in the list of references, readers have to waste some time to identify correctly what the abbreviation stands for. See the example:

KAU [Kerala Agricultural University] 2016. Package of Practices Recommendations: Crops (15th Ed.) Kerala Agricultural University, Thrissur, 392p.

In-text citation: (KAU 2016)

Sources with No Given Author or Editor

If no author or editor's name is mentioned, look for the organization's name, and check, if it can be listed as a case of corporate authorship. However, if the name of the organization is unclear, unknown, or authorship cannot be determined, do not use anonymous, as it is not allowed under CSE citation style. Instead, you may include the first word or a few words of the title to identify the source apart from others followed by an ellipsis (…) as shown in the example.

End reference

WARASA-Jan Sahbhagitha: Guidelines for National Watershed Development Project for Rainfed Areas (NWRDPA). 2003. Ministry of Agriculture, Department of Agriculture and Co-operation, Govt. of India, 97p.

In text reference: The Ministry of Agriculture, Govt. of India has issued the guidelines for NWDPRA (WARASA… 2003).

Compiled or Edited Works

In the case of a compilation or edited book, the surname with initials of the chapter author with year of publication is given first, followed by the title of the chapter. Then put 'In' with a colon (In:) to indicate that the article is from a particular book and provide the name of editor/s along with (ed.) or (eds) in parenthesis and a comma. Then, the name of the book is given. Remember that in the case of a compiled or edited book, the title of the chapter is written in sentence case only, but the title of the book must be in title case. Instead of total pages, only the pages referred are indicated. List the editor first, if your in-text citation is to the work of the editor. See the examples:

Gregory, P.J. 1988. Crop growth and development. In: Wild, A. (ed.), Russell's Soil Conditions and Plant Growth (11th Ed.). ELBS/Longman, London, pp. 31–68.

Jenkins, S.R. 1987. Searching the literature. In: Hawkins, C. and Sorgi, M. (eds), Research: How to Plan, Speak and Write about It (4th Indian reprint, 1993). Narosa Publishing House, New Delhi, pp. 29–59.

Wild, A. (ed.). 1988. Russell's Soil Conditions and Plant Growth (11th Ed.). ELBS/Longman, 991p.

A Translation

Zonn, S.V. 1986. Tropical and Subtropical Soil Science (transl. Russian, Victoriova, M.). Mir Publishers, Moscow, 422p.

Book with Author, Editor, and Translator

Popov, A. S.1985.Cryoconservation and plant cell bank. In: Butenko, R.G. (ed.), Plant Cell Culture (transl. Russian, Chernilovskaya, P.E and Degtyareva, G.Y.).MIR Publishers, Moscow, pp. 175–198.

Lecture/Course Notes

The sequence is author, year, title, name of institution, and place. Indicate that these are lecture/course notes in square brackets.

Thomas, C.G. 2005. Research: Planning, Implementation, and Reporting [Lecture notes]. College of Horticulture, Vellanikkara, 179p.

A Work Published in Several Volumes

Gamble, J. S. 1928–1932. Flora of the Presidency of Madras State (3 volumes). Allard and Son, 21, Hart Street, W.C.

Note: When all the volumes are indicated, page numbers are not given.

One Volume Out of Several Volumes

Randhawa, M.S. 1982. A History of Agriculture in India: Vol. 2. Eight to Eighteenth Century. Indian Council of Agricultural Research, New Delhi, 358p.

Article in an Encyclopaedia or Dictionary

Pimental, D. and Pimental, M. 2008. Human population growth. In: Encyclopaedia of Ecology, Vol .3, Elsevier, Amsterdam, pp. 1907–1912.

Journal Article by One Author

Connolly, J. 1988. Experimental methods in plant competition research in crop weed systems. Weed Res. 28: 431–436.

In-text citation: Conolly (1988)

Journal Article by Two Authors

Antony, S. and Thomas, C. G. 2014. Performance of hybrid napier cultivars under rainfed ecosystems in humid tropics. Range Manag. Agrofor. 35(1):169–172. *In-text citation*: Antony and Thomas (2014)

Journal Article by Multiple Authors Up to 10

Seufert, V., Ramankutty, N., and Foley, J.A. 2012. Comparing the yields of organic and conventional agriculture. Nature 485:229–232.

In-text citation: Seufert et al. (2012)

Note: The name of the journal 'Nature' is written without abbreviation as the title of the journal has one word only.

Journal Article by Multiple Authors more than 10

Sometimes, we may have to deal with articles having numerous authors. If their number is less than 10, CSE style specifies to list all the authors at the end. However, if the number of authors of an article is more than 10, list only the first 10 in the list of references followed by et al. For in-text citation, use the surname of the first author only followed by et al. See the example of an article, which appeared in PLOS Biology with 164 authors! In the example, the names of only 10 authors have been included.

Thomson, S.A., Pyle, R.L., Ahyong, S.T., Alonso-Zarazaga, M., Ammirati, J., Araya, J. F., Ascher, J. S., Audisio, T.L., Azevedo-Santos, V.M., Bailly, N., et al. 2018. Taxonomy based on science is necessary for global conservation. PLOS Biol. 16(3): e2005075. https://doi.org/10.1371/journal.pbio.2005075. [Accessed 10 July 2018].

In-text citation: Thomson et al. (2018)

Journal Article Accepted for Publication (but Not yet Published)

Journal articles or other works accepted for publication but not published should be referred to as 'in press'. The 'in press' category includes manuscripts about which the authors have received written notification from the journal that they have been accepted for publication following the journal's pre-publication procedures such as peer review. For example:

Shylaja, P.V and Thomas, C, G. 2004. Efficacy of pre-emergence herbicides for weed control in cocoa seedling nursery. J. Plant. Crops 32(2) (in press).

Note: Do not forget to update the publication status of the work before your thesis is submitted.

A Thesis

Renu, S. 1999. Emergence and competition of 'Polla' (*Sacciolepis interrupta* (Willd.) Stapf.) in semi-dry rice. MSc (Ag) thesis, Kerala Agricultural University, Thrissur, 69p.

Conference Proceedings and Conference Papers

Generally, conference proceedings are published in book format, but many are now published as a part of journal issues or as a supplement to journal issues. Those published in book format share many characteristics of compiled books with editors. Conference papers are cited in the same way as contributions to edited books, but the major difference in citing papers lies in their titles and in the provision for giving the date and place of the conference.

Proceedings with a Specific Title

Reynolds, C. and Atta-Krah, A.N. 1986. Alley farming with livestock. In: Kang, B.T. and Reynolds, L. (eds), Alley Farming in the Humid and Sub humid Tropics. Proceedings of an international workshop, Ibadan, Nigeria. International Institute of Tropical Agriculture, Nigeria, pp. 27–36.

Proceedings Without a Specific Title

Shylaja, P.V., Thomas, C.G., and Abraham, K. 2002. Effect of pre-emergence herbicides on soil microorganisms. In: Das, M.R. (ed.), Proceedings of the Fourteenth

Kerala Science Congress, 29-31 January 2002, Kochi. Kerala State Committee on Science, Technology, and Environment, Government of Kerala, pp. 312–314.

Note 1: If the 'proceedings' is with a specific title, the date of conference is not mandatory. However, if the proceedings have no specific title, and it starts simply with 'Proceedings of ……', then the dates of conference should also be mentioned as a part of the title, and this shall be put in title cases as shown in the examples. Note that in the former case, only the specific title has been put in title cases.

Note 2: There are certain journals, the names of which start with 'Proceedings'. In such cases, the titles of journals shall be abbreviated starting with 'Proc'.

Conference Abstract

Prusty, J.C. and Behera, B. 1992. Integrated weed managemt in rainfed direct seeded rice [abstract]. In: Abstracts, Annual Weed Science Conference; 3–4, March 1992, Hisar. Indian Society of Weed Science, Hisar, p.34. Abstract No. 5.1.5.

Note: Provide abstract number also, if given, along with page numbers.

Technical Bulletin/Series/Report

Evans, D.O. and Rotar, P.P. 1987. Sesbania in Agriculture. Westview Tropical Agriculture Series No. 8, Westview Press/Boulder, London, 192p.

Annual Report

These are not periodicals in the usual sense, but are important regular publications of government institutions, research centres, universities, companies, and other organizations. A reference for an annual report should include the name of organization, date of publication, short descriptive title, if applicable, years covered, and total pages.

IGFRI [Indian Grassland and Fodder Research Institute] 2000. Annual Report 1999–2000. Indian Grassland and Fodder Research Institute, Jhansi, 101p.

A Patent

The order is name/s of inventor/s, year of issue, title of patent, code number of patent including country of issue, and source.

Quinn, R.V.1991. Tree trunk-smoothing device. United States Patent No. 5056258. Available: http://www.freepatentsonline.com/5056258.html [Accessed 10 June 2011].

Article from a Weekly

Sen, A. 2009. An unequal country. India Today, 24 Aug.2009, pp. 18–20.

Note: In this case, you may also add the date of issue. Fortnightly periodicals will also be listed and cited like the above.

Article from a Newspaper with Author

Articles in newspapers are handled almost similarly as those for journal articles. However, names of newspapers are never abbreviated, and column location is added. Omit any volume and issue numbers while listing newspaper articles.

The sequence is as: Author(s). Date. Title of article. Title of newspaper (edition). Section: beginning page of article (column no.).

Shelar, J. 2018 Oct. 14. Plotting transplant outcomes. The Hindu (Kerala Edition): 15(col.2).

In text reference: (Shelar 2018)

Article from a Newspaper Without Author

If the article is without author, begin the entry with the title of the article. Note that the use of 'anonymous' is not permitted. The date follows the title.

A walk to conserve Killi River. 2018 Oct. 14. The Hindu (Kerala Edition): 3(col.2).

In text reference: (A walk to conserve... 2018)

15.9 Websites and Other Online Sources

Online sources of information are increasingly used for scientific works. Sometimes, a printed book, part of a printed book, or a published journal may be available online. For example, most of the FAO publications are available for free download from the Internet. However, beware of the risks involved as anybody can open a web page and post misinformation on it. Therefore, before using a source, ensure it is reliable and trustworthy. Do not cite doubtful or misleading information posted by interest groups on the Internet. In general, websites maintained by reputed universities, institutes, and organizations can be considered authentic sources. However, personal judgement is highly essential in this regard.

The particulars required for citing and listing electronic sources are the same as those required for printed documents. Additional information you need to obtain are the Internet address or Uniform Resource Locater (URL), digital object identifier (DOI), if available, and the date of access of the information from the Internet in square brackets.

Although many people depend upon Internet for up-to-date information on a variety of topics, unlike printed sources, many of these are transient in nature, and may go offline or be revised at any time. Therefore, when you cite and list information from an Internet source, do not forget to include the date of access.

Article in an Online Journal with DOI

Fried, H.G., Narayanan, S., and Fallen. B. 2018. Characterization of a soybean (*Glycine max* L. Merr.) germplasm collection for root traits. PLOS ONE 13(7): e0200463. Available: https://doi.org/10.1371/journal.pone.0200463 [Accessed 31 July 2018].

Article in an Online Journal, No DOI

Tonukari, N. J. 2004. Cassava and the future of starch. Electr. J. Biotechnol. 7(1). ISSN: 0717-3458. Available: http://www.ejbiotechnology.info/index.php/ejbiotechnology/article/view/v7n1-i02/492 [Accessed 21 March 2007].

Printed Journal Article Freely Available on the Internet

Include usual reference details as for a printed article followed by details of the Internet site and date of access. For example, an article from *Journal of Biosciences* published by the Indian Academy of Sciences, which is available free on the Internet, can be listed as:

Ranganathan, S. Kundu, D., and Vudayagiri, S. D. 2003. Protein evolution: intrinsic preferences in peptide bond formation: a computational and experimental analysis. J. Biosci. 28(6): 683–690. Available: http://www.ias.ac.in/jbiosci/dec2003/683.pdf. [Accessed 08 Dec.2003].

A Thesis from Shodganga

Sumithra, K. 2007. Physiological, biochemical, and molecular responses of five V*igna radiata* (L.) Wilczek cultivars under salinity stress. Pondicherry University, Puducherry, 300p. Available: http://hdl.handle.net/10603/1016 [Accessed 11 Oct.2018].

Note: Shodhganga is a digital repository of theses and dissertations submitted to Indian universities, maintained by the Information and Library Network (INFLIBNET), an autonomous Inter-University Centre of the UGC (University Grants Commission), India. Note that the last number in the URL (1016) is thesis number.

A Thesis from Krishikosh

Sereena, J. 2004. Hybridization and molecular characterization of papaya (*Carica papaya* L.) varieties. PhD thesis, Kerala Agricultural University, Thrissur, 261p. Available: http://krishikosh.egranth.ac.in/handle/1/5810015064. [Accessed 08 Oct. 2018]

Note: KrishiKosh is an institutional repository of National Agricultural Research System (NARS) of India.

E-Book

Jarrard, R.D. 2001. Scientific Methods [E-book]. http://www.emotionalcompetency.com/sci/sm_all.pdf.236p. [Accessed 10 June 2018].

Printed Book Freely Available on the Internet as HTML Document

In the case of a printed book, which is also available on the Internet as HyperText Markup Language (HTML) document, include the usual reference details as for a book followed by details of the Internet site. As these are given as HTML document, page numbers cannot be shown.

Roose, E. 1996. Land Husbandry-Components and Strategy. FAO Soils Bulletin No.70. Food and Agriculture Organization, Rome. Available: http://www.fao.org/docrep/t1765e/t1765e00.htm [Accessed 02 June 2008].

Printed Book Freely Available on the Internet as PDF Document

Several printed books are available as Portable Document Format (PDF) documents on the Internet. Include all the usual reference details as for a book including total number of pages followed by details of the Internet site and date of access.

Heuperman, A.F., Kapoor, A.S., and Denecke, H. W. 2002. Biodrainage: Principles, Experiences and Applications. Food and Agriculture Organization of the United Nations, Rome, 79p. Available: ftp://ftp.fao.org/agl/iptrid/KSR_6.pdf [Accessed 11 Nov. 2011].

Conference Proceedings from a Website

Bruinsma, J. 2009. The resource outlook to 2050: by how much do land, water, and crop yields need to increase by 2050? In: FAO Expert Meeting on 'How to Feed the World in 2050', 24–26 June 2009, Rome. Available: ftp://ftp.fao.org/docrep/fao/012/ak542e/ak542e06.pdf [Accessed 02 Feb. 2012].

Government Report from a Website

GOK [Government of Kerala]. 2017. Agricultural Statistics 2016–17. Department of Economics & Statistics, Government of Kerala, Thiruvananthapuram, 220p. Available: http://www.ecostat.kerala.gov.in/images/pdf/publications/Agriculture/data/2016-17/rep_agristat_1617.pdf. [Accessed 10 May 2018]

Web Document with Author

Pretty, J. 2006. Agroecological approaches to agricultural development [on-line]. Available: http://www.rimisp.org/getdoc.php?docid=6440 [Accessed 05 Nov.2011].

Web Document: No Author

Referencing Why, When and How. 2013. Lincoln University, New Zealand, 14p. Available: https://library2.lincoln.ac.nz/documents/ReferencingWhyWhenHow. [Accessed 19 Oct.2018]

Professional Site

IRRI [International Rice Research Institute]. 2018. IRRI home page. Available: http://irri.org/ [Accessed 26 Sept.2018]

Web Document: No Publication Date

Sometimes, publication date may not be mentioned in the web document. In that case, the order of listing can be: the name of the author, 'date unknown' in square brackets, title, version number (if applicable), name and place of the sponsor of the source, the words 'Available:' and then Internet address, and the date accessed in square brackets.

E-Mail Message

Documentation of an e-mail message requires the following information: author's name, e-mail address, date of sending, subject line, type of communication (personal e-mail, distribution list, office communication, etc.) in square brackets, and date of access. When you cite the e-mail in the text, you begin the sentence with the name of the author and after completing the sentence write 'through e-mail' in parenthesis.

Thomas, C.G. (gtcgthomas@gmail.com). 2008, Jan.30. Controversy over organic farming [personal e-mail, accessed 2 Feb.2008].

In-text citation: Thomas (2008) asserted that………. (through e-mail).

Note: E-mail addresses should never be cited without the permission of the owner of the address.

CD-ROMs

The details required are the same as for a book with the form of the item indicated. Give the bibliographic details in the order of authors (director or producer, if available), year (of recording), title, format (e.g. CD-ROM, video recording, etc.), publisher, and place of recording. Any special credits and other information that might be useful can be noted after the citation. Use the same format for films, videos, DVDs, and television and radio programmes. For example:

Sudheeesh, M. V. 2005. Natural Resource Management. CD-ROM, Lal Creations, Thrissur.

If the producer or director's name is not mentioned, then the listing can be:

Natural Resource Management 2005. CD-ROM, Lal Creations, Thrissur.

In-text reference: In the first example, 'Sudheesh (2005) reported that …' or at the end the sentence, '… (Sudheesh 2005)'. In the second example, cite at the end of the sentence as: '… (Natural Resource Management 2005)'.

15.10 Reference Management Software

Although the Council of Science Editors Manual (CSE 2014) recommends three methods for citing references, namely, the Name-Year (N-Y) system, the Citation-Sequence (C-S) system, and the Citation-Name (C-N) system, the N-Y system is the most popular one. A major advantage of the system is that manual citation and listing are easy. In the USA and certain European countries, they follow C-S or C-N system, where they number the citations. However, as additions and deletions demand renumbering, manual reference management with C-S or C-N system is cumbersome for the authors. In these cases, it would be a good idea to go for reference management software.

At present, several software packages are available for reference management, either priced or freely downloadable from Internet sources. For example, **Mendeley** is a free reference management software from Elsevier. **EndNote** is a popular commercial reference management software from Clarivate Analytics (previously, Thomson Reuters). Prominent priced software packages include:

EndNote (https://www.endnote.com/)
Papers (http://www.readcube.com/papers/)
Bookends (www. sonnysoftware.com/)
SciRef (http://sci-progs.com/)
Citavi (https://citavi.com/)

Most commercial software packages provide trial versions free. If you want to experiment with such software, first follow any of the website addresses given above in parenthesis, and click 'download a free trial'. Try it, and if satisfied, order for the original software. However, note that these software packages go for revisions at regular intervals, and each time you have to shell out some money for an upgraded version! You can also try free software packages. Some of these perform excellently well on par or a bit better than commercial packages. Popular among them include:

Mendeley (https://www.mendeley.com)
Zotero (https://www.zotero.org/)
JabRef (http://jabref.sourceforge.net/)
Qiqqa (http://www.qiqqa.com/)
RefWorks (https://www.proquest.com/)
CiteULike (http://www.citeulike.org),
Bibus (http://www.sourceforage.net/projects/bibus-biblio)
Refbase (http://www.refbase.net)
BibSonomy (https://www.bibsonomy.org/), and
Wikindx (https://sourceforge.net/ projects/wikindx/)

A warning! Whether you use reference management software or intend to do it manually, you should first try to have a clear idea about various documentation styles and their rules of citation and listing!

Annexure

Some common abbreviations used for journal titles

Terms	Abbreviations	Terms	Abbreviations
Abstract	Abstr.	Academy	Acad.
Acta	*No abbreviation*	Advancement	Advmt
Advances	Adv.	Africa/African	Afr.
Agribusiness	Agribus.	Agricultural	Agric.
Agriculture	Agric.	Agriculturist	Agric.
Agrobiologia	Agrobiol.	Agroforestry	Agrofor.
Agronomica	Agron.	Agronomy	Agron.
Alabama	Ala.	Alteration	Alter.
Alternative	Altern..	America/American	Am.
Anales	An.	Analysis	Anal.
Analytical	Anal.	Anatomy/Anatomical	Anat.
Animal	Anim.	Annals	Ann.

(continued)

(continued)

Terms	Abbreviations	Terms	Abbreviations
Annual	Annu.	Antibiotics	Antibiot.
Antimicrobial	Antimicrob.	Applications	Appl.
Applied	Appl.	Aquaculture	Aquac.
Aquatic	Aquat.	Archives	Arch.
Arid	*No abbreviation*	Arkansas	Ark.
Aromatic	Aromat.	Artificial	Artif.
Asian	*No abbreviation*	Aspects	Asp.
Assessment	Assess.	Association	Assoc.
Astronomical	Astron.	Astronomy	Astron.
Atlantic	Atl.	Atmospheric	Atmos.
Atomic	At.	Audiovisual	Audiov.
Australia/ Australian	Aust.	Austria	*No abbreviation*
Avian	*No abbreviation*	Banking	Bank.
Behaviour	Behav.	Better	*No abbreviation*
Bibliography	Bibliogr.	Bimonthly	Bimon.
Biochemistry	Biochem.	Biodiversity	Biodivers.
Bioenergetics	Bioenerg.	Bioengineering	Bioeng.
Bioethics	Bioeth.	Bioinformatics	Bioinforma.
Biophysics	Biophys.	Biosafety	Biosaf.
Bioscience	Biosci.	biweekly	biwkly
Board	*No abbreviation*	Botany/botanical	Bot.
Breeding	Breed.	British	Br.
Bulletin	Bull.	Bureau	Bur.
California	Calif.	Cambridge	Camb.
Canada/Canadian	Can.	Carbohydrate	Carbohyd.
Carolina	Carol.	Cell	*No abbreviation*
Celluar	Cell.	Centre/Central	Cent.
Chemical/ Chemicals	Chem.	Chemist	Chem.
Chemistry	Chem.	Chemotherapy	Chemother.
Chronicle	Chron.	Circular	Circ.
Climate	Clim.	Clinical	Clin.
Coconut	*No abbreviation*	College	Coll.
Colorado	Colo.	Commission	Commn
Committee	Comm.	Commonwealth	Commw.
Communication	Commun.	Community	*No abbreviation*
Comparative	Comp.	Complementary	Complement.

(continued)

(continued)

Terms	Abbreviations	Terms	Abbreviations
Compost	*No abbreviation*	Composting	Compost.
Computer	Comput.	Conference	Conf.
Congress	Cong.	Conservation	Conserv.
Control	*No abbreviation*	Cooperation	Coop.
Coordinated	Coord.	Copenhagen	Cph.
Cotton	*No abbreviation*	Critical	Crit.
Crop/crops	*No abbreviation*	Culture	Cult.
Current	Curr.	Cytology	Cytol.
Dairy	*No abbreviation*	Dairying	Dairy.
Defence	Def.	Delaware	Del.
Dental	Dent.	Department	Dep.
Design	Des.	Development	Dev.
Dakota	Dak.	Disease	Dis.
Dissertation	Diss.	Division	Div.
Drainage	Drain.	Drug	*No abbreviation*
Dynamics	Dyn.	East	E.
Eastern	East.	Economics	Econ.
Ecosystem	Ecosyst.	Edition	Ed.
Education	Educ.	Electric	Electr.
Electrical	Electr.	Electronic	Electr.
Empire	Emp.	Engineering	Eng.
Engineer/s	Eng.	England	Engl.
English	Engl.	Environment	Environ.
Environmental	Environ.	Equipment	Equip.
Ethnobotany	Ethnobot.	European	Eur.
Evolution	Evol.	Evaluation	Eval.
Excerpta	*No abbreviation*	Experiment	Exp.
Experimental	Exp.	Extension	Ext.
External	Extern.	Faculty	Fac.
Farm/Farmer	*No abbreviation*	Farming	*No abbreviation*
Fauna	*No abbreviation*	Federal	Fed.
Federation	Fed.	Fertigation	Fertig
Fertility	Fertil.	Fertilizer	Fertil.
Field	*No abbreviation*	Fish	*No abbreviation*
Fisheries	Fish.	Flora	*No abbreviation*
Florida	Fla	Fodder	Fodd.
Food	*No abbreviation*	Forage	*No abbreviation*

(continued)

(continued)

Terms	Abbreviations	Terms	Abbreviations
Forecasting	Forecast.	Forest	For.
Forester	For.	Forestry	For.
Freshwater	Freshw.	Functional	Funct.
Gazette	Gaz.	General	Gen.
Genetics	Genet.	Geography	Geogr.
Geophysical	Geophys.	Geophysics	Geophys.
Georgia	Ga	Global	Glob.
Grain	*No abbreviation*	Grassland	Grassl.
Groundwater	Groundw.	Growing	Grow.
Handbook	Handb.	Health	*No abbreviation*
Healthcare	Healthc.	Herbage	Herb.
Highway	Highw.	History	Hist.
Homeopathy	Homeopath.	Homestead	Homestd.
Horticultural	Hortic.	Horticulture	Hortic.
Horticulturist	Hortic.	Hormone	Horm.
Human	Hum.	Hungarian	Hung.
Husbandry	Husb.	Hydromechanics	Hydromech.
Hygine	Hyg.	Immunity	Immun.
Improvement	Improv.	Index	*No abbreviation*
India/Indian	*No abbreviation*	Industrial	Ind.
Industry	Ind.	Information	Inf.
Institute	Inst.	Institution	Inst.
Instruments	Instrum.	Intelligence	Intell.
International	Int.	Ireland	Irel.
Irish	Ir.	Irrigation	Irrig.
Island	Isl.	Japan/Japanese	Jpn.
Joint	Jt	Journal	J.
Kansas	Kans.	Kentucky	Ky
Knowldge	Knowl.	Laboratory	Lab.
Land	*No abbreviation*	Leaflet	Leafl.
Letter	Lett.	Liberty	Lib.
Library	Libr.	Life	*No abbreviation*
Lifesaving	Lifesav.	London	Lond.
Magazine	Mag.	Malayan	Malay.
Management	Manag.	Marketbook	Markb.
Marketing	Mark.	Material/s	Mater.
Mathematics	Math.	Mechanical	Mech.

(continued)

(continued)

Terms	Abbreviations	Terms	Abbreviations
Mechanics	Mech.	Medical	Med.
Medicine/ Medicinal	Med.	Memoirs	Mem.
Methods	*No abbreviation*	Mexico	Mex.
Michigan	Mich.	Microscopy	Microsc.
Miscellaneous	Misc.	Modern	Mod.
Molecular	Mol.	Monograph	Monogr.
Monthly	Mon.	Mutation	Mutat.
National	Natl	Natural	Nat.
Nature	Nat.	Nebraska	Neb.
Netherlands	Neth.	New Zealand	N.Z.
News Letter	Newsl.	North	N.
Northern	North.	Nottingham	Nottm
Nuclear	Nucl.	Nutrition	Nutr.
Obstetrical	Obstet.	Obstetrics	Obstet.
Occupational	Occup.	Oceanography	Oceanogr.
Official	Off.	Oklahoma	Okla.
Optics	Opt.	Optimization	Optim.
Oregon	Or.	Organic	Org.
Organization	Organ.	Outline	Outl.
Outlook	*No abbreviation*	Pacific	Pac.
Paediatrics	Paediatr.	Pakistan	Pakist.
Paper	Pap.	Petrol	Pet.
Petrology	Petrol.	Pharmaceutical	Pharma.
Phenomena	Phenom.	Phenomenon	Phenom.
Philippines	Philipp.	Philosophy	Philos.
Photosynthesis	Photosynth.	Physical	Phys.
Physics	Phys.	Plant	*No abbreviation*
Plantarum	Plant.	Plantation	Plant.
Planter/s	Plant.	Plastics	Plast.
Political	Polit.	Pollution	Pollut.
Polymer	Polym.	Postgraduate	Postgrad.
Poultry	Poult.	Power	*No abbreviation*
Practical	Pr.	Preview	Prev.
Proceedings	Proc.	Production	Prod.
Progress	Prog.	Protection	Prot.
Publication/s	Publ.	Quality	Qual.
Quarter	Quart.	Quarterly	Q.

(continued)

(continued)

Terms	Abbreviations	Terms	Abbreviations
Queensland	Qld	Radiation	Radiat.
Range	*No abbreviation*	Rangeland	Rangel.
Reclamation	Reclam.	Record	Rec.
Regulation	Reg.	Rehabilitation	Rehabil.
Relative	Relat.	Report	Rep.
Republic	Repub.	Research	Res.
Resource	Resour.	Review	Rev.
Rivisita	Riv.	Root	*No abbreviation*
Royal	R.	Rubber	*No abbreviation*
Ruminants	Rumin.	Scandinavia	Scand.
School	Sch.	Science	Sci.
Scientific	Sci.	Scottish	Scott.
Sedimentary	Sedim.	Sericulture	Seric.
Series	Ser.	Service	Serv.
Silviculture	Silvic.	Situation	Situ.
Society	Soc.	Soil	*No abbreviation*
South	S.	Southern	South.
Special	Spec.	Spectroscopy	Spectrosc.
State	St.	Station	Stn
Statistics	Statist.	Sterility	Steril.
Structural	Struct.	Studies	Stud.
Sugar	Sug.	Surface	Surf.
Surgery	Surg.	Survey	Surv.
Sustainable	Sustain.	Symposium	Symp.
Synthetic	Synth.	Systematic	Syst.
Systems	Syst.	Taxonomy	Taxon.
Technical	Tech.	Technique	Tech.
Tennessee	Tenn.	Texas	Tex.
Theoretical	Theor.	Therapeutics	Ther.
Tobacco	Tob.	Tradition/traditional	Tradit.
Transactions	Trans	Trinidad	Trin.
Tropical/tropicale	Trop.	United States	U.S.
University	Univ.	Utilisation	Utilis.
Vegetable	Veg.	Veterinary	Vet.
Vitamin	Vitam.	Viticulture	Vitic.
Washington	Wash.	Water	*No abbreviation*
Weekly	Wkly	Welding	Weld.

(continued)

(continued)

Terms	Abbreviations	Terms	Abbreviations
West	W.	Western	West.
Weston	West.	Wild	*No abbreviation*
Wildlife	Wildl.	Wisconsin	Wis.
World	World	Worldtrade	Worldt.
Worldwide	Worldw.	Yearbook	Yearb.
Yearly	Yrly	Zone	*No abbreviation*

Some frequently used abbreviations of title words of journals as approved by the International Standard Serial Number (ISSN) are given in this Annexure. For a complete list of abbreviations, visit the site: http://www.issn.org/services/onlineservices/access-to-the-ltwa/. Please refer page No. 379 of this book for additional sources.

Bibliography

AMA [American Medical Association] 2007. AMA Manual of Style: A Guide for Authors and Editors (10th Ed.). Oxford University Press, New York, 1024p.

APA [American Psychological Association] 2013. Publication Manual of the American Psychological Association (6th Ed.). American Psychological Association, Washington DC, 272p.

Barrass, R. 2002. Scientists Must Write: A Guide to Better Writing for Scientists, Engineers and Students (2nd Ed.). Routledge, London, 204p.

BSI [British Standards Institution] 1990. Recommendations for Citing and Referencing Published Material (2nd Ed.). British Standards Institution (BS 5605:1990), 8p.

Cargill, M. and O'Connor, P. 2009. Writing Scientific Research Articles: Strategy and Steps. Wiley-Blackwell, 173p.

COA [Commonwealth of Australia] 2002. Style Manual for Authors, Editors and Printers (6th Ed.). John Wiley & Sons, Brisbane, Australia, 550 p.

Cousens, R. 1987. Theory and reality of weed control thresholds. Plant Prot. Q. 2: 13–20

Craswell, G. 2004. Writing for Academic Success: A Postgraduate Guide. Sage Publications, London, 271p.

CSE [Council of Science Editors] 2014. Scientific Style and Format: The CSE Manual for Authors, Editors, and Publishers (8th Ed.): The University of Chicago Press, 722p.

Davis, M. 2005. Scientific Papers and Presentations. Academic Press, Massachusetts, 356p.

Day, R.A. and Gastel, B. 2006. How to Write and Publish a Scientific Paper (6th Ed.). Cambridge University Press, Cambridge, 320p.

Gustavii, B. 2008. How to Write and Illustrate a Scientific Paper (2nd Ed.). Cambridge University press, Cambridge, 168p.

Hakansson, S. 1983. Competition and Production in Short-lived Crop Weed Stands: Density Effects. Report 127, Swedish University of Agricultural Sciences, Uppasala, 85p.

ISSN [International Standards Serial Number].2017. The List of title word abbreviations [on line]. Available: http://www.issn.org/services/onlineservices/access-to-the-ltwa/ [Accessed 06 May 2018].

Leggett, G., Mead, C.D., Kramer, M.G., and Beal, R.S. 1985. Handbook for Writers. Prentice- Hall, New Jersey, 558p.

Lester, J.D. and Lester, J.D. Jr. 2009. Writing Research Papers: A Complete Guide (13th Ed.), Longman, 416p.

MLA [Modern Language Association of America] 2016. MLA Handbook for Writers of Research Papers (8th Ed.) Modern Language Association of America, 146p

Neville, C. 2010. The Complete Guide to Referencing and Avoiding Plagiarism (2nd Ed.). McGraw-Hill Education, 288 p.
Renu, S. and Thomas, C.G. 2000. Stale seedbed technique for the management of Sacciolepis interrupta in semi-dry rice. Indian J. Weed Sci. 32(3&4):146-149
Rubens, P. 2004. Science and Technical writing: A Manual of Style (2nd Ed.). Routledge, New York, 427p.
Sheng, T.C. 1989. Soil Conservation for Small Farmers in the Humid Tropics. FAO Soils Bulletin No. 60. Food and Agriculture Organization of the United Nations, Rome, 104p
Thomas, C. G. 2008. Forage Crop Production in the Tropics. Kalyani Publishers, Ludhiana, 333p.
Thomas, C. G., Abraham, C. T., and Sreedevi, P. 1997. Weed flora and their relative dominance in semi-dry rice culture. J. Trop. Agric. 35:51–53.
Thomas, C. G. 2010. Land Husbandry and Watershed Management. Kalyani Publishers, Ludhiana, 716p.
Turabian, K.L. 2007. A Manual for Writers of Term Papers, Theses, and Dissertations (9th Ed.). The University of Chicago Press, 461p.
UChicago [University of Chicago] 2017. Chicago Manual of Style [CMOS].
University of Chicago Press, 1144p.

Weblinks

AMA style- http://www.amamanualofstyle.com/oso/public/index.html.
APA style- https://www.apastyle.org/manual
BibSonomy-https://www.bibsonomy.org/
Bibus -http://www.sourceforage.net/projects/bibus-biblio
Bookends-www. sonnysoftware.com/
CAS Source index search tool- http://cassi.cas.org/search.jsp
Citavi-https://citavi.com/
CiteULike-http://www.citeulike.org
EndNote- https://www.endnote.com/
FAO [Food and Agriculture Organization of the United Nations]. 2018. The State of Food Security and Nutrition in the World 2018. Building Climate Resilience for Food Security and Nutrition. FAO, Rome, 181p. Available: http://www.fao.org/3/I9553EN/i9553en.pdf [Accessed 10 Oct. 2018].
Genamics JournalSeek- http://journalseek.net/
JabRef -http://jabref.sourceforge.net/
Journal database, NCBI- http://www.ncbi.nlm.nih.gov/nlmcatalog/journals ISSN, LTWA-http://www.issn.org/2-22661-LTWA-online.php.
Mendeley-https://www.mendeley.com
Papers-http://www.readcube.com/papers/
Qiqqa -http://www.qiqqa.com/
Refbase-http://www.refbase.net
RefWorks-https://www.proquest.com/
Scientific Style and Format (CSE) - https://www.scientificstyleandformat.org/Home.html
SciRef-http://sci-progs.com/
Web of Science journal abbreviations- https://images.webofknowledge.com/images/help/WOS/A_abrvjt.html
Wikindx -https://sourceforge.net/ projects/wikindx/
Zotero-https://www.zotero.org/

Chapter 16
Improve Your Writing Skills

> *Information is a source of learning. But unless it is organized, processed, and available to the right people in a format for decision making, it is a burden, not a benefit.*
> *William G. Pollard* (1911–1989), American Physicist

Researchers and research students are expected to write various kinds of scientific reports. These include assignments, term papers, project reports, seminar reports, project proposals, theses, abstracts, research reports, research papers, reviews, and popular articles. All these come under the purview of 'scientific writing' or 'technical writing'. Scientists should accept that writing is part of their work, and they must be able to write using good English. Many wrongly feel that English is not important for science and technology, and chances are that such persons may end up as poor communicators. If you do not have reasonable command over English, you may sometimes reproduce exactly what others have reported in literature, inviting trouble for committing *plagiarism* (Chap. 24). *Paraphrasing* is a major technique employed by scientists to avoid plagiarism (Sect. 24.3) for which a thorough mastery over the language is indispensable.

16.1 Use Plain English

Scientific writing is highly practical, and its main purpose is to convey information unlike most literary works that entertain, amuse, or stir up emotions. Scientific publications, therefore, demands a simple and straightforward style of writing. A major requisite for scientific writing is brevity. There are many simple ways to economize space while writing. Eliminating redundant words and modifying phrases to make them simple and short are widely accepted techniques. For writing in an unambiguous way, you have to select correct words. Wherever possible, use short words instead of long words and short phrases instead of long phrases. It would be wise to choose a short word with the correct meaning rather than using a long word with the wrong meaning.

C. G. Thomas, *Research Methodology and Scientific Writing*,
https://doi.org/10.1007/978-3-030-64865-7_16

The length of sentences can be pruned through restructuring. Wherever possible, using active voice instead of passive voice can save space substantially. Yet, another alternative to save space is to combine ideas of several sentences into one, as many short, simple sentences waste words. However, do not use long, complex, and confusing sentences.

Unlike everyday writing, in technical writing, you should follow some additional rules and conventions with respect to words, numbers, units, abbreviations, and grammar. Make sure that there is no ambiguity in meaning due to improper use of words. Before finalizing your manuscript, the language should be scrutinized for usage, flow, and clarity by checking for awkward phrasing, biased language, clichés, and jargons, which are inappropriate for a scientific report. Care must also be taken to avoid grammatical and stylistic mistakes. In this chapter, an attempt has been made to highlight some specific cases with examples to improve your skill in writing scientific reports.

The message conveyed in the articles must be clearly defined and not surrounded by distractions. A *distraction* is anything in the writing that unintentionally raises a question or irritation in the reader's mind. The reader then pauses and looks for an answer to this question instead of reading further. There are many types of distraction, which slow down and irritate the reader. Sometimes, the reader may not realize exactly the distraction. If an article is well written, the reader simply does not notice the writing style, because the clearly presented subject matter attracts him. If the article is hard to read and not reader friendly, the reader's attention may be diverted from the subject and focused on the defects of the writing itself. Sometimes, technical writing is intended to convince someone to do something. For example, an application for a research grant must be prepared in such a way as to convince the granting agency to fund the proposed work.

16.2 Use Paragraphs Properly

The ideas presented in a scientific report must be arranged in proper paragraphs. A paragraph is a unit of thought or information. Hence, the idea expressed in the paragraph should be homogenous in relation to the subject matter. It should be homogeneous in content and should treat the content in logical and sequential order. All the sentences in the paragraphs must provide information to the topic, and the first and the last sentence should help to link paragraphs so that readers understand instantly how your thoughts about one aspect lead onto others.

Paragraphing gives the readers a mental and visual break and helps them to know that one thing has been over and that it is time to think of the other. Try to avoid long paragraphs. If the paragraph is long, some writing may not get the required attention it deserves. Short paragraphs are easier to read allowing efficient communication with the readers. However, forming clusters of short or single sentence paragraphs is not appropriate. In short, avoid very short and long paragraphs, which may confuse the readers and are indications of poorly organised writing. Therefore, arrange the ideas into paragraphs of appropriate length after analyzing the order of sentences and checking their connections.

When you begin to write a paragraph, you must have a clear concept of what it is going to be about. A paragraph is difficult to understand, if it is not organized around a defined topic. By using a topic sentence, you can organize your thoughts in each paragraph. The topic sentence creates the expectation of what the paragraph will be about and the supporting sentences fulfil that expectation. Once the topic sentence has been correctly framed, the paragraph is completed with supporting sentences that give all the remaining information that the reader needs to know. If a paragraph begins with a clear topic sentence, you immediately get a good idea of what information will follow.

Sentence Checking in Paragraphs

Examine rigorously the sentences and words in a paragraph to locate ambiguities and distractions. Many of these distractions are examples of wrong usages of the eight parts of speech (*noun*, *pronoun*, *adjective*, *verb*, *adverb*, *preposition*, *conjunction*, and *article*) or of their positions in sentences. Check every sentence to make sure that it has a subject and a verb, and that they agree. For example, many readers may not easily understand what is wrong with the following sentence:

- The new collection of flowering plants *are* superb.

In the example, the subject is 'collection' and not the subject compliment 'flowering plants'. As the noun 'collection' is singular, the verb should be 'is' and not 'are'. This error occurs mainly because of the nearness of the subject compliment (here, flowering plants) to the verb. The sentence should be corrected as:

- The new collection of flowering plants *is* superb.

See some other examples:

- The number of scientists participating in the workshop *is* small.
- A number of scientists *were* asked to report for the investigation.

In the first sentence, 'number' preceded by 'the' takes singular verb as it refers to a unit of scientists. In the second sentence, however, 'number' preceded by 'a' takes a plural verb as it refers to individual scientists.

- Whatever you says, five million rupees *are* a big amount.

The above sentence is not correct. In the sentence, the plural noun 'five million rupees' denotes some specific amount considered as a whole, hence the verb must take singular form. The correct sentence is:

- Whatever you says, five million rupees *is* a big amount.

When two or more subjects are connected by 'or', 'nor', 'neither…nor', or 'either…or', they take singular verbs. It is wrong to write the sentences as given below:

- Neither digging the fields nor hand weeding *are* easy.
- Hand weeding or sickle weeding *are* cumbersome.

The above sentences should be corrected as:

- Neither digging the fields nor hand weeding *is* easy.
- Hand weeding or sickle weeding *is* cumbersome.

See some other examples, where the verb form is correctly used.

- Every one *was* engaged in animated discussion.
- Each one of these students *is* intelligent and clever.
- Each of the members *has* one vote.
- Every teacher and every student *is* expected to know the basics of computer applications.

Singular indefinite pronouns such as 'anyone', 'each', 'each one', 'everyone', 'either', 'neither', 'no one', 'one', and 'someone' as indicated in the above examples, take singular verbs (*Indefinite pronouns* are pronouns that replace nouns without specifying which noun they replace). In addition, along with such indefinite pronouns, singular personal pronouns only are used. Examine the following sentence:

- One of the candidates gave up *their* seat and left.

The above sentence is not correct, as 'one of the candidates' is singular, and must take singular personal pronoun. The sentence should be corrected as:

- One of the candidates gave up *her* seat and left.

In the same manner, when you use relative pronouns, make sure that the pronoun and the antecedent agree in all respects (Wren and Martin 2005; More details in Sect. 16.9).

Chances are that similar mistakes as shown above may go unnoticed. If you are using a word processor with grammar and spell-check facility, it may sometimes detect the error. However, do remember that it is a machine; and in many cases, it may not detect the error at all, and your judgment will be final.

16.3 Avoid Gender Specific or Discriminatory Language

In scientific writing, it is important to use a language that includes both men and women making no distinction between the genders. Unfortunately, the English language has many words with discriminatory overtones such as chairman, housewife, businessman, and mankind. Until recently, it was perfectly agreeable to use 'man' as such or as a prefix or suffix to certain words in contexts where man represents both man and women. However, in the modern world, when the context includes both man and women, it is improper and discriminatory to use the word 'man' to denote both males and females. If your intention is to represent human beings as a whole, you have to use appropriate alternatives such as humans, people, humankind, or humanity.

Although gender-specific words may be applicable in some situations, it is better to use words that include both men and woman. Check for the inadvertent bias. For example, the term 'housewife' is no more acceptable; use 'homemaker' instead. The word 'chairman' is out, use 'chair', 'chairperson' or 'president.' Similarly, the feminine forms such as poetess, hostess, and actress are also on the way out; use gender-neutral words such as poet, host, steward, and actor (for a list of typical discriminatory words and alternatives, see Table 16.1).

Find Alternatives to Masculine Pronouns to Represent Both Genders

You may find out alternatives to pronouns to represent both men and women. In English, third person singular pronouns change form to indicate gender (he, she). Until recently, when the pronoun could refer to either male or female or when the antecedent's gender was unknown, it was grammatically correct to use the singular masculine pronoun. For example, nobody could find any fault with the sentence: *The farmer must manage 'his' farm following scientific practices.* However, gender-sensitive people have raised objections to this grammatical rule blaming it as sexist. It is incorrect to imply that all farmers are men! The objections have been widely debated, and it is a preferred practice now to avoid the masculine pronoun when the antecedent includes both genders. Sometimes, the antecedent can be set in the plural to denote both genders. For example, the statement in the above example can be made gender-neutral by a minor change as: *Farmers must manage 'their' farms following scientific practices.*

In writing, make it a habit to use gender-neutral plural antecedents as students, researchers, scientists, teachers, doctors, nurses, personnel, patrons, people, humans, farmers, peasants, participants, and the like according to the context. A better way out is to change the wording of sentences. For example, the sentence, *The researcher must learn to maintain 'his' laboratory clean,* can be changed to, *The researcher must learn to maintain the laboratory clean.*

Although some people use 'he or she', 'he/she', or 's/he' to represent both man and woman, the construction seems awkward. Adopt these only as a last resort; often, you can recast the whole sentence to avoid such a usage.

There are several pronouns in English, which are conventionally given a gender. The pronoun 'she' is sometimes used to refer to things, which can hold people such as countries, ships, or vehicles. For example, cars are frequently called 'she'. Use 'it' instead. Similarly, do not write 'sister centre' or 'sister institute'; instead use 'related centre' or 'sibling centre', or change the sentence.

Girl, Lady, Female, Woman, and Spouse

You must be vigilant when using these words. The term 'girl' has hints of immaturity and dependence. It is not proper to use the term girl to refer to young women scientists or supporting staff. Similarly, it is not correct to use the term 'lady' as a synonym for 'woman'. 'Lady' is better used to suggest certain standard of maturity, decorum, or sophistication. Note that in 'ladies and gentlemen', the usage is alright. For example,

Table 16.1 Examples for gender specific or discriminatory terms and alternatives

Terms	Alternatives
Actress	Actor
Businessman	Business manager, executive, business people (plural)
Cameraman	Photographer; cinematographer
Chairman	Chairperson; chair; president
Delivery man	Courier
Domestics, maids, servants	Domestic workers
Draftsman	Drafter
Forefathers	Ancestors
Foreman	Supervisor
Frenchmen	The French
Freshmen	First year students
Gentleman's agreement	Unwritten agreement, agreement based on trust
Girl Friday, man Friday	Aide, key aide, assistant, helper
He, himself	He or she, he/she; s/he; himself/herself; one; they; their, or recast the sentence to represent both man and woman
Headmaster/Headmistress	Head teacher, principal
Hostess	Host
Housewife	Homemaker
Laymen	Non-professionals; lay persons
Line men	Line crew
Lower class	Working class; people with lower earnings
Maid	Domestic worker, house cleaner
Man	Humans (in the sense of either man or woman)
[To] man (verb)	To operate, to work, staff serve at (or on, or in)
Man-hours	Work hours, labour time
Mankind	Humankind; humanity
Man-made	Hand-made, hand built, human-made, synthetic, manufactured, fabricated
Manpower	Staff, labour, workforce, workers, human power
Man-to-man	One-to-one, person-to-person
Media man	Media person
Middleman	Intermediary, trader
Mother tongue	First language
Mentally retarded	Mentally challenged; persons mental with disabilities
Physically handicapped	Person with physical disabilities
Poetess	Poet
Policemen	Police officers

(continued)

Table 16.1 (continued)

Terms	Alternatives
Postman	Mail carrier; postal worker
Salesgirls	Sales clerks; sales assistants
Sister centre	Sibling centre, related centre
Spokesman	Spokesperson; representative
Statesman	Diplomat; political leader; public official
Stewardess	Flight attendant; steward
The black	The black people; African; Afro-American; AfroCaribbean; Indian-American
The blind	People with little or no sight; someone with visual impairment
The disabled	The disabled people; people with disabilities
Waiter/Waitress	Waiting staff
Workman	Worker

you may address: Good morning ladies and gentleman; *not*: Good morning men and women.

The term, 'female' is appropriate when the corresponding choice for the other sex is male, for example, male members and female members. However, 'woman' is the most common word for referring to an adult female person. For example, you can say, a team of four women and six men. Remember that the term 'woman' indicates independence, competence, seriousness, and sexual maturity compared to a 'girl'.

'Spouse' is a gender-neutral word. For example, if the husband is mentioned, then his spouse is his wife, and if the wife is referred to, then her spouse is her husband.

Avoid Discriminatory Words

Similar to gender-specific words, discriminatory words that stereotype any person, regardless of gender, race, nationality, creed, age, or handicap are also unacceptable for scientific literature. Unless you are not specific, readers might make assumptions about race, age, sex, or handicap.

Use alternate expressions for phrases considered derogatory such as 'physically handicapped' (use 'person with physical disabilities'), 'mentally retarded' (use 'mentally challenged' or 'persons with mental disabilities'), and 'lower class' (use 'working class' or 'people with lower earnings'). Avoid using 'the disabled' in a depersonalized form. Use 'the disabled people', or 'people with disabilities'. Similarly, 'the blind' is vague. Use 'people with little or no sight', or 'someone with visual impairment'. Saying 'the blacks' in a depersonalized form may also be considered derogatory. Use alternate expressions such as 'the black people', 'African', 'Afro-American', 'Afro-Caribbean', or 'Indian-American' according to the situations.

16.4 Titles

We use titles such as Mr, Mrs, Ms, and Miss for addressing men or women in a formal context. In India, Sri. (*Sriman*), Smt. (*Shrimati*), and Kum. (*Kumari*) are the frequently used titles for addressing people. Other titles in common use are Dr and Professor. Gen. (General), Lt (lieutenant), Capt. (captain), and Col (colonel) are titles used by the Army. Generally, first mention of a person requires the full name with appropriate title, but thereafter, only the surname is sufficient along with the title.

Mr (mist (r)): Mr (Mister), derived from Master, is a commonly used title to address men, whether married or unmarried. The title 'Master' is now used for boys only. Mr is sometimes combined with certain positional names, for example: Mr President, Mr Speaker, Mr Vice Chancellor, and Mr Dean. Note that in these cases, the female equivalent is Madam and *not* Mrs or Ms, for example: Madam President, Madam Speaker, and Madam Dean. The plural form of Mr is Messieurs, abbreviated as Messrs (from the traditional French title).

Mrs (misiz): The title Mrs, derived from the archaic "Mistress", is used to address married women.

Miss (miss): This title is commonly used to address unmarried women. However, the present trend is to replace it by 'Ms', restricting 'Miss' to address female children above the age of 13 probably up to 19, the so called teenagers. For girls younger than 13, normally, titles such as 'Miss' or 'Ms' is not used. Note that 'Miss' is also used by school going students to address their female teachers, even if they are married.

Ms (miz): Ms is a recent term advanced and propagated by gender sensitive people to refer to a woman who does not wish to have to say whether she is married or not, and when the marital status of the woman is unknown or irrelevant. In fact, Ms is a merged version of Mrs and Miss, which can be used instead of Mrs or Miss. The title Ms is getting wide acceptance now, replacing both Mrs and Miss.

Sri. (Sriman): This is a common title in India to address men, whether married or unmarried.

Smt. (Shrimati): This title is used to address married women.

Kum. (Kumari): A common title in India to address unmarried young women.

Dr (Doctor): Doctor as a title originated from the Latin word *Doctoris*, meaning teacher. Abbreviated as Dr or Dr., this title is widely used to designate a person who has obtained a doctorate degree in any discipline. Doctorates may be research doctorates (e.g. PhD, DSc) or professional doctorates (e.g. MD). In many countries, physicians with bachelors degree are also addressed by the title Dr. Note that if someone has a doctorate, it is impolite to disregard that title knowingly and address with ordinary titles such as Mr and Mrs.

Professor: 'Professor' is generally considered as a job title. A person holding any of the ranks of professor such as *professor, associate professor, or assistant professor* may be addressed orally as *Professor* or *Professor… (with person's name)* as a courtesy. However, only serving professors use it as a title.

Please note that a retired professor with a doctorate would continue to be Dr… (*name*), but identified as a former professor. Normally, a retired professor would not present himself/herself as Professor… (*name*); however, a former student can greet her/him as Professor… (*name*).

Remember that the full form 'Professor' shall be used instead of 'Prof.', especially when writing about somebody who is a professor or for addressing in letters and e-mails, for example, Professor Choudhary and *not* Prof. Choudhary. Addressing somebody with the abbreviated form (Prof.) is considered discourteous.

Use of Periods in Titles

Some titles in use are actually contractions formed by the first and last letters of a word as in Dr (Doctor), Mr (Mister), Mrs (Mistress), and Ms. Please note that in British English, when the last letter of the abbreviation and the last letter in the original word without the abbreviation are the same, a period or full stop is omitted after the abbreviation. However, in US English, these titles are written with full stops as, Dr., Mr., Mrs., and Ms.

Avoid Awkward Combination of Titles

Avoid using weird combination of titles such as Professor (Dr) Thomas, Prof. (Dr) Thomas, or Dr (Mrs) Thomas as these are considered pompous and awkward.

If you are very particular to show that one is a professor and has doctorate degree, you may use, for example, as Professor C. G. Thomas, PhD. Another alternative is to write Dr. C. G. Thomas, Professor. However, there is no justification for addressing a woman with an odd double title as Dr (Mrs) Thomas! It is also gender discriminatory, as you may not address a man with the title Dr (Mr) Thomas! You may further note that the wives of doctors or professors do not share their husbands' titles. For example, we may say Mrs Singh and *not* Mrs Dr Singh or Mrs Professor Singh.

Titles in Technical Writing

In technical writing, avoid formal titles such as Dr, Professor, Mr, Mrs, Ms, and their equivalents in other languages such as Sri., Smt., Kum., Mme, Dame, and Monsieur. However, you can use formal titles in the 'acknowledgments' section of a thesis or research paper.

16.5 Verbs: Care with Tenses

Use of correct tense forms of verbs is important in academic and scientific writing as its wrong usage is irritating to the reader and reflects poorly on the author's writing skills. Generally, five tense forms of verbs are used in scientific writing. These include the present simple tense, the present continuous tense, the present perfect tense, the future simple tense, and the past simple tense.

Present simple tense: In technical writing, remember to put generally accepted facts and statements, which do not specify any particular period and always true in the present simple tense. For example, the statements, *Seeds germinate under favourable conditions* and *Water boils at* 100 °C, are not restricted to a particular period and that the statements are true in all circumstances. Similar are the cases with many recommendations on agricultural practices. In technical writing, the present simple tense predominates in almost every type of writing situations except those explicitly set in the past (e.g. historical reviews, experimental results) or in the future (e.g. proposals, recommendations). In these exceptional cases too, the simple present can be used to indicate generalization that is not restricted to the past or to the future.

Present continuous tense: The present continuous tense is used to express an action in the present, something that is currently happening. See the sentence: *Students are studying lessons.* It indicates an action (*studying*) currently happening. This form should be reserved for situations where you are describing an event in the process of occurring, that is, an incomplete, ongoing event. It is a useful form in progress reports and in letters.

Future simple tense: The future simple tense is used to express the future, especially for predicting something as in: *Students will study lessons.* This is simply a prediction, which may or may not occur in the future.

Present continuous tense with 'going to': Instead of future simple tense, the present continuous tense with 'going to' is often used to indicate events that have already been decided and certain to happen shortly. This type of usage is common in lectures and conversations as in: *I am going to present a short summary.* Note the meaning carefully here; it indicates an impending action which is certain to happen without much delay.

Past simple tense: The past simple tense is used to express completed actions that occurred at some time in the past but no longer occurs. For researchers, such events include the methods or the procedures that they have conducted and the results they observed. Similarly, when you refer to your present results, you should use the past simple. 'Abstract' is another section, where past simple is frequently used because you are referring to your own present results. When you write, all those occurred explicitly in the past must be set in the simple past, but general statements, which are always true, in the present tense. For example, the following sentence is set in the simple past: *When the hormone was sprayed, the most striking changes were rapid growth of stem and a hastening of leaf fall.* The sentence means that the statement was true in the particular conditions of the reported experiment, and that we do not know if the statement is always true. However, the sentence in the present simple tense: *When the hormone is sprayed, the most striking changes are rapid growth of stem and a hastening of leaf fall* implies that the statement is true always.

Present perfect tense: Occasionally, we may also use the present perfect tense for actions began in the past but still going on. Note that the simple past tense is used to express a completed action in the past that no longer occurs. For example, in the 'Review of Literature' section of a scientific paper, single isolated studies are usually referred to in the simple past tense (e.g., Day (1983)) reported that…). Whereas, multiple studies suggesting an ongoing sequence of studies are usually referred to in the present prefect tense (e.g., Day (1983); Legget et al. (1986); Rubens (2004); and Gustavii (2008) have reported that…).

In a thesis or paper, most of the 'Introduction' and 'Discussion' should be in the present simple tense because you are mostly talking about previously established knowledge. The Introduction usually includes background materials, which are generally accepted as facts. The author also explains why the research in question is important. Sometimes, the author needs to refer to literature relevant to the present work, and indicates his/her opinion of the research being reported. When the present tense is used, indication is that the research findings are still true and relevant, although the original research may have been conducted some time ago.

In 'Discussion' section, the significance of the results is explained, and the present tense is apt for this. However, one may use the past tense to summarise the findings, in combination with the present tense to explain or interpret what the results mean.

For writing the 'Results' section, generally the past tense is used. However, when you refer to figures, tables, and graphs, use the present tense only.

For writing the 'Methods', simple past tense must be used as we are describing what was done. For describing methods, past tense in passive voice is often used.

For summarizing the main findings and the major implications of the study and to offer suggestions for future research, you may use a combination of tenses. However, as the 'Abstract' portion refers to results, simple past tense is often used.

May and Might

'Might' is the past tense of 'may', but when used in sentences, both words apply to the present or the future showing probability. 'May' simply indicates a possibility, but 'might' adds a greater degree of uncertainty to the possibility. This shade of difference could be seen in the following sentence:

- Farmers who adopt plant protection measures in time *may* be rewarded by way of good crop, and those who are complacent *might* face difficulties.

Note that the use of 'may' and 'might' implies only a difference in intended meaning. For example, when we say, *You had better plant the seedling now or the rainy season 'may' be over*, we suggest a real possibility. Instead, if we say, *You had better plant the seedling now or the rainy season 'might' be over*; the possibility is still there but the chances are comparatively less.

In addition to the above usage, '*might*' can be used as the past tense of '*may*' if situation warrants. In the present tense, we usually say, *Thomas thinks that he 'may' go to New Delhi for higher studies.* In the past tense, we say, *Thomas thought that he 'might' go to New Delhi for higher studies.*

In the 'Discussion' part of a thesis, you may find excessive use of might as: *This might be due to…* Try to avoid this kind of vagueness, and be specific in your explanations and conclusions.

16.6 The Active or Passive Voice Dilemma

It is common to use both active and passive voices in everyday use, but the situations in which active or passive voice is suitable must be clearly understood. In sentences with active voice, the subject does something by the verb upon the object as in the sentence: *Insects attacked the crop.* When the sentence is set in the passive voice, the order of the words is reversed, and the sentence becomes: *The crop was attacked by insects.* The object becomes the subject and something is done to the subject. In active voice, the subject is 'active' as it acts; however, in passive voice, the subject is 'passive', as it suffers or receives some action of the verb. Although both examples mentioned above convey the same meaning, the sentence in the passive voice requires two extra words!

In normal writing, active voice is preferable to passive voice. In active voice, the stress is on who is 'acting' or 'doing'. This is especially so in cases where the actor needs to be mentioned. If you write in passive voice mentioning the actor, the sentence becomes verbose and slow, as it requires more words than the active form. Compare the following two sentences and decide which is appropriate.

- The sugarcane fields were destroyed by wild elephants.
- Wild elephants destroyed the sugarcane fields.

The message is clear; always use the active voice unless you have better explanations to use the passive voice. If you feel that the active voice is better for the context, you may convert the sentence in the passive to active form. For this, you have to locate the true subject, verb, and drop the form of *to be*. Using these details, convert the passive to active voice as shown below:

Passive: Nutritive quality of hybrid napier was studied by Antony and Thomas (2014).
Active: Antony and Thomas (2014) studied nutritive quality of hybrid napier.

The passive voice is useful to know what had happened when the actor cannot be identified as in the sentence: *The crop was damaged*. The passive voice is also useful to describe what was done when the actor need not be mentioned as in the 'Materials and Methods' section of a research paper. See the following sentences:

- The fertilizers were applied basally.
- The observations were made from the middle row.

The above sentences in passive form are acceptable, because what was done is more important in such cases rather than who did it. Similarly, the passive voice is proper in the following examples too:

- Shade tolerant weeds are found in coconut gardens.
- Wetlands should not be converted.
- About 2000 plants are used for food purposes.

In some cases, especially for writing definitions, passive voice is appropriate. Examine the following examples:

- The embryo is defined as the part of a flower that develops into another plant.
- The part of a flower that develops into another plant is called the embryo.
- The embryo is the part of a flower that develops into another plant.

Among the three, the first two set in the passive voice are better, although they require more words than the third sentence.

16.7 Be Definite About the Use of the 'Articles'

The adjectives '*a*', '*an*', and '*the*' are the articles. Among these, 'a' and 'an' are the indefinite articles, because they leave indefinite the person or thing spoken of (e.g. a book, meaning any book). The indefinite article 'a' precedes a word beginning with a sounded consonant, and 'an' precedes a word beginning with a vowel sound. 'The' is called the definite article, because it points out a definite person or thing (e.g. he read *the* book, meaning some particular book). 'The' has a 'definite' meaning. If your sentences are definite (in other words, if your reader knows exactly which ones you mean), you normally use 'the'. However, if you are talking about indefinite things, which your reader does not know anything about, you use the articles differently.

Writers in science are somewhat better in the use of indefinite articles. However, they may go wrong while using the definite articles. Typical mistakes are omission of 'the' when it is actually needed and using 'the' when it is not required. Some important rules governing the use of articles are listed below with examples. When in doubt, consult a good book on grammar.

Case 1. When we use a singular noun to talk about things in general, we do not use any articles. Examples:

- Light is essential for photosynthesis (*not* the light)
- Cardamom is grown in the high ranges (*not* the cardamom)

Case 2. The definite article *the* must be used when a singular noun is meant to represent a whole class.

- *The* cow is worshipped in India.
- Marconi invented *the* radio

Case 3. We do not use *the* before plural nouns when used in a general sense.

- Green manure plants provide good amount of green matter (green manure plants in general).
- Scientists are studying the big problem of climate change (scientists in general).

However, if the reference is to particular ones, then *the* must be used before plural nouns.

- *The* green manure plants in the farm are not ready for incorporation.
- *The* scientists who are working on the climate project are not happy.

Case 4. Use ‘a’ the first time you speak about something. The second time, third time, and so on, you must use ‘the’ as shown in the examples.

- We used *a* new method of seed scarification. *The* method was very useful.
- Multiple cropping is *a* system of intensification of cropping in time and space by growing two or more crops on the same field in a calendar year. *The* system is followed in most tropical countries.

Case 5. When writing about particular things or when there is only one of something, *the* must be used. Examples:

- *The* capital of India is New Delhi.
- Which fruit is called *the* king of fruit?

Case 6. Generally, no articles must be used with continents, countries, states, villages, panchayats, towns, streets and lakes. However, when the name of the country has a common name in it, or the name suggests that the country is made up of smaller units, *the* must be used. Examples:

Canada, India, Kerala, Thrissur, and Vellanikkara, but, *the* USA, *the* UAE, and *the* Netherlands (also *the* Hague).

Case 7. ‘*The*’ must be used with names of seas, rivers, oceans, island groups, mountain groups, areas, deserts, etc. Examples:

the Indian Ocean, *the* Himalayas, *the* Andamans, *the* Middle East, *the* Ganges, and *the* Sahara.

Case 8. Do not use ‘*the*’ for acronyms, which are sounded like words, but use ‘*the*’ when the letters of the acronym are spelt out or when the full name is used. Examples:

IRRI (pronounced ‘iree’), CADA (‘kada’), and ICRISAT (icri-sat), but the UN (‘the you-en’), the IMF (‘the ai-em-ef’), the International Rice Research Institute, and the India Agricultural Research Institute.

Case 9. Use ‘the’ before the names of municipal or government departments and before the names of shops, business houses, banks, foundation, etc. Examples:

the Ministry of Education, *the* Public Library, *the* Hindu, *the* Department of Agronomy, the Ford Motors, and the Rockefeller Foundation.

Please note the expressions: Kerala University, Calicut University, etc., but *the* University of Kerala, *the* University of Calicut, etc. Similarly, Agricultural College, Horticultural College, etc., but *the* College of Agriculture, *the* College of Horticulture, etc.

Case 10. Do not forget to put the definite article with superlatives. The only important exception is when ‘most’ is used to mean ‘very’ or ‘the majority of’.

Example: *The* highest yield was from plot A.
Example with most: That is most kind of you (means very kind of you).
Most birds can fly (means the majority of…).

16.8 Prepositions

Prepositions are used to mark a relationship between a noun, pronoun, or noun equivalent and another word. Choose the correct preposition and use it for your writing. Your task become difficult as the same preposition can have different uses. In the same way, different prepositions can have very similar uses (e.g. in the evening, on Monday evening, at night), and it is easy to make mistakes.

Prepositional idioms, prepositional verbs followed by a gerund, and prepositional adjectives followed by a gerund are considered here.

Prepositional Idioms

Many nouns, verbs, and adjectives are used with particular prepositions forming prepositional idioms. A prepositional idiom consists of a noun, verb, or adjective followed by a preposition forming an expression with an idiomatic meaning. Unlike the original word, the meaning of a prepositional idiom is largely determined by the preposition that comes after the word. In several instances, a single verb can give more than one meanings depending on the preposition that comes after it. Choose the correct preposition in such usages. Some examples of prepositional idioms are given in Table 16.2.

Examples with prepositional idioms:

- Delhi is *famous for* ancient monuments.
- The teacher *commented on* her presentation.

Table 16.2 Examples of prepositional idioms

Acquaint with	Adapt to	Addict to	Afflicted with
Agreement with	Alternative to	Analogous to	Apprise of
Attempt (n.) at	Attempt (v.) to	Averse to	Back down
Back away	Back out	Back up	Bear with
Break down	Bring about	Characteristic of	Comment on
Compliments of	Conducive to	Confine to	Conformity with
Congratulate on	Contemporary of	Contrasted with	Correlation with
Derive from	Different from	Divide into	Endowed with
Experienced in	Get across	Get around	Famous for
Fond of	Ignorant of	Implicit in	Infected with
Inference from	Infested with	Interested in	Involved in
Irrespective of	Keep at	Keep off	Keep back
Liable to	Make up	Make out	Marry to
Originate in	Pass off	Pass out	Proof of
Resemblance to	Respite from	Result of	Run across
Succeed in	Supplement to	Susceptible to	Teeming with
Theorize about	Tolerant of	Variance with	Want of

Table 16.3 Examples of prepositional verbs followed by a gerund

Addict to	Depend on	Prevent from
Aim at	Excel in	Prohibit from
Approve of	Give up	Put off
Assist in	Insist on	Refrain from
Be better of	Justify in	Rely on
Believe in	Keep on	Succeed in
Bend on	Look forward to	Think about
Count on	Object to	Think of
Despair of	Persist in	Worry about

- The weather is *conducive to* disease development in crops.

If you are not sure of prepositions, you may end up using wrong prepositions like 'with compliments *from*' or 'marriage *with*', especially in marriage invitation cards. If the invitee detects the error, it will be embarrassing for the host (proper usage is 'with the best compliments of' and 'marry to').

Prepositional Verbs Followed by a Gerund
Some verbs always combine with a preposition and show specific meaning. These 'prepositional verbs', must be followed by a *gerund* (A gerund is that form of the verb, which ends in—*ing*, and does the work of a noun). A short list of prepositional verbs, which are always followed by a gerund is given in Table 16.3.

Examples with prepositional verbs:

- Most post-modern intellectuals *insist on promoting* organic farming.
- Drought hit farmers *objected to increasing* the cost of seeds and fertilisers.
- Agricultural officers should not forget the *duty of helping* poor farmers.

Prepositional Adjectives Followed by a Gerund
Just as prepositional verbs as described above, there are certain prepositional adjectives too which are always followed by a gerund. Mistakes can occur while using prepositional adjectives. See Table 16.4 for a few examples of prepositional adjectives.

Examples with prepositional adjectives:

- The *practice of cutting* firewood for fuel is disastrous for the forests.
- The *custom of keeping* a few cattle is part of the homestead farming system.
- Rice farmers were *fortunate in getting* a hike in procurement prices.

Table 16.4 Examples of prepositional adjectives followed by a gerund

Accustomed to	Duty of	Necessity of
Afraid of	Excuse for	Power of
Averse to	Expert in	Practice of
Capable of	Fond of	Pretext for
Chance of	Fortunate in	Privilege of
Confident of	Humiliation of	Right in
Custom of	Intent on	Satisfaction in
Desirous of	Interested in	Successful in
Difficulty in	Knack of	Tired of

16.9 Relative Pronouns: Use the Most Appropriate

Pronouns are simple words that substitute for a noun or noun phrase, which would otherwise have to be repeated. Remember that pronoun is one among the eight parts of speech, which means *for-a-noun*. Normally, when speaking of oneself (the first person), the meaning of a pronoun such as 'I', 'me,' and 'we' is clear without stating the noun the pronoun represents. In the case of second person too, it is easy to identify the intention of using a pronoun such as 'you' or 'yours'. In contrast, when a pronoun represents a person or thing spoken about (the third person: he, she, it, they, etc.), then problems can arise to correctly identify the person or antecedent.

Certain pronouns in addition to acting as a third person pronoun similar to the above do another job of relating or joining words, phrases, or clauses together like conjunctions. Such pronouns are called *relative pronouns*. The referred noun in the sentence, which goes before the relative pronoun, is called its *antecedent*. Commonly used relative pronouns are:

- that (used for things)
- which (used for things)
- who (used for people)
- whom (used for people)
- whose (usually used for people to show possession)
- where (used for places)
- what (for things, meaning the thing(s) that).

When you use relative pronouns, make sure that the pronoun and the antecedent agree in all respects. Agreement problems may surface especially when the antecedent is unclear, vague, ambiguous, or even absent. Pronouns should also agree in number with their antecedents. The pronoun–antecedent agreement normally does not occur in the case of simple pronouns, as the antecedents are clear. If you know the subject well, sometimes you may be able to understand the meaning of vague antecedents.

Examine the sentence: 'Some plants on the northern side of the plot had few flowers, but they were plentiful on the southern side'.

In the sentence, 'they' might refer to plants or flowers. The problem is that even if you know the subject, you have to spend some time to identify the correct antecedent. You may recast the sentence to avoid ambiguity:

> Some plants on the northern side of the plot had few flowers, but on the southern side of the plot, all the plants had plentiful flowers.

In some cases, you may have used a relative pronoun but without an antecedent! See the sentence: 'If strict sanitation measures are not followed, the crop may be infested with several diseases, *which* ultimately reduces yield'. The antecedent for 'which' is not clear from the sentence; in fact, it is absent! The sentence may be revised without 'which' as: "If strict sanitation measures are not followed, the crop may be infested with several diseases ultimately reducing yield".

Restrictive and Non-restrictive Clauses with Pronouns

Some sentences may have clauses introduced with the help of relative pronouns (a clause is a group of words containing a subject and a verb). A *restrictive relative clause* is a clause containing essential information about the noun that comes before it. If you leave out the restrictive element in such a sentence, the entire meaning of the sentence is affected. Restrictive relative clauses can be introduced by the relative pronouns, *that*, *whose*, *who*, or *whom.*

See an example: *Scientists who publish research papers in peer-reviewed journals only are recognized.* The clause, '*who publish research papers in peer-reviewed journals*' restrict the kind of scientists you were talking about. Without that clause, the meaning of the sentence would change. Note that restrictive clauses or phrases are not separated off with commas.

On the other hand, non-restrictive clauses contain some more information that could be easily dropped without affecting the meaning or structure of the sentence. Often, non-restrictive clauses are introduced by the relative pronouns, *which, whose, who,* or *whom.* Please note that we cannot use 'that' to form a non-restrictive clause. The non-restrictive element provides additional information without restricting the overall meaning of the sentence. See the example: *My research guide, who had undergone overseas training, is a well-known scientist.* In the example, leaving out the clause '*who had undergone overseas training*' does not change the meaning of the resultant sentence, '*My research guide is a well-known scientist*'. A non-restrictive clause is usually surrounded by commas or in certain cases, parentheses.

The Relative Pronoun, Whose

The relative pronoun 'whose' is usually used for people to show possession, but sometimes also used for plants, animals, and even inanimate objects as in the sentence: *All the plants whose lower branches were removed grew taller*. By putting commas, the meanings of certain sentences with 'whose' can be made non-restrictive: *All the plants, whose lower branches were removed, grew taller*. The first sentence, '*All the plants whose lower branches were removed grew taller*', contains a restrictive clause explaining that only those plants whose lower branches were removed grew taller (there were many plants; the lower branches of only some plants were removed). On the contrary, the second sentence, '*All the plants, whose leaves had been removed, grew taller*' means that all the plants grew taller, and the clause surrounded by commas, *whose leaves had been removed*, gives extra information that all the plants had their leaves removed. See some other sentences where whose has been used in a restrictive sense:

- It was a small sleepy village *whose* importance I did not recognize when I visited it.
- When I was walking through the village, I met a farmer *whose* ingenuity attracted me.

Instead of 'whose', 'of which' can be used to refer to inanimate things, and it is preferred. See the example.

- It was a small sleepy village the importance *of which* was not recognized when I visited it.

When to Use 'Which' or 'That'

'That' and 'which' are two common relative pronouns used in technical writing. The distinction between 'that' and 'which' must be clearly understood for unambiguous expressions. Authors, who are not sure to use 'which' or 'that', have a tendency to use 'which' for everything in place of 'that'. It is wrong to use 'which' instead of 'that' in restrictive clauses as explained earlier. It appears that the choice was optional in the past, and at that time, 'that' and 'which' were used almost synonymously! The approach nowadays, however, is towards clear distinction between them. You should particularly note the differences in meaning of sentences when *that* or *which* is used. Use *that* with restrictive clauses without commas and *which* with non-restrictive clauses with commas. Consider the following two sentences:

- The hormone treatments that were applied before flowering were effective.
- The hormone treatments, which were applied before flowering, were effective.

The first sentence is restrictive, and the second sentence is non-restrictive. The first sentence implies that the treatments with hormones were applied in more than one way, but only those applied before flowering were effective. *That were applied before flowering* is a restrictive clause because it describes only a restricted part of the subject it modifies. In the second sentence, hormone treatments were applied in only one way, and a clause is added to indicate what that was. In this sentence, *which were applied before flowering* is a non-restrictive clause. This kind of clauses applies to the entire subject providing additional information. This is just extra information only, and even if you delete the central clause surrounded by the commas, and write: '*The hormone treatments were effective*', it will not affect the meaning of the sentence. See another example written in two forms:

- The rice crop that was transplanted in the last week of June was not affected for want of irrigation during its maturity stages.
- The rice crop, which was transplanted in the last week of June, was not affected for want of irrigation during its maturity stages.

From the first sentence with the restrictive clause, we could understand that the rice crop was transplanted at more than one instance, but only the crop transplanted in the last week of June was not affected for want of irrigation during its maturity stages. Probably, rainfall received during the critical stages was enough. In the second non-restrictive sentence, all the crop plants were transplanted only once, and the clause gives additional information about their time of transplantation (last week of June).

16.10 Transitional Words and Phrases

Transitional words and phrases act as connecting links between ideas, allowing the reader to move from one point to the next smoothly. Transitions improve the logical connections between thoughts. In writing, transitions are used with many objectives such as to generalize, to introduce an example, to add something, to compare or contrast something, to demonstrate cause and effect, to summarize, or to conclude. It is essential to understand how transitional words and phrases can be used to combine ideas in writing. You may find the use of transitions more frequently in reviews and discussions compared to other forms of technical writing. From among the many widely used transitions with almost synonymous meanings, selection of an appropriate one with less number of words is also an issue for researchers. Transitional words and phrases include conjunctive adverbs, coordinating conjunctions, and phrases. Some commonly used transitions as applied to scientific writing are discussed in this section.

1. **Contradiction/contrast**

Use of transitional words and phrases for contrasting or contradicting ideas is quite common in scientific writing. Some transitional words convey the presence of evidences to the contrary or point out alternatives, and thus introduce a change in the line of reasoning. Several words can be used synonymously to indicate the same meaning, but the sentence structure may vary according to the words or phrases chosen. See the list in Table 16.5.

Some of the most common transitional words and phrases used to indicate contrasting or contradictory points in theses and research articles are shown through examples below.

Although, Though, Even Though, But
Consider the sentence: *Although there are several types of water conservation measures, only a few are suitable for the humid tropics.* Compound sentences as

Table 16.5 Transitional words and phrases for contradiction/contrast

Though	Although	Even though	Otherwise
On the contrary	On the other hand	In contrast to comparison	In
Notwithstanding	Even so	Whereas	All the same
At the same time	In spite of	Despite	Instead
Conversely	Above all	In reality	In fact
Unlike	While	Albeit	After all
Yet	Still	Regardless	Rather than

shown in this example contain a minimum of two verbs and sometimes two subjects to express two contrasting ideas.

Compound sentences with contrasting ideas can be made in various ways. A common approach is to use *although*, *though*, *even though*, or *but* to connect two sentences. '*Although*' is similar to '*but*', which can be used at the start of sentences or between phrases. Although '*but*' and '*though*' are frequently used in spoken English or informal letters to start a sentence, in formal writing, these should not be used to begin a sentence. Examples:

- *Although* there are several types of water conservation measures, only a few are suitable for the humid tropics.
- There are several types of water conservation measures, *although* only a few are suitable for the humid tropics.
- *Even though* there are several types of water conservation measures, only a few are suitable for the humid tropics.
- There are several types of water conservation measures, *but* only a few are suitable for the humid tropics.
- There are several types of water conservation measures, *though* only a few are suitable for the humid tropics.

Please note the use of comma (,) in the middle of the compound sentence. '*Although*' can be replaced with '*though*' or '*even though*'. However, '*though*' and '*but*' should not be used to start a sentence in formal writing.

However, Nevertheless, Nonetheless, Even So, Still, Yet

In the above examples, the compound sentences could be split into two sentences, and begin the second sentence using '*however*', '*nevertheless*', '*nonetheless*', or '*yet*'.

- There are several types of water conservation measures. *However*, only a few are suitable for the humid tropics.
- There are several types of water conservation measures. *Nevertheless*, only a few are suitable for the humid tropics.
- There are several types of water conservation measures. *Nonetheless*, only a few are suitable for the humid tropics.
- There are several types of water conservation measures. *Even so*, only a few are suitable for the humid tropics.

- There are several types of water conservation measures. *Yet*, only a few are suitable for the humid tropics.

On the Other Hand, on the Contrary

While writing, do not confuse between '*on the other hand*' and '*on the contrary*'. We use the phrase '*on the other hand*' to express the other side of a question. At the same time, '*on the contrary*' is used to deny some earlier statement. '*On the other hand*' is used to present another point of view while agreeing to the statement in the preceding sentence, and '*On the contrary*' is used to express an alternate view. See the following examples:

- In the UK, 'watershed' is the term for the boundary or ridgeline of the catchment that separates water on either side. *On the contrary*, in the USA, 'watershed' means the entire catchment including the ridgeline.
- Readers hate abbreviations as they have to keep searching back through the document to know what each abbreviation means. *On the contrary*, authors love abbreviations, as they can avoid writing long phrases or terms in full.
- Command area development approach is pursued in irrigated areas for its comprehensive development. *On the other hand*, watershed area development approach is adopted for rainfed areas where availability of water is dependent on rainfall.

Whereas, Compared to, In Contrast to

Consider the sentence: *Herbaceous perennial crops like cardamom may not survive summer season without irrigation,* ***whereas*** *tree crops like rubber can tolerate drought.*

In the above sentence, '*whereas*' is used to compare two divergent statements. In this case, '*whereas*' is used to compare a disadvantage of herbaceous perennial crops to an advantage of tree crops. The above sentence could be written using '*compared to*' or '*in contrast to*' as shown below. However, when using '*compared to*' or '*in contrast to*', '*which*' should also be added.

- Herbaceous perennial crops like cardamom may not survive summer season without irrigation *compared to* tree crops like rubber, *which* can tolerate drought.
- Herbaceous perennial crops like cardamom may not survive summer season without irrigation *in contrast to* tree crops like rubber, *which* can tolerate drought.

In certain sentences with contrasting information, instead of using '*whereas*', '*but*' or '*however*' can be used as shown below:

- Herbaceous perennial crops like cardamom may not survive summer season without irrigation, *but* tree crops like rubber can tolerate drought.
- Herbaceous perennial crops like cardamom may not survive summer season without irrigation. *However*, tree crops like rubber can tolerate drought.

Please note that in the above sentences with '*but*' or '*however*', '*which*' is not required unlike '*compared to*' or '*in contrast to*'.

In Spite of, Despite

Examine the following two sentences.

- *In spite of* the requirement of additional food production, more and more farmers are deserting farming.
- *Despite* the requirement of additional food production, more and more farmers are deserting farming.

There is no difference between the sentences in meaning, but the second sentence is more common in a formal context. A major attraction with '*despite*' is reduction in number of words with the same intended meaning as that of '*in spite of*'.

The above sentences could also be written using '*but*' or '*although*', but some changes in the structure are necessary.

- The requirement of additional food production is great, *but* more and more farmers are deserting farming.
- *Although* the requirement of additional food production is great, more and more farmers are deserting farming.

See the following examples too wherein *in spite of*, *although* and *however* are used with the same meaning:

- *In spite of* the delay in the onset of monsoons, farmers are expecting good harvest.
- *Although* there was a delay in the onset of monsoons, farmers are expecting good harvest.
- There was a delay in the onset of monsoons. *However*, farmers are expecting good harvest.

The phrase, '*in spite of*' is used as the opposite of '*because of*'. See the examples, and note carefully the differences in meaning.

- *Because of* the recession, export of spices has declined (recession is the cause).
- *In spite of* the recession, stock prices improved (recession has no effect).

2. **Introducing examples**

For introducing an example, we may use certain transitional words or phrases (Table 16.6). These transitions are used to introduce examples as support, to indicate importance, or as illustrations so that an idea hit the reader.

Commonly used transitions for introducing examples are: *for example*, *for instance*, *such as*, *like*, and *including*.

Consider the example: *In the tropics, many cultivated grasses such as napier, guinea, para, and congo signal are grown.*

Note that '*such as*' used in the above sentence has the same meaning as '*for example*', '*for instance*', '*including*', and '*like*'. In written English, '*such as*' and '*including*' have the same usage, but '*like*' is more common in spoken English. The

Table 16.6 Transitional words and phrases for introducing examples

For example (e.g.)	For instance	Such as	Like
Including	Thus	To illustrate	To demonstrate
Especially	Specifically	Notably	Particularly
Markedly	Chiefly	Namely	In particular

phrase, '*for example*' and '*for instance*', are interchangeable, but *for example* is more common. It is also possible to begin a sentence with '*for example*' or '*for instance*' when it is followed by a complete idea.

When '*for example*' or '*for instance*' is used to list something, it is usually placed between a comma (,) and a colon (:). However, when a sentence begins with '*for example*', a comma is normally put after 'for example' to set it off from the rest of the sentence. However, when *including*, *such as* and *like* are used, commas or colons should not be used.

- In the tropics, many cultivated grasses are grown, *for example*: napier, guinea, para, and congo signal.
- In biological sciences, there is no clear distinction between theory and law. *For example*, it is still the 'theory' of evolution and not the 'law' of evolution, although it does not mean that biologists consider it is any less than a law.
- Cereals *like* wheat and rice form the major food of the masses.
- In the tropics, many cultivated grasses *including* napier, guinea, para, and congo signal are grown.
- Many spices are grown in Kerala *such as* pepper, cardamom, and ginger.

Note that '*such as*' and '*for example*' introduce a partial account only and not a complete one. More spices are grown in Kerala than the three mentioned in the example. When you provide the complete list, you cannot use '*such as*', '*including*', or '*for example*'.

3. **Adding an idea**

Some transitional words are used to add something to what has previously been indicated, to reinforce ideas, and to show continuation of an idea noted earlier (Table 16.7).

Among these, '*furthermore*', '*moreover*', *in addition*', and '*besides*' are frequently used in scientific writing especially in discussion part of a thesis or article. All these four transitions have similar meaning. Consider the sentence: *2,4D is a*

Table 16.7 Transitional words and phrases for adding an idea

Furthermore	Moreover	In addition	Besides
And	Also	Or	As well as
Likewise	Similarly	Too	Again

herbicide capable of controlling broadleaf weeds in grasses. Furthermore, it acts as a hormone at low doses.

In the above sentence, '*furthermore*' can be replaced with '*in addition*', '*moreover*', '*besides*', '*and*', or '*also*'. All the six words have exactly the same meaning and usage, and give extra information. They must be followed by a comma (,).

- 2, 4-D is a herbicide capable of controlling broadleaf weeds in grasses. *In addition*, it acts as a hormone at low doses.
- 2, 4-D is a herbicide capable of controlling broadleaf weeds in grasses. *Moreover*, it acts as a hormone at low doses.
- 2, 4-D is a herbicide capable of controlling broadleaf weeds in grasses. *Besides*, it acts as a hormone at low doses.
- 2, 4-D is a herbicide capable of controlling broadleaf weeds in grasses. *Also*, it acts as a hormone at low doses.
- 2, 4-D is a herbicide capable of controlling broadleaf weeds in grasses. *And*, it acts as a hormone at low doses.

Note that in formal writing, '*also*' and '*and*' should not be used to begin a sentence and consider revision of sentences with the above words. Only '*furthermore*', '*in addition*', '*moreover*' and '*besides*' are used to start sentences in formal writing.

4. **Consequence/effect**

These transitions state the consequences or effects from what is already written in the preceding sentence. The commonly used words and phrases to show consequence or effect are given in Table 16.8.

Because, Because of, Due to, As a Result of, Owing to, Since

All the above words or phrases are used to introduce a reason or a cause. However, they often cause confusion, leading to incorrect usage. Take the case of '*because*', a subordinating conjunction, which always indicates a cause. The important point to remember is the construction of sentences with noun phrases and verb phrases.

- This is *because of* limited supply of irrigation water (includes a noun phrase).
- This is *because* supply of irrigation water is limited (includes noun phrase plus verb phrase).

Table 16.8 Transitional words and phrases for stating consequences or effects

Because	Because of	Due to	As a result of
Owing to	Since	Thereby	Thus
Therefore	So	So that	For this reason
Accordingly	Consequently	Hence	For this purpose
Hence	Under the circumstances	Provided that	In case
As	For	While	Unless

Note that '*due to*', '*as a result of*,' or '*owing to*' can be used instead of '*because of*' (but not instead of '*because*'). For instance, "This is *due to* the paucity of funds".

There are subtle differences between '*due to*' and '*because of*'. In the example, '*The harvest was delayed because of rains*', '*because of*' is correctly used but '*due to*' is not appropriate. However, '*due to*' can be used as an adjective as in the sentence, '*The crop failure was due to drought*'. Although '*on account of*' can be used instead of '*because of*', avoid its use as the former has an additional word.

The transitional word, '*since*' may indicate either cause or time as in the sentences:

- *Since* you are not interested to join, I'll do the job myself (indicates cause).
- He has been ploughing *since* yesterday (indicates time).

Try to avoid using '*since*' in sentences where it could indicate either cause or time, and thus is ambiguous. For example, in the sentence: *Since we visited the place, we have been receiving frequent calls*, it is difficult to understand whether '*because we visited the place*' or '*from the time we visited the place*' is correct. In such sentences, if we are meaning cause, it is better to use '*because*' to avoid confusion.

Thereby and Thus

'*Thereby*' and '*thus*' are two transitional words often seen in scientific writing. In most cases, you can replace them with '*because*', '*consequently*', '*as a result of*', and '*for this reason*'. Consider the sentence:

- The diverted water from the canal was applied through drip irrigation system, *thereby* saving much water and improving water application efficiency.

In the above sentence, there is a cause and an effect connected by '*thereby*' with the meaning 'in this way'. The cause is, '*The diverted water from the canal was applied through drip irrigation system*' and the effect is, '*saving much water and improving water application efficiency*'. If you put the effect before the cause, the sentence can be constructed with '*because*'.

- There was much saving in water and improved water use efficiency, *because* the diverted water from the canal was applied through drip irrigation system.

'*Thus*' can also be used instead of '*thereby*' with some minor changes. Note that '*thus*' and '*thereby*' must be followed by a noun or gerund. Some more examples:

- Desertification continues unabated, *thereby* making huge chunks of land unproductive.
- Desertification continues unabated, *thus* huge chunks of land are made unproductive.
- Continued felling of trees is taking a heavy toll on our environment, *thereby* affecting the health of many hilly watersheds.

In the above sentences, you can use '*consequently*', '*as a result of*', '*for this reason*', '*because of*' as conjunctive adverbs in the same sense as '*thereby*' or '*thus*'.

- Desertification continues unabated, *consequently* huge chunks of land are made unproductive.
- *As a result of* unabated and continued desertification, huge chunks of land are made unproductive.
- *Because of* unabated and continued desertification, huge chunks of land are made unproductive.
- Desertification continues unabated, *for this reason*, huge chunks of land are made unproductive.

Unless and Otherwise

'*Unless*' and '*otherwise*' are used to express similar ideas meaning '*if... not*'. However, they are placed differently. See an example, where '*unless*' is appropriately used.

- Unless watershed management is done with proper planning and programmes, they will not yield the desired results.

  This sentence could also be written with '*otherwise*':

- Watershed management should be implemented with proper planning and programmes, *otherwise*, they will not yield the desired results.

Note that '*unless*' is used in the first phrase of the compound sentence, and '*otherwise*' is used in the second phrase of the compound sentence. When you use '*otherwise*' in the second phrase, the first phrase usually contains '*should*' in the sense 'If… not…' See an example with '*if …not…*' to express the same meaning.

- *If* watershed management is implemented without proper planning and programmes, it will *not* yield the desired results.

5. **Adding a comparison**

These transitional words or phrases help to introduce similarity to what has been stated already (Table 16.9).

'*Alternatively*', '*likewise*', and '*similarly*' are commonly used in writing. These expressions can be used to split compound sentences. Sometimes, you may feel that separate sentences are better instead of going for sentences with '*either …or…*' or 'neither…*nor…*' If the choices are written in positive simple sentences, the second sentence must start with '*alternatively.*' If the choices are written in negative simple sentences, the second sentence usually starts with '*likewise*', or '*similarly*'. If two

Table 16.9 Transitional words and phrases for adding a comparison

Alternatively	By the same reason	Similarly	Likewise
In similar fashion	Comparatively	Coupled with	Together with
In like manner	Correspondingly	Identically way	In the same way

choices are possible, '*both*' can be used. The following examples illustrate these points better.

- Rice seeds can be directly sown in the main field. *Alternatively,* they can be sown in a nursery and the seedlings transplanted.
- Digging is not an easy farm operation. *Likewise*, hand weeding is also not easy.
- Digging is not an easy farm operation. *Similarly,* hand weeding is also not easy.
- The results from *both* treatments are not encouraging.

The first example could also be written as a compound sentence with *either... or...*:

- Rice seeds can be *either* directly sown in the main field *or* sown in a nursery and the seedlings transplanted.

The second example could also be written as follows with *neither...nor...*:

- *Neither* digging the fields *nor* hand weeding is easy.

Note that '*either...or...*' is used in positive sentences to describe a choice between two or more possibilities, where only one choice is possible; whereas, in negative sentences, '*neither...nor...*' is used.

Sometimes, *whether...or* is more appropriate than *either ... or...* to combine sentences and form compound sentences. See the following three sentences.

- Keep the fields weed free until the critical period of weed competition is completed. This can be done by hand weeding. This can also be done by the use of herbicides.

These sentences can be combined into a compound sentence using *whether... or...*

- Keep the fields weed free until the critical period of weed competition is completed, *whether* by hand weeding *or* by the use of herbicides.

The strength of the sentence is improved; and in the process, you were able to reduce the number of words from 31 to 24.

In the above example, '*whether...or...*' is used to indicate that two conditions produce the same result. In this respect, '*whether...or...*' is similar to '*either...or...*' that is often used to construct compound sentences combining two phrases.

- Keep the fields weed free until the critical period of weed competition is completed, *either by* hand weeding *or* by the use of herbicides.

See another example with three sentences: '*The area under food crops is decreasing. This may be due to rapid urbanisation. It may also be due to low profitability margins*'.

The above three sentences could be combined and written as: '*The area under food crops is decreasing,* **whether** due to *rapid urbanisation* **or** *low profitability margins*'.

Table 16.10 Transitional words and phrases for establishing time

As soon as	In the meantime	In the first place	Immediately
Quickly	Finally	After	Later
Until	Since	Then	Before
Hence	Formerly	Suddenly	Shortly
Eventually	Meanwhile	During	Forthwith
Now	Instantly	Currently	Occasionally
At the moment	Nowadays	At present	Presently

6. **Establishing time**

These transitional words and phrases have the function of limiting, restricting, and defining time (Table 16.10).

Many of these transitions can be used synonymously. For example, '*now*', '*nowadays*', '*presently*', '*at present*', '*at the moment*', and '*currently*' indicate the same meaning. See the sentence below:

- Library catalogues come in many shapes and sizes, usually on cards or in sheaves. *Presently*, in most of the libraries, it is in digital form.

In the above sentence, '*presently*' can be replaced with '*now*', '*nowadays*', '*at present*', '*at the moment*', and '*currently*'. Note how '*nowadays*' is written. Earlier it was '*now a days*', but now the accepted usage is '*nowadays*'.

7. **Generalizing**

Certain transitional words and phrases are used to generalize statements (Table 16.11). Generalizing transitions are frequently seen in textbooks and introduction part of research articles.

Usually, *often*, *generally*, and *normally* are the more common ones, which are almost synonymous in meaning. See the examples below.

- A brook is a small stream, *usually* a first order stream.
- A brook is a small stream, *often* a first order stream.
- A brook is a small stream, *generally* a first order stream.
- A brook is a small stream, *normally* a first order stream.

Table 16.11 Transitional words and phrases for generalizing statements

Often	Usually	Normally	Generally
Generally speaking	Ordinarily	As a rule	As usual
Broadly speaking	On the whole	As usual	In general

Table 16.12 Transitional words and phrases for qualifying statements

Perhaps	Probably	Possibly	Never
Always	Frequently	Nearly	May be
Although	Almost	For the most part	Although

8. **Qualification**

Certain words and phrases are useful to qualify another point to the preceding statement (Table 16.12).

See the following examples in which *probably*, *possibly*, and *perhaps* are synonymously used.

- A creek is a slightly bigger brook, *perhaps* a second order stream.
- A creek is a slightly bigger brook, *probably* a second order stream.
- A creek is a slightly bigger brook, *possibly* a second order stream.

Bibliography

AMA [American Medical Association] 2007. AMA Manual of Style: A Guide for Authors and Editors (10th Ed.). Oxford University Press, New York, 1024p.

Antony, S. and Thomas, C.G. 2014. Nutritive quality of hybrid napier cultivars grown under rainfed ecosystem. J. Trop. Agric. 52 (1): 90–93.

APA [American Psychological Association] 2013. Publication Manual of the American Psychological Association (6th Ed). American Psychological Association, Washington DC, 272p.

Axelrod, R.B. and Cooper, C.R. 1997. The St. Martin's Guide to Writing (5th Ed.). St. Martin's Press, New York, 791p.

Barrass, R. 2002. Scientists Must Write: A Guide to Better Writing for Scientists, Engineers and Students (2nd Ed.). Routledge, London, 204p.

CSE [Council of Science Editors]. 2014. Scientific Style and Format: The CSE Manual for Authors, Editors, and Publishers (8th Ed.): The University of Chicago Press, 722p.

Day, R.A. 1983. How to Write and Publish a Scientific Paper (2nd Ed.; Reprint, 1997). Vikas Publishing House, New Delhi, 160p.

Day, R.A. and Gastel, B. 2006. How to Write and Publish a Scientific Paper (6th Ed.). Cambridge University Press, Cambridge, 320p.

Glasman-Deal, H. 2010. Science Research Writing for Non-Native Speakers of English. Imperial College Press, London, 257p.

Gustavii, B. 2008. How to Write and Illustrate a Scientific Paper (2nd Ed.). Cambridge University press, Cambridge, 168p.

Huckin, T.N. and Olsen, L.A. 1983. English for Science and Technology: A Handbook for Non-native Speakers. McGraw–Hill Book Company, 549p.

Leggett, G., Mead, C.D., Kramer, M.G., and Beal, R.S. 1986. Handbook for Writers. Prentice- Hall, New Jersey, 558p.

Malmfors, B., Garnsworthy, P. and Grossman, M. 2004. Writing and Presenting Scientific Papers (2nd Ed; Indian Edition, 2011). Viva Books, New Delhi, 153p.

McCaskill, M.K. 1990. Grammar, Punctuation, and Capitalization: A handbook for Technical Writers and Editors. NASA SP-7084, National Aeronautics and Space Administration, Washington DC, 106p. Available: https://ntrs.nasa.gov/archive/nasa/casi.ntrs.nasa.gov/19900017394_1990017394.pdf [Accessed 10 Dec. 2003].

MLA [Modern Language Association of America] 2016. MLA Handbook for Writers of Research Papers (8th Ed.) Modern Language Association of America, 146p.

Mountford, A. 1977. English in Focus: English in Agriculture. Oxford University Press, 156p.

Rubens, P. 2004. Science and Technical writing: A Manual of Style (2nd Ed.). Routledge, New York, 427p.

Swan, M. 1983. Practical English Usage. The English Language Book Society and Oxford University Press (639 entries).

Turabian, K.L. 2007. A Manual for Writers of Term Papers, Theses, and Dissertations (9th Ed.). The University of Chicago Press, 461p.

UChicago [University of Chicago] 2017. Chicago Manual of Style [CMOS]. University of Chicago Press, 1144p.

Whimster, W. F. 1985. What the critical reader looks for in an original article: A guide for writers. In: Hawkins, C. and Sorgi, M. (ed.), Research: How to Plan, Speak and Write about It (4th Indian Reprint, 1993). Narosa Publishing House, New Delhi, pp. 85–109.

Wren, P.C. and Martin, H. 2005. High School English Grammar and Composition (Revised, Prasada Rao, N.D.V.), S. Chand & Co, New Delhi, 416p.

Chapter 17
Use Appropriate Words and Phrases

Better than a thousand hollow words is one word that brings peace.
Lord Buddha

In scientific writing, in addition to grammatical mistakes, chances of wrong usages of words and phrases are many. In this chapter, some strategies to cut down words and improve the overall writing style have been included followed by lists of singular and plural words, commonly confused words, unnecessary use of several words instead of a single word, unnecessary qualification of words, unnecessary phrases, and typical clichés (stale idiomatic expressions) along with alternatives compiled from various sources (e.g., Hawkins and Sorgi 1985; Barras 2002; Malmfors et al. 2004; Day and Gastel 2006; Mathews and Mathews 2007). In most cases, the alternatives are helpful to reduce the number of words. Always prefer the alternatives suggested in the right column.

17.1 Economy in Words

Serious writers employ several methods to reduce the number of words in sentences. This is especially important in scientific writing where brevity is valued. Most journals restrict the maximum number of words for a journal article and abstract. Some common strategies to cut down words and improve the overall style are included in this section.

Avoid Nouns Derived from Verbs as Subjects

There are several nouns in English derived from verbs. Although these nouns can be useful in many situations, avoid using them as subjects in technical writing, as the sentence becomes long, and sometimes difficult to understand. When such nouns from verbs are used as subjects of sentences, they combine with additional verbs such as *occur*, *take place*, *make*, *perform*, and *result*. Go through the following examples,

C. G. Thomas, *Research Methodology and Scientific Writing*,
https://doi.org/10.1007/978-3-030-64865-7_17

and find out how many extra words are required when certain verbs have been changed to nouns and used as subjects. Learn to improve similar sentences as indicated.

Written: You may *take into consideration* the prevailing prices (8 words).
Improved: You may *consider* the prevailing prices (6 words).
Written: *Development* of flowers *occurs* under cool weather (7 words).
Improved: Flowers develop under cool weather (5 words).
Written: An *increase* in nutrient uptake *took place* (7 words).
Improved: Nutrient uptake *increased* (3 words).

Some routine use of nouns derived from verbs that you must avoid in scientific writing are: *make an adjustment* (adjust), *make an assumption* (assume), *reach a conclusion* (conclude), *make a selection* (select), *perform an analysis* (analyse), *take into consideration* (consider) and *conduct an investigation* (investigate).

Avoid Adjectives Derived from Verbs

Adjective means 'added to'. An *adjective* is a word used with a noun to add something to its meaning. Adjectives are generally formed from nouns, verbs, and other adjectives. When adjectives derived from verbs are used in writing, some precautions are needed. Verbs lose their effectiveness when needlessly converted to adjectives. Moreover, extra verbs such as *produce*, *result*, and *exert* are needed for such sentences. Readability also suffers. If required, improve the sentence by converting the verb-derived adjective back to the original verb.

In the following examples, verbs were made into adjectives. These sentences can be improved by converting them into verbs again. By doing so, we can also reduce the number of words in the sentences.

Written: The treatment produced a *stimulatory* effect on the crop (9 words).
Improved: The treatment *stimulated* the crop (5 words).
Written: The herbicide was *injurious* to rice seedlings (8 words).
Improved: The herbicide *injured* rice seedlings (6 words).
Written: The hormone was *inhibitory* to cell division (7 words).
Improved: The hormone *inhibited* cell division (5 words).

Avoid 'Result in'

Most researchers and students have a tendency to use the phrase '*result in*' repeatedly without actually thinking about the vagueness or ambiguity it creates. This usage indicates that it links a cause and effect. Something happens whereby the effect is the result of the cause. In fact, the term does not mean this and causes only vagueness, ambiguity, and wordiness in a thesis or paper. In most science literature, '*to result in*' is used as an unnecessary verb whose object is a noun made from a verb. Try to avoid its use, and by doing so you can also economize on words. Consider the following sentence:

- If done properly, mulching can *result in* the *suppression of* weeds (11 words).

This sentence can be improved by removing *result in* as follows:

- If done properly, mulching can *suppress* weeds (7 words).

Use the improved version, as the force of the sentence is increased by changing the noun 'suppression' to the verb 'suppress'. In the process, the number of words is also reduced from 11 to 7. Another example with an improved version is given:

- *Adoption of* water conservation measures *resulted in* higher yield (9 words).
- Water conservation measures *caused* higher yield (6 words).

In the example, 'adoption of' and 'result in' are redundant. By eliminating unnecessary words, substantial reduction of words is achieved without forfeiting information.

Nouns as Adjectives

As mentioned earlier, the role of adjectives is to modify nouns, thus adding something to its meaning. Often, they are single words like '*live* specimen'. Sometimes, a preposition is used to link two nouns, and the resultant prepositional phrase acts as an adjective, for example, '*tree of* mango'. However, the second noun can be used directly as an adjective without the use of a preposition. For example, 'mango tree' means the same as 'tree of mango'. You can use nouns as adjectives like this to reduce the number of words without affecting the proposed meaning. However, you must make sure that the proposed meaning is clear eliminating vagueness, if any.

The adjectives and nouns are combined into compound words in different ways. Certain words have a space between the words such as living room, post office, middle class, interest rate, root hairs, pesticide residue, full moon, half-sister, real estate, application rate, and associate professor. These are *open compound words.* Once an open compound word is formed, it sometimes transforms over time. For example, *keyword* was originally *key word.* These are *closed compound words.* They do not have a space between them, and examples include: second hand, makeup, farmhouse, textbook, headache, sunflower, railroad, earthquake, downstream, and grasshopper.

Whenever a noun becomes an adjective, a preposition is eliminated, but the resultant noun should show its original meaning. However, uncertainty may occur when more than one preposition could have involved. The meanings of each of the nouns used as an adjective are clear and their uses are proper in the following examples: *interest rate* (the rate of interest), *fertilizer application* (the application of fertiliser), *fungus disease* (a disease caused by fungus), *root hairs* (the hairs on the roots), and *pesticide residue* (the residue of pesticide). However, this may not be the case always. Certain compound words may have at least two possible meanings for each noun used as an adjective. In such cases, make clear what was really meant. For example, the compound word, *weed removal* has two meanings; 'removal of the weeds' (from lands) or 'removal by the weeds' (of water, nutrients, etc.). Similarly, *soil application* can be 'application in the soil' or 'application of the soil'. In such cases of ambiguity, amend the sentences so that the intended meaning is clear.

Do not add on several noun adjectives one after the other, as they may create problems in understanding. Probably, words that can act as adjectives only can be used in a sequence without losing clarity. For example, when you write, 'the dwarf, high yielding, non-lodging rice', it is clear that all the adjectives modify 'rice'. At the same time, when a long sequence of nouns modifies the noun at the end of the sequence, it may confuse you. For example, it is difficult to understand the meaning of 'cucumber mosaic virus resistance mechanism'. For clarity, revise the sentence by including some prepositions as 'resistance mechanism *for* cucumber mosaic virus'.

Inserting Phrases Using Punctuation Marks

Look at the sentence: '*Small farmers, who are resource poor, cultivate as many crops as possible in their homesteads*'.

In the example, the phrase '*who are resource poor*' is an inserted phrase. In fact, we are combining sentences by inserting information as a part of compound sentences. It is grammatically a '*non-restrictive appositive*' (When one noun or phrase follows another to describe it, the noun or phrase that follows is said to be in *apposition,* and such noun or phrase is the *appositive.* Apposition means *placing near*). It gives additional information. If you take out the inserted phrase, the sentence still has meaning and is still grammatically correct as in: *Small farmers cultivate as many crops as possible in their homesteads.*

Inserted phrases with 'which', 'whose', and 'who' have already been discussed in Sect. 16.9. Inserted phrases must begin and end with commas (,…,), em dashes (—…—) or parentheses (.….) as shown in the following examples.

- My daughter, a scientist by profession, wants to study martial arts.
- My daughter—a scientist by profession—wants to study martial arts.
- My daughter (a scientist by profession) wants to study martial arts.

Among the three, the most commonly used are commas. However, the choice of commas, dashes, or parentheses depends on the relation of the phrase to the rest of the sentence and on the required emphasis. The following rules of thumb may be used.

- When a slight division in thought from the rest of the sentence is meant, commas may be used.
- When the middle phrase contains internal commas, dashes are appropriate to emphasize the element enclosed and clarify its meaning.
- When parentheses are used to enclose phrase, it means that the phrase is only loosely connected to the rest of the sentence.

17.2 Singular and Plural Words

Generally, the plural of nouns is formed by adding 's' or '*es*' to the singular. In the case of some nouns, the plurals are formed by changing the inside vowel of the singular as in *man* and *men*, *foot* and *feet*, or *mouse* and *mice.* For some nouns, the singular and the plural forms are alike. Certain nouns do not follow any of these conventions. Some frequently used words that create difficulties in scientific writing in the use of singular and plural forms are listed below.

Singular	Plural
Alga	Algae
Analysis	Analyses
Antenna	Antennae
Appendix	Appendices (appendixes)
Axis	Axes
Bacterium	Bacteria
Basis	Bases
Brother	Brothers (sons of the same parent)
	Brethren (members of a society)
Crisis	Crises
Criterion	Criteria
Curriculum	Curricula
Datum	Data
Equipment	Equipment
Erratum	Errata
Focus	Foci or focuses
Formula	Formulae (scientific usage)
	Formulas (slang)
Fungus	Fungi
Furniture	Furniture
Genius	Genii (spirits)
	Geniuses (persons of great talents)
Genus	Genera
Glassware	Glassware
Head of the Department	Heads of the Department
Hypothesis	Hypotheses
Index	Indices (signs used in algebra)
	Indexes (tabular contents given in a book at the end)
Information	Information
Locus	Loci
Matrix	Matrices (matrixes) in a book
Medium	Media (mediums)
Mouse	Mice
Nucleus	Nuclei
Ovum	Ova

(continued)

(continued)

Singular	Plural
Parenthesis	Parentheses
People[a]	Peoples[b]
Phenomenon	Phenomena
Radius	Radii
Serum	Sera
Sheep	Sheep
Species	Species
Spectrum	Spectra
Stimulus	Stimuli
Stomata	Stoma
Stratum	Strata
Swine	Swine
Syllabus	Syllabi
Thesis	Theses
Vertebra	Vertebrae

[a]Certain collective nouns such as people, poultry, and cattle although singular in form are always used as plurals (e.g. 'These cattle are mine').
[b]The plural form 'peoples' is sometimes used to mean a body of persons that are united by a common culture, for example, peoples organization.

17.3 Commonly Confused Words

Some words are often confused, and you may wonder which one is correct. Many words in English appear or sound alike but have different meanings. The most common pairs of words that frequently cause problems in scientific writing are listed below as a quick-reference guide. Try to use these words in a proper way after understanding their correct spellings and meanings.

access/assess

access: right to use, right of entry
assess: appraise; make a judgment

advice/advise

advice (noun): recommendation about what to do; guidance
advise (verb): to recommend something; to give an opinion

affect/effect

affect: to have an effect on; to influence

effect: to bring about a result; outcome

affluent/effluent

affluent: wealthy
effluent: something that flows out, usually waste matter

all together/altogether

all together: all in one place; all at once
altogether: completely; in total

alternate/alternately/alternatively

alternate: to perform by turns; interchange
alternately: first one thing then another in a repeated fashion
alternatively: one thing as an alternative to another; otherwise; on the other hand

among/between

among: in the middle of (used with more than two choices)
between: in the middle of (used with only two choices)

amount/number

amount: a sum, indicates something that can be measured
number: a quantity, indicates something that can be counted

appraise/apprise

appraise: to assess
apprise: to inform someone

assent/ascent

assent: agree; agreement; approval
ascent: climb; rise

bare/bear

bare: to uncover; naked
bear: to carry; to tolerate

berth/birth

berth: a bunk in a train or ship
birth: the emergence of a baby from the womb; beginning; origin

born/borne

born: natural; having started life
borne: carried (past tense of bear)

brake/break

brake: a device for stopping a vehicle; to stop a vehicle
break: to separate into pieces; a pause

canvas/canvass

canvas: a type of strong cloth
canvass: to seek people's votes

cite/sight/site

cite (verb): quote; refer to
sight (noun): vision; view; prospect
site (noun): location; position

chord/cord

chord: harmony, musical tones
cord: part of a string, twine or, rope

climactic/climatic

climactic: forming a climax as in plant succession
climatic: something related to climate

compared with/compared to

compared with: used in the analysis of similarities and differences
compared to: used to liken one thing to other as in a simile

complement/compliment

complement (verb): to go together; an addition that improves something
compliment (verb): to praise or express approval; an admiring comment

complementary/complimentary

complementary (adjective): balancing; serving to complete
complimentary (adjective): admiring; free of charge

continuous/continual

continuous: non-stop, incessant
continual: repeatedly; frequently

council/counsel

council: a group of people who manage or advise
counsel: to advise; to guide

decrease/reduce

decrease: lessen in number
reduce: lessen in amount

defuse/diffuse

defuse: to make a situation less tense
diffuse: to spread over a wide area

dependant/dependent

dependent (adjective): rely on someone or something
dependant (noun): a person who rely on somebody for support

discreet/discrete

discreet: prudent; careful not to attract attention
discrete: separate and distinct

disinterested/uninterested

disinterested: impartial
uninterested: not interested

dose/dosage

dose: a dose is a measure of something to be administered at one time
dosage: the regulated administration of doses, usually as quantity per unit of time

ensure/insure

ensure: to make sure; guarantee
insure: indemnify; underwrite; to give compensation for death or damage

envelop/envelope

envelop: encircle; surround
envelope: a paper cover or packet usually to put a letter

farther/further

farther: beyond, used with distance
further: additional, used with time or quantity; supplementary

foreword/forward

foreword: an introduction to a book
forward: onwards; ahead

fewer/less

fewer: a smaller number, used with objects that can be counted by number
less: a smaller amount, used with objects that can be measured by amount

imply/infer

imply: to suggest or indicate without saying it directly
infer: to assume or deduce from something

infect/infest

infect: some organisms such as bacteria *infect* to produce an *infection*
infest: some external organisms such as fleas *infest* and produce an *infestation*

lay/lie

lay: to place or put something down
ie: to recline, to lie down

lead/led

lead: a soft metal (noun); to guide (verb)
led: past participle of lead

loose/lose

loose: to unfasten; to set free
lose: to be deprived of; to be unable to find

majority/most

majority: greater part; bulk; the excess of one number over another (e.g. in an election)
most: nearly all

many/much

many: great in number (countable)
much: great in amount (measurable)

method/methodology

method: how to perform a task, a technique
methodology: the study of methods, a tactic

meter/metre

meter: a measuring device
metre: a metric unit; rhythm in verse

objective/rationale

objective: purpose; goal
rationale: underlying principle; justification

practice/practise

practice (noun): the use of an idea or method; the work or business of a lawyer, medical practitioner, dentist, etc.
practise (verb): to do something regularly; to do something repeatedly to gain skill

precede/proceed/proceeds

precede: to go before
proceed: to continue, to keep on

proceeds: income; earnings

prescribe/proscribe

prescribe: to order authoritatively; to authorize use of medicine
proscribe: to officially forbid something; disallow

principal/principle

principal: the most important; the head of a school or college
principle: a fundamental rule or belief

quite/quiet

quite: somewhat; rather
quiet: calm; not noisy

regime/regimen

regime: establishment; management; political organization
regimen: rules followed

respectively/respectfully

respectively: in that order; correspondingly
respectfully: reverentially; showing politeness

sceptic (skeptic)/septic

sceptic: a person inclined to doubt
septic: infected with bacteria

stationary/stationery

stationary: static; not moving
stationery: writing materials

storey/story

storey: a level of a building
story: a tale or account

unique/unusual

unique: only one of its kind; without equal; sole
unusual: strange; not common

varying/various

varying: changing
various: means of several kinds

verbal/oral

verbal: using words; unwritten
oral: spoken; by word of mouth; relating to the mouth

while/whilst/whereas

while: at the same time as
whilst: even as; although
whereas: in contrast to; on the contrary.

17.4 Avoid Using More Words Instead of a Single Word

In most scientific reports such as thesis or research paper, word limits are usually prescribed. Develop the habit of using short single words instead of several words to mean the same thing. This not only saves space but also readability. You can also reduce printing charges!

Avoid	Prefer
A considerable number of	Many
A decreased number of	Fewer
A greater length of time	Longer
A large number of	Many
A large proportion of	Much
A majority of	Most
A number of	Many, several
A sufficient number of	Enough
Adjacent to	Near
An adequate amount of	Enough
An increased amount of	More
Are found to be in agreement	Agree
Are of the same opinion	Agree
As a consequence of	Because
As of this date	Today
At an early date	Soon
At the present time	Now
At this point in time	Now
A variety of	Various
Based on the fact that	Because
Be of the same opinion	Agree
Bring about	Cause
By means of	By, with
Come to the same conclusion	Agree
Decreased in length	Shortened
Decreased in width	Narrower

(continued)

(continued)

Avoid	Prefer
Despite the fact that	Although
Due to the fact that	Because
For the most part	Mainly
For the reason that	Because
From the point of view of	Because
Give an account of	Describe
Give rise to	Cause
Has a tendency to	May
Has the capability of	Can
Having regard to	About
In a large number of cases	Often
In a number of cases	Some
In a position to	Can, may
In as much as	For, as
In close proximity	Near
In connection with	About, concerning
In considerable quantities	Abundant
In few cases	Rarely
In most cases	Usually
In some instances	Sometimes
In terms of	About
In the absence of	Without
In the affirmative	Yes
In the event that	If
In the order of	About
In the possession of	Has, have
In the vicinity of	Near
In view of	Because, since
Increased in length	Lengthened
Increased in size	Enlarged
Is aware of	Knows
Is characterized by	Shows
It is doubtful that	Possibly
It would appear that	Apparently
Not in accordance with the facts	False
Of considerable size	Large
Of long standing	Old

(continued)

(continued)

Avoid	Prefer
On behalf of	For
On no occasion	Never
On numerous occasions	Often
On one occasion	Once
On the basis of	By
On the grounds that	Because
On two occasions	Twice
Owing to the fact that	Because, as
Presents a picture similar to	Resembles
Quite a large quantity of	Much
Referred to as	Called
Subsequent to	After
Sufficient number of'	Enough
The great majority of	Most
The vast majority of	Most
Through the use of	By, with
Unanimity of opinion	Agreement
Undergo transformation	Change
Was of the opinion that	Believed
With reference to	About
With regard to	Concerning, about
With respect to	About
With the possible exception of	Except
With the result that	So that

17.5 Avoid Unnecessary Qualification of Words

Tautology is defined as needless repetition of an idea in a different word, phrase, or sentence. A tautological statement uses more than one word to say the same thing as in 'definitely proved'. If a part of the meaning is repeated and appears as unintentional or awkward, then it is a case of tautology, unnecessary qualification of words. However, a repetition of meaning for improving the style of writing may not necessarily be tautological. Wordiness can also occur, when absolute words are mistakenly modified. For example, there is no difference between 'absolutely perfect' and 'perfect'. Many tautologies are unintentional, but often hinder reader comprehension and weaken the writer's credibility.

Avoid	Prefer
About 5–6 samples	5–6 samples
Absolutely essential	Essential
Absolutely perfect	Perfect
Actual number	Number
Aggressive in character	Aggressive
Almost imperfect	Imperfect
Almost unique	Unique
Analyses were made	Analysed
Are found to be in agreement	Agree
At a later date	Later
At a rapid rate	Rapidly
Categorically denied	Denied
Causal factor	Cause
Circular in outline	Circular
Close proximity	Proximity (or write *near*)
Come to the conclusion	Conclude
Completely full	Full
Completely surrounded	Surrounded
Conclusive proof	Proof
Cylindrical in shape	Cylindrical
Definitely proved	Proved
Deliberately chosen	Chosen
Disappear from sight	Disappear
During the course of	During, while
Each and every	Each
Enclosed herewith	Enclosed
End result	Result
Eradicate completely	Eradicate
Every individual one	Everyone
Fewer in number	Fewer
First of all	First
For the purpose of	For
Give consideration to	Consider
Give positive encouragement	Encourage
Green in colour	Green
Grouped together	Grouped
Hard evidence	Evidence
Has knowledge of	Knows

(continued)

(continued)

Avoid	Prefer
In a satisfactory manner	Satisfactorily
In order to	To
In the range of 10 to 15	From 10 to 15
In two equal halves	In halves
Is indicative of	Indicates
Is suggestive of	Suggests
It is apparent that	Apparently
It is clear that	Clearly
It is often the case that	Often
Large in size	Large
Make adjustments to	Adjust
Make an examination of	Examine
Not actually true	Not true
Of a mild nature	Mild
Of common occurrence	Common
Of the chronic type	Chronic
On a regular basis	Regularly
Percolate down	Percolate
Pooled together	Pooled
Quite impossible	Impossible
Quite obvious	Obvious
Quite unique	Unique
Rather interesting	Interesting
Real problems	Problems
Really dangerous	Dangerous
Related to each other	Related
Reverted back	Reverted
Shabby in appearance	Shabby
Small in size	Small
Smaller in size	Smaller
Stunted growth	Stunted
Superimposed over each other	Superimposed
Take into consideration	Consider
Take measurements of	Measure
Very relevant	Relevant
Ways and means	Ways, means (*never use both*)
Whether or not	Whether

(continued)

(continued)

Avoid	Prefer
Wholly new	New
With a view to	To
With the exception of	Except

17.6 Remove Unnecessary Phrases

It is common to use many words and phrases just as ‘fillers’ rather than for its actual meaning in spoken English. They do not have much meaning than simply clearing one’s throat or buying some more time. Probably, neither the speaker nor the audience is even aware of them unless they exceed the limit. Nevertheless, writing in the same way as speaking may invite troubles. Some authors use unnecessary qualifying words, which do not have any specific meaning. Similarly, certain vague introductory statements may ruin some sentences or even paragraphs.

In scientific writing, where each word matters, take care to remove all unnecessary and meaningless phrases and ‘fillers’ for clarity and brevity. For example, most ‘it... that…’ phrases are unnecessary. Although most of us know that the phrases, ‘it must be recalled that…’ or ‘it is interesting to note that…’ are pointless fillers, we commit the errors repeatedly! When equivalent alternatives exist, choose the best one that takes the least space. Nevertheless, when doing with these kinds of editing, remember that clarity is more important than brevity. The following lists contain several of such unnecessary expressions.

Short Phrases Instead of Long Phrases

Avoid	Prefer
30 s in duration	30 s
30 cm in breadth	30 cm broad
30 m in length	30 m long
A relationship to	Related to
A small number of	A few
Adverse climatic conditions	Bad weather
An excessive amount of	Too much
An inadequate amount of	Too little
As a matter of fact	In fact
As is the case	As happens
At a distance from	Away from
Cases of short duration	Short cases

(continued)

(continued)

Avoid	Prefer
Commonly occurs	Is common
Created the possibility	Made possible
Diversity of opinion	Many views
During the month of January	During January
For a further period of 10 years	Coming 10 years
Has been engaged in a study of	Has studied
Have the appearance of	Look like
I am in the process of	I am
In a very real sense	In a sense
In almost all instances	Nearly always
In black and white	clearly
It consists essentially of …	It has…
It has been reported by Gill	Gill reported
It is generally believed. …	Many think
It is possible that the cause is…	The cause may be
Of constant occurrence	Always
On a regular basis	Regularly
On a temporary basis	As temporary
On a theoretical level	Theoretically, in theory
On account of	Because of
On a practical level	Practically, in practice
Over the period of a decade	Over a decade
The general opinion is	Many think
The majority of authors	Most authors
The present author believes	I think
There is no reason to believe that	I think
This result would seem to indicate	This result indicates
We are in the process of making	We are making

Needless Expressions, Which Can Be Dropped

For the record…
I'd be more than happy…
I have no hesitation in saying…
In the case when…
It has been proposed that…
It has long been known that…
It is a well-known fact that…

It is clear that much additional work will be required …
It is interesting to note that…
It is recognized that …
It is the purpose of this study…
It might be stated that …
It must be recalled that…
It should be emphasized that…
Needless to emphasize…
Needless to say…
Not many studies on this aspect have been reported…
The fact of the matter is…
There are relatively few studies reported …
Without any hesitation, I can state that …

17.7 Avoid Clichés

You should also try to avoid stale idiomatic expressions or clichés and stereotyped phrases. A *cliché* is an expression that has been worn out by constant use. Many clichés are examples of figurative language, which have lost their influence as we hear them too often. As clichés have lost their vigour, they will reduce the effectiveness of our writing. *Jargons* also come under this category. Jargons are words or expressions used by a particular group of people but difficult for others to understand. Although you may not be able to avoid technical jargons altogether, try to avoid using too many jargons.

Some idiomatic expressions, which lost their punch and became clichés or stereotyped are given in the list. Try to use alternate expressions to make the sentence effective.

Clichés	Alternate expression
A different ball game	Another matter
Break new ground	Start something new
Burning the midnight oil	Study or work hard
Facts and figures	Data, facts, or figures
Finishing touches	Final changes, final additions
In black and white	Clearly
In the pipeline	In preparation
Last but no means least	Lastly, *or simply omit the expression*
Last but not the least	Lastly, *or simply omit the expression*
Leave no stone turned	Make every effort
Paradigm shift	Change in pattern
Raving beauties	Ravishing or unusually attractive beauties

(continued)

(continued)

Clichés	Alternate expression
See the light at the end of the tunnel	Making progress
Sweeping statement	Over-generalization
Take on board	Note
Throw more light	Provide more information
Time and again	Repeatedly
Time frame	Period
Trump card	A final recourse
Vested interests	An action for selfish ends

Bibliography

Axelrod, R.B. and Cooper, C.R. 1997. The St. Martin's Guide to Writing (5th Ed.). St. Martin's Press, New York, 791p.

Barrass, R. 2002. Scientists Must Write: A Guide to Better Writing for Scientists, Engineers and Students (2nd Ed.). Routledge, London, 204p.

CSE [Council of Science Editors]. 2014. Scientific Style and Format: The CSE Manual for Authors, Editors, and Publishers (8th Ed.): The University of Chicago Press, 722p.

Davis, M. 2005. Scientific Papers and Presentations. Academic Press, Massachusetts, 356p.

Day, R.A. and Gastel, B. 2006. How to Write and Publish a Scientific Paper (6th Ed.). Cambridge University Press, Cambridge, 320p.

Glasman-Deal, H. 2010. Science Research Writing for Non-Native Speakers of English. Imperial College Press, London, 257p.

Gustavii, B. 2008. How to Write and Illustrate a Scientific Paper (2nd Ed.). Cambridge University press, Cambridge, 168p.

Hawkins, C. and Sorgi, M. (eds) 1985. Research: How to Plan, Speak and Write about It (4th Indian Reprint, 1993). Narosa Publishing House, New Delhi, 184p.

Huckin, T.N. and Olsen, L.A. 1983. English for Science and Technology: A Handbook for Non-native Speakers. McGraw–Hill Book Company, 549p.

Leggett, G., Mead, C.D., Kramer, M.G., and Beal, R.S. 1985. Handbook for Writers. Prentice- Hall, New Jersey, 558p.

Lester, J.D. and Lester, J.D. Jr. 2009. Writing Research Papers: A Complete Guide (13th Ed.), Longman, 416p.

Malmfors, B., Garnsworthy, P. and Grossman, M. 2004. Writing and Presenting Scientific Papers (2nd Ed; Indian Edition, 2011). Viva Books, New Delhi, 153p.

Mathews, J.R. and Mathews, R.W. 2007. Successful Scientific Writing: A Step by-Step Guide for the Biological and Medical Sciences (3rd Ed.). Cambridge University Press, 256p.

McCaskill, M.K. 1990. Grammar, Punctuation, and Capitalization: A handbook for Technical Writers and Editors. NASA SP-7084, National Aeronautics and Space Administration, Washington DC, 106p. Available: https://ntrs.nasa.gov/archive/nasa/casi.ntrs.nasa.gov/19900017394_1990017394.pdf [Accessed 10 Dec. 2003].

Mountford, A. 1977. English in Focus: English in Agriculture. Oxford University Press, 156p.

Oliver, P. 2008. Writing Your Thesis (2nd Ed). Sage Publications, New Delhi, 174p.

Rubens, P. 2004. Science and Technical Writing: A Manual of Style (2nd Ed.). Routledge, New York, 427p.
Swan, M. 1983. Practical English Usage. The English Language Book Society and Oxford University Press (639 entries).

Chapter 18
Punctuation Marks and Abbreviations

Arise, awake, and stop not till the goal is reached.
Swami Vivekananda (1863–1902)

Punctuation marks are symbols used in writing to make the meaning of sentences clear and to make reading easier. There are four types of punctuation. In *end punctuation*, symbols are used to indicate the ending of a sentence, which include periods or full stops, question marks, and exclamations marks. *Internal punctuation* marks, which occur within sentences, are used to indicate relationship of certain words or group of words to the rest of the sentence. These include commas, semicolons, colons, dashes, and parentheses. *Direct quotation punctuation* includes quotation marks and brackets. The fourth category is *word punctuation*, which includes capitals, italics, apostrophes, and hyphens, often used to indicate that certain words have a special character.

The use of periods or full stops are well known, and there may not be much confusion in its usage. The most commonly used and misused punctuation marks in scientific writing are the comma (,), the semicolon (;), the colon (:), the parentheses (), the apostrophe ('), the dashes (—, –), and the hyphen (-). Style manuals such as AMA (2007), APA (2013), CSE (2014), and UChicago (2017) have recommendations on punctuation. An overview of commonly used punctuation marks in scientific writing is attempted here.

18.1 Periods

Periods or full stops usually signal the end of a sentence. Its primary purpose is to separate complete thoughts, and to mark the end of declarative and imperative sentences (Remember that interrogative sentences end with a question mark and exclamatory, with an exclamation point). A period or full stop should be used only after a sentence completes with subject and predicate. Ordinarily, a period should not be used after headings on separate lines, after titles of tables, and after items in

C. G. Thomas, *Research Methodology and Scientific Writing*,
https://doi.org/10.1007/978-3-030-64865-7_18

a numbered list except in cases where one or more items are complete sentences. However, a period or colon is normally put after run-in headings to separate it from the main text. See some uncommon uses of periods:

- The tradition is to put a period after figure captions whether they are complete sentences or otherwise:
  Fig. 1. Stream bank erosion.
- Put a period after certain abbreviations and contractions as in: a.m., i.e., e.g., and Fig.
- Spaced periods. Put three dots (...) to indicate the omission of a word or a group of words. However, to indicate omission of one or more paragraphs or sentences, use an entire line with spaced periods (..).

18.2 Comma

In a sentence, to represent the shortest pause, comma is used. A common usage is to separate a series of more than two words, phrases, or clauses in the same construction, as:

- The job was exhaustive, tiresome, and boring.
- After land preparation, you may sow, dibble, or drill the seeds.

The comma used here is called *serial comma* or *series comma*, as a comma is used immediately before a coordinating conjunction such as '*and*' and '*or*' preceding the last item in a list comprising of three or more items.

A typical series take the form *a*, *b*, and *c* or the form *a*, *b*, or *c*. To illustrate, a list of three fruits can be written in a series as 'apple, mango, and pine apple'. In scientific writing, routine use of serial commas helps to prevent ambiguity. Note that many style guides, for example, *The CSE Manual for Authors, Editors, and Publishers* (CSE 2014) recommends the use of serial commas. AMA (2007) and APA (2013) also recommended the use of serial comma. This serial comma is also called *Oxford comma* or *Harvard comma* because the Oxford University Press and the Harvard University Press have traditionally been following this type of comma usage in their publications.

Note that in ordinary writing, some writers omit the comma in simple series as in: 'bread, butter and jam'; or 'leaves, stems or roots'. However, as the comma is sometimes vital for clarity, it is preferable to include it even in ordinary writing. See the example:

- The new college building has a spacious library hall, four classrooms, a canteen with waiting room and computer centre.

Without the comma after waiting room, it may lead to misreading as 'a canteen with waiting room and computer centre', and you may wonder whether the computer centre is part of the canteen or the college! Therefore, write the sentence with the comma as:

- The new college building has a spacious library hall, four classrooms, a canteen with waiting room, and computer centre.

Uses of Comma in Writing

1. Put a comma before a coordinating conjunction such as '*and*', '*but*', '*for*', '*or*', '*nor*', '*so*', and '*yet*', when they connect two independent clauses (that is, sentences that can stand-alone).

   - Watershed management strives to conserve natural resources, and promotes programmes to ensure livelihood security of people.

   In fact, these are two sentences. (1) Watershed management strives to conserve natural resources. (2) It promotes programmes to ensure livelihood security of people. These two are joined using 'and'. Sometimes, a comma is omitted when one or both of the independent clauses are very short as in the sentence

   - The spirit is willing but the flesh is weak.

2. A comma is used to break up introductory phrases and clauses from a main clause.

   - According to Barrass (2002), writing is part of science, and young scientists and engineers should be taught to write well.

3. Commas help to set off adverbial clauses that precede or interrupt the main clause.

   - When the student noticed some problems with the new equipment, she immediately reported the matter to her research guide.
   - The thesis, after the evaluation of the advisory committee, was submitted to the University for fixing a date for final examination.

4. When transitional words and phrases such as 'on the contrary' and 'on the other hand'; conjunctive adverbs such as 'consequently', 'furthermore', and 'however'; or phrases that introduce an example such as 'for example' and 'for instance' are used, commas are needed to set them off from the rest of the sentences (see Sect. 16.10).
5. Commas are used to separate coordinate adjectives modifying a noun.

   - The harsh, hot weather is affecting all agricultural activities.

6. Commas set off from the rest of the sentence 'inserted cases' such as non-restrictive modifiers with 'which', 'who', and 'whose' and non-restrictive appositives (An appositive is a noun or a group of words used as noun that describes or renames another noun).

   - The herbicide, which was applied post-emergence, controlled nut sedge.

- Agronomy, the oldest agricultural discipline, is taught in all agricultural colleges.

7. When contrasting and opposing expressions come within sentences, commas must be used.
   - The teacher changed his approaches, not his principles.
8. Commas are used before introducing a direct quotation, after finishing a direct quotation, and for putting split quotations.

   Arun said, 'I am going'.
   'I am going', Arun said.
   'I won't go', Priyanka said with conviction, 'even if they expel me'.
9. In direct addresses and mild interjections, commas are used.

   'You can write the examination, Ankitha, if you want'.
   'Oh, that's terrible!'.
10. Use commas in dates. However, when just the month and year are given, the comma is usually omitted. Similarly, if a date is written as day–month–year, do not use commas. See examples:
    - The experiment was laid out on Monday, Aug. 16, 2010.
    - I began my career in Oct. 1980.
    - The experiment was laid out on 16 Aug. 2010.

18.3 Semicolon

The semicolon is used to represent a greater pause that that of comma. Usually, semicolons are used instead of full stops, in situations where sentences are grammatically independent, but the meaning is closely connected.

- The proposal is good; let us hope that they will approve it.

The main uses of semicolon are indicated below:

1. A semicolon is used to break up closely related main clauses, if they are not joined by a coordinating conjunction. Example:
   - Some farmers come forward to adopt new technologies quickly; others do not.
2. Semicolons link main clauses joined by conjunctive adverbs such as *furthermore*, *therefore*, *indeed*, *consequently*, and *however.*
   - There are innumerable plants on earth; *however*, only a few are grown for food.

   Note that the above sentence can also be written as two sentences.

- There are innumerable plants on earth. *However*, only a few are grown for food.

3. A semicolon is used to separate phrases and clauses, which contains internal comas.

   - In academic fields, research involves activities such as defining, redefining, and solving problems; observing facts, detecting patterns, and their interpretation; formulation of hypotheses, conducting experiments, and analysis of data; revision of existing theories and laws; and practical application of information already generated.

4. Use a semicolon before phrases such as *that is* and *namely* when they are used to introduce an independent clause.

   - Soil conservation activities must be planned 'bottom up' rather than 'top down'; that is, these must be undertaken with the active cooperation and involvement of the farmers involved rather than thrusting upon them by the outside experts.

5. When the main clauses joined by a coordinating conjunction are long or internally punctuated, use a semicolon before the coordinating conjunction.
6. Semicolons are used to separate phrases and clauses which themselves contain commas, or if the clauses are long.

   - The following may be the reasons for rejecting a proposal for funding: (1) The researchable idea is not important; (2) The applicant has not taken sufficient care to plan the project; (3) …

18.4 Colon

Both scientists and students often get confused about the use of colons. Colons should not be used in place of commas or semicolons.

1. When a second independent clause explains a preceding independent clause, a colon can be used to separate them. Please note that in this case, a colon is used after a complete sentence.

   - I will always remember the first day of my office: It was a raining, chilly Monday.

2. A colon is used to separate parts in a title comprising of general and specific information.

   - Drought management: Importance of drought avoidance strategies in crop plants.

3. Use a colon to direct attention to an appositive.
   - The research station is famous for one thing: good rapport with farmers.
4. Use a colon to introduce a series or list that augments an introductory sentence.
   - Three persons attended the interview: Joseph, Ahmed, and Sruthi.
5. Use a colon to introduce a list after the introductory statement which contains the words 'as follows' or 'the following'. Note that the example below contains a colon and used semicolons instead of commas as the clauses are long.
   - The following may be the reasons for rejecting a proposal for funding: (1) The researchable idea is not important; (2) The applicant has not taken sufficient care to plan the project; (3) …
6. Use a colon to indicate time and ratio.
   - The function will start at 10:00 a.m. and end by 11:30 a.m.
   - The C:N ratio of moist soils is about 10:1.
7. A colon is normally put after run-in headings to separate it from the main text. **Method of science**: The method of science is a practical methodology of acquiring knowledge by formulating specific questions and systematically finding answers.
8. Colons are used to introduce quotations. Bertrand Russell (1872–1970), a renowned philosopher of the twentieth century, said: 'We know very little, and yet it is astonishing that we know so much, and still more astonishing that so little knowledge can give us so much power'.

18.5 Quotation Marks

Inverted commas are used to indicate double and single quotation marks. These are generally used to enclose a quotation or the exact words of a speaker. In fact, there is no specific rule about which type of inverted commas to use (double or single) in quotations; but you should be consistent in one or the other throughout a piece of writing. Single inverted commas are more common in British English while the Americans seem to prefer double inverted commas. However, if you want to enclose a quotation within another quotation, use the style you have not used already. For example, if you have been using double inverted commas in your thesis or paper, put any further quoted material in such quotations within single inverted commas and vice versa.

Please note that in British English, the custom is to use single quotation marks to enclose direct quotations and to use double quotation marks, if a quotation occurs within another quotation. If you want to enclose a quotation within quotation, the sentence must be; as,

The teacher advised, 'In the Academic Handbook, refer the section on "Reference Writing" for more details'

18.6 Apostrophe

1. The main use of an apostrophe is to form the possessive case to show ownership. The possessive form of a singular or plural noun not ending in 's' is formed by adding *'s*, and for those ending in 's' just by adding an apostrophe only. See the examples:

Singular	Plural
Owner's	Owners'
Animal's	Animals'
Friend's	Friends'
Man's	Men's
Thomas'	–

2. In the case of a compound noun, form the possessive by adding *'s* to the end of the compound noun as sister-in-law's home and professor-incharge's report.
3. For showing joint possession, add *'s* to the last item of a series; but if you want to indicate individual possession, then add *'s* to each item:
   - Geo and Jim's laboratory (one laboratory)
   - Student's and guide's efforts (separate efforts)
4. Earlier, the possessive case (*'s*) was not permitted for inanimate nouns. Instead, the preposition '*of*' was used along with the item, for example, the strength *of* the steel wire (*not* steel wire's strength). At that time too, there were some exceptions with inanimate words representing a collection of animate beings (e.g., company's profits, university's curriculum) and words expressing measure or time (e.g., 2 h work).
   Note that the current practice is to dispense with both *'s* and '*of*' as in *company profits, university curriculum*, and *two hours work*.
   Similarly, we will not say systems' analyst, but *systems analyst* or table's top, but *tabletop*.
5. If the object has some lifelike qualities, we may use *'s* on an inanimate object, for example: computer programme's name; earth's rotation.
6. Never use apostrophes to indicate plurals of abbreviations consisting of capital letters, for example, POPs and *not* POP's. Similarly, periods are also written without apostrophe, for example: 1960s and *not* 1960's.
7. The apostrophe is also used to indicate letters left out of contractions, for example: *it's* (it is), *Gov't*, and *Nat'l*. Note that as contractions are unusual in technical writing, these types of apostrophe usage are also rare in technical writing.

18.7 Brackets

Brackets or square brackets [] are tall punctuation marks used in matched pairs in sentences and for other purposes. The following are some of their common uses:

1. In ordinary writing, square brackets are generally used to enclose words added by someone other than the original writer, in order to clarify some aspects as shown in the example.
   - She [the forensic expert] could not produce sufficient evidences.
2. Square brackets are also used to indicate missing material added by someone other than the original author, especially in quoted text.
   - 'I welcome the decision [ban on smoking]'
3. The Latin term [sic], meaning, 'thus was it written' is used after a reprinted text to indicate that it appears exactly as given in the original source (sometimes, its spelling may be incorrect or grammatically incorrect).
4. In scientific writing, square brackets are used in reference citations (see Chap. 15).
5. Brackets are also used in proof reading for some specific functions.
6. In mathematics, square brackets have a specific role, especially in writing equations.

18.8 Parentheses

Parentheses or round brackets ( ) are used to indicate some additional information about words or sentences. The use of parenthesis in scientific writing is indicated below:

1. Use parentheses in sentences to enclose a phrase, which give additional information.
   - Smallholder farmers (the farmers who own small areas) indulge in many farming activities for their own consumption or for sale to customers.
2. Use parentheses for references and abbreviations.
   - Lester and Lester (2009) reported that …
   - Kerala Agricultural University (KAU) …
3. Use parentheses for definitions and for providing subsidiary information that may be of interest only to certain readers.
   - Leguminous plants fix nitrogen in their roots (some non-legumes too).
4. Use parentheses in pairs ( ) when numbering items in a list within running text.

- The following may be the reasons for rejecting a proposal for funding: (1) The researchable idea is not important; (2) The applicant has not taken sufficient care to plan the project; (3) ...

5. In mathematics, parenthesis or round brackets are part of equations or formulae.

18.9 Dashes

A dash takes the form of an em dash/long dash (—) and en dash/short dash (–). The em dash has approximately the width of the letter 'M', and the en dash has the width of 'N' (slightly longer than the hyphen).

1. Use em dash or long dash for emphasis. It is used to mark an abrupt shift in sentence or afterthoughts.
   - 'Could I—should I even try to—experiment with this new device?'
   - The experiment is over—the result is discouraging.
2. Use the long dash to set off introductory lists or a series of words or phrases.
   - Data collection, tabulation, classification, and interpretation—the process seems endless for a researcher.
3. Use the em dash or long dash to isolate or separate parenthetical matter instead of putting it in parenthesis.
   - Smallholder farmers—the farmers who own small areas—indulge in many farming activities for their own consumption or for sale to customers.
4. An en dash is used to show a period similar to the meaning as *from...to....*
   - The experiment was conducted during the years 2009–2010.
5. An en dash can be used to show a range of numbers. Examples:
   - Choose 10 random numbers in the range of 1–200.
   - You may refer pages 50–55 in your textbook for the procedure.
6. Use an en dash instead of a hyphen for combining open compound words when one or two of them are made up of two words (An 'open compound word' is formed when a modifying adjective is used with its noun to create a new noun). See the examples:
   - The speaker is an African American–Nobel laureate.
   - It was a middle class–upper class conflict.

   Please note that when using the en dash, the em dash, or the hyphen, you should not put a space either before or after them.

18.10 Hyphens

A hyphen (-) can be considered as a very short dash generally used to combine two words, and to form a compound word (e.g., knee-high, counter-productive, self-confidence, ex-minister). However, hyphen is omitted in commonly used words (e.g., coeducational, coordinate, cooperation).

1. Use a hyphen to join two or more words serving as a single adjective before a noun, for example: a well-known scientist (The scientist is well known).
2. Use a hyphen to avoid an ambiguous or awkward union of letters, for example: re-create, for create again (not recreate) and bell-like (not bellike).
3. Use a hyphen to form compound numbers from twenty-one through ninety-nine (e.g., twenty-five, thirty-six).
4. Do not use a hyphen to replace the preposition 'to' between numerals except in graphs, tables, and brackets, for example, 20 m to 30 m (not 20 m-30 m).
5. Hyphens should never be used interchangeably with dashes (em dash or en dash), which are strikingly longer. In addition, remember there should not be spaces around hyphens as their main purpose is to stick words together. For example, write 2017-18 (*not* 2017—18; 2017 –18; or 2017- 18).
6. Avoid using hyphens in 'closed compound words', which looks like one word. These words were originally used as two words; however, because of frequent use, they are now accepted as single words. Closed compound words are usually made up of only two words. Examples include honeybee, notebook, cooperation, coordination, bookshop, online, keyword, makeup, farmhouse, textbook, headache, sunflower, railroad, earthquake, downstream, and grasshopper.

18.11 Virgule

Virgule, also called solidus or slash, is the slanting mark (/).

1. A virgule is used to separate alternatives such as and/or, he/she, s/he, his/her, him/her, himself/herself, man/woman, and wife/husband.
2. It often represents 'per' in units, for example: 10 g/L (10 gram per litre) and 5 Mg/ha (5 megagram per hectare).
3. It is also sometimes used as a mathematical sign for division as in 6/10, meaning six divided by ten.

18.12 Italics

Often, the purpose of using italics is to make the meaning of certain words or phrases clear and reading easier. Italics help us to differentiate words or phrases from the rest of the sentence, so that the implication of the italicized words is clear. In addition, there are several conventional uses for italics. In mechanical typewritten text or hand-written text, we underscore the words or phrases as putting them in italics is not possible. The following are some of the common uses of italics:

1. An important word in a discussion on first use or a technical term followed by its definition is often italicized.
   - A *paraphrase* is a restatement of the written matter of another person in your own words and style.
2. Letters used as letters are italicized:
   - You just press the letter *n* to indicate 'no' and the letter *y* to indicate 'yes'.
3. Most mathematical symbols and letter symbols representing physical concepts are written in italics. However, chemical symbols, computer symbols, and abbreviations are not italics.
4. Use italics in citing titles of books, reports, compilations, journals, and newspapers, but not titles of research articles, conference papers, theses, papers in compilations, or patents.
5. Words from languages other than English used in sentences, which are unfamiliar to readers (except proper names), are usually written in italics.
6. Scientific names of plants and animals including genus, species, and variety are italicized.

18.13 Capitals

1. **Capitalize proper nouns, their derivatives and abbreviations, and common names used as proper nouns.**
   Certain nouns, derivatives, and abbreviations must be capitalized always. See Table 18.1 for some specific examples under different categories.
2. **Headline style capitalization.**
   Headline style calls for all principal words to be capitalized (***title case*** style.). First, decide the principal words in the headline, and then follow the guidelines given below.
   - Often, the articles—*a*, *an*, and *the*; the prepositions or adverbs—*at*, *by*, *for*, *of*, *in*, *up*, *on*, and *to*; and the conjunctions—*and*, *as*, *but*, *if*, *or*, and *nor* are not subjected to title case capitalization even if they are part of headlines. However, capitalize the first and last words in a headline even if they are articles or prepositions.
   - Generally, words with four letters or more are considered principal words and are capitalized.
   - Capitalize both elements of a two-element hyphenated compound word except the second element of a compound numeral, for example: GeoHydrologic Unit, but Thirty-nine.

Table 18.1 Examples for capitalization of words

Items	Examples
Personal names	Geethu George
Races	Jews
Nationalities	Indian
Languages	Malayalam
Computer languages	Basic, Fortran, Pascal
Specific places	New Delhi, London
Specific organizations	College of Horticulture, Thrissur
Planets	Jupiter, Venus, Mars (but the earth, the sun, and the moon)
Historical events	The French Revolution
Documents	Declaration of Independence
Days of the week	Monday
Months	January
Eras	the Eighties, the Twentieth Century
Holidays	Christmas
Religious terms	The Virgin, Lord Shiva, The Almighty
Titles of books	Freedom at Midnight
Plays	Waiting for Goethe
Magazines	Outlook
News papers	The Hindu
Journals	Agronomy Journal
Titles, when they precede a proper noun	Professor Thomas, Ms Alice, Dr George
Personal titles	Project Officer, Professor of Agronomy
Major sections of a book, report, article, or thesis	Preface, Introduction, Results, Index
Common nouns used as an essential part of a proper noun	the Pacific Ocean, the Indus River
The words north, south, east and west when used in place names	The Far East, North Africa, South East Asia
For emphasis	Use PENCIL only

3. **Proper nouns need to be put in title case style.**
   The derivatives of proper nouns when used in a proper sense are capitalized as shown in the following examples:

Arabia	Arabian Sea
China	Chinese food
Italy	Italian culture

(continued)

(continued)

Newton	Newtonian physics
Rome	Roman law

However, derivatives of proper nouns that have acquired an independent meaning are not capitalized. Examples:

Arabia	arabic numbers
China	bone china
Italy	italics/italicize
Kelvin	kelvin (unit of temperature)
Manila	manila envelope
Newton	newton (unit of force)
Paris	plaster of paris
Pascal	pascal (unit of pressure)
Pasteur	pasteurize
Rome	roman (numbers/fonts)

4. **Stylistic uses for sentence style capitalization.**
   In scientific writing, table entries, captions, and footnotes are not written as full sentences, but the first letter of these non-sentence elements are capitalized as a matter of style just like sentences. This style of capitalization is applicable in the following cases.
   - Table entries and table subtitles
   - Run-in headings, head notes, box heads, etc.
   - Footnotes (text as well as table)
   - Figure captions and figure labels

5. **Avoid unnecessary capitalisation of words.**
   - When writing about directions, capitalize north, south, east, and west only when they come at the beginning of a sentence or when used in place names; otherwise, use only lower case letters. Similarly, names of seasons need not be capitalized (e.g., fall, autumn, winter, midwinter, spring, and summer).
   - Only when the word 'the' is part of an official name, it is capitalized, for example: The Hague, but the Indian Agricultural Research Institute.
   - Acronyms like 'radar' and 'laser' that have become accepted as common nouns are written in lower case letters instead of capital letters.
   - Common names of plants and animals are written in lower case letters (e.g., rice, cow). However, first letter of the name will be in capitals, if it was derived from a proper noun (e.g., American bollworm, Job's tears).
   - Common names of elements are never capitalized (e.g., carbon, hydrogen).

18.14 Abbreviation, Initialism, and Acronym

An *abbreviation* is a shortened form of a word or a group of words. It is normally written with a period or full stop at the end, for example, abstr. for abstract. However, if the abbreviation consists of only capital letters, it is written without periods, for example, IARI. Sometimes, certain abbreviations are pronounced as a single word, which are called *acronyms*. An abbreviation created by joining the first letters of each of the successive words or major words of a long compound term to pronounce it as a single word is an acronym. An example is ICRISAT for 'International Crops Research Institute for the Semi-Arid Tropics'. ICRISAT is pronounced as ic-ri-sat. Acronyms are pronounced as words and written without periods. Note that commonly used acronyms like 'radar' (Radio Detecting and Ranging) and 'laser' (Light Amplification by Stimulated Emission of Radiation) are written in lower case letters instead of capital letters as these terms are now accepted as common nouns.

Another type of abbreviation is *initialism* produced by joining the initial letters of all the words in a long multiword term. Initialisms are pronounced as separate letters, for example, ICAR (Indian Council of Agricultural Research) is pronounced as i-c-a-r. Plural of an acronym or initialism can be formed by adding an 's' as in PCRs, CRTs, etc. Note that apostrophe is not used here.

In general, abbreviation is an obstacle in reading, as the readers have to keep searching back through the document to remind them what each abbreviation means. On the contrary, authors love to use abbreviations, as they can avoid repeating long phrases or terms in full. Therefore, take some extra care while using abbreviations. Try to avoid abbreviations altogether or use them sparingly. However, if an acronym or initialism is popular than its proper term or full form, use the abbreviated form only. For example, the acronyms, radar and laser are more popular and common than their expanded forms. Similarly, the initialism AD is much more common than the original form *Anno Domini*.

Abbreviations shall be written in full at least for the first time they are used in the document, followed immediately by the abbreviated form in parentheses. Full form of the abbreviation shall be repeated, if it is used again in the document after a lengthy gap. It is also a good idea to list and explain them at the beginning of a document.

Abbreviations of Phrases from Other Languages

Many abbreviations of phrases from other languages, especially Latin, are commonly used in writing. A list of common abbreviations used in writing inclusive of Latin abbreviations is given in Annexure.

The abbreviation, etc. (*et cetera*, meaning 'and so forth' or 'and other things') is frequently used in writing to avoid the continuation of a series of descriptions. See the example: *Pulses include pigeon pea, green gram, black gram, cowpea, etc.* The 'etc.' in this sentence stands for 'and other types of pulses'. Please note that the abbreviated version of *et cetera* (etc.) should always be followed by a period (full stop), and it is customary, both in American and in British English, that 'etc.' always be preceded by a comma. For example, it must be a, b, c, etc., and *not* a, b, c etc. While using 'etc.', the extension of ideas implied by it must be clear. Consider the example: *Surface water drains to a surface water body—lake, pond, river, estuary, etc.* In this example, the reader has no difficulty in mentally listing other possible surface water bodies. However, in the sentence, '*We walked, ate, etc.*', the reader has no clues to the implied idea. The sentence should be, '*We walked, ate, played, and danced*', for making sense. It is also incorrect to say or write 'and etc.' as the 'and' here would be redundant. The present trend, however, is to dispense with the use of 'etc.' as it conveys no additional information except that the list is incomplete. Instead, use 'including', 'such as', or 'like' immediately before the list, which can convey the same meaning.

In formal writing, instead of phrases and abbreviations from other languages, the English equivalents are increasingly used now. For example, it is better to avoid abbreviations such as *loc. cit* (in the place cited), *op. cit* (in the work cited), and *ibid.* (in the same work) in technical writing as these words often contribute to ambiguity. The Latin abbreviations such as *i.e.* (*id est* = that is), *e.g.* (*exempli gratia* = for example), and *viz.* (namely) are also prone to misuse, and therefore, their English equivalents are recommended in technical writing. Foreign words are usually written in italics. However, widely used words or abbreviations from other languages such as 'viz.' and 'etc.' need not be italicized.

Use of Capital Letters and Periods in Abbreviations

In general, acronyms are written in capital letters without periods as in IRRI and CIMMYT. Nevertheless, acronyms like 'radar' and 'laser' are written in lower case letters only without periods as these terms are now accepted as common nouns. Initialisms may be written in either upper case or lower case letters. Periods are not used when all are uppercase letters as in WHO, but are used when they are lower case letters as in a.m. When writing academic degrees, whether composed of upper case letters alone or combined with upper and lower case letters, periods are not used between letters as in BA, BSc, MPhil, and PhD. Similarly, periods should not be used after the symbols for an SI unit such as g (gramme or gram), s (second), and m (metre). For more details on SI units, see Chap. 19.

In British English, when the last letter of the abbreviation and the last letter in the original word without abbreviation are the same, a period or full stop is omitted after the abbreviation. For example, put periods after ed. (edition), No. (number), and Vol. (volume), but omit periods after eds (editions), Vols (volumes), and Nos (numbers). Similarly, certain honorific titles are actually contractions formed by the first and last letters of a word as in Dr (Doctor), Mr (Mister), Mrs (Mistress), and Ms (Ms is a blend of Mrs and Miss, which can be used instead of Miss or Mrs). Please note that while the above abbreviations are written without full stops as Dr, Mr, Mrs, and Ms in most Commonwealth countries, in the USA, it is written with full stops as Dr., Mr., Mrs., and Ms.

Annexure: Abbreviations Commonly Used in Writing

Abbreviations	Expansion	Meaning
AD	*Anno Domini*	In the year of the Lord, precedes numerals with no space between letters; as in 'AD 200'
ad lib.	ad libitum	At pleasure
ad loc.	*ad locum*	At the place
ad val.	*ad valorum*	According to value
Amt	Amount	Amount
anon.	*Anonymous*	Not named or identified
Approx.	Approximately	Approximately
BC	Before Christ	Follows numerals with no space between letters, as in 500 BC (cf. AD)
Ca	*Circa*	About, approximately; used to indicate an approximate date, as in 'ca.1452'
ch., chap.	Chapter	Chapter
chs, chaps	Chapters	Chapters
cent.	*Centum*	A hundred
cf.	*Confer*	Compare (one source with another)
Co.	Company	Company
C/o	Care of	Care of
col.	Column	Column
cols	Columns	Columns
contd	Continued	Continued
cv.	Cultivar	Cultivated variety
diss.	Dissertation	Dissertation
do.	*Ditto*	The same
doc.	Document	Document
Ed.	Editor	Editor
eds	Editors	Editors
Ed.	Edition	Edition
ed. cit.	*Edition citato*	The edition cited
e.g.	*Exempli gratia*	For example
esp.	Especially	Especially
et al.	*et alia*	And others; Thomas et al. means 'Thomas and others'
etc.	*et cetera*	And so forth
et seq.	*et sequentia*	And the following
ex.div.	*Extra dividendum*	Without dividend

(continued)

(continued)

Abbreviations	Expansion	Meaning
f.	*Folio*	A leaf of a book
Fig.	Figure	Figure
ff.	*Folgen folio*	Following pages
f.v.	*Folio versa*	On the back of the page
ibid.	*ibidem*	In the same place (book, chapter, page), i.e., in the immediately preceding source, normally capitalized and not underlined as in 'Ibid., p.45'
id.	*idem*	The same
i.e.	*id est*	That is
illus.	Illustrations, illustrator illustrated by	Illustrations, illustrator, illustrated by
in loc.	*In loco*	In its place
in loc.cit.	*In loco citato*	In the place cited
loc.cit.	*Loco citato*	In the place (passage) cited
MS, M.S.	*Mano sinistro*	Written by hand or typed, manuscript
NB, N.B.	*Nota bene*	Note well
n.d.	*No date*	No date; in a book's title or copyright pages
No.	*Numero*	Number
n.p.	*No place*	No place (of publication)
Ns	*New series*	New series
op.cit.	*Opere citato*	In the work cited
q.e.	*Quod est*	Which is
q.s.	*Quantum sufficit*	A sufficient quantity
q.v.	*Quod vide*	Which see
p.	*Pagina* (page)	Page
pp.	*Paginae* (pages)	Plural of page (For some words, doubling the letters in the abbreviation is used to indicate plural forms)
per an	*Per annum*	Per year
per cent	*Per centum*	By the hundred
Proc.	Proceedings	Proceedings
PS, P.S	*Post scriptum*	Written after something, a post script
pseud.	Pseudonym	Pseudonym
qty	Quantity	Quantity
rev.	Revised, revision, review, reviewed by	Revised, revision, review, reviewed by
sec.	Section	Section
secs	Sections	Sections

(continued)

(continued)

Abbreviations	Expansion	Meaning
sic	Thus	Written intentionally in parenthesis to indicate that the writer is quoting the author accurately along with the error
sp.	Species	Species (a taxonomic unit under genus)
spp.	Species	Plural of species
suppl.	Supplement	Supplement
trans., tr.		Translator, translated, translated by, translation
u.s	*ut supra*	As above
var.	Variety	Variety (a taxonomic unit under species)
vid.	*Vide*	See
v.i.	*Vide infra*	See below
viz.	*Videlicet*	Namely (z = medieval Latin symbol of contraction)
vs, v.	*Versus*	'Against' as used in citing legal cases
v.s.	*Vide supra*	See above

Bibliography

AMA [American Medical Association] 2007. AMA Manual of Style: A Guide for Authors and Editors (10th Ed.). Oxford University Press, New York, 1024p.

APA [American Psychological Association] 2013. Publication Manual of the American Psychological Association (6th Ed.). American Psychological Association, Washington DC, 272p.

Barrass, R. 2002. Scientists Must Write: A Guide to Better Writing for Scientists, Engineers and Students (2nd Ed.). Routledge, London, 204p.

CSE [Council of Science Editors] 2014. Scientific Style and Format: The CSE Manual for Authors, Editors, and Publishers (8th Ed.): The University of Chicago Press, 722p.

Day, R.A. and Gastel, B. 2006. How to Write and Publish a Scientific Paper (6th Ed.). Cambridge University Press, Cambridge, 320p.

Huckin, T.N. and Olsen, L.A. 1983. English for Science and Technology: A Handbook for Non-native Speakers. McGraw–Hill Book Company, 549p.

Leggett, G., Mead, C.D., Kramer, M.G., and Beal, R.S. 1985. Handbook for Writers. Prentice- Hall, New Jersey, 558p.

Lester, J.D. and Lester, J.D. Jr. 2009. Writing Research Papers: A Complete Guide (13th Ed.), Longman, 416p.

Malmfors, B., Garnsworthy, P. and Grossman, M. 2004. Writing and Presenting Scientific Papers (2nd Ed; Indian Edition, 2011). Viva Books, New Delhi, 153p.

McCaskill, M.K. 1990. Grammar, Punctuation, and Capitalization: A Handbook for Technical Writers and Editors. NASA SP-7084, National Aeronautics and Space Administration, Washington DC, 106p. Available: https://ntrs.nasa.gov/archive/nasa/casi.ntrs.nasa.gov/19900017394_1990017394.pdf [Accessed 10 Dec. 2003].

MLA [Modern Language Association of America] 2016. MLA Handbook for Writers of Research Papers (8th Ed.) Modern Language Association of America, 146p.

Penn, J. 2018. The Punctuation Guide. Available: https://www.thepunctuationguide.com/index.html [Accessed 20 Sep. 2018].

Rubens, P. 2004. Science and Technical writing: A Manual of Style (2nd Ed.). Routledge, New York, 427p.

Swan, M. 1983. Practical English Usage. The English Language Book Society and Oxford University Press (639 entries).

Turabian, K.L. 2007. A Manual for Writers of Term Papers, Theses, and Dissertations (9th Ed.). The University of Chicago Press, 461p.

UChicago [University of Chicago] 2017. Chicago Manual of Style [CMOS]. University of Chicago Press, 1144p.

Wren, P.C. and Martin, H. 2005. High School English Grammar and Composition (Revised, Prasada Rao, N.D.V.), S. Chand & Co, New Delhi, 416p.

Chapter 19
Units and Numbers

An experiment is a question, which science poses to nature, and a measurement is the recording of nature's answer.
Max Planck (1858–1947)

Humans recognized the importance of measurements long ago. Many of the traditional units of measurements evolved naturally. By the first half of the twentieth century, there were two well-established systems of weights and measures—the *imperial system* and the *metric system.* The imperial system, also called the British system or the FPS system, was derived from the old Anglo-Saxon measurements evolved as a natural way of expression of measures. This system was prevalent in the British Empire and its colonies. However, the metric system was a well-thought-out system of weights and measures developed during the French revolution period. The metric system was followed in France and other European countries.

Metrology is the science of measurements, which includes both experimental and theoretical aspects of measurements in any field of science and technology.

The imperial system of measurements evolved based on standards that were readily available in nature. For instance, the '*inch*' was the 'knuckle of the thumb' and the '*foot*', the length of the human foot. An '*acre*' was the land area that could be ploughed in a day by a pair of oxen. The '*mile*' came from the Roman legionaries. Their '*milli*' was 1000 paces or about 1618 yards. The mile was later standardised at 1760 yards. Although these units are now precisely defined, their conversion into larger or smaller units is cumbersome, especially for scientific works.

19.1 Units and Dimensions

Measurement includes units and dimensions. For measurement, physical quantities must be organized in a system of dimensions. For example, each of the seven base quantities used in the SI is regarded as having its own dimension. The dimensions of a quantity refer to the basic nature of the quantity, that is, how the quantity is related

C. G. Thomas, *Research Methodology and Scientific Writing*,
https://doi.org/10.1007/978-3-030-64865-7_19

to length, mass, time, or charge. A particular quantity can be reported in different kinds of units, but it will always have the same dimension.

Units provide standards to measure the quantities defined by physical laws called *dimensions*. Unit is defined as any measure used as a standard; and dimension is any measurable quantity or extent. For example, length (l) is a dimension and metre (m) is a unit. There is a special relationship between dimension and unit. A unit belongs to one and only one dimension. The metre is a unit used to measure the dimension length and length only. It cannot be used to measure any other dimensions, for example, mass or time. While the dimension of a unit is unique, the units allowed to describe a dimension are not. For example, metre and foot are just two of the many alternate units available to describe the dimension, length.

Primary and Derived Dimensions

There must be enough number of dimensions to account for all of the quantities to be measured. These required dimensions are called *primary* or *fundamental dimensions*. Traditionally, the primary dimensions accepted are mass, length, time, and temperature.

From the primary dimensions, *derived dimensions* can be constructed as combinations of the primary set. For example, velocity can be defined as the time rate of change of distance. The dimension of 'velocity' (V) can be written as length divided by time (L/t). The quantity 'acceleration' (A) can be defined as the time rate of change of velocity, and could be written as V/t (velocity divided by time), or written in terms of the accepted primary dimensions, L/t^2.

The dimensional formula of a derived dimension will be the dimension written in terms of the accepted primary dimensions. The dimension of a unit and the dimensional formulae of a unit are the same as the derived dimension containing the unit. For example, Joule (J) is a unit used to describe heat, energy, and work. Therefore, the dimensional formula of Joule is the same as for the derived dimension, work, ML^2/t^2.

19.2 A Brief History of the Units

There were many traditional systems of units of measures in different parts of the world. Growing trade between countries, however, necessitated the inevitability of common standards for measurements. Further, some of the units were not found appropriate for scientific works. Lack of common standards led to confusion and significant inefficiencies in trade and research. The French government wanted to alleviate these problems by formulating a system of measurement that could be followed in all the countries.

The metric system developed during the French revolution period (1791–1795) was the forerunner of the now International System of Units or the SI system in short. In fact, it started with Louis XVI, who charged a group of savants to develop a new system of measurement. Their work laid the foundation for the 'decimal metric system', which has evolved into the modern SI. After the revolution, the new republican government took over the idea of the metric system, and made some significant changes. The French National Assembly in 1790 entrusted the Academy of Science to devise a simple, decimal-based system of units. The system they formulated began to be called the *metric system* (metric is from metre, the first standard of measurement under this system). The system was adopted by France in 1790, and propagated in other European countries by Napoleon.

Gauss (Johann Karl Friedrich Gauss, 1777–1855), a German mathematician, strongly supported the use of metric system for the physical sciences. In 1832, Gauss measured the earth's magnetic force in terms of a decimal system involving the units, the *millimetre*, the *gram*, and the *second* for the quantities, *length, mass,* and *time* respectively. Later, these measurements were extended to include electrical phenomena. Use of metric system in the field of electricity and magnetism received a boost in the 1860s through the British Association for the Advancement of Science (BAAS). The BAAS was trying for a rational system of units with *base units* and *derived units*, and in 1874, they introduced the *CGS system*, based on the three basic units, *centimetre, gram* and *second*, using prefixes such as micro to mega to denote decimal submultiples and multiples.

In 1870, France took the lead in organizing a convention to evolve a unified metric system acceptable to all nations. In 1875, the 'Metre Convention' (*Convention du Mètre*) or the 'Treaty of the Metre' was signed in Paris by the representatives of 17 nations. The Treaty established a 'General Conference on Weights and Measures' (CGPM, *Conférence Générale des Poids et Mesures*), an intergovernmental treaty organization, which would meet periodically to adopt new definitions as the need arises. The first CGPM in 1889 authorized the international prototypes for the 'metre' and the 'kilogram'. Along with the astronomical second as the unit of time, these three units comprised a three-dimensional mechanical unit system similar to the CGS system, but the sanctioned base units were *metre*, *kilogram*, and *second*. This system began to be called the MKS system.

The above 'Metre Convention' remains the basis of all later international agreement on units of measurement. There are now 54 member States in CGPM, including all the major industrialized countries. The CGPM is the international authority on SI, and they are responsible for modifying the SI as and when necessary to reflect the latest advances in science and technology. The 'General Conference on Weights and Measures' meets in Paris once every four years (the 26th meeting of the CGPM was held in Nov. 2018) and is made up of representatives of the governments of the member States.

The 'Metre Convention' also created the 'International Bureau of Weights and Measures' (BIPM, *Bureau International des Poids et Mesures*) and the 'International Committee for Weights and Measures' (CIPM, *Comité International des Poids et Mesures*). The BIPM, located in Sèvres, a suburb of Paris in France, has the responsibility of continuing unification of measurements. It operates directly under CIPM, which functions something like an executive committee of CGPM. The CGPM elects the members of the CIPM. Currently, the CIPM, which has 18 members, each belonging to a different member state, meets every year at the BIPM.

As already discussed, the first CGPM in 1889 sanctioned the international prototypes for the *metre* and the *kilogram.* Along with the astronomical *second* as the unit of time, these three units comprised a three-dimensional mechanical unit system (the MKS system) similar to the CGS system. The 10th CGPM, in 1954, approved three more base units, the *ampere*, the *kelvin* and the *candela* for electric current, thermodynamic temperature, and luminous intensity respectively, and declared the MKS system as an internationally suitable system. The 11th CGPM in 1960 adopted the name *International System of Units* (*Le Systeme Internationale de Unites*) for the system. The abbreviation *SI* (*Systeme Internationale*) is same in all languages. The 14th CGPM in 1971 adopted one more basic unit, the *mole* (mol) for the amount of substance, thus bringing the tally of base units to seven. The 14th CGPM also adopted the pascal (Pa) as a derived unit of pressure equal to 1 N per square metre.

The 26th General Conference on Weights and Measures (CGPM) held from 13 to 16 November 2018, voted unanimously to accept the revised definitions of four SI base units, namely, kilogram, ampere, kelvin, and mole, which the International Committee for Weights and Measures (CIPM) had proposed. The new definitions are effective from 20 May 2019, the World Metrology Day.

19.3 The SI System

The 11th CGPM in 1960 adopted the name *International System of Units* (*Le Systeme Internationale de Unites*) or the SI system for the MKS system with some additional rules and regulations. As seen already, the SI system is a decimal system; each unit is 10 times larger than the previous unit. Any measurement given in one SI unit can be changed to another SI unit by just moving the decimal place. For simplicity, very large and very small measurements are expressed as multiples of 10 of the base unit. All these subunits have names, and they are useful for measuring very large or very small things.

The SI system has devised specific rules for the prefixes, the derived units, and related affairs concerned with units. The system is created on seven clearly defined *base units* for seven *base quantities*, the *metre*, the *kilogram*, the *second*, the *ampere*, the *kelvin*, the *mole*, and the *candela* (Table 19.1). There is also provision for *derived quantities*, which are defined in terms of the seven base quantities through a system of algebraic relations linking the corresponding quantities (Table 19.2). Derived names are often given special names and symbols, for example: degree Celsius, coulomb, siemens, volt, ohm, joule, newton, lux, hertz, radian, watt, and pascal (Table 19.4). Researchers must be abreast with the latest trends in SI. Each SI base unit has a history of its own, and as the need arises, it would become essential to redefine the units. The story of metre and kilogram is worth to mention as examples.

The French Academy of Sciences in 1791 accepted the definition of the *metre*, the base unit of length, as 'equal to one-ten millionth of a quadrant (one-fourth the circumference of earth, from the pole to the equator) of the earth's meridian'. Later on, it was observed that the first prototype of metre was less by 0.2 mm, because of

Table 19.1 Base units of the SI system

Base quantity	Name of the base unit	Symbol
Length (l)	Metre	m
Mass (not weight) (m)	Kilogram	kg
Time (t)	Second	s
Thermodynamic temperature (T)	Kelvin	K
Electric current	Ampere	A
Amount of substance (n, Q)	Mole	mol
Luminous intensity	Candela	cd

Table 19.2 Examples of SI derived units from the basic units

Derived quantity	Name	Symbol
Area	Square metre	m^2
Volume	Cubic metre	m^3
Speed, velocity	Metre per second	m/s
Acceleration	Metre per second squared	m/s^2
Density	Kilogram per cubic metre	kg/m^3
Specific volume	Cubic metre per kilogram	m^3/kg
Amount-of-substance	Mole per cubic metre *or* mole charge per litre	mol/m^3 mol_c/L
Luminance	Candela per square metre	cd/m^2

an error in calculation due to earth's rotation. In 1889, a new prototype made of an alloy of platinum with 10% iridium was accepted, and the metre was redefined as the 'length of the prototype of platinum-iridium bar at the melting point of ice'. In 1960, a new definition was accepted based on the wavelength of radiation from krypton 86 atom. In 1983, scientific advances necessitated once again to redefine the metre as 'the length of the path travelled by light during a time interval of 1/299,792,458 of a second'. The definition of metre has been slightly changed by 26th CGPM held during Nov. 2018, which will be effective from May 2019. The revised definition is: 'the metre, the unit of length, is defined by taking the fixed numerical value of the speed of light in vacuum c to be 299,792,458 when expressed in the unit m s^{-1}'.

Similarly, the *kilogram* was originally defined as the mass of a cubic decimetre of water at the ice point, and it was like that way until the end of the eighteenth century. In 1889, the first CGPM sanctioned the international prototype of the kilogram, made of platinum-iridium. Yet, there was some ambiguity in the use of the words, *weight* and *mass*. The third CGPM (1901), in a declaration confirmed that, 'The kilogram is the unit of mass; it is equal to the mass of the international prototype of the kilogram'. The 26th CGPM drastically changed the definition of the kilogram. The new definition effective from May 2019 is: 'The kilogram is the unit of mass, and is defined by taking the fixed numerical value of the Planck constant h to be

$6.62607015 \times 10^{-34}$, when expressed in the unit J s, which is equal to kg m^2 s^{-1}, where the metre and the second are defined in terms of c and $\Delta\nu$Cs'.

Along with the popularization of SI units, detailed rules and style conventions for using the SI units for scientific use have also been devised. This is important to ensure that scientific literature is not hindered by ambiguity. It is true that because of continued usage, people are reluctant to abandon certain commonly used non-SI units and accept equivalent SI units in their place. Some salient points in relation to SI style conventions that must be followed while writing are attempted here. For a detailed account on SI units, readers may visit BIPM website (https://www.bipm.org/) or refer to Thompson and Taylor (2008).

In agriculture and allied fields too, the International System of Units must be followed. This means sometimes giving up familiar units, and in some cases, the new units may look unreasonable and even illogical. However, with the passage of time, several SI units once seemed unacceptable became accepted. It is hoped that greater familiarity with the system would lead to frequent use and widespread acceptance.

Detailed information on the SI system is given in Tables 19.1, 19.2, 19.3, 19.4, 19.5, 19.6 and 19.7. The base units with name and symbols are shown in Table 19.1. A few examples of SI derived units from the basic units are given in Table 19.2, and for ease of understanding and convenience, important derived units have been given special names and symbols (Table 19.3). For the units, multiples of 1000 are suggested and are named by the use of prefixes to the basic units (Table 19.4).

Some special units, which are outside the SI system, but accepted for use along with the SI are shown in Table 19.5. There is still another group of units outside the SI system, which are currently accepted for use with the SI, but subject to further review are indicated in Table 19.6. In Table 19.7, factors for conversion of non-SI units to acceptable SI units with abbreviations are given.

Definitions of the SI Base Units

There are seven well-defined base units, the *metre*, the *kilogram*, the *second*, the *ampere*, the *kelvin*, the *mole*, and the *candela*. Universally accepted definitions of the units are given below.

Length: The *metre* is the length of the path travelled by light in vacuum during a time interval of 1/299,792,458 of a second. The symbol is m.

Mass: The kilogram is equal to the mass of the international prototype of the *kilogram*, made of platinum-iridium. The symbol is kg.

Time: The *second* is the duration of 9,192,631,770 periods of the radiation corresponding to the transition between the two hyperfine levels of the ground state of the cesium 133 atom. The symbol is s.

Thermodynamic temperature: The *kelvin* is defined as the fraction 1/273.16 of the thermodynamic temperature of the triple point of water. The symbol is K.

Electric current: The *ampere* is the current required to produce, in vacuum, a force equal to 2×10^{-7} N/m between two parallel conductors of infinite length and negligible cross section, 1.0 m apart. The symbol is A.

Table 19.3 Important derived units, with special names and symbols

Derived quantity	Name	Symbol	Expression in terms of other SI units	Expression in terms of SI base units
Activity (of a radionuclide)	Becquerel	Bq	–	s^{-1}
Celsius temperature	Degree celsius	°C		K
Electric charge (Q) quantity of electricity	Coulomb	C	–	s A
Electric conductance (G)	Siemens	S	A/V	m^{-2} kg^{-1} s^3 A^2
Electric potential difference (ψ), electromotive force	Volt	V	W/A	m^2 kg s^{-3} A^{-1}
Electric resistance (R)	Ohm	Ω	V/A	m^2 kg s^{-3} A^{-2}
Energy (E), work (W), quantity of heat	Joule	J	N m	m^2 kg s^{-2}
Force (F)	Newton	N	–	m kg s^{-2}
Frequency (v)	Hertz	Hz	–	s^{-1}
Illuminance	Lux	lx	lm/m^2	m^2 m^{-4} cd = m^{-2} cd
Inductance	Henry	H	Wb/A	m^2 kg s^{-2} A^{-2}
Luminous flux	Lumen	lm	cd sr	cd
Plane angle	Radian	rad	1	m m^{-1} = 1
Power, radiant flux	Watt	W	J/s	m^2 kg s^{-3}
Pressure, stress (P)	Pascal	Pa	N/m^2	m^{-1} kg s^{-2}
Solid angle	Steradian	sr	1	m^2 m^{-2} = 1

Table 19.4 Prefixes to indicate multiples and submultiples of the base or derived units

Prefix	Multiple	Symbol	Prefix	Multiple	Symbol
yotta	10^{24}	Y	deci	10^{-1}	d
zetta	10^{21}	Z	centi	10^{-2}	c
exa	10^{18}	E	milli	10^{-3}	m
peta	10^{15}	P	micro	10^{-6}	μ
tera	10^{12}	T	nano	10^{-9}	n
giga	10^9	G	pico	10^{-12}	p
mega	10^6	M	femto	10^{-15}	f
kilo	10^3	k	atto	10^{-18}	a
hecto	10^2	h	zepto	10^{-21}	z
deca	10^1	da	yocto	10^{-24}	y

Table 19.5 Non-SI units that are accepted for use with the SI units

Name	Symbol	Value in SI units
Minute (time)	min	1 min = 60 s
Hour	h	1 h = 60 min = 3600 s
Day	d	1 d = 24 h = 86 400 s
Degree (angle)	°	$1° = (\pi/180)$ rad
Minute (angle)	′	$1' = (1/60)° = (\pi/10\,800)$ rad
Second (angle)	″	$1'' = (1/60) = (\pi/648\,000)$ rad
Litre	L	$1\ \text{L} = 1\ \text{dm}^3 = 10^{-3}\ \text{m}^3$
Metric ton/tonne	t	$1\ \text{t} = 10^3\ \text{kg} = 1\ \text{Mg}$
Bel	B	B = (1/2) ln 10 Np

Table 19.6 Other units outside the SI, which are currently accepted for use with the SI, subject to further review

Name	Symbol	Value in SI units
Nautical mile		1 nautical mile = 1852 m
Knot		1 nautical mile per hour = (1852/3600) m/s
Are	a	$1\ \text{a} = 1\ \text{dam}^2 = 100\ \text{m}^2$
Hectare	ha	$1\ \text{ha} = 1\ \text{hm}^2 = 10\,000\ \text{m}^2$
Bar	bar	$1\ \text{bar} = 0.1\ \text{MPa} = 100\ \text{kPa} = 1000\ \text{hPa} = 10^5\ \text{Pa}$
Millimetre of mercury	mmHg	1 mmHg ≈ 133.322 Pa
Ångström	Å	$1\ \text{Å} = 0.1\ \text{nm} = 10^{-10}\ \text{m}$
Barn	b	$1\ \text{b} = 100\ \text{fm}^2 = 10^{-28}\ \text{m}^2$
Curie	Ci	$1\ \text{Ci} = 3.7 \times 10^{10}\ \text{Bq}$
Roentgen	R	$1\ \text{R} = 2.58 \times 10^{-4}\ \text{C/kg}$
Rad	rad	$1\ \text{rad} = 1\ \text{cGy} = 10^{-2}\ \text{Gy}$
Rem	rem	$1\ \text{rem} = 1\ \text{cSv} = 10^{-2}\ \text{Sv}$

Amount of substance: The *mole* is the amount of substance of a system, which contains as many elementary entities as there are atoms in 0.012 kg of carbon 12. The symbol of mole is 'mol'.

When the 'mole' is used to express the amount of substance, the elementary entities must be specified. The entities may be atoms, molecules, ions, electrons, or specified groups of particles. Photons are also included. One mole of a substance contains Avogadro's number of particles (the number of atoms in 0.012 kg of carbon 12 is equivalent to 6.022045×10^{23} particles).

Luminous intensity: The candela (symbol, cd) is the unit of luminous intensity in a given direction. The *candela* is defined as the luminous intensity in a given direction of a source that emits monochromatic radiation of frequency 540×10^{12} Hz, and that has a radiant intensity in that direction of 1/683 W per steradian.

Table 19.7 Factors for converting non-SI units to acceptable SI units with standard abbreviations

(1) Non SI units	(2) Multiplication factor $(F)^{a}$ (Col. 1 × F = Col. 3)	(3) SI units
Length		
Angstrom unit, Å	0.1	Nanometre, nm (10^{-9} m)
Chain	20.12	Metre, m
Foot, ft	0.3048	Metre, m
Furlong (660 feet)	201.168	Metre, m
Link	0.2012	Metre, m
Inch, in.	25.40	Millimetre, mm (10^{-3} m)
Micron, μ	1	Micrometre, μm (10^{-6} m)
Millimicron (mμ)	1	Nanometre, nm (10^{-9} m)
Mile	1.60934	Kilometre, km (10^{3} m)
Nautical mile	1.852	Kilometre, km (10^{3} m)
Yard	0.9144	Metre, m
Mass		
Dram, dr	1.772	Gram, g (10^{-3} kg)
Grain, gr	0.0648	Gram, g (10^{-3} kg)
Ounce, oz	28.3495	Gram, g (10^{-3} kg)
Pound, lb	453.592	Gram, g (10^{-3} kg)
Quintal, q	10^{2}	Kilogram, kg
Tonne or metric ton, t	10^{3}	Kilogram, kg
Tonnc or mctric ton, t	1	Megagram, Mg
Ton, short (2000 lb), t	907.185	Kilogram, kg
Ton, long (2240 lb)	1016.5	Kilogram, kg
Tons (2000 lb) per acre, t/a	2.242	Megagram per hectare, Mg ha^{-1}
Volume		
Gallon, gal (US)	3.78531	Litre, L (10^{-3} m^{3})
Ounce, fluid, oz.	0.029573	Litre, L (10^{-3} m^{3})
Bushel, bu (U.S.)	35.2381	Litre, L (10^{-3} m^{3})
Cubic feet, cu.ft.	0.028317	Cubic metre, m^{3}
Cubic feet, cu.ft	28.3165	Litre, L (10^{-3} m^{3})
Pint	0.437	Litre, L (10^{-3} m^{3})
Quart	0.946	Litre, L (10^{-3} m^{3})
Area		
Acre, a	0.404685	10^{4} m^{2} (= 1 ha)
Acre, a	4046.85	Square metre, m^{2}

(continued)

Table 19.7 (continued)

(1) Non SI units	(2) Multiplication factor (F)[a] (Col. 1 × F = Col. 3)	(3) SI units
Are, a	1	Square decametre, dam^2 (10^2 m^2)
Are, a	10^2	Square metre, m^2
Hectare, ha	10,000	Square metre, m^2
Hectare, ha	0.01	Square kilometre, km^2
Square feet	0.0929	Square metre, m^2
Square inch	645	Square millimetre, mm^2
Square mile	2.59	Square kilometre, km^2
Others		
Atmosphere	1.01325	10^5 Pa (= 1 bar)
Bar	10^5	Pascal, Pa
Bar	10^2	Kilopascal, kPa
Bar	0.1	Megapascal, MPa
Calorie, C	4.184	Joule, J
Calorie per square centimetre minute (irradiance)	698	Watt per square metre, W m^{-2}
Calorie per square centimetre (Langley)	418,40	Joules per square metre, J m^{-1}
Curie	3.70×10^{-10}	Becquerel, Bq
Degree (angle)	1.75×10^{-2}	Radian, rad
Degree centigrade	1	Degree Celsius (°C)
Dyne	10^{-5}	Newton, N
Erg	10^{-7}	Joule, J
Gram per cubic centimetre	1	Megagram per cubic metre, Mg m^{-2}
Gauss	10^{-4}	Tesla, T
Foot pound, ft-lb	1.356	Joule, J
Einstein(E)	1	Mole of photons or quanta (mol)
Metre head of water	9.8	kPa
mm mercury	133.0	Pa
Micromhos/cm (μmhos/cm)	0.1	Millisiemen per metre, mS m^{-1}
Mile per hour	0.477	Metre per second, m s^{-1}
Millimhos/cm (mmhos/ cm)	1.0	Decisiemen per metre, dS m^{-1}

(continued)

Table 19.7 (continued)

(1) Non SI units	(2) Multiplication factor $(F)^{a}$ (Col. 1 × F = Col. 3)	(3) SI units
Pound per acre, lb/a	1.121	Kilogram per hectare, kg ha^{-1}
Pounds per square foot, lb/ft^2	47.88	Pascal, Pa
Parts per million (ppm)	1	mg kg^{-1}; μL L^{-1} (liquids); μmol mol^{-1} (gases)
Parts per billion (ppb)	1	μg kg^{-1}; nL L^{-1} (liquids); nmol mol^{-1} (gases)
Temperature (°F − 32°)	0.5555	Temperature, °C
Temperature (°C + 273.16)	1	Temperature, K

[a]Dividing the SI units in column 3 by the factor in column 2 will give non-SI Units in column 1

Please note that the 26th General Conference on Weights and Measures (CGPM) held from 13 to 16 November 2018, voted unanimously to accept the revised definitions of four SI base units, namely, kilogram, ampere, kelvin, and mole, which the International Committee for Weights and Measures (CIPM) had proposed. The new definitions are effective from May 20, 2019. The definitions of kilogram, ampere, kelvin, and mole have been revised by setting precise numerical values for the Planck constant (h), the elementary electric charge (e), the Boltzmann constant (k), and the Avogadro constant (NA), respectively. The metre, second, and candela have already been defined by physical constants, subject to correction to their present definitions. The objective of revised definitions is to improve the SI system without changing the size of any units, thus ensuring continuity with existing measurements. Common people, consumers, and most trades and industries may not feel any immediate impacts, but scientists and researchers expect the changes to inspire new technologies and reduce the cost of calibrating several processes and instruments. For information on revised definitions, please visit BIPM website: https://www.bipm.org/en/measurement units/rev-si/.

For historical reasons, the gram is not the SI base unit for mass. The kilogram is the only base unit with a prefix. One gram equals 0.001 kg. Further, note that weight is technically a measure of the force produced by gravity whereas kilogram is a unit of mass. The proper unit of weight is the Newton.

19.4 SI Unit Rules and Style Conventions

The SI system stipulates specific rules for writing units and symbols. Separate style conventions have also been proposed. Important among them are indicated here.

Name of Units

Begin all unit names in lower case letters, except at the start of a sentence or in titles in which all main words are capitalised, for example: metre, kilogram, and second. Exception to the rule is 'degree Celsius' (note that 'degree centigrade' is obsolete and no longer used).

Names of units are considered plural with numerical values greater than 1, equal to 0, and less than -1. All other numerical values get the singular form of the unit name. Examples: 200 metres, 1.6 metres, 0 degrees Celsius, -6 degrees Celsius, 0.6 metre, 1 metre, and -0.5 degree Celsius. However, do not use unit symbols in plural form. The symbols must be used only in singular form, for example, 45 g and *not* 45 gms.

For compound units involving division, use a slash or solidus (/) to form the symbol with no space before or after the slash (e.g. kg/m^2) or use negative exponents (e.g. kg m^{-2}). If a compound unit involving division is spelled out, the word *per* is used, for example, metres per second. However, only one *per* (spelled form of slash) is permitted in a written unit name. Using two or more 'pers' or slashes in the same expression is not appropriate because they cause ambiguity. To avoid this problem, negative superscripts can be used. For example, instead of writing kg/m^2/s, write either kg/m^{2}s (as all symbols to the right of the slash belong to the denominator) or kg m^{-2} s^{-1}.

Tonne and Mega Gram

Take care when the unit 'ton' is used. In SI system, it must be either 'tonne' or 'metric ton' to avoid confusion with 'short ton' and 'long ton'. The old 'short ton' (2000 pounds) and 'metric ton' are with the same symbol, 't'. Another 'long ton' (2240 pounds) was also in use. Use 'megagram' (Mg) instead of 'tonne' (t) to avoid confusion with these units.

Kelvin and Celsius

Use of degree Celsius and the symbol °C require some caution. The base unit of thermodynamic temperature (T) is measured with respect to its difference from the reference temperature at the ice point, $T_0 = 273.16$ K. For ordinary use, however, the temperature difference with symbol t ($t = T - T_0$) is measured in degree Celsius, a derived unit of temperature. In other words, the numerical value of temperature difference expressed in degree Celsius (°C) is equal to the numerical value of the same difference when expressed in kelvin (note that it is K and *not* °K).

Please note that 0 K is equal to -273.16 °C; therefore, thermodynamic temperature, K = (°C + 273.16). For example, 30 °C is equal to 303.16 K. The unit, degree

Celsius, may be used in combination with SI prefixes, for example, millidegree Celsius (m °C).

Litre and Cubic Metre

'Litre' is a non-SI unit accepted for use with the SI units. It is widely used for common volume measurements. Take care when using the litre and its symbol 'l', which were originally adopted by the CIPM in 1879. In 1979, the CGPM adopted an alternative symbol for the litre, L, to avoid confusion between the English letter 'l' and the number '1'. The litre is now defined as the volume of 1 kg of pure water at its maximum density (which is about 4 °C) and is equal to 1.000028 dm^3, although it is often defined in general books as 1 dm^3. Note that cubic metre is the accepted SI unit for volume, which is equal to 1000 L.

Candela, Lumens, and Lux

The SI unit *candela* (cd) indicates how bright the light source is, or it measures luminous intensity. Candela is from candle, which is roughly equivalent to the light from a single candle. The candela is based on the sensitivity of human eye. The *lumen* (lm) is a measurement of luminous flux or the total amount of visible light. Lux (lx) measures illuminance, which is the amount of light on a surface per unit area. A single lux is equal to one lumen per square meter. Remember candela measures the visible intensity from the light source, lux is how bright your surface will be, and lumen is how much light is given off from a source.

$$1\,\mathrm{lm} = 1\,\mathrm{cd\,sr}$$

$$1\,\mathrm{lx} = 1\,\mathrm{lm/m^2} = 1\,\mathrm{cd\,sr/m^2}.$$

Pascal and Bar

Pascal is a derived unit of pressure under SI with a special name. Bar is a unit outside SI, but currently accepted for use with the SI subject to further review. 1 bar = 100 kPa = 10^5 Pa. In meteorology, bar is widely used. For measuring water potential, multiples of pascal (kPa and MPa) are preferred than bar. Old units such as atmosphere or pF should not be used for measuring pressure or water potential.

Use of Prefixes

Use only accepted prefixes by CIPM for denoting multiples or submultiples of base or derived units (Table 19.4).

Apply only one prefix to a unit name (e.g. nm, *not* mμ m). The prefix and unit name are joined together without a hyphen or space as in 10 Mg.

Symbols for prefixes greater than kilo are capitalized. This rule is important because some letters are the same as some symbols or prefixes, for example, G for giga and g for gram; and M for mega and m for both milli and metre.

Unit Symbols

Use symbols, when units are used along with numerals, for example, 20 g and 20 m. Do not use unit symbols in plural form. They must be used only in singular form, for example, 45 g and *not* 45 gms. Periods or full stops should not be used after unit symbols unless at the end of a sentence.

Wrong form: The vine is 95 cm. long.
Correct forms: The vine is 95 cm long.
The length of the vine is 95 cm.

For writing or printing unit symbols, use lower-case letters only, for example, m (metre) and s (second). However, symbols for units named after individuals have the first letter capitalized, but the name of the unit is written in lower case, whereas most other symbols are not capitalized. Examples include kelvin, K (from Lord William Thomson Kelvin (1824–1907)) and pascal, Pa (from Blaise Pascal (1623–1662)). However, degree Celsius, °C, is an exception to this rule. A recent exception is the litre symbolized with a capital 'L' to avoid confusion with numeral 1.

Do not begin a sentence with a symbol. When decimal values less than 1.0 are used, they should have a zero to the left of the decimal, for example, 0.85 g, and not .85 g.

For printing unit symbols, only roman fonts (upright) are to be used regardless of the typeface used in the surrounding text.

Use numerical superscripts (^{2} and ^{3}) to indicate squares and cubes; and never use sq., cu., etc. along with units.

Leave a space between the last digit of a numeral and its unit symbol and between unit symbols when more than one is used, for example, 40 kg iron rod and *not* 40kg iron rod or 40-kg iron rod. However, if you are using the spelled out name of a unit, follow the normal rules of English, for example, it is correct to say; 'a 30-m measuring tape'.

Do not mix information with unit symbols or names. For example, instead of writing, 'It contains 25.5 mL H_2O/kg' or 'It contains 25.5 mL of water/kg', write 'The water content is 25.5 mL/kg'.

Do not mix unit symbols and spelled out unit names, and never mix SI units with units of another system.

Wrong forms: kilogram/m^3, kg/cubic metre, kilogram/cubic metre, kg per m^3, kilogram per metre^3, or kg per cubic metre.
Correct forms: kg/m^3, kg m^{-3}, or kilogram per cubic metre.

Use of Abbreviations

Use only standard abbreviations for unit names, prefix names, unit symbols, and prefix symbols. For example, non-standard abbreviations such as sec, sq.cm, cc, and mps are not acceptable for second, cubic centimetre, and metre per second. Instead, use s, cm^2, cm^3, and m/s.

SI system do not encourage the use of 'part per million', 'part per billion', or their equivalents ('ppm', 'ppb', etc.) to depict the values of quantities. Instead, express clearly the values as mg/kg or μL/L for ppm and μg/kg or nL/L for ppb.

Mathematical Notation

When using unit symbols along with mathematical notation, care must be taken to write them unambiguously. Avoid wrong forms as indicated below.

Wrong forms	*Correct forms*
80 ± 5%	80% ± 5%; (80 ± 5)%
100 ± 5 g	100 g ± 5 g; (100 ± 5) g
55 × 32 cm	55 cm × 32 cm
15 °C–25 °C; 15 to 25 °C	15 °C to 25 °C; (15 to 25) °C.

Digit Spacing

In writing SI units, divide the digits which occur on either side of the decimal point into groups of three using a single space counting from both the left and right of the decimal point (remember, in popular writing, instead of spaces, commas are used for this purpose). However, if there are only four digits, this rule is not insisted.

Wrong forms: 45682.12785; 45,682.1278; 56,000,000; 6 000
Correct forms: 45 682.127 85; 56 000 000; 6000.

Object and Quantity

Differentiate clearly an *object* and any *quantity* describing the object. For example, understand clearly the differences between 'surface' and 'area', 'body' and 'mass', 'resistor' and 'resistance', and 'coil' and 'inductance'. Accordingly, for example, instead of 'an area of 100 m^2', write, 'a surface area of 100 m^2'.

When using a quotient quantity, write it unambiguously. For example, instead of writing 'mass per unit volume', write 'mass divided by volume'.

Multiplication and Division of Units

Show the symbols for combined units formed by multiplication using either a centred dot or a space. Spaces are normally preferred, and dots are used only if confusion occurs without their use, for example, m · s or m s.

For showing symbols for units formed from other units by division, use a solidus or oblique stroke (/), a horizontal line, or a negative exponent. Examples: m/s, m.s^{-1}, or m s^{-1}.

Do not repeat the solidus on the same line to avoid vagueness unless parentheses are used. For example, it is correct to use m/s^2 or m s^{-2} but m/s/s is not appropriate.

19.5 Non SI Units Accepted for Use

Certain non-SI units are still widely used due to many reasons. The units in this group that are accepted for use along with the SI units are given in Table 19.5.

Although 'litre' is outside the SI, it is widely used for common measurements. Take care when using the litre and its symbol 'l', which were originally adopted by the CIPM in 1879. In 1979, the CGPM adopted an alternative symbol for the litre, L, to avoid confusion between the English letter 'l' and the number '1'. The litre is now defined as the volume of 1 kg of pure water at its maximum density (which is about 4 °C) and is equal to 1.000,028 dm^3, although it is often defined in general books as 1 dm^3.

Note the spelling of ton. In SI system, it must be either tonne or metric ton to avoid confusion with short and long tons. However, the old short ton (2000 pounds) and metric ton are with the same symbol, 't'. Another long ton (2240 pounds) was also in use. To avoid confusion with these units, use of megagram (Mg) rather than tonne (t) is encouraged.

Note that the unit bel is most commonly used with the SI prefix deci: 1 dB = 0.1 B.

Please note that plant physiologists and biochemists usually use dalton (Da) as the unit of atomic mass (1 Da = 1 g mol^{-1}), although it is not an SI unit. However, it has been accepted for use by the International Union of Pure and Applied Chemistry (IUPAC).

Some popular units which are outside the SI system, but currently adopted for use along with SI subject to further review includes *hectare, are, bar, nautical mile, knot, angstrom, barn,* and *millimetre of mercury* (Table 19.6). The units *are* (1 a = 1 da m^2 = 100 m^2) and *hectare* (1 ha = 100 a = 10,000 m^2) are widely used in agriculture for measuring land areas. You may also find widespread use of *acre* and *cent* among farmers, but never use them for scientific purposes. Similarly, the *bar* (bar) is widely used as a unit of pressure but multiples of the *pascal* (Pa) are preferred.

You may find several non-SI units still in use for ordinary transactions. It may take some time to phase out such well-established units. The factors given in Table 19.7 are helpful to convert non-SI units to acceptable SI units and vice versa.

Please note that the units such as *atmosphere, micron, calories, degree centigrade, Einstein, parts per million (ppm),* and *parts per billion (ppb)* are discarded units. Use acceptable SI units using the multiplication factor.

19.6 Care with Numerals

There are several issues to be considered while using numerals in scientific reports. Common errors with numerals and the correct way of presenting them are indicated here.

Writing Small Numbers
Examine the following sentences:

- The flower has 5 petals.
- Out of 300 patients admitted, eleven died.

The above sentences need some corrections. In writing, always spell out one digit numbers, but numbers of two or more digits should be expressed as numerals. Change the sentences to:

- The flower has five petals.
- Out of 300 patients admitted, 11 died.

There are some exceptions to the above convention, which include:

1. With SI units, even if they are one-digit numbers, do not spell them out (e.g. 8 g, 6 m, and 5 m^2)
2. If the number is a decimal, you need not spell it (e.g. Guinea grass has 9.0 percent crude protein in its herbage)
3. For writing dates, use simple numbers only (e.g. April 4)
4. For numbers after labels (e.g. Chapter 7)
5. For numbers in a list (e.g. 5 mangoes, 2 bananas, and 3 oranges)
6. When the context includes larger numbers and small numbers (e.g. five samples; but 6 and 12 samples and *not* six and 12 samples)
7. Use numerals for all pages, days, figures, and diagrams even if they are less than 10.
8. Do not write two numbers together either as figures or as numbers as they contribute to ambiguity. For example, write 'three 100 ml bottles,' and *not* '3 100 ml bottles' or 'three hundred ml bottles.'

Spell Out Numbers Occurring at the Beginning of a Sentence
Consider the sentence: 72 ha *of land were developed for rubber*. Do not start a sentence with a numeral like this way. Always spell out numbers that occur at the beginning of a sentence. Amend the above sentence as: *Seventy-two hectares of land were developed for rubber*. If necessary, recast a sentence, especially those involving SI units, to eliminate numerals at the beginning as in the example: A *land of* 72 ha *was developed for rubber*.

Percentage
Percent (sometimes written per cent) and percentage denote rate per hundred. Percent is used with numbers and percentage without numbers, for example: a small

percentage (*not* small percent). Look at the sentence: *The new rice cultivar has 8.5% protein in its grains.* It is recommended to spell out '%' in between sentences. The above sentence shall be corrected to: *The new rice cultivar has 8.5 percent protein in its grains.* The symbol for percentage (%) is most appropriate for cases in parenthesis and tabular data, for example: *The fodder contains high amount of crude protein* (30%).

Wherever possible, the percentage should be defined to avoid ambiguities. For example, distinguish clearly 'percentage by mass' from 'percentage by volume'.

You may exercise some discretion when using percentages in scientific reports. Gustavii (2008) highlighted some conventions of scientific writing regarding their use.

1. When the total number is less than 25, avoid using percentages.
2. When the total number is between 25 and 100, percentages should be expressed without decimals (8%, *not* 8.4%).
3. When the total number is between 100 and 10,000, add one decimal (8.4%, *not* 8.42%).
4. If the total number exceeds 10,000, add two decimals (8.42%).
5. Include the original data along with percentages for clarity, for example: *Out of 200 plants exposed, wilting occurred in 80 plants* (40%).

Fractions

Wherever possible, use decimals instead of fractions in writing. For values less than one, always insert a zero before the decimal, for example: 0.73% (*not* 0.73%). When fractions must be used, write them out using the hyphen, for example one-third, but half-filled (*not* one-second).

Dates

- Use numerals for the year, but name the month to avoid misrepresentation of a date. For example, write 08 June 2018 (or June 8, 2018) instead of 08.06.2018 or 06.08.2018.
- When date is given, use a comma between the date and the year as in May 12, 2012.
- If a date is written as day–month–year, do not use commas as in 12 May 2012.
- If the month and year only are given, comma should not be used as in May 2012.
- We write AD 500, but 500BC (*not* 500AD or BC 500).

Abbreviations of Days and Months

While writing, abbreviations of months and days can be used. Months with more than five or more letters can have abbreviations. Use the following recognized abbreviations:

Days: Sun., Mon., Tues., Wed., Thurs., Fri., Sat.
Months: Jan., Feb., Mar., Apr., May, June, July, Aug., Sep., Oct., Nov., Dec.

Writing Digits

In ordinary writing, the digits of numerical values having more than three digits are separated into groups of three (thousands) using commas. However, comma is not usually put when there are only four digits. Examples: 15,689,000, 15,000, and 1550. A modern trend is to insert spaces instead of commas, for example: 15 689 000, 15 000, and 21 000.

In SI system of units, instead of commas, the digits of values having more than three digits are separated into groups of three using a single space counting from both the left and right of the decimal point. As in the case of commas, omission of the space is preferred when there are only four digits. Examples: 26 000 000, 48 730.062 81, and 4500.

19.7 Roman and Arabic Numerals

Roman numerals are those used by the ancient Romans. They are letters converted into numbers, for instance $I = 1$, $V = 5$, $X = 10$, and so on. They do not have any zeroes and do not follow the digital system of arabic numerals. In fact, the concept of zero and the digital system are Indian origin, later adopted by the Arabs. This system of numerals came to be called arabic numerals, as the Europeans came to know of this system from the Arabs. Although roman numerals are no longer used in mathematical connotations or scientific purposes, it is still used for certain purposes such as numbering chapters in a book or numbering items. The general rules of roman numerals with examples are given below.

- If a letter is repeated, its value is also repeated (XX = 10 + 10 = 20)
- A letter or letters placed after one of greater value adds thereto (VII = 5 + 1 + 1 = 7)
- A letter placed before another of great value subtracts there from (IX = 10 − 1 = 9)
- A dash over a numeral multiplies its value by thousand (X = $\overline{10}$ × 1000 = 10,000) (Table 19.8).

Higher Arabic Numerals

There is some confusion in the terminology of numerals of higher values. Up to one hundred million, there seems to be no problems. However, 10^9 is called billion in the USA and France and 10^{12} is called billion (one million million) in the UK and European countries. However, the latest trend is to follow the Americans and in scientific literature, people are increasingly using American style. Note that 'lakh' and 'crore' are used only in the Indian subcontinent. The higher arabic numerals from 10^5 to 10^{18} are shown in Table 19.9 for a comparison.

Table 19.8 Roman equivalent of some arabic numerals

Arabic	Roman	Arabic	Roman	Arabic	Roman
1	I	16	XVI	500	D
2	II	17	XVII	600	DC
3	III	18	XVIII	700	DCC
4	IV	19	XIX	800	DCCC
5	V	20	XX	900	CM
6	VI	30	XXX	1000	M
7	VII	40	XL	2000	MM
8	VIII	50	L	5000	V
9	IX	60	LX	6000	VM
10	X	70	LXX	10,000	$\overline{\text{X}}$
11	XI	80	LXXX	20,000	$\overline{\text{XX}}$
12	XII	90	XC	50,000	$\overline{\text{L}}$
13	XIII	100	C	100,000	$\overline{\text{C}}$
14	XIV	200	CC	500,000	$\overline{\text{D}}$
15	XV	400	CD	100,000	$\overline{\text{M}}$

Table 19.9 Higher arabic numerals from 10^5 to 10^{18}

Number	USA and France	UK and other European countries	India
10^5	Hundred thousand	Hundred thousand	One lakh
10^6	Million	Million	Ten lakhs
10^7	Ten million	Ten million	One crore
10^8	One hundred million	One hundred million	Ten crores
10^9	Billion	Milliard (thousand million)	Hundred crores
10^{12}	Trillion	Billion	–
10^{15}	Quadrillion	Thousand billion	–
10^{18}	Quintillion	Trillion	–

Bibliography

Barrass, R. 2002. Scientists Must Write: A Guide to Better Writing for Scientists, Engineers and Students (2nd Ed.). Routledge, London, 204p.

BIPM [Bureau International des Poids et Mesures] 2018. The International System of Units (SI) [homepage]. International Bureau of Weights and Measures (Bureau International des Poids et Mesures, BIPM). Available: https://www.bipm.org/en/measurement-units/ [Accessed 18 Nov. 2018].

BIPM [Bureau International des Poids et Mesures] 2006. SI Brochure: The International System of Units (SI) (8th Ed.) Text in English, 88p. Available: https://www.bipm.org/utils/common/pdf/si_brochure_8_en. [Accessed 12 Jan. 2013].

BIPM [Bureau International des Poids et Mesures] 2006. The International System of Units Supplement 2014: Updates to the 8th edition (2006) of the SI Brochure. Available: https://www.bipm.org/utils/common/pdf/si_supplement_2014.pdf. [Accessed 12 Jan. 2013]

CSE [Council of Science Editors] 2014. Scientific Style and Format: The CSE Manual for Authors, Editors, and Publishers (8th Ed.): The University of Chicago Press, 722p.

Gustavii, B. 2008. How to Write and Illustrate a Scientific Paper (2nd Ed.). Cambridge University press, Cambridge, 168p.

McCaskill, M.K. 1990. Grammar, Punctuation, and Capitalization: A handbook for Technical Writers and Editors. NASA SP-7084, National Aeronautics and Space Administration, Washington DC, 106p. Available: https://ntrs.nasa.gov/archive/nasa/casi.ntrs.nasa.gov/19900017394_1990017394.pdf [Accessed 10 Dec. 2003].

NIST [National Institute of Standards and Technology] 2018. International System of Units (SI). The NST Reference on constants, units and uncertainty. Available: https://physics.nist.gov/cuu/Units/index.html [Accessed 22 Oct. 2018].

NIST [National Institute of Standards and Technology] 2018. Historic vote ties kilogram and other units to natural constants. News: 16 Nov. 2018. Available: https://www.nist.gov/news-events/news/2018/11/historic-voteties-kilogram-and-other-units-natural-constants [Accessed 18 Nov. 2018].

Salisbury, F.B. and Ross, C.W. 1992. Plant Physiology (4th Ed.) Wadsworth Publishing Company, California, 682p.

Thompson, A. and Taylor, B.N. 2008. Guide for the Use of the International System of Units (SI). NIST Special Publication 811. National Institute of Standards and Technology, Gaithersburg (USA), 85p. Available: https://physics.nist.gov/cuu/pdf/sp811.pdf. [Accessed 10 Jan. 2018].

Chapter 20
Conference Presentations and Posters

If I have seen further than others, it is by standing upon the shoulders of giants.
Sir Isaac Newton (1642–1726)

Not only writing about research but also how to speak about research and research findings are important issues for a researcher. Scientists get opportunities to speak about their works in various capacities. Presenting papers on ongoing or completed research works in scientific meetings or conferences is such an opportunity. Sometimes, as an expert in the field, you may be invited to speak or lecture on the status of a certain subject, which would in all probability be a review of work done on the subject until then and future prospects. 'Keynote address' is another form of presentation, wherein, the invited expert gives a scholarly speech on the theme topic. In most institutes of higher learning, teaching and research go together. In such institutes, in addition to research works, scientists may also be giving lectures to students on advanced topics.

20.1 Scientific Meetings

There are several kinds and types of scientific meetings depending upon the purpose and the players involved. Scientific meetings may take many names—conference, seminar, symposium, congress, workshop, colloquium, panel discussion, lecture, or brainstorming. There are some differences between these meetings. Considerable overlaps between these meetings are also obvious.

Conference: Conference is a general term for a scheduled meeting for consultation, exchange of information, or discussion with a formal agenda. A scientific conference is a conference of researchers to present and discuss their works, and it is often known by names as symposium, congress, or conference itself. Most scientific societies regularly convene annual or biennial conferences.

Seminar: A seminar is a discussion-based meeting on a defined topic wherein an expert leads a small group of participants. The presenter often gives a lecture

C. G. Thomas, *Research Methodology and Scientific Writing*,
https://doi.org/10.1007/978-3-030-64865-7_20

followed by discussions. The word 'seminar' is from a Latin term *Seminarium,* meaning nursery or seed plot. A seminar has the function of bringing together small groups for frequent meetings, focusing each time on a particular subject.

Symposium: Symposium, referred originally to a drinking party after a banquet (the word 'symposium' is from a Greek word *Sympotein*, meaning, 'to drink together'), has now come to refer to an academic conference. A symposium is a prearranged meeting of a number of experts and research workers where different papers on a defined subject are presented and discussed; and at the end, some recommendations on the discussed problems are made. In a symposium, all the participants including presenters form the audience.

Congress: It is a formal gathering of delegates for discussion and generally action on some issues. A congress is larger than a symposium with many defined subjects. An example is Indian Science Congress, which is conducted every year.

Workshop: A workshop is an intensive meet by a small group of people to concentrate on a defined area of concern. In a workshop, all the participants are supposed to be involved in the discussion. The purpose of workshops may vary; it may be for information sharing, problem solving, formulation of new policies, or training.

Colloquium: Colloquium refers to an academic meeting at which specialists deliver speeches on related topics and then answer questions related to them. The word 'colloquial' means familiar conversation and the speakers are expected to speak in familiar language.

Convention: A convention is a large companionship meeting of a closely linked fraternal group such as an organization, a company, a society, or a political group for various purposes. It may not be a scientific meeting.

Lecture: Lecturing is a form of academic teaching, which involves larger student groups with less active participation. Invited lectures on various topics of interest by known personalities or experts are also common in research establishments.

Brainstorming: Brainstorming is a process of eliciting divergent ideas on a specific topic or problem from a group of people. Brainstorming with scientists is often conducted to solve certain contingency issues. Brainstorming can extract multiple ideas on a given topic, and the discussion that follows helps the group to explore and compare a variety of possible solutions to the specific problem.

20.2 General Features of a Scientific Conference

Conferences in academic fields are regular events under the auspices of government institutions, scientific societies, or organizations in the form of congress, symposium, or as conference itself. Presentation in meetings may take the form of oral or poster presentation. The organizers often identify a theme area, and invites applications from researchers working in the identified area for presentation of their research works in the conference.

Conferences are expensive events, and the organizers have to find out sources of fund for various expenses in connection with the conference such as food and accommodation, decoration and stage arrangements, transport, rent of premises, travel expenses of dignitaries, and publication expenses. The organizers usually seek the help of a funding agency for partly financing the event. They may also collect a fixed amount as registration fees from the participants to cover a part of the expenses.

Validity of Conference Presentations

Most researchers feel that presentation is a more rapid way of reporting research works than writing papers or articles. These meetings serve as an excellent forum for presentation and exchange of new ideas. You may also get an opportunity to mingle with experts and fellow workers in your specialization. However, presentation of a piece of scientific work in a conference in oral or poster form cannot be equated with a written paper on the subject in a reputed journal. Scientists and research students are often attracted to conferences because of the lure of easy acceptance of their papers. Before sending papers to conferences, you should check some important aspects. First, you have to judge the credibility of the conference. You should particularly ascertain the antecedents of the organizers and check whether it is simply a fund raising strategy of the organizers. Many 'conference proceedings' are normally not subjected to peer review, and because of any real quality control, 'proceedings paper' gets a low status. If the conference reports are not in full proceedings form (papers with the IMRAD structure, peer reviewed, and with the questions and answers at the conference), it is not considered as validly published primary literature. In any case, conference paper fetches less credit compared to a regular journal article. If the proceedings contain only 'abstracts' or 'extended abstracts', it is clear that these are not validly published, and therefore, once the conference is over, you should publish the data immediately in a regular primary journal.

You can approach the conference in a different way. Conferences, if conducted properly, serve as an excellent forum for presentation and exchange of new ideas. Both oral and poster presentations in such conferences can be used to report ongoing works. This gives the researcher an opportunity to disseminate new findings and to get valuable feedback from the audience. This kind of presentation at a meeting is also useful for the preparation of the paper for later publication because the audience's questions may probe weaknesses in the work or the researcher's account of it, and shows where s/he needs to expand and explain.

Submission of Abstracts and Papers

The participants are supposed to submit an abstract of their presentation along with registration form, which will be reviewed before finally accepting it for inclusion in the meeting. At the time of registration itself, the researchers have to indicate whether they intend to make oral presentation or poster presentation.

The abstract submitted for the conference must be short and concise confining to the word limit prescribed. Some conferences require presenters to submit a paper of about 6–15 pages, which will be peer reviewed by a group of referees or committee chosen by the organizers. Once the abstract is accepted for oral presentation, the author has to prepare for the presentation and present it at the chosen venue and time.

The script of all presentations (full papers of oral presentations and posters) are usually pooled together along with questions and answers and published as 'proceedings' in addition to the compilation of abstracts distributed before the conference. Sometimes, the organizers publish only a compilation of 'abstracts' or 'extended abstracts' for distribution before the start of the programme.

Conference Sessions

A conference usually consists of an *inaugural session*, *technical session*, and *plenary session*. Technical session may be conducted in one session or in different sessions. Usually, sessions are conducted one after the other so that the participants get a chance to attend all the sessions. However, in big symposia or congress, the sessions may be simultaneously conducted at different venues. Select those sessions, which are more relevant to you. Sometimes, for each technical session, there may be a 'keynote address' by a famous scientist, which will be followed by presentations from participants (Fig. 20.1). The keynote lecture is often longer, lasting up to an hour or sometimes more. A plenary session is a part of conference where all the participants cutting across different technical groups are invited to attend. Plenary sessions are usually scheduled at the start of a conference and in the latter half, during which prominent speakers give plenary speeches. Sometimes, keynote addresses are also arranged as a part of plenary sessions. The inaugural session is usually conducted as a plenary session.

Fig. 20.1 Keynote address is a common feature of many conferences

The technical session is controlled by a chairperson, often a senior and well-known scientist from among the participants. For each session, rapporteurs should also be arranged. Rapporteurs are also competent scientists, probably from among the participants or from the organizing institute itself, who are expected to prepare a report of the proceedings after each session. They have to write down the questions and answers that come up during the discussion. Usually, there will not be any discussion on the 'keynote address'.

While scheduling technical sessions, time for discussion should be allotted after each presentation. The speakers are expected to deliver a talk with the help of presentation aids, displaying key figures and research results in about 10 to 30 min. However, exact time limit is decided based on the total number of papers for presentation and total time at the disposal of the organizers. Often, each paper is discussed immediately after its presentation so that the participants will not forget what they want to ask. Sometimes, all the papers would be discussed after they were presented to save time, although it may create confusion and losing touch with the details presented in the earlier papers. During discussion, questions can be raised for clarification and for further details. In the concluding session, which is often conducted as a plenary, chairpersons of various sessions present a gist of the proceedings along with their comments and conclusions.

Poster presentations are integral part of conferences, and separate timings may be set apart for poster sessions. Participants who have submitted requests for poster presentation have to set up posters at the scheduled time and venue by following the already intimated guidelines for poster preparation and display by the organizers. For a general account on preparation of posters and display, read Sect. 20.5 in this chapter.

In addition to oral and poster presentations, sometimes, conferences also feature panel discussions and workshops on various issues. In most conferences, cultural programmes, tours, business meetings, and similar events are also arranged as part of the programme. Large conferences may also have stalls for displaying the advances made in the theme area.

Oral Presentation of Research Papers

Presentation of research papers is a routine affair in scientific conferences. You can see a set pattern for presentation in such meetings. As already discussed, it may take the form of oral presentation or poster presentation depending upon the importance and preference of the researchers. Booth (1996), Alley (2003), and Davis (2005) discussed various strategies to succeed in scientific presentations.

Oral presentation of research papers (see also Sect. 20.3) is an important part of technical sessions (Fig. 20.2). Try to present your results in a concise manner in the conference. Do not attempt to present too much data and information in a short presentation. Convey only the main points and stimulate the interest of the audience so that they can seek out additional information from the speaker in the course of subsequent discussion. In many cases, however, the speaker takes the first few minutes for introducing the subject thus wasting precious time. It is more important to use the valuable time allotted to you for the presentation of results and the message.

Fig. 20.2 Oral presentation of papers is an important feature of a conference

Similarly, a detailed account of materials and methods is also not required; only the major points are needed. Interested persons will have the opportunity to learn about the techniques by referring to the published work or through discussion with the speaker.

The presentation of results should receive maximum attention. Resist the temptation to present tables after tables of voluminous data. Even the most attentive listener may not have the patience or the capacity to absorb all that is being presented. Therefore, subject the voluminous data collected in the course of the investigation to careful analysis and project only the core of the findings.

It is very difficult to convey new information we obtained from our research works through speech alone. If you speak with the support of Liquid crystal display (LCD) projector, it would be more appealing to the participants. Presently, it almost became a norm to use an LCD projector to present salient findings, especially tables, charts, and photographs. It is relatively easy to prepare colourful slides using computer programmes, which can be projected through LCD projectors. Make good and effective slides to catch the attention of the viewers. For more details on preparation of presentation slides, see Sect. 20.4.

20.3 Planning and Delivering a Presentation

Similar to writing, scientists must also learn to speak and present well. Presentation is an important part of conferences and lectures. Normally, the reader of an article can choose his own speed, sometimes skipping unimportant paragraphs or coming back to the article later. However, a listener of a speech or lecture must accept the speed chosen by the speaker. You must note that while delivering a speech, there is no time to pause, to hang around difficult passages, or to study an illustration again. For example, one unknown word, which makes the listener stumble, may cause loss of contact with the speaker. The speaker must speak clearly in simple language and give clear indications when s/he moves on to the next point. This is important because while listening to a speech, the audience do not have the privilege of ideas separated by punctuations or paragraphs as in a written article. Continuous attention of the listeners is needed to make the speech effective. Understand that an irritated or antagonized listener will leave you mentally and will never come back to you until your lecture is over!

We do not call all speeches presentations. When we say ***presentation***, we mean a talk illustrated by visual contents. Of course, a speech can be made without any visual aids, but not presentations! In a presentation, an individual makes an effort to share his/her ideas, information, and knowledge with another individual or individuals through visual medium. Most scientists now uses a computer aided presentation, and therefore, the nuances of the old 'oral presentation' taught in communication or extension classes could not be applied fully.

The idea of lecturing or presentation in public is worrying to many, but thorough planning can obviate most of the fears of the presenter. In addition to confronting an audience, some other problems may also perturb the speaker, for example, the strict time limit, which may be 10 or 15 min, if it is a conference or symposium. It is an arduous task to combine and put into a brief format the results of research works over months or years. Probably because of these worries, some presentations become incomprehensible and ineffective. Nevertheless, with sufficient planning and preparation, anybody can succeed in presenting even complicated matters in an understandable form in the stipulated time.

Individuals differ in oratory abilities. Some are borne speakers and have the natural talent to captivate an audience. Even if you do not have this natural talent, there is no reason to worry; anyone can become a great speaker by hard work with good preparation and meticulous planning. Because of lack of preparation, certain individuals make an interesting subject dull or a simple one complicated! How you deliver the messages is important; it must be unambiguous and clear. For an effective presentation, it is important to give due attention to the following 15 guiding principles.

1. **The form of talk**

Plan the speech in such a way that it has an introduction, body of talk, and conclusion. A short introduction is always desirable for clarity. While introducing the topic, mention the context and motivation for presenting the topic. It is important to make the audience understand that your talk is important to them. Take care to avoid repeating the same type of introduction, especially when the conference theme is a single subject. Many novices spend considerable time to introduce the subject and may not get enough time to present the most important part of the presentation, the body of talk. In the case of a symposium presentation, the body of talk is composed of the results and discussion. In other cases such as seminars or lectures, you can divide the body of talk into various sections and subsections. Although body is the most important part of a speech or presentation, unless you take special interest, audience interest may go down during this part. Therefore, structure the content to be both logical and interesting using various techniques such as good visual aids, body language, and regulation of voice. You must also set apart some time for concluding your talk. The conclusion should be clear and the audience should understand that the end of the talk has reached. When you conclude, give emphasis on main points and their impacts and finally come to your recommendations including some future lines of work.

2. **Preparation and rehearsal**

Adequate planning and preparation are essential for successful presentation. Nervousness and anxiety of success are real blocks, which prevent the speakers from delivering effective presentations. Bear in mind that if you have prepared thoroughly, you can overcome most of your problems of nervousness and anxiety easily. In university departments, rehearsals are routine for speeches and presentations. Proper rehearsal exudes confidence, and during such rehearsals, time is checked and incomprehensible slides are eliminated. If there is a time limit for the presentation, adjust the presentation to this time limit when you rehearse. Remember that time taken by rehearsals should be shorter than that actually allowed at the meetings. If the time allotted is 15 min, plan to do it in 13 min, because at the meeting, some more time may be required. Prepare your notes neatly in large letters on $3'' \times 5''$ cards or small sized sheets. Cards are handy as they are thicker than ordinary paper. In addition to the points you want to discuss, you may also write on note cards important quotations or hints for anecdotes, jokes, etc., which you propose to describe for greater impact. Give each note-card a number and a sub-heading.

If possible, conduct a mock presentation before your colleagues, who critically evaluate your shortcomings and defects of your slides. It is also a good idea to record your speech, especially to find out the time taken for completing the speech and your shortcomings.

3. **Take care of your audience**

It is impossible to deliver a presentation without considering the audience. Considering the audience and their expectations will help you to choose the most suitable approach, the structure, the content, and the details of presentation. You first consider the knowledge level of your audience and assess what they already know about the subject. You should also take into account other factors such as likes and dislikes, attitudes and values, and the benefit to the audience by listening to your presentation. Try to gather as much information as possible about the audience prior to giving a talk so that you present the most useful and relevant information possible keeping away from offending or embarrassing remarks.

In certain scientific meetings, all the participants may belong to a specialty discipline as that of the presenter, who can understand all the deliberations including the jargons used. In some cases, the group is composed of specialists from different disciplines with a general background. A third category may have a general background only, for example: generalists, farmers, and students. Presenting a talk before a specialist group as that of the first category is relatively easy, as they show more enthusiasm and can understand the contents easily. If the audience belongs to the second and third categories, the presentation has to be structured taking into consideration their expectations and weaknesses.

While making the presentation, take care not to offend the value system of any member of the audience, for they may get alienated during your presentation. Similarly, if you fail to recognize and acknowledge their existing knowledge and experience, they may feel insulted. The audience may get bored, if you fail to relate to their needs and aspirations. Listeners may also be given an opportunity to question or contribute to the discussion; otherwise, they may feel frustrated.

The success of a presentation can be judged easily by the reaction of audience. Try your best to raise and maintain audience's interest so that the audience understand and remember your messages. Some disturbing factors in the room, for example: strange noise, noise of the equipments, and movement of people in the hall, may annoy the audience diverting their attention. Normally, audience attention varies during presentation, higher at the beginning, lower in the middle, and again higher towards the end. Because of expectations and curiosity, audience attention is usually high at the beginning. Again, interest increases towards the end because of anticipation of your conclusion. Taking a cue from audience reaction, amend and improve your style of presentation.

4. **Avoid too much details**

Many scientists attempt to present too much data and information in a short time. Any attempt to cram all the data into a short communication will obscure the message. Remember that the oral presentation of a scientific work cannot be equated with a written research paper. The material may consist of work that has taken one or more years to complete, and it may be impossible to condense this much information into a few minutes. The purpose of the talk, therefore, is to bring out a few focal points that can be made in the allotted time. Condense and convey to the audience the main findings and stimulate their interest so that they seek out additional information from the speaker during the time allotted for discussion.

5. **Avoid clichés and use jargons only sparingly**

Plan the speech in simple English using clear unambiguous short words. Find better words by eliminating those clichés and phrases, which through constant use have ceased to convey any meaning to the listeners. A typical cliché you often hear while concluding the talk is 'last but not the least…' or 'last but no means least…'. It is irritating to hear this often! Some other typical clichés in talks, which you should avoid, are 'to be perfectly honest…', 'with all due respect…', 'I'd be more than happy…', and 'for the record…'.

Similarly, avoid jargons as far as possible. Each branch of science has some jargons, which is understood only by those routinely using them. Occasionally, this specialized language may become necessary for easy communication as in information technology and medicine. However, if the speech is intended for the public or scientists of other disciplines, technical jargons make the speech hard to grasp. Moreover, many new words are introduced each year, and existing dictionaries are unable to explain them. Similarly, use of abbreviations repeatedly without explanations is frustrating to those unfamiliar with them.

6. **Use humour and wit**

Humour provides brief reprieve from complex subjects and revives interest. It can be used to reinforce serious ideas. However, humour must be good, as a distasteful humour will spoil the quality of the lecture and may embarrass the audience. Do not crack jokes simply for its sake. The speakers should learn the art of using good humour successfully. Bear in mind that too many jokes and hilarious overtones may cause annoyance and irritate the audience instead of reviving interest. It may also take away your valuable time.

7. **Timing of speech**

The speaker must adhere to the allotted time. Reputation of the speaker is at risk by speaking for too long. Those who exceed the allotted time without permission from the Chair invite scorns from the audience and upset the entire plan of a meeting. You should have a rough estimate of time needed for showing slides. Do not attempt to draw on blackboard or white board as the time needed for it is unpredictable. If you want to become a good speaker, set apart enough time for practice and rehearsal. Some organizers issue guidelines on timing and presentation, and these should be studied carefully for compliance.

The chairperson of the meeting should be firm in dealing issues like exceeding time limits. Many methods are in vogue for warning a speaker. One way is by using lights, often a green warning light followed by a red one. In most meetings, a warning bell is given just one minute before the scheduled time limit.

8. **Speak instead of reading**

Some speakers prefer to ‘read’ than speak! A written speech is monotonous and seldom makes an impact on the audience. Only in exceptional cases, reading is allowed, for example, when a speaker has to give a paper in another language or those ritualistic occasions when a speaker is invited to deliver a routine speech. When we speak, we repeat ourselves to make a point; but when a paper is read, as it is done mechanically, important points will not be repeated. If a speaker must read, the speed limit should not exceed 120 words a minute (approximately 1.5 min for each double-spaced typewritten page); and pauses are essential after each definite point to allow the audience to think about it. Even in this case, you should acquaint yourself with the script thoroughly. While reading from a prepared speech too, as in a regular speech, you should lift your head and look at the audience at short intervals.

The main reason for speakers ‘reading’ their papers rather than ‘speaking’ is that they are not confident at speaking. This happens mainly with non-native speakers of

English, especially whose medium of instruction were not English. A compromise is, however, possible. You first script the talk and practice it repeatedly so that the script gradually becomes unnecessary. If needed, write and use a few key points as prompters. Predictably, the result is usually a memorized speech with the help of some important points. Using slides as a prompter is the best way out.

9. **Speak slowly**

You have to control your speed of talk and deliver the talk slowly. Often, fast delivery of talk betrays lack of confidence on the part of the speaker. Choose a moderate speed as speaking fast confuses the audience. Normal speed of talk is 125–150 words per minute. Understand that speaking slowly is a courtesy to listeners, especially if the official language of the meeting is not the mother tongue of majority of participants. Pronounce the words properly, putting stresses at the right places. Speak with enough volume so that every one in the audience can hear you clearly.

While delivering the talk, vary the pace of speaking and make it slower occasionally for emphasis. Make pauses between phrases and sentences. Speakers who deliver the speech in a monotonous manner may put the audience to sleep! Bear in mind that variation of speech delivery is helpful to tide over this problem. You should also have a control over your sound. Do not scream at the audience, but speak in a moderate loud voice. Similarly, avoid nasalization and vocalized pauses such as 'er', 'um', and 'uh'. Take care not to use phrases such as 'you see', 'like', 'you know', 'I mean', and 'is it clear?' Most speakers often vocalize pauses like the above because they are uncomfortable with even a bit of silence when standing on the podium facing an audience. Do not bother much about unfilled pauses, as these are acceptable in a presentation.

10. **Regulation of voice**

A good voice is a gift but anyone can improve the quality of speech with enthusiasm and proper training. The speech organs are the lips, the jaws, and the tongue, and the speech organs too need exercise! Reading aloud every day for a few minutes usually helps. It is also a good idea to watch how effective speakers vary the pitch of their voice to match the thought, idea, or feelings they convey. There are certain generalized set of rules applicable to talks and presentations regarding regulation of voice. For regulating your speech, the following six elements—pitch, volume, rate, quality, animation, and pause—are important.

Pitch: Pitch means the tone of sounds depending upon the rate of vibration of vocal chords. When you talk, there is continuous variation in the levels at which your voice is pitched. Pitch ranges from high or shrill tones of soprano to low or deep tones of bass as in an opera.

Volume: Volume of speech refers to fullness or power of the sound, and it ranges form very soft to very loud.

Rate: It indicates the speed at which you speak words. The rate of most people is 120–150 words per minute, almost equal to one-half to two-thirds of a double-spaced, typewritten page.

Quality: By the term quality, we mean the characteristic tone of a speaking voice. Sometimes, a voice may be termed as 'creaky' or 'jarring' to the ears or it may be termed 'pleasant'.

Animation: The liveliness of speech is known as animation. Some speakers are talented, and they make the speech lively.

Pause: It refers to a temporary stop or a juncture in a speech.

A good speaker makes use of the potential of all the six vocal elements mentioned above. If possible, record your speech, critically examine it, and try to amend it in subsequent efforts based on your self-criticism.

11. **Personal elements of the speaker**

Personal elements of the speaker include the attire and looks of the speaker and the body language. The first impression of the speaker on the audience is created by the appearance of the speaker. Body language is a form of nonverbal communication consisting of body posture, gestures, eye contact, facial expressions, and movements. While making a presentation or other forms of communication, humans send and interpret many signals through body language almost subconsciously.

Attire and looks: The first impression of the speaker on the audience is created by the appearance of the speaker, especially the dress s/he worn, the colour of the dress, and hairstyle. You can significantly influence the formality of an occasion by what you wear although most scientists do not have the reputation for being well dressed! Although the dress worn by the speaker seems unimportant for a talk, it is better to wear only dresses in which you are comfortable. Avoid any bizarre or unusual dress. A professional appearance creates a good first impression. By the looks and posture of the speaker, the audience begins to form an opinion about your personality, and may even visualize the way the speaker is going to deliver the talk!

Posture: Posture is an important part of body language and refers to the way one stands and walks. You need to adopt a posture that conveys confidence to the audience and that makes you comfortable. Normally, body movements and position of hands, legs, and head give an indication of the personality of an individual. We can form an opinion whether the speaker is competent, vibrant, dynamic, nervous, and self-assured. Keep the heels together, stand erect, chest high, and rest one hand on the podium but do not lean on it. Do not clench the podium, as it conveys a defensive posture.

Gestures: It is natural for humans to use gestures while speaking. Hand gestures are powerful means of communication, because when you use your hands to illustrate points, the audience not only hears what you are saying but also sees what you are saying. While making a planned speech or presentation, gestures should come out naturally. Avoid noticeable and awkward gestures, for example, playing with buttons, tie, belt, and wristwatch. Take care to avoid repetitive gestures such as opening and closing a pointer, jumping from one foot to the other, or some dancing gestures. Only small gestures are needed in scientific meetings.

Eye contact: Eye contact with the audience is important because it tells whether the speaker is sincere and believable and whether the listener is interested. In fact, it is a way to judge audience feedback, enabling the speaker ready to act, adjust, and reframe the message while it is being passed on. If you take on the audience with your eyes, the audience will return the look and will concentrate more on what you have to say. Most often, this process is automatic, and eye contact with the audience takes place unconsciously. A good speaker looks at all sections of audience and not on some specific individuals. If the audience is large, make sure that you have looked several times at every section of people in the room and that you have made eye contact with individuals in those sections before concluding the presentation. Looking at the floor, the ceiling, the door, or somewhere else is distracting; sometimes, taking a cue from you, the audience may also look to those structures instead of you!

Facial expression: Facial expression of the speaker must change while the speech is progressing. They can use facial expressions to convey their feelings, attitudes, and emotions. Appropriate facial expressions make speakers more interesting to listen to, which also enhances their credibility.

Movement: Some speakers do not move at all and looks glued to the pedestal! This happens mainly when the speaker has to operate the LCD projector. In big conferences, there will be arrangements for operating projecting equipments, and the speaker is free to move. While making the speech, the speaker should move about the room occasionally rather than standing behind a podium for the entire presentation. However, do not over do the movement; only a moderate amount of movement is necessary. Movement makes the presentation more interesting to listen. In general, movement produces a more congenial environment in which the physical and psychological distance between the speaker and the audience is narrowed. While making the speech, speakers often move with purpose contributing to the overall impact of the presentations. For example, some speakers walk towards the audience to emphasis a point. Sometimes, your movements may be to operate the projection equipment. You should be able to operate the projection devices so that you can avoid awkward movements and gestures.

12. **Enhance the credibility of the speaker**

The audience usually judge the speaker based on his/her credibility. This is an impor tant issue for the success of presentation and feedback from the audience. Audience usually judge the credibility of speakers based on three components—***competence***, ***character***, and ***dynamism***. Competence includes the speaker's intelligence and expertise; character includes sincerity and trustworthiness; and dynamism includes energy and enthusiasm. The audience should feel that you are competent, dynamic, and with good character. If you are able to deliver your speech fluently, expressively, and with conviction, your credibility is enhanced!

13. **Abstracts and proceedings**

In most conferences, 'abstracts', 'extended abstracts', or full length 'proceedings' are given to the participants at the time of registration itself for reference. In occasional seminars too, abstracts or full lecture notes are usually distributed. There are differences of opinion among scientists in the timing of distribution of abstracts for seminars. Some says it is better to distribute abstracts or full-length notes at the end of the talk; otherwise, attention may be lost while they are being presented. However, listeners prefer to have it at the beginning of presentation. As this is convenient for the organisers too, distribution of abstracts at the beginning of a presentation is the usually accepted norm. Ordinarily, an abstract consists of a summary of the talk, some ideas to stimulate thinking, and useful references.

14. **Use presentation aids effectively**

In this era of information explosion, we cannot think of a presentation without presentation aids or visual aids. You should use presentation aids effectively. An object that relates to the subject being lectured can be called a ***presentation aid*** or ***visual aid***. Ineffective use of presentation aid or visual aid is very common in scientific conferences and lectures. Good presentation visuals applied in the correct manner help your listeners stay attentive. There are a number of aids that can be used individually or in combination to enhance the quality of a presentation as well as to increase audience interest and information retention. Some commonly used presentation aids are:

- Computer-based slides
- overhead transparencies
- 35 mm slides
- photographs
- video
- flip charts
- white boards
- black boards
- handouts
- models
- real objects.

Presentation aids are an important factor for a successful presentation, and as a speaker, you should give enough thought to such visual aids. These visual aids give the audience something tangible to look at facilitating a better understanding of what you are talking. It is normal for the audience to learn more of what they are seeing and listening rather than just listening. Because of strict time limit in academic conferences like symposia, speakers may not be able to use many presentation aids. Therefore, they go for the most convenient and contemporary aid, the computer-based slides using the software 'PowerPoint'. See Sect. 20.4 for more information on preparation of presentation slides.

Many presenters just click through slides without directly mentioning and discussing the contents. Try to familiarize yourself with the equipment and ensure that you know how to operate your slide equipment. It is distracting to the audience to see a scene wherein you fumble with the equipment. You should also practice with your slides. Note that it may damage your reputation as a presenter if you go on clicking here and there for the correct slide. You should know precisely where and when each slide appears in the presentation. In computer-based presentations, a laser pointer is often used to point at important points. Use it with caution. For example, if you have shaky hands, the laser pointer only shows your nervousness to the audience. You can use your mouse as a pointer. Some people use slide software for animation and highlighting. The traditional pointing stick is also good.

Some speakers rely completely on presentation aids to deliver their speech. Note that slides are not always necessary for the success of a speech! You should be able to turn them off when it is not needed. Simply projecting slides one after other is not a good practice. Remember that as a speaker you must be your most important presentation aid! The audience has assembled there to hear you and interact with you. The power of your words and the strength of your speech delivery should hold the attention of all. Presentation aids should be in a supportive role only. You should ask yourself whether this particular slide is actually needed. For example, if you are discussing a personal experience or observation, it is more effectively communicated without a slide. Similarly, when you are introducing a research paper or project proposal, it is awkward to project some related sentences through slides. You should learn to discuss these without text slides. Avoid projecting a slide unless they are relevant to what you are currently discussing in your talk. After you finish discussing the points depicted in a slide, see that the screen is empty so that the audience focuses their attention entirely on you.

15. **Never read slides to the audience**

Slides have an important place in oral presentation; but it is important to maintain a judicious balance with regard to the number of slides. The slides are intended to be an aid to what one has to say; but should not be a substitute for it. Some speakers, the moment they are at the speaker's desk, start projecting slides, and they base their entire presentation on the projected slides. This makes the presentation boring and monotonous. The impression gathers in such cases is that the speaker is trying to impress the audience with beautiful slides, which may not have much substance, and is unable to face the audience in the absence of slides.

Some speakers read the written slide word by word as though the audience cannot read! You must allow the listeners to read the slides; but you need to highlight any particular points of importance. Similarly, try to avoid speaking in one language while the slides are written in another, as this is confusing, except perhaps for captions on graphs or diagrams.

Some General Tips for Success

Make sure that the speaker is seen and heard while giving lectures or presentation. If it is a large gathering, a microphone and sound system must be arranged. Try to follow the 15 guiding principles mentioned already for a successful speech delivery. Always look at the audience and not elsewhere. Looking at the screen or slides too long is awkward. Stand still and address the audience. Vary the tone of your voice and the pace of your talk. Personality of the speaker is also important. Although the dress worn by the speaker is unimportant, wear only dresses in which you are comfortable. Avoid any bizarre or unusual dress. Be your natural self. It is often said that research workers, who has no opportunity to teach, are unable to express their thoughts effectively in public. This may not be true. If you are serious, you can convey your thoughts in simple language with an analytic mind. Remember that perfection in speaking is mainly an acquired characteristic.

A faulty communication is often due to the defects of technique used, but this can be corrected.

If you are a first timer to a conference, it is quite natural that you may feel tense when you are waiting for the turn. If you have prepared well to present the topic, the delivery of talk takes place without problems. Always think that the listeners are your friends. If you feel tense while waiting for your turn to deliver the talk, try to diffuse it by talking to somebody or by stretching your hands. Be confident and firm in what you say, but learn to accept criticisms and suggestions. Finally, strive to complete your speech within the specified time.

20.4 Preparing Good Presentation Slides

Preparation of good quality presentation slides involves skills. Slides are made for devices such as overhead projectors, slide projectors, or computer-based LCD projectors. These are excellent devices to project text-based aids prepared in the format of overhead transparencies, 35 mm slides, or computer-based virtual slides. In the past, before the spread and universal acceptance of these equipments, the common devices were black boards, white boards, or flip charts. Presently, in most conferences, the organizers arrange facilities for presentation of computer-based slides only because of its huge populaity and easiness in preparing virtual slides.

Overhead Projectors

A projector for projecting magnified images of line drawings, tables, graphs, write-ups, etc., prepared on transparent sheets (illuminated from below) onto a vertical screen is called an *overhead projector*. It can be used in classroom lectures, seminars, and conferences. Overhead projectors are on the way out after the onslaught of computer-based LCD projectors.

Slide Projectors

Until recently, slide projectors that project 35 mm slides were very common in meetings. The slides for such projectors were specially made as photographic transparencies consisting of texts, diagrams, and photographs, mounted on a film holder. Slide projectors are seldom seen now.

LCD Projectors

At present, almost all the meetings use liquid crystal display (LCD) projectors of various hues for projecting slides directly from a computer. These have almost replaced slide projectors and overhead projectors. LCD projector is a type of video projector for displaying video, photographs, diagrams, or computer data on a screen or other flat surface. It can be considered as a modern analogue of the slide projector or overhead projector. An additional advantage of LCD projector is facilities for video projection, which is lacking in an overhead projector or slide projector.

Preparation of Slides

With personal computers, especially with programmes like PowerPoint, it is quite easy to prepare quality slides. Although these virtual slides have a vital role in oral presentation, it is important to maintain a sensible balance between the speech and the number of slides. The main purpose of a slide is to show certain aspects of a talk which are more easily understood by illustrations. The slides can also act as visual prompters to remind the speaker as well as the audience of the major points of the presentation. The speaker may then explain the subject by simply commenting on the slides. This is preferable to reading from a script. It is a good idea to keep prints of the slides, as this practice saves the speaker from looking at the slides except for just checking by a glance that the right slide has been projected.

Slides can be prepared to show texts, data, drawings, and pictures. Remember that copyright rules are applicable to visual presentation as with other scientific materials. Therefore, the sources must be shown when you are using the data of others. If data are modified, write 'modified from' or 'from' and quote name and source.

Legibility is an important requirement of a slide. The data for slides have to be prepared in a special manner. Graphs, tables, and figures prepared for research papers or thesis should not be used as such for a slide. In the case of publication, readers will spend as much time as they needs to understand each illustration. However, in the case of a slide, it is the speaker, who decides how much time to use for explaining the data. Follow the steps mentioned below to deliver clear messages through slides.

Generate ideas for presentation: Select the most important ideas or information, which should be shown through slides. Show the salient ideas through bulleted texts, charts, graphs, diagrams, or photographs.

Determine the best way of presentation: All the materials for presentation should be scrutinized to find the best way of showing it, whether by text, table, chart, graph, diagram, or photograph. In most cases, tables can be conveniently converted into bar charts or diagrams showing relationships and trends. However, think twice which is more appealing and comprehensible—table or diagram. Definitely, the visual appeal of the slide matters.

Keep the number of slides to minimum: Do not consider slides as a substitute for your speech. They should be used only to reinforce key ideas and basic concepts. You should plan to have not more than one slide for every 1–2 min of presentation time. For example, for a 20 min talk, approximately, 15 slides are recommended. If your slides exceed this, you are not providing your audience enough time to absorb the information passed on by you through the slides.

Select only pertinent information: It is not required to cram all the information in a slide, as the speaker will explain the details. Give a title to each visual aid, which acts as a headline. Make the headline short and punchy.

Simplify the data: Do not show raw data, which may confuse the audience. If you want to show tables, make them simple. Often, statistical data in tables can be converted in to a diagram showing relationships and trends.

Include one central idea per slide: Each slide should have only one central idea or message. Slides with multiple ideas may confuse the audience.

Edit the contents: Stringent editing of unnecessary or complex data is essential before presenting them before the audience through slides.

Converting Materials into Slides

We present text, data, diagrams, pictures, and photographs through slides. We have to bestow extreme care when these are made to order for slides.

Text Material

If the slides contain only text material, limit them to about three or four lines and never exceed seven per frame. If more lines are to be accommodated, split the slide into two. Use bullets to present the idea.

It is always recommended to use '*sans serif*' typeface (a typeface with no 'serifs', that is, the short flairs and finishing strokes at the top or bottom of letters, as in 'Times New Roman' fonts) for visual display (Kops and Worth 1996). A few appealing sans serif typefaces are Ariel, Helvetica, Verdana, Tahoma, Optima, Universal, and Microsoft Sans Serif. Use typefaces above 18–24 points for text matter and 32 point or more for titles depending upon the fonts (Type sizes are measured in points; 72 points = 1 in.). Do not go for fancy fonts, as the audience may think about the fancy style instead of concentrating in your speech!

To give emphasis to specific words, use of fonts with boldface, italics, or underlining can create a powerful effect. Colour can also create similar effects. However, use them sparingly; if overused, they lose their impact. Use different colours for title and text matter. Avoid using very bright colours. Slides with a coloured background are preferred. Although many design templates are possible through 'PowerPoint', those with blue are often preferred. Slides seem to work best when a dark background is combined with lighter type, for example, yellow or white type on a navy or dark green background. Colour of bullets can be varied.

Present the headlines in all 'upper case' or 'title case'. If you select this pattern, be consistent through out the slides. The body text of a visual is presented in 'sentence case', as this is the most familiar format.

Slides with text are most effective when lines are kept short. Use consistent phrasing, for example, all commands, all statements, or all adjectives and nouns (see Fig. 20.3).

Treatment of Data

While presenting, data are shown mostly to stimulate the interest of the audience rather than to give specific details. Data can be presented in various ways as tables and charts. For more information on tables and charts, refer to Chap. 13. When you use them for presentations through slides, you have to follow some additional precautions.

Tables: Tables are generally used to show comparisons between different groups. Limit the columns and rows to four. You can use shading or appropriate colours to differentiate one column from another.

How to write 'Review of Literature'

- A comprehensive review of the works done on the subject selected for the study and the lacunae in the subjects are presented.
- The review must be divided in to sections with sub headings.
- Present the findings of various authors interconnecting them with your observations.
- An inexperienced author may simply list the findings in a chronological order and the quoted references may not have any relationships.

This slide can be improved by converting all the statements into commands. Improved slide (after modification) is shown below:

Writing 'Review of Literature'

- Include a comprehensive review of the previous works
- Divide the review into sections with sub headings
- Present various authors' findings interconnecting them
- Avoid listing the findings simply in a chronological order

Fig. 20.3 How to improve a slide by converting all statements into commands

Bar charts: Bar charts are used to illustrate comparisons between groups. Sometimes, instead of tables, you may choose bar charts. Limit the number of bars to five or maximum seven.

Pie charts: Pie charts illustrate the division of a whole into different parts. A useful rule of thumb is to limit the number of slices to five and to place percentages and numbers inside the pie and the labels and descriptions of each one of the slices outside.

Line graphs: Line graphs are often used to express changing relations when they occur in relation to time. Put the scale on the left margin and the time scale on the bottom. Limit the number of curves to three or four, and distinguish each by different colours or symbols.

Scatter diagram: This is used to illustrate the degree of correlation or distribution in groups.

Flow charts: Flow charts are ideal to describe successive stages of a process.

Drawings

Line drawing or tonal drawings can be used for presentation. Sketches will be useful to explain parts of plants such as flowers. The parts of a machine can also be explained easily with a line drawing.

Pictures and Photographs

Nowadays, inserting pictures in slides is very easy and inexpensive. Using a digital camera, you can directly insert images on the slides. Give suitable titles in 'text boxes'. Do not overcrowd pictures with information. Another option is to scan pictures and use the scanned images on slides.

20.5 Poster Presentation in Meetings

Poster presentations are integral part of scientific meetings, and the organizers encourage posters because of several reasons, the prime being time constraints (Fig. 20.4). Some participants submit their papers for 'poster' presentation only. Sometimes, the organizers decide whether to admit the paper for oral presentation or poster presentation. Papers that contain results of just one aspect or results from some ongoing study may be accepted for poster displays only. When a huge number of scientists send their papers for the conference, there is no other option but to send some papers to poster sessions.

A poster is different from a research paper or an oral presentation, and different techniques are used for its presentation. A major advantage of poster, according to Mathews (1990), is that it allows the audience to view the material at a reader-selected rate leading to better viewer comprehension and retention. However, poster sessions in most conferences usually have numerous posters at display, and competition to attract the attention of viewers is intense. Therefore, it is utmost important that a poster should be designed in a unique manner to make the onlookers curious and stimulate their interest. The title and its wording are also important. As the viewers may not spend much time before a poster, the message should be presented briefly

Fig. 20.4 Most conferences may have sizable number of posters on display. A view of posters displayed at a conference

so that the viewers can quickly understand the message. You will get good tips for poster making and presentation from Mathews (1990), Malmfors et al. (2004), and Day and Gastel (2006).

Planning and Preparation

Before you start preparing the poster and arranging the exhibits, make sure that you have received details such as the type of display board, stand, dimensions for length and height, and method of fixing exhibits. You should note carefully the size specified for the poster. In most conferences, the size specified is 90 cm × 120 cm or A0 size (84.1 cm × 118.9 cm). Note the dimensions properly and check whether the dimensions meant are for portrait or landscape orientation. For example, 90 cm × 120 cm means portrait orientation with 90 cm length and 120 cm height. However, if the dimension is 120 cm × 90 cm, it means landscape orientation with 120 cm length and 90 cm height. At present, it is relatively easier to prepare attractive single sheet posters. Instead of one single poster, you can also prepare multiple sheet posters, for example, two A1 size sheets (59.4 cm × 84.1 cm), four A2 size sheets (42 cm × 59.4 cm), or eight A3 size sheets (29.7 cm × 42 cm) to fit into the size of A0 (84.1 cm × 118.9 cm).

Poster Designing

Poster designing is an art. A poster can be prepared adopting a variety of layouts depending on the form of graphics and photographs. However, the text and the graphics must be as concise and condensed as possible to fit into the specified dimension of a poster. Do not make thickly packed, overcrowded posters with too much information, as the reader may get tired of reading them and may sometimes skip the poster altogether. Therefore, select and include only the key points and arrange them legibly with plenty of space between the exhibits. Selection of colours and the combinations of typefaces, tables, and graphics can be important (colour of display paper and pins too!); use them logically to attract the viewer. However, keep the use of warm, bright colours to the minimum as it may mask the significance of the scientific information being displayed.

The advent of computers and Desk Top Publishing (DTP) softwares have revolutionized poster preparation and presentation. You can now prepare a poster easily on a display paper of poster size (A0 size) and get it laminated with all the attractiveness, brevity, and clarity (ABC) of posters! Please note that majority of organizers will not allow posters made on flex sheets. Usually, posters are printed on photopaper, vinyl, or glossy paper.

The Subject Matter and the Main Headings

The title of the poster must be concise and to the point. The names of authors should be in slightly smaller lettering, and include their addresses too. Just like a research paper, the brief report exhibited through poster should also follow the IMRAD structure. Organize the subject matter into sections such as Introduction, Objectives, Materials and Methods, Results and Conclusion. However, the matter under each section should be brief. Arrange these sections in sequence, usually from left to right. Include graphs, tables, and photographs as illustrating materials for the Results. The References, not more than six, may also be given.

Try to arrange the text and data in an appealing and concise manner. The information can be presented as text, tables, charts, histograms, curves, drawings, cartoons, halftones, photographs, or paintings. While presenting, you can split the information into conceptual units that are not essentially paragraphs. Grouping the information into comprehensible units gives a sense of categorical organization to the poster. It is not necessary to write in complete sentences. You can use sentence fragments in a bulleted format as bulleted phrases convey information quickly and effectively.

Use the Right Letters

Letters used must be large enough to allow people to read without much strain. You can check this from a distance of 1.2 m. Lettering should be bold and appropriately coloured. The size of letters will depend on the typeface selected and background; but typefaces with 72 points can be considered for the title, 36 points for the names of authors and other headings, and 24 points for the rest of the text (1 point = 1/72 in. = 0.35 mm). The typeface you select helps to create the overall effect of your poster. Lettering can be made by a word processor. Use different typefaces for main headings and text. A common practice is to use sans serif typefaces for title and headings (e.g. Helvetica, Ariel) and serif typefaces (e.g. Times New Roman, Palatino) for the text. However, do not select more than two typefaces; limit one for the headings and the other for the text. You should also be consistent in using them throughout the poster.

Transporting Poster Materials

Transporting poster materials to the venue of the meeting is a problem. A common method is to put each illustration on separate bits of cardboard. This makes the exhibit easily portable in a carry case and easier to fit on the stand. It may also allow a more attractive layout. However, it needs more time to put up the poster. You can also put your poster inside a cardboard cylinder after making the poster in a spool form. The cardboard cylinder will keep the poster intact. While reaching the meeting venue, take the poster out, and straighten it.

Method of Fixing

The poster should be set up at a reasonable eye level. Mark the number as given in the schedule of programmes on the poster. There are several kinds of fixing methods, but these must be checked in advance. Usually, panel pins or drawing pins are used. Double stick tape, sellotape, sticky fixers, and the like are also used. It is always advisable to take scissors, sellotape, and extra pins with you as a precaution in case the materials provided by the organizers may not be enough.

During the Meeting

The author/s of the poster shall be available at the conference venue during the poster session to answer questions. Separate timings may be set apart for poster sessions. The authors are supposed to stand near the display at the specified day and time. Participants can question the authors when they pass through the posters.

If the visitor is enthusiastic about the poster and the subject, handouts providing details of the subject covered can also be supplied. However, this is not needed, if the background paper of the poster is already given in the 'Proceedings' or 'Abstracts'.

20.6 Students' Seminar

In most universities, seminars are offered as credit courses (credit seminars) to teach students how to conduct a presentation. This is often accomplished through a formal presentation before teachers and students with a seminar leader. In fact, a seminar is based on 'Socratic method of teaching'. According to Socrates, it is more important to facilitate students to think and find answers themselves than to merely fill their brain with answers. Therefore, Socrates regularly engaged his students in dialogues by responding to their questions with questions instead of answers. Instead of convergent thinking, this seminar method promotes divergent thinking. A seminar will be open to discussion; the participants can raise questions and indulge in debates.

As a part of credit seminar, students will be given a topic from their discipline or they can choose one; and the seminar paper should be an excellent review on the chosen topic along with references, listed according to the name-year sequence (or the system your institute follows). The credit seminar is also an opportunity for the students to learn how to prepare a review. Preparation time given to students might be about one month. Normally, the student will be allowed to conduct the seminar, only if s/he submits the draft review before the due date. The student is also supposed to provide an abstract to the listeners before starting the seminar. Students may prepare for presentation and speech delivery as detailed already in this chapter. One or two rehearsals improve the quality of seminar. The delivery of speech and presentation are done in the presence of an evaluation committee of teachers led by the seminar leader. The time allotted is usually 30 min for presentation and 10 min for discussion. Once the oral presentation is over, the draft will be returned to the student, and s/he is expected to submit the review paper along with the questions raised during the discussion time and the answers given. Plagiarism is a problem in such seminars. Students should be taught well how to avoid plagiarism, how to do paraphrasing, and how to cite and list sources (refer Chaps. 15 and 24).

For assessing student's seminars, universities must have formulated certain broad guidelines. The seminars have to be assessed properly and given grades. A broad outline for assessment is given below.

Coverage of topic (write up)

Classification of information
Review of literature
Citation and listing of references
Knowledge of the topic
Writing style including grammar.

Presentation

Structure and overall content of the presentation
Speech delivery (*not* reading slides)
Use of visual aids
Design and content of visual aids
Ability to explain properly
Enthusiasm and contact with the audience
Distribution of time and keeping the time limit.

Discussion

Ability to defend points presented
Handling questions and mode of answering.

Hand out

The content and overall layout of abstract or handout.

Research students, in addition to the credit seminar, usually conducts three or more seminars on their research topic, one before the start of the programme highlighting the importance and methodology of the programme, one in the midway of their research showing the progress of the work, and one after the entire work is over just before thesis submission. The idea behind such seminars is to confront students with the methodology of their chosen subject and to familiarize them with practical problems that might crop up during their research work.

20.7 Group and Panel Discussion

In general, group and panel discussions are arranged to formulate certain strategies or implement certain programmes. The group should have a leader and a rapporteur. In a typical ***group discussion***, the group is arranged in a circle so that the participating members can see each other. Ensure that all are seated during discussion. Initially, the gathering should go round the circle each one introducing him /her. When newcomers join the group, they should introduce themselves to the group leader and to the group. Start the session in time and close at the prearranged time. Do not allow the monopoly of any particular speaker. Strict control of timing is important. Keep seated when you speak. The discussion meeting is not a place for long speeches. Informality is the rule to be followed. Make your points in a few words and pass the ball to some one across the group. If the discussion lags, help the leader by putting questions that will draw others out. Come prepared for the discussion with some salient points and questions, if the subject is already known to you.

A ***panel discussion*** is slightly different from a group discussion. It is a discussion among selected group of persons under a leader in front of an audience, which joins them later. Panel discussion is a good idea when the crowd is big, for example, if it is more than 40–50. A good panel may have members ranging from six to eight including the leader. The purpose of panel discussion is to get important facts and viewpoints, stimulate the audience, and put a basis for wide participation of the audience later.

Bibliography

Alley, M. 2003. The Craft of Scientific Presentations: Critical Steps to Succeed and Critical Errors to Avoid. Springer, New York, 241p.

Barrass, R. 2002. Scientists Must Write: A Guide to Better Writing for Scientists, Engineers and Students (2nd Ed.). Routledge, London, 204p.

Beebe, S.A. and Beebe, S.J. 2009. Public Speaking: An Audience-Centered Approach. (7th Ed.), Pearson international, 491p.

Booth, V. 1996. Communicating in Science: Writing a Scientific Paper and Speaking at Scientific Meetings (2nd Ed.), Cambridge University Press, 96p.

Davis, M. 2005. Scientific Papers and Presentations. Academic Press, Massachusetts, 356p.

Day, R.A. and Gastel, B. 2006. How to Write and Publish a Scientific Paper (6th Ed.). Cambridge University Press, Cambridge, 320p.

Hamlin, S. 1988. How to Talk So People Listen: The Real Key to Job Success (2nd Indian Reprint, 1989). Universal Book Stall, New Delhi, 265p.

Hawkins, C.1985. Speaking at meetings. In: Hawkins, C. and Sorgi, M. (ed.), Research: How to Plan, Speak and Write about It (4th Indian Reprint, 1993). Narosa Publishing House, New Delhi, pp. 60–84.

Kops, G. and Worth, R. 1996. Great Speakers Aren't Born: The Complete Guide to Winning Presentations. Lifetime Books, Hollywood, 188p.

Malmfors, B., Garnsworthy, P., and Grossman, M. 2004. Writing and Presenting Scientific Papers (2nd Ed.). Viva Books, New Delhi, 153p.

Mathews, D.L. 1990. The scientific poster: Guidelines for effective visual communication. Tech. Commun. 37(3): 225–232.

Matthews, J.R. and Matthews, R.W. 2007. Successful Scientific Writing: A Stepby-Step Guide for the Biological and Medical Sciences (3rd Ed.) . Cambridge University Press, 256p.

Richmond, V.P. and McCroskey, J.C. 1997. Communication: Apprehension, Avoidance, and Effectiveness (5th Ed.). Allyn and Bacon, Boston, 152p.

Chapter 21
Guidelines for Successful Lecturing

The mediocre teacher tells. The good teacher explains. The superior teacher demonstrates. The great teacher inspires.
William Arthur Ward (1921–1994)

Lectures are the most common method of teaching and learning in universities. Lecture method is just one of several teaching methods, but it is predominantly used in classroom teaching involving large students groups. The noun 'lecture', meaning 'the action of reading,' is from the Latin, *lectus* (to read). Presently, the word is used to refer to an oral treatise on a particular subject before a group of students or others for teaching. Note that the word 'lectern' is the reading desk commonly used by lecturers. In a college or university, during a lecture class, a professor (assistant, associate, or full professor) gives an oral presentation of subjects included in the syllabus, most often accompanied by visual elements.

Mark Twain (1835–1910), the author of the much acclaimed '*The Adventures of Tom Sawyer and The Adventures of Huckleberry Finn*' (he worked as a lecturer too!) is often credited with a hilarious description of college lectures: 'Lecture is a talk by which the notes of the professor become the notes of the students, without passing through the brains of either'.

You may agree that there are some truths behind the above joke! Lecturing should be concerned with the active transfer of knowledge, understanding, skills, and attitudes. Therefore, deliberate actions are required by both lecturers and students for the successful transfer of knowledge. Otherwise, it would simply be a mechanical transfer of lecture notes! The lecturer needs to plan, structure, and communicate clearly and interestingly. On their part, students need to listen, observe, take notes, and subsequently study and apply the knowledge thus gained.

For teaching in universities or colleges, although you do not need a separate teaching qualification, it is imperative that you have acquired the competence required for a teacher. However, most universities seldom take any serious steps to train their lecturers, in contrast to school teachers who need to have a degree in education or teachers training certificate. Of late, some universities have started providing some avenues to acquire teaching skills during M.Phil. or Ph.D.

C. G. Thomas, *Research Methodology and Scientific Writing*,
https://doi.org/10.1007/978-3-030-64865-7_21

Even if you are not a regular 'teacher' but engaged in research activities only, still you should be able to lecture! In most advanced research institutes and centres of higher learning, in addition to normal research activities, scientists may also be collaborating with teaching activities through research guidance and lecturing on advanced topics. Although this chapter has been designed for those whose main activity is teaching, it would be equally applicable for those who attempt to venture occasional lectures.

Most lecturers take teaching casually. Teaching is both an art and a science. Unfortunately, the general impression in colleges and universities is that if you have subject knowledge and a Master's degree, you can lecture! This notion is changing and you may find many books to enhance the art of lecturing and learning. Some seminal books are those of Race (2007), Petty (2009), Davis (2009), and Pauk and Owens (2011). These books have gone through several editions and you may go through the latest editions. Friedmann (1990) narrated 56 laws of good teaching. In this chapter, general guidelines are proposed for good teaching and effective learning under nine heads.

1.	Qualities of a good lecturer
2.	Planning for lecture
3.	The process of lecturing
4.	Use visual aids effectively
5.	Sustain the interest of students
6.	Improve motivation of students
7.	Build rapport with the students
8.	Plan and prepare for group work
9.	Proper conduct of examinations

21.1 Qualities of a Good Lecturer

Most teachers learn the art of lecturing by experience or by watching others lecture. Some may ask a friend or colleague to listen and criticize their lectures. Remember that a lecture, especially to a large audience, is analogous to a stage performance, which requires great skill and knowledge. Not all those recruited for lecturing may become good teachers. The ideal qualities required for a good lecturer are mentioned below. While selecting lecturers for university level teaching, the recruitment body must have some mechanism to ascertain that the new incumbents have these qualities.

Consider the following 10 qualities, and do an introspection about you as a teacher in terms of these qualities.

Passion for Teaching

One of the essential qualities of a good teacher is a passion for teaching. Some are born teachers with great passion for teaching, and they require no instruction on how to teach or lecture. Their success may be due to their craving for teaching, enthusiasm, originality, sincerity, and feeling for showmanship as well as personal qualities. These qualities enable them to attract and sway the students throughout the lecture.

Love of Students

The teacher should be a gentle, kind, benevolent, and magnanimous person. Teachers must be friendly, and they should be able to love their students. Students can share their problems with a loving and caring teacher without any fear or hesitation. A situation in which teachers are seen as enemies must not prevail in the campus. Students should love their teachers and when they do love them, they idolize such teachers. Eventually, they will respect the teacher, will do the homework, and ultimately will bring greater outputs and results.

How does a teacher become 'unapproachable'? Do an introspection, and find out whether you are an 'approachable' teacher. This is not just a matter of personality and dedication, a number of techniques and skills can help you. Most of all, students should have faith in you to become approachable. In the classroom, students should not hesitate to ask questions. Remember that questions from students is a vital part of the learning process; but some students, especially shy ones, will not ask questions at all, especially to an over-formal or unfriendly teacher.

Communication and Oratory Skills

All teachers must be able to communicate easily with good oratory skills. If you feel that your skills are not up to the mark, you can improve them and become good speakers by training. If your communication skills are good, you can deliver the lectures with better results. Undoubtedly, students love to hear a teacher who can speak well.

Good Personality

A good teacher must have good personality. Students like teachers with good personality, and this ensures good communication, care, and understanding, eventually leading to good results. Do not equate personality with beauty or handsomeness. It is more related to your attitude. In fact, personality is a combination of characteristics that shape an individual's unique character, and is manifested in each individual's characteristic patterns of thinking, feeling, and behaving. It involves moods, attitudes, and opinions, and therefore, most clearly expressed in interactions with other people. Anyone can have a good, decent, friendly, and presentable personality by some efforts. Some tips are to dress reasonably well, behave properly, smell good, be gentle, and compassionate.

Deep Knowledge
You must love your discipline and the subjects coming under your specialization, and must have deep knowledge in these fields. If you do not have enough knowledge of a certain subject that you teach, you are never going to make a name for yourself. It is better to teach only those subjects in which you have good expertise. Proficiency in the subject makes you morally strong and confident.

Be a Good Listener
Apart from the passion for teaching and oratory talents, a good teacher should also be a good listener. To be a good listener, you should have patience and compassion. Students adore teachers who can listen to them as well. When teachers develop this listening quality in themselves, they start to become great teachers.

Sense of Humour
A teacher with a good sense of humour can sway the students. This is essential to carry the class with you without boredom or disinterest. Good teachers with their sense of humour, strong communication skills, and personality can maintain the discipline and coherence of the class.

Dynamism
A good teacher must be dynamic, which means that as your students evolve, so do you and so does your teaching. Being dynamic is a way of being a good teacher, leader, and sometimes a student for learning new concepts and ideas in the classroom as well as outside.

Motivational Skills
The motive for participation and recognition is evident among students. Acceptance and approval are strong personal needs in most people. Many students tend to respond to a caring teacher who conveys interest in them, who listen carefully to them, and who respect them as individuals. More details on motivation are given elsewhere.

Good Language
The teacher must be proficient in the medium of instruction. In India, the medium of instruction at the university level is English. Naturally, college and university teachers must have proficiency in both written and spoken English. Some teachers may not be confidant in giving lectures in English, and may end up mixing English with regional languages producing what is sometimes known as 'Manglish' or 'Hinglish'. This is not fair, but if they wish, they can improve their speech by listening to radio, TV, English movies, and reading English newspapers. If they are still hesitant, try to improve the language by joining some short-term spoken English classes.

21.2 Planning for Lecture

Remember that students are not as informed or aware of how a course operates as everybody thinks. If the students are coming to a system hitherto unfamiliar to them, they are particularly vulnerable to uncertainty and do not know what is going on. For example, students new to semester systems and internal evaluations, require some time for adjustments. A fresher may not always remember what is said at orientation or introductory classes. They must be provided comprehensive course information, and if possible, materials like academic rulebook so that they will be able to conform to details.

Teachers should not take lecturing casually, but must plan for it thoroughly. You may start with the course outline. The following are some stages in the planning process.

Be Ready with the Course Outline

Prepare and be ready with the course outline before the start of the semester itself. Devise the course outline covering objectives, lecture and practical schedules, assignments, assessment, contact hours, main references books as well as an outline of expectations of students. Explain the course outline at the first lecture itself, and inform the students where and how to get help, facilities, helpful web sites, learning support, and other general information.

Do Your Homework Well

Prepare well for the lecture and make an outline of the topic to be taught in the class. Do not attempt to give a lecture unless your knowledge far exceeds the content of your lecture topic. Therefore, read and learn the material thoroughly, going through the textbook or additional reference materials. For advanced courses like M.Sc., M.Tech., M.Phil., MD, and Ph.D., collect information from recent reviews and research journals. For those figures that change frequently, for example, area and production of crops, provide only the latest data. In this era of information explosion, it is easy to access latest information from several sources including web-based sources.

Read and Prepare

Often, the information being given to the students through textbooks is at least 5–8 years old. In olden days, whatever we learnt remained fresh for a considerable time, at least the lifetime of an individual! Now, within a few years, the information becomes stale and needs updating. Even great professors or scientists, if not in touch with the current developments in their respective fields, become outdated, and they would be a laughing stock among the students. Try to keep abreast with the growth of knowledge. Similar to the habit of reading newspapers daily, scientists should also inculcate the habit of reading journal articles regularly. Reading journal articles and reviews should form a regular feature of the life of a successful lecturer.

It is not fair on your part to expect the students to learn or understand anything that you cannot learn or understand. Each time you prepare for the class, try to re-experience your subject; and in this way, it will remain forever fresh and spontaneous.

Decide the Portions to Be Covered

Decide the portions to be covered during each lecture class. After reviewing the guidelines or curriculum, set your broad goals for the course. The goals of an introductory course might include stimulating students' interest in the field and providing them with sufficient foundation to pursue that interest. Estimate the time required to address the topics, and then increase this estimated time by 25 percent to allow time for questions and interruptions from students and for the inevitable slippage in large groups.

Recommend Textbooks
You must recommend a good textbook for the course, but do not follow it as such in lectures. In certain fields such as agriculture, there is a problem of good textbooks, which can be prescribed or followed. Most often, sufficient copies of the books prescribed may not be available for purchase or may not be available in the library. In such situations, it is quite natural that the students rely heavily on the notes, which they are able to take while lecturing. Sometimes, instructions may be required on note taking (Chap. 22). The Cornell method seems to be ideal for university-level lectures. However, students should not neglect textbooks, and you can avoid this by making textbooks an integral part of teaching and learning strategies.

Prepare Lecture Notes and Outlines
Preparation of lecture notes is a major task of lecturers, and you should do it judiciously. You must find time for arranging facts, definitions, examples, equations, and answering critical questions. Some teachers prepare their lectures well in advance after taking into account students' reactions to previous lectures. While preparing notes, you may use colour codes to highlight difficult points, distinctions between major examples, and important information. You may also include notations that indicate times to pause and asking questions. It is also a good idea to note in the margin, for example, when to write on the board, questions to be asked, humour to be cracked, examples, and similar techniques to enhance the effectiveness of the lecture. Use clear examples, which reinforce the points you are trying to make.

Some teachers lecture directly from notes. It is not a good practice, but you may carry the lecture notes to reassure you. The correct practice must be to prepare and use brief mnemonic aids using the notes to enhance lecture delivery. By lecturing from this outline, you can produce sentences more for the ear than for the eye, thereby making it easier for students to grasp the material.

Whether distribution of handouts, lecture notes, or lecture summaries among students is necessary for academic teaching is a debatable issue. Some teachers reject handouts altogether saying these promote 'spoon feeding'. However, most students like to get handouts as these materials free them from regular note taking. Active concentration on two activities at the same time, understanding the lecture and noting it on paper, is difficult for the students. A handout usually consists of a summary of the lecture with or without a record of some of the details, some ideas to stimulate thinking, and useful references. Handouts are helpful for the lecturers too, for even if he/she missed something while lecturing, it would be there in the handouts. Similarly, if the lecturer cannot prescribe a proper textbook for the course, full lecture notes or handouts will be in much demand.

Structure the Lecture

Structure your lectures to help students retain the most important material. In general, students' retention is greater at the beginning of the class, decreases as the period goes on, and then increases slightly in anticipation of the end! Plan your classes so that the main points come at a time when students are most attentive.

Design Your Lectures in 15-min Blocks

Do not plan to lecture for a full period. The average student's attention span is between 10 and 20 min (Penner 1984). After this time, students may have difficulty in concentrating on the speaker. For each lecture, plan to change the pace every 15 min or so to relieve the monotony and recapture students' interest. For example, ask students to solve a problem at their seats or in groups of two or three, give a demonstration, use an audiovisual aid, ask a question, or tell a story or anecdote. When you design your blocks, ensure that each one cover a single point with examples. Ideally, it should end with a brief summary. If you are sure that you cannot complete a block as decided, leave an entire block for the ensuing class or shorten it for the present class.

21.3 The Process of Lecturing

Successful lecturing is a formidable task. In the classroom teaching process, the teacher sends a verbal message containing some information to the students, which they integrate into their existing knowledge. However, this process of converting information into knowledge is not as simple as it seems. The clearer the message the easier it is to be understood by students. Strengthening the messages given orally by supplementary means, especially by visual aids, helps the listeners to receive the message through the ear and the eye, thus enhancing receptivity.

There are some issues, which you should take care of while starting the lecture, during lecturing, and winding up the class.

Begin the Lecture in Time

Punctuality is an essential virtue required for both the teacher and the student. Always begin the lecture on time and stop on time. You may start each lecture with an outline of what you are going to cover in that period. It is not a good practice to start the lecture fast and end slowly, the indication is that you have exhausted your material! Similarly, never start slowly and end fast exhausting your students. Start the lecture steadily with calm and ease, and complete the lecture within the allotted time. It is always useful to end the presentation by drawing attention to the major conclusions, but this should not result in a repetition of what has already been said.

Before the students arrive for the class, make sure you have everything you need. For practical classes, you should already have rehearsed any experimental procedure you or your class will be carrying out. Make sure that the audio-visual equipment works properly. You need not wait for late coming students, as it is unfair to those who have arrived on time. Moreover, it may encourage late arrival in the future. Going to the class a little early and talking informally with the students create good ambience. When the students arrive, you may like to greet them at the door and watch them as they take their places.

Start the Lecture
You may start the class confidently in a relaxed mood keeping your tone conversational. In the beginning, some amount of nervousness is normal, but you can overcome this easily. An easy tip is to relax yourself by taking deep breaths or tightening and then releasing the muscles of your body from toes to jaw. Once you start the lecture and under way, your nervousness will decrease. Seize students' attention with your opening remarks. It is a good idea to open with a provocative question, startling statement, unusual analogy, personal anecdote, dramatic contrast, powerful quote, short questionnaire, demonstration, or mention of a recent news event.

Deliver the Lecture with Ease
Avoid dull and monotonous lecturing. The students' interest should be directed to the subject you are handling, and not to your voice or mannerisms. Understand that the mere one-time use of a word or idea does not expose the students to it. Just intimation will not help; you should substantiate your arguments and points. A lecturer's job is not to tell students all that they need to know. Instead, they must provide a review of the essentials of the subjects supported by examples. It is a mistake to cover the whole thing written in the prepared note or textbook. Leave such elaborate details to textbooks or distributed handouts. Discuss problems, explain difficult points, clarify concepts and principles, answer questions, and coax the students to refer to sources for further information. A lecturer should act as a trendsetter for students, who by listening can move forward more quickly than they could by reading alone.

Do not make the class a show of your erudition; the students are far less interested in what you know than in what they can learn. Consider competence as the best way to gain popularity among students than any gimmicks.

During the lecture, never crack jokes for their own sake, but only in the service of what you teach. Similarly, do not tease your students in the class; it may affect their self-respect and enthusiasm. Losing your temper in the class also makes you unpopular. Maintain good eye contact with the students. While lecturing, always look at the students, and not at the ceiling, the floor, or the black board. Take care of your body language too as discussed in Sect. 20.3. When you introduce a new term or concept, you need to reinforce it by writing it on the board, and by repeating it in such a way that the students have a chance to note it down.

However, repeating every sentence is a waste of time.

Avoid Reading Instead of Lecturing
Students hate 'readers' and 'dictators'. Never read or dictate notes in the class instead of lecturing. Lecturing verbatim from a script such as your lecture note tarnishes your

image as a teacher. Only incompetent or lazy teachers resort to simple reading. If the purpose of your 'dictation' is to allow the students to take down notes as given in your prepared notes, then a better way is to ask the students to take photocopies of the notes! If you simply read from a text, you will find yourself detached from the material, and your students will feel disengaged. Understand that lecturing and dictation are different. Lecturing is a creative process, because students enthusiastically receive and learn what is being lectured and make notes on salient points. At the same time, dictation is purely mechanical; students just note down passively what is read to them for later understanding.

Moreover, reading from notes reduces your opportunities to engage your class in conversation and prevents you from maintaining eye contact. It also casts your voice down toward your notes instead of up and out towards the students. Lecturing must resemble a natural, spontaneous conversation between the instructor and the students, with each student feeling as though the instructor is speaking to him/her.

There are some lecturers, who speak almost by heart as in dictation, and the students note down almost every word. From an ideal lecture, students make only pertinent notes with carefully selected headings and subheadings, definitions, phrases, words, numbers, and abbreviations to serve as memoranda. Students are likely to learn most from such lectures. In such lectures, they are not fully occupied in writing, and have time to think, select, and note only the essential points.

A recent trend among new generation teachers is to use 'presentations' for everything! Some lecturers base their entire lecture on projected slides made through 'PowerPoint'. 'PowerPoint' can be a very effective tool for enlivening the lecture, if used properly. Unfortunately, PowerPoint presentations frequently end up as long sequences of slides, each containing a list of bullet points. Sometimes, it may be a reproduction of lecture notes as such. Crowded slides with too much material will not hold students' interest. This is compounded when the lecturer just read out the slides! The impression gathers in such instances is that the lecturer is lazy and not well versed with the subject, and that he/she is unable to face the students in the absence of slide display. Understand that speaking the same words that are written decreases the ability to understand what is being presented. Never make the lecture dull and boring by resorting to these practices. Remember students hate teachers who read from slides just as they do to those who read notes instead of lecturing!

Sometimes, lecturers 'read' the material, if they are not proficient in English language. This happens mainly with non-native speakers of English. There is no other way out, but to improve your language proficiency.

Restate and Periodically Summarize Key Points

You have to ensure that the students are always in an active learning mood. Occasionally students may become simply passive listeners. To prevent students from sinking into passive listening, involve students' active listening skills by interspersing questions throughout your lecture. Allocate enough time for students to interact with you by asking questions or demanding additional explanations. Begin and close the lecture with a summary statement. Students need to see how each new topic relates to what they have already learned and what they will learn in the coming days.

Look for the Feedback
The students should receive the messages well. An easy method to check for this is to look for feedback from the students. Feedback from students can be in different forms, verbal—when students answer questions, or in non-verbal signs—when students express in body language their reaction to what they have received from the message; they nod with their heads, they laugh after a joke, they look bored, and the like. Incomprehensible and unclear messages are not easily stored in the memory, and they are quickly forgotten. Sometimes, you may notice that some students are showing signs of sleepiness. This does not always mean that they are bored because of your lecturing! They may be tired due to something else. Another usual mistake is to assume that the students' silence is related to understanding on their part. This need not be correct; sometimes, they might be confused!

You should have a policy for encouraging questions and try to answer all questions as best as possible. Remind them of the Chinese proverb attributed to Confucius: 'He who asks a question is a fool for five minutes; he who does not ask a question remains a fool forever'. There should be a reasonable chance for students to formulate and ask questions. It is not enough simply asking 'Any questions?' with your back turned to the audience. Reword it as an honest gesture such as, 'Should I repeat this once again?' or 'Which portion is still not clear to you'. In a large class, you must repeat the student's question so that all the students know what question you are going to answer. It is a good idea to reserve two-to three-minute blocks for questions at transition points in your lecture(after every 15 min or so). Allow the students to utilize this time to think, even if nobody asks a question. This will encourage students to review the material recently covered, and if they desire, can ask questions.

Admit Ignorance
Occasionally, you may feel peeved at your ignorance, especially when some bright students ask some unexpected questions. If you do not know the answer, it is better to admit ignorance rather than becoming angry or annoyed, but you should know enough to tell whether your ignorance is just your own or everyone else's as well! If you are unable to answer the question properly, tell them politely that you would find out the answer and inform them in the subsequent lecture class. You can also allow the student know that the question goes well beyond what you can address in lecture, and allow the particular student or others to find the answer and report back.

Similarly, equating ignorance or lack of knowledge on the part of the students with stupidity when answering questions is wrong; instead, encourage them to put some more hard work and study with earnestness.

Manage the Class Effectively
Establish reasonable rules for student behaviour. Tackling late comers is a big issue. If the students keep on coming during the lecture, it will be a huge nuisance. You can form a policy regarding late coming and announce it in the class. Tell the students that you expect them to arrive promptly. However, most teachers allow five minutes, and you may announce in the class that after five minutes, the door would be closed. Use the first couple of minutes to discuss a related issue to take account of stragglers.

Similarly, you should have some rules to manage with talking during the class, not doing homework, and other disruptive behaviour.

Explain your rules early and stress the value of cooperation and consideration.

21.4 Use Visual Aids Effectively

Visual aids increase the effectiveness of a lecture and help to increase the interest of the students. These devices introduce variety and help to avoid monotony for better participation by the students, and sometimes are helpful to overcome language barriers. Visual aids fall into two categories, ***projected*** or ***non-projected***. Non-projected aids include chalkboards, white boards, charts of various kinds, cards, and flannel graphs. These are simple and effective, and can be used alone or along with projected aids. These visual aids not only add clarity but also help to present complex ideas with simpler ones.

Presently, there are several new generation gadgets available for lecturers to make the class more effective, for example, LCD or LED projector and white board. In the past, black board was the only visual aid available for the lecturer. You may select one or more type of visual aids according to availability and ease in handling. The ubiquitous slide projector and overhead projection are on the way out; and now, the norm is computer slides projected through LCD system ('PowerPoint' slides). Occasionally, maps, models, photographs, drawings, etc. are also used.

Blackboards

Black board or chalkboard is extensively used in academic institutions, as it is an effective and cheaper way of giving a visual appeal to the materials presented in the class. The black board can be effectively used for a group of 30–40 students, and it is commonly used to present new words, terms, rules, definitions, and classification. It can also be used to illustrate certain facts and ideas by maps, diagrams, and sketches.

You should learn to use a black board properly. Before you start your class, clean the black board by erasing, if required. Use a good duster to clean the board. Always clean the board by holding the duster horizontally and wiping the board in the vertical direction top to bottom. You may start erasing from the top, and when you reach the bottom in the vertical direction, go to the top and start again erasing from top to bottom. This way you can complete the cleaning process. When you write, make letters and drawings large enough to be easily seen by all the students. A distance of 10 m away from the board would need lettering of at least 6 cm height. Similarly, when you lecture, do not cover the material on the board by standing in front of it.

If a drawing or chart is complex, put it on the board before your class or you may draw a faint line in advance and finish the drawing during the presentation. You need not put elaborate details, as too much accuracy is not needed. Similarly, try to avoid putting too many materials on the board at a time. Use the whole area of the board rather than a corner of it. At the end of the class, do not forget to clean the board by wiping out whatever you have written.

White Boards
A recent trend is to replace black boards with white boards. You can use a white board in the same way as that of a black board. Use good-quality white boards; otherwise, erasing written matter from the board becomes difficult. Similarly, use good-quality white board pens for writing. All the other instructions given in the case of black board are equally applicable for the white board.

Models
Models are very effective for teaching certain specific items. Many kinds of models can be made and used. For example, models of equipment, structures, animals, microorganisms, human body parts, DNA strand, etc. make communication very easy. Similarly, models of irrigation system, soil erosion structures, water measuring devices, engines, etc. can be very effectively used.

Charts
Charts are made to represent diagrammatically the relationships or trends, ingredients of a product, steps in a process, or sequence of an event. There are various forms of charts as already discussed in Chap. 13. Select and use the most appropriate among them.

Overhead Projectors
It is a projector for projecting onto a vertical screen magnified images of tables, charts, graphs, diagrams, or text materials prepared on transparent sheets. It can also be used in classroom lectures. However, overhead projectors are on the way out after the onslaught of computer-aided projectors.

Computer Based Projectors
For classroom teaching, LCD or LED projectors can be conveniently used, especially for showing photographs, diagrams, tables, and terms that require emphasis. A lecturer can prepare good-quality slides easily through the help of computer programmes like 'PowerPoint' (see also Sect. 20.4). Show the slides as and when the lecturer explains the contents. Bear in mind that the slides are intended to be an aid to what one has to say; but should not be a substitute for it.

PowerPoint can be a very effective tool for enlivening the lecture. It is easy to import graphics, photographs, charts, graphs, audio and video clips, and to insert live web links. Used well, it can generate interest and provide rich and varied information. Unfortunately, PowerPoint presentations frequently exclude these features, and simply end up as long sequences of slides, each containing lists of bullet points through which the lecturer works in pedestrian fashion. Slides with bullet points, interspersed with other types of material, can work exceptionally well. The following 12 guidelines are helpful to use PowerPoint effectively in lectures:

1. Keep the number of slides to a minimum.

2. Do not cram all the information in a slide, as you will explain the details.
3. Each slide should have only one central idea or message.
4. Give a title to each slide, which acts as a headline.
5. The use of slides is to enhance and illustrate the presentation. If a slide does not really add anything, do not include it.
6. Avoid using complex background images, which detract attention.
7. Ensure good colour contrast between text and background.
8. Use sans serif typefaces such as Arial, Helvetica, or Verdana for writing texts on slides.
9. Use typefaces above 18–24 points for text matter and 32 point or more for titles depending upon the fonts.
10. Never use slides with bullet points only.
11. Consider use of animations within presentations to build graphic explanations of complex ideas.
12. You may import and use images and sound or video clips as appropriate, but while doing so, ensure that they are compliant with copyright laws.

Students now expect all lecturers to provide copies of the PowerPoint presentations from their lectures (if lecture notes are not given). The concern often expressed by lecturers is that students may see handouts, whether provided in hard copy or electronically, as a replacement for attending the lectures. The reality is that students may indeed think so, if attending the lecture gives them no benefit over and above the PowerPoint presentation! The lecturer must warn the students from the outset itself that simply taking the handouts is not going to give them the best learning experience.

21.5 Sustain the Interest of Students

Sustaining the interest of students throughout the lecture is a formidable task. Hawkins (1985) stressed on several techniques to sustain the interest of the audience while lecturing. Some general tips to arouse and sustain the interest of the students are provided here. All the lecturers can master these tips, if they take lecturing a bit seriously.

Remember that attention in lectures usually declines after about 15 min, so vary activities or switch the focus of the lecture.

Rhetoric: Develop skill in the mode of expression and speech delivery. Some teachers hold sway over the students through their oratory skills. Although this ability is often inborn, anyone can improve his/her speech by practice. Choose the most appropriate words, avoid worn out words and clichés, and speak emphatically with the correct gestures and body language.

Quoting examples: Include sufficient examples to prove your points and for clarification. Interesting cases will hold the attention of students, and they will easily understand difficult concepts.

Humour: Occasional use of humour in the class revives interest and gives light relief. However, not all lecturers may be proficient in the use of humour, and they should learn the art of using humour successfully. Too many jokes and hilarious overtones may annoy and irritate the students.

Setting problems: This is a good idea to take the class with you. You present a problem to the students so that they can take part in thinking of various solutions to it.

Provocation: Provoke the students to think in different ways. Attacking accepted hypotheses or traditional ideas will definitely provoke the listeners.

Mentioning contemporary problems: You can mention some current topics of interest. Certain topics may have already aroused interest and have been headlines in the daily press or in the TV channels.

Allowing interruptions: Instead of continuing one-way delivery of lecture, allowing interruption from the students is a good practice to sustain interest. Encourage the listeners to take part in the lecture by asking questions. However, when the audience is large, the questioner may irritate the rest of them who feel frustrated because they may think that the progress of the lecture is being delayed.

21.6 Improve Motivation of Students

In colleges and universities, teachers often complain that students' motivation is not what it used to be. Competent and talented teachers try to identify the motives that activate different groups of students, and change their teaching techniques to enhance students' motivation. Teachers should learn to praise their students for their accomplishments. Take your students as seriously as they take you. When you become more familiar with the students, they may take some freedom with you, but do not confuse this familiarity on the part of the students with lack of respect. Similarly, students may disagree with you on many counts, but do not equate intellectual disagreement with personal animosity.

Some lecturers behave as though theirs is the only course the students undertake during the semester! Understand that there is a limit for students' learning capacities. You should show compassion and understanding in students' problems.

Some symptoms of failing motivation are easily discernible as in-class behaviour or out-of-class behaviour. Race (2007) listed in-class and out-of-class symptoms of low motivation.

Some in-class symptoms of low motivation:

- Late coming to class and leaving early
- Talking to friends in class about other things
- Looking out of the window, scribbling, drawing, doodling, writing letters to friends, sending text messages on mobile phones
- Lack of engagement, not asking questions, not being willing to answer questions, nor volunteering responses when invited

- Diverting lecturers from the main issues
- Not coming in equipped with pens, paper, books, calculators, and so on
- Taking a longer break than is intended during long sessions, or failing to return at all, yawning, looking disinterested, and avoiding eye contact.

Someout-of-class symptoms of low motivation:

- Regular absence without reason
- Inadequate preparation towards class work
- Handing in scribbled last-minute work—botched, or not handing in any work
- Low quality individual and group work
- Damaging each other's attitude
- Work avoidance strategies—giving in too easily to doing only unimportant tasks and putting off doing important ones
- Ignoring lecturers out of class
- Doing only what's necessary for coursework marks, but not doing other things
- Not buying books nor using library resources
- Maintaining poor folders and disorganized collections of handouts.

The reasons for low motivation must be identified and tackled well. You may use some tactics. Learners require some reward or 'reinforcement' for learning.

Students' motivation can be ***intrinsic*** or ***extrinsic***. When students are motivated from within, it is intrinsic motivation. For example, some students are driven by intellectual curiosity. By entrusting students with some responsibilities, the teacher can maintain their motivation at a high level. They should be encouraged to find out answers to open questions and queries by themselves.

Similarly, students' motivation is enhanced when they understand that the subject taught is important because of its practical significance. Certain students try to get satisfaction in attaining competence and proficiency in a specific field when they understand that this gives them a better standing or enhanced influence in their community. This can be called ***competence motivation.***

The motive for participation and recognition is also evident among students. Acceptance and approval are strong personal needs in most people. An expected reward of some kind such as praise or satisfied curiosity can be a strong motivation. Many students tend to respond to a caring teacher who conveys interest in them, who listen carefully to them, and who respect them as individuals. Students may do well, if they are motivated extrinsically (***extrinsic motivation***) too. For example, the expectation of receiving a degree certificate for its premium value of job guarantee rather than for the competencies it certifies is a strong motivating force. Experienced teachers know the importance of rewarding their students with praise, attention, and other types of encouragement.

21.7 Build Rapport with the Students

Good teacher–student relationship is based on mutual respect. Students respect teachers for their teaching skills, personal qualities, knowledge, and professionalism; and in turn, teachers must respect students as individuals and their sincere attempts to learn. It is important to realize that respecting each individual student is not the same as some kind of generalized respect for the class as a whole; and that the teacher's respect must be shown as well as felt, otherwise the student will be unaware of it. There must be good rapport between the student and the teacher, which creates some sort of formal and personal authority.

Rapport and Formal Authority

Rapport between the teacher and students takes time to establish, usually passing through two phases. In the first phase, you achieve a position of power by virtue of your role as a teacher. Do not expect students to like you from the first class itself. You normally start in a formal relationship with your students. The position of being a teacher confers on you certain duties, but with these duties, some rights are also inherent. Your students accept your authority, if you apply it with confidence, and you begin to feel confident.

Formal authority is established through non-verbal methods. For example, when facing the students, stand up straight with shoulders back. Give instructions with a confident tone of voice, expecting that they obey you. If the student has not done something that you asked them to do in reasonable time, be confident and emphatic; you might show a puzzled surprise at the impertinence, but never show that you are perturbed. Authority is conveyed principally with body language. Proximity and eye contact can show commanding posture. The closer you are with the students, the greater your effect, especially if you invade the student's 'personal space' and adopt a commanding posture. Sustained eye contact while you are speaking is a good tool.

Rapport and Personal Authority

The phase two in developing student–teacher relations is a gradual shift from the formal authority towards teacher's personal authority. A teacher who wields formal authority with some skill in teaching and make the students learn will earn the respect of their students. If everything goes well as expected, in due course, the relationship may shift so that it will be based on personalities. The source of a teacher's power must become the desire of students to please the teacher, and to build their own self-image through the teacher's approval. This is called 'personal authority'.

21.8 Plan and Prepare for Group Work

Group work is common in some practical and field works. When planning to give group work, think carefully about what you wish your students to achieve and adjust the size accordingly. Large groups mean fewer clusters to monitor and assess; however, certain tasks may require smaller, more interactive clusters. Figure out expectations from each group, and offer guidance on how to coordinate the group project. If two or more groups are assigned the job of exploring the same topic or project, arrange to have the groups give feedback to one another. If the students are supposed to work outside class hours or on holidays, remind the groups to get contact details for one another to enable consultations. You may develop some measures to increase accountability on individuals working within the groups. For example, before finalizing grades, get the students agree on a percentage for individual contributions.

Suitably planned group work, especially when groups are given clearly defined and achievable goals, stimulates deeper and wider learning outcomes for most students. Such effective group works allow the students to clarify ideas through discussions and debates, and improve interpersonal and communication skills.

Group activities allow teachers to encourage the use of problem-solving and critical thinking skills. Students often consider teachers the sole source of knowledge and problem solvers. However, through group works, students get an opportunity to gather knowledge through their own efforts. Some students may use group work as an occasion to loaf and wander; therefore, accountability measures are needed. A strategy to curb this tendency is to give marks for individual inputs and participation. Understand that group work tasks should relate to course objectives, and must play a role in assessment, rather than being used in an isolated manner.

Assignments are part of university teaching, and sometimes you may also give homework. If properly done, assignments would be of immense value as an additional learning instrument. Short essay assignments and long essays called term papers are the usual types of assignments given to the students. Apart from these, there are many other useful types, which can also be given to the students, depending upon the need and leaning situations (see Chap. 22 for more details).

It is a good idea to stagger due dates for submission of assignments, especially when the number of students are large. Divide the students randomly into convenient groups, and announce the dates of submission of papers. All students must be given their assignment topics at least two weeks before their due date. This approach frees the lecturer from reading and responding to all the assignments at one occasion.

21.9 Conduct of Examinations

Examinations are part of the system. The main purpose of examinations is to ensure that the students have learned well the material taught in the class. When you start lecturing, you should imagine that your students are there to learn and not just for passing examinations. Examination-oriented teaching or reminding them always about impending examinations are not healthy practices. In the recent past, all the examinations were public examinations, questions being set by external examiners.

Most academic programmes in India involve internal and external evaluation. Continuous evaluation is the core of any internal evaluation system. In many universities, although external component is still part of the system, a part of the examination is conducted as internal. When setting question papers, go for a balanced mix of easy and difficult questions. For 'quiz' type examinations, objective type questions, including multiple choice questions, are preferred. If the system has the provision of only one internal examination (probably, a midterm examination), include both objective type and descriptive type questions. Avoid asking questions on topics that you did not ask students to learn or you could not understand.

Remember that in examinations, you cannot simply 'give' marks or grades, but students have to 'earn' grades by studying and properly answering. You should be impartial in valuing answer papers of internal examinations, and students should feel that you are impartial and unbiased in awarding marks. Good teaching and good examination should go together, but grading must be based on the performance of students in the examination.

There must be a meticulous and foolproof mechanism in the university for conducting external examinations and valuing answer papers. Teachers and students frequently make complaints on several issues related to the conduct of examinations such as out of syllabus questions, skewed questions concentrating only on some parts of the syllabus, below standard questions, questions on outdated concepts ignoring the present ones, and so on. For setting questions, in addition to the syllabus, a copy of the lecture outline must also be given to the examiner. Clear instructions must be given to the external examiner on the pattern of questions.

Conduct the examination in a proper way so that the temptation of students to indulge in copying and unscrupulous practices is eliminated. Fixing CCTV camera, employing sufficient number of invigilators, comfortable seating arrangement, and similar measures act as deterrents in copying. Do not allow the students to take mobile phones or such electronic gadgets with them. You can allow calculators depending upon the subject.

Bibliography

Barker, L. and Watson, K. 2000. Listen Up: How to Improve Relationships, Reduce Stress, and Be More Productive by Using the Power of Listening, St. Martin's Press, 203p.

Bligh, D. 2000. What's the Use of Lectures? Jossey Bass, San Francisco, 368p.

Boch, F. 2005. Note taking and learning: A summary of research. The WAC J. 16: 101–113.

Brown, G. 1987. Lores and laws of lecturing. Phys. Bull. 38: 305–307.

Davis, B.G. 2009. Tools for Teaching (2nd Ed.). John Wiley & Sons, 608p.

Erickson, B.L., Peters, C.B., and Strommer, D.W. 2006. Teaching First-Year College Students Jossey-Bass, San Francisco, 204p.

Friedmann, H.C. 1990. Fifty-six laws of good teaching. J. Chem. Edu. 67(5):490.

Hawkins, C.1985. Speaking at meetings. In: Hawkins, C. and Sorgi, M. (eds), Research: How to Plan, Speak and Write about It (4th Indian Reprint, 1993). Narosa Publishing House, New Delhi, pp.60–84.

Hoskins, S.L. and Newstead, S.E. 2009. Encouraging student motivation. In: Fry, H., Ketteridge, S. and Marshall, S. (eds.), A Handbook for Teaching and Learning in Higher Education Enhancing Academic Practice (3rd Ed.). Routledge, Oxon, UK, pp 27–39.

Morton, A. 2009. Lecturing to large groups. In: Fry, H., Ketteridge, S. and Marshall, S. (eds), A Handbook for Teaching and Learning in Higher Education Enhancing Academic Practice (3rd ed.). Routledge, Oxon, UK, pp 58–71.

Nichols, R.G. and Stevens, L.A. 1957. Are you listening? New York, McGrawHill, New York, 235p.

Pauk, W. and Owens, R.J.Q. 2011. How to Study in College (10th Ed.). Wadsworth, Boston, USA, 406p.

Penner, J.G. 1984. Why Many College Teachers Cannot Lecture: How to Avoid Communication Breakdown in the Classroom. CC Thomas, Springfield, USA, 212p.

Petty, G. 2009. Teaching Today: A Practical Guide (4th Ed.). Nelson Thornes, UK, 614p.

Race, P. 2007. The Lecturer's Toolkit: A Resource for Developing Assessment, Learning and Teaching (3rd Ed.). Routledge, Oxon, 264p.

Robertson, A.K.1994. Listen for Success: A Guide to Effective Listening. Irwin Professional Pub, Australia, 197p.

Chapter 22
Note Taking, Note Making, and Assignments

Live as if you were to die tomorrow. Learn as if you were to live forever.
Mahatma Gandhi (1869–1948)

Note taking is a normal practice when students attend lectures, seminars, and discussions. Noting down important points is also routine when reading books, journals, and other reference materials for preparing essays, seminars, presentations, theses, and examinations. These activities are important for researchers and teachers too, if they want to excel in their profession. We use the phrases, *note taking* and *note making*, to distinguish the above two processes of hearing and reading. 'Note taking' is a passive process, mostly done by students or listeners during lectures, but 'note making' is a more active and focused activity where one note down relevant and required information after reading a book or article. Note taking and note making represent two related key academic skills students and researchers must nurture, but one has to approach them prudently for greater utility.

A Chinese proverb says, 'Tell me, I'll forget. Show me, I'll remember. Involve me, I'll understand'. To put it differently, hearing is forgetting, seeing is believing, and doing is learning! It is often said that people remember 90 percent of what they do, 75 percent of what they see, and 20 percent of what they hear (Ellis 2013). The pre-eminent position of experiential learning is because of these facts. However, experiential learning is impossible in all situations, and lectures are the preferred method of teaching when large students' groups are involved.

The major difference between note taking in lectures and note making from reading is the absence of control one has over the process, because lectures happen in real time. This means that when you take down notes from lectures you cannot slow down or pause the teacher, spool back, and then replay. Probably, the maximum you can do is to ask the teacher to be a little bit slow! Nevertheless, during the process of note making from reading, you can easily go over again to something you could not understand. You can easily stop and reread something again, if you want. Later, if you are reminded of some information you want to look up, you will be able to do it promptly by going through the book and make notes again. Thus, you can make sure that your notes are complete.

C. G. Thomas, *Research Methodology and Scientific Writing*,
https://doi.org/10.1007/978-3-030-64865-7_22

In most academic institutions, assignments are routine, and they act as additional learning instrument. Many types of assignments can be given to the students such as essays, literature reviews, annotated bibliographies, critical reviews, reflective journals, and case studies depending upon the need and leaning situations. Essay assignment is the most common among them. Essay writing increases the expertise of students in some particular subject area and is helpful to refine their analytic and writing skills. Note making is an integral part of assignments.

22.1 The Issue of Forgetting and Recall

Lectures are the most common method of teaching and learning in universities. During a lecture class, the instructor gives an oral presentation of subjects included in the syllabus, most often accompanied by visual elements. Soon after initial learning of an information, our recall is almost 100 percent. However, after time lapses, we begin to forget what we have learned.

Most of us are not sure how we remember events and facts, and why we forget them quickly. Psychologists believe that the process of remembering involves information passing from our short-term memory into our long-term memory. Initially, information may be stored in the short-term memory for a brief period, probably a few seconds. However, the long-term memory can store information for a great period, but practically most of it is quickly forgotten. The content of the short-term memory is short-lived and is easily displaced by new information. If students are given new information one by one in quick succession, they will not have enough time to process it suitably in the short-term memory; hence, the information will not be retained. For retention in the long-term memory, some measures such as recitation have to be done deliberately. Once the information is passed into the long-term memory, it will be retained.

Although the information you read from textbooks too is forgotten quickly, the rate of forgetting for things you hear from lectures occurs even faster. This has to do something with the way the short-term or working memory is designed. Forgetting and remembering occur automatically; these are not conscious events. What we call 'forgetting' is the brain's built-in mechanism to ensure that it is not overburdened with useless knowledge. The brain wants to remember useful information only. However, our brain tends to consider an idea useful in the long term only, and if it is not used regularly, it may be forgotten. Repetition or frequent use ensures that something is remembered. Therefore, teachers must make sure to recall and use frequently any information you want your students to remember.

The famous forgetting curve of Hermann Ebbinghaus shows how information is lost over time when you do not try to retain it (Murre and Dros 2015). A typical forgetting curve shows that humans tend to lose 50 percent of their memory of newly learned knowledge in a matter of days unless they consciously review the learned material. The level at which we retain information depends on the strength of our memory and the amount of time passed since learning. The more frequently one repeat something, the more likely it is to stick. Each recall helps to fix learning more firmly in the long-term memory. Consequently, it takes increasingly longer time for forgetting the learned idea after each recall. Note taking and subsequent recall make the information fixed in the long-term memory. This is the rationale behind the suggestion to review the notes taken during the lecture regularly to improve memory retention. Reviewing at regular intervals increase retention, and over time, less frequent review is needed.

22.2 Note Taking

Note taking is the process of recording information when somebody is speaking. Often, notes are drawn from a transient source such as an oral discussion at a meeting (notes of a meeting are called *minutes*) or a lecture, in which case, the scribbled notes may be the only record. Note taking is an important skill for students, especially at the university level. It is a battle against time; and it can be a challenge, if the content of the lecture is predominantly factual, and you want to record all the facts but the lecturer is going through it fast. The person who takes notes will be under severe time pressure, and therefore, different note taking styles and techniques have evolved to make the best use of time. Listening can be active or passive, and active listening occurs when you listen carefully to make sure that you understand and learn the information being conveyed to you. By taking notes, you can be an active listener and a critical thinker. It helps us to keep our body active and involved. On the contrary, when you listen to a radio or simply speak to a person, you may be doing it passively.

During note taking, you record the essence of the material lectured, freeing your mind from having to recall everything. As a learning tool, it helps you to clarify thinking, forcing you to be selective in picking ideas, organizing ideas, and finally, assisting you to remember. Most probably, the teacher may have given a copy of lecture slides but you still need to take down notes! Sometimes, the teacher would offer more explanation and examples than that appear on the slides. In most lectures, teachers may also share information not readily available in textbooks. If you are not an avid note taker, you may miss out such information. It also helps you to link classroom learning to textbook readings easily. Moreover, scribbling notes in the class helps you to avoid drowsiness or distraction.

How to Improve Note Taking Skills

Consider the following seven guidelines seriously for effective note taking and learning from classroom lectures.

1. **Preparing to take notes**
   Note taking will be easier, if you bestow attention to certain things before the lecture class. Always bring paper, pen, pencil, ruler, and such other things needed for writing notes to the class. Take with you an organized notebook or enough sheets of paper, which can be filed neatly. Sit near or at the front of the classroom

so that you can engage with the teacher and ask questions more easily. For this purpose, you have to come to the class early and occupy the coveted seat much before the lecture!

2. **Review the course syllabus**
   Look at your course outline to see what would be the topic of the lecture so that you could anticipate the discussion points and mentally prepare for it. You may check the relationship between the lecture class and the prescribed textbook, lecture notes, or handouts already given to you. In certain classes, the textbook would be the primary source of information, with the result that the class time is exclusively used for clarifying and applying the concepts. In certain other courses, materials being lectured and noted down in the class would be more important than the textbook. If the lecture is based on textbook material or lecture notes already given, read the assigned chapters before the lecture class. This makes you receptive to the new ideas and concepts.
3. **Review the previous class**
   Go through the notes from the previous class to refresh your memory and get mentally warmed-up for the new topic. Every teacher has a different lecturing style, and therefore, you find out a style of note taking that will be effective for each particular course. Courses that are heavily based on the textbook may require you to focus on the main ideas in class while filling in the details afterwards from your text. In certain courses, the lecturer may be delivering separate material not given in the text. In such lecture classes, you have to take detailed notes, as you cannot use your text as a reference.
4. **Note taking materials**
   For classroom note taking, it is better to use A4 size papers. For occasional lectures, notepads are commonly used. Store your notes in a loose leaf folder so that you can move the pages around as well as add any new material such as handouts and other reference materials. Pens, pencils, and coloured highlighters are familiar tools for handwritten notes.
   In recent times, there are other effective tools for recording lecture material and for making notes electronically on a computer, for example, tablet, iPad, or smartphone. Some students try audio recording of lectures for later hearing, and they leisurely make notes. At present, most digital recorders are affordable, and you can transfer files easily to a computer. For recording, smart phones and devices such as iPads and tablets are popular choices, which allow both audio and visual recording. Bear in mind that you must have the permission of the lecturer to make such digital recordings, as intellectual copyright is a crucial issue to consider here. Some lecturers may provide a copy of their lecture notes or presentation slides to the students.
5. **Note down important points during lectures**
   From a lecture, you may wonder what need to be noted and in how much details. This depends on the subject of lecture and your purpose of note taking. If the information passed on by the lecturer is not available from usual channels or very difficult to obtain from other sources, then the notes can be as detailed as possible.

However, if the information is readily available from common textbooks, then you can focus on the main points highlighted by the lecturer.
Because of differences in speed of lecturing and writing, you may not be able to go for verbatim noting of lectures; it is not desirable either. Write only the main points. The indicators that trigger note taking are the following (Lynch 2004; Boch 2005)

- Main points as emphasized by the lecturer
- Writing on the board/PowerPoint slides: This is a very powerful indicator for note taking. Students identify the cue and write them down.
- Dictation: Sometimes, the teacher acts as if s/he is dictating the information, for example, slow delivery of speech so that students can write.
- Examples
- Diagrams
- Formulae and their derivations
- Calculations
- Listing of information: projected or written on the board
- Specific terms, catch phrases, key words, and definitions
- Relevant questions and answers
- Look for cues.

Listen and look for cues from lecturers as given below for writing notes. These cues indicate that something important is forthcoming.

- Non-verbal cues such as facial expressions and hand and body signals
- Visual cues including references to names, sources, events
- Phonological cues such as changes in voice volume, speed, etc.
- When information that is more important is passed on, often, the speaker speaks slower, louder, and more directly
- Signposts and linking expressions.

As we know, signposts on roads show the direction and distance to nearby places. Signposts in lectures indicate the direction of talk. Listen for the words and phrases that might direct you as:

• Sequencing words:	Firstly, secondly, next, finally
• Summary words:	In conclusion, to sum up, in summary
• Illustration words:	For example, to illustrate, that is
• Reason and result words:	Consequently, obviously, because, therefore
• Contrast words:	However, but, despite, on the other hand, conversely
• Concession words:	Although, even though, in the light of, given that
• Addition words:	Also, too, in addition, furthermore, moreover
• Emphasis words:	Importantly, specifically, especially, obviously

Immediately after the lecture or after returning to your room, review your notes. This enables you to identify any gaps in the information. You can also formulate questions that need further reading. You can highlight key points and add relevant references and link them to the main points of the lecture.

It is a good idea to revise your notes on a regular, ongoing basis, for example, at least for five to ten minutes once a week. Each time you skim through your notes the material will become a little more firmly embedded in your memory.

6. **Abbreviations for making notes**

   The average writing speed of a student is around 18–24 words/min, whereas a lecturer speaks at a rate of around 125 words/min. Unless everything is said at dictation speed, or students develop exceptional short hand skills, you cannot write everything as teachers never speak slowly enough for students to write down everything that is said. You can use abbreviations and shortened versions of commonly used words. Some common abbreviations and symbols are listed in Tables 22.1 and 22.2.

7. **Using notes for revision**

   Use the notes from lectures as a key part of your revision strategies when preparing for examinations or for participation in tutorials. Organize and review your notes in a logical way so that you can make connections between individual ideas and an overview of the whole subject. Read the notes again, highlighting

Table 22.1 Shortened versions of common words

And so on	- etc	Against/Versus	- vs
Association	- assoc	Before	- b4
Behaviour	- behav	Cannot	- cx
Characteristics	- chx	Could	- cd
Curriculum	- curri	Definition	- def
Department	- dept	Education	- edu
Especially	- esp	Evidence	- evid
Example	- eg	Exercise	- exer
Experiment	- expt	Formula	- fmla
Frequency	- freq	Function	- funct
Government	- govt	Homework	- hmwk
Important	- imp	That is	- i.e
Information	- info	Maximum	- max
Note	- nb	Organization	- org
Paragraph	-para	Probably	- prob
Question	- quest	Should be	- s/b
Solutions	- soln	Society	- soc
Strategy	- strat	Support	- sup
With	- w/	Without	- w/o
With respect to	- wrt	Would	- wd

Table 22.2 Symbols instead of words

And	&/+	Approximately	≈
At, at the rate of	@	Because	∵
Change	Δ	Clarify	?
Copyright	©	Decreasing	↓
Each way	↔	Equal to or greater than	≥
Equal to or less than	≤	Female(s)	♀
Greater than	>	Infinity, always	∞
Increasing	↑	Less than	<
Male(s)	♀	Not the same as	≠
Parallel	‖	Percent	%
Plus or minus	±	Plus, in addition	+
Equals, identical	=	Sum of	S
Therefore	\	That is	ie
Unequal	≠	Very important	**

keywords and ideas. Revision must be a continuous process to be done throughout the semester, not just when examinations are scheduled.

22.3 Note Taking Systems

Traditionally, most students take *linear notes*, that is, write notes from the upper left corner of the page, and then, going from left to right and from top to bottom until the page is full. Often, students take notes without understanding anything on note taking systems. When they take verbatim notes as in linear notes, they try to write down everything the teacher says or as much of it as they can in no particular format. Such notes are often hard to decipher and are usually incomplete because a teacher may be speaking more rapidly than students can write. Just because majority of students adopt this pattern does not mean that it is a good note taking method.

Linear note taking is laborious for students because it bores them mentally and exhausts them physically. Moreover, writing whole sentences is time-consuming and results in a pile of information, of which, only a small percentage is actually important. Students often have to waste valuable time in rereading large chunks of their notes to find the information they are looking for. Once a page has been filled, it is hard to add additional information to it. Often, students are forced to squeeze it in between the lines or add it at the bottom of the page with an asterisk. Presently, there are more efficient approaches than this kind of verbatim notes. The following are some improved methods of note taking:

- The Cornell note taking method
- Outline method

- Mind mapping
- Charting

Among these, the Cornell method is the most popular and useful for classroom note taking.

The Cornell Method

The 'Cornell method' is probably the most useful method of note taking for university students. It was devised about 50 years ago by Walter Pauk, a lecturer at Cornell University in the USA (Pauk and Owens 2011). It has been designed for classroom note taking, which includes some post-processing too. The Cornell system is both a note taking and a study system, having a systematic format of condensing and organizing notes without the need for laborious recopying. This method can be used in any lecture situation, and it employs an easy format to distinguish major concepts and ideas. There are five steps (RQ3R) to it—record, question, reduce (summarize), recite, and review.

Step 1. Record

Prepare your writing sheets in a two-column table format as shown in Fig. 22.1. The left-hand column, the cue column, should take up 2.5 in. (6.25 cm) space, leaving the remaining 6 in. (15 cm) for recording information (the record column). Provide 2 in. (5 cm) at the bottom portion of the paper to write summaries. Use only one side of each sheet of paper. Note the main points and restate in your own words the

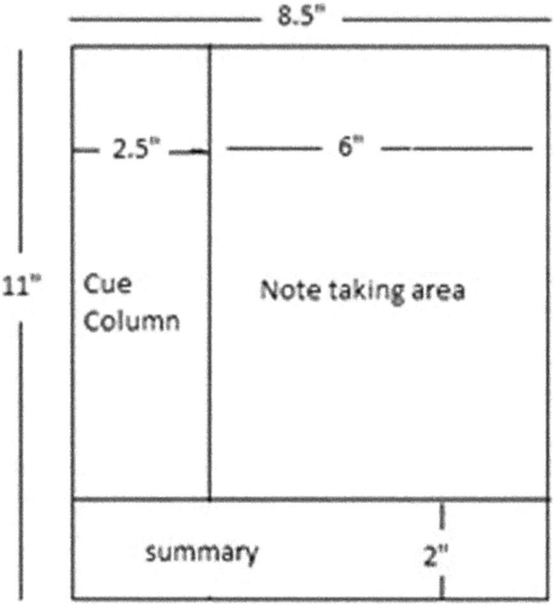

Fig. 22.1 A typical two-column note taking sheet

facts and ideas presented. Record definitions as stated or written. Changes in topic may be noted with headings. Key ideas are written with each topic using numbers or in bullet form. Use telegraphic or brief sentences, abbreviations, and symbols to increase the note taking speed. Write legibly so that your notes make sense to you while reviewing later.

You can take notes in sentence or paragraph form, as lists, as definitions, by adding drawings, or by using a combination of these formats.

Sentences. Take key ideas from a lecture and write them down in your own words. You probably may not have enough time to write out complete sentences. Therefore, write telegraphically, leaving out articles such as 'a', 'an', and 'the', and abbreviate frequently used words. Examples have already been given.

Paragraphs. In this approach, related sentences are clustered together in a block of text, often under a heading or label that serves to tie them together. Note that these are not traditional paragraphs where one complete sentence flows smoothly to the next. The sentences in paragraph-style notes will usually be telegraphic, and smooth transitions are not important. However, these paragraphs share one important thing with traditional paragraphs. All the sentences should relate to the same main idea.

Definitions. A term or phrase is defined here, by adding a dash or colon, and then providing a concise explanation or elaboration.

Lists. Start with a topic, name, term, or process, and then list phrases or telegraphic sentences related to it. Avoid numbering the items unless the numbers are relevant to the list. You may begin each with an asterisk or a bullet point.

Drawings. Note drawings and diagrams as such. A sketch can often convey locations or relationships more effectively instead of description.

Step 2. Question

Frame questions based on the information recorded in notes and write them in the recall cues column on the left-hand side of notes. Try to predict questions on an examination point of view. Consider these questions for guidance when you revise and study.

Step 3. Reduce (summarize)

After each lecture class, without losing much time, summarize the ideas and facts in a few words or sentences in the reduce column provided at the bottom of the page. The summary helps to show relationships between the main points and help to strengthen our memory.

Step 4. Recite

Recitation means explaining yourself the information in the notes in a louder voice. This should be triggered by the test questions you have noted in the cues column. Recitation is helpful to improve learning, understanding, and for retrieval for useful information.

Step 5. Review

Compare your notes with a classmate to check accuracy and understanding. Review and edit your notes to fill in any missing details. You may also check your notes against the textbook or handout. For better results, review your notes within 24 h but before the next class. Keep your notes in a secure place. By the end of the week, you shall do another review. Giving a few days' time to forget some points recorded, and then purposefully recalling again stimulates your long-term memory.

Some general tips

- Put dates on your lecture notes and number all pages
- Keep separate notes for each course
- Begin taking notes for each lecture on a new page
- Never use a sentence when you can use a phrase, or a phrase when you can use a word
- Use indentations to distinguish between major and minor points
- Write most notes in your own words, but formulae, definitions, and specific facts should be noted as such
- Use abbreviations and symbols wherever possible
- Note down unfamiliar vocabulary and unclear areas.

If you should miss something completely, leave a blank space and fill it later.

Mind Mapping Method

The mind map format is an alternative to linear note taking and can be used during lectures, presentations, or when reading a text. A mind map is a graphic method to represent ideas and concepts by relating each fact or idea to every other fact or idea. Information is presented around a central idea. Instead of writing whole sentences, individual keywords or short phrases are noted, which are connected to the central idea with lines.

The mind map begins with a main concept or idea that the rest of the map revolves around it; therefore, choosing the central idea is the first important step. Draw an image or write a keyword at the centre of a blank piece of paper. Decide on important words or short phrases relating to the central idea. Add these and connect them to the central idea with thick lines, the so-called branches. Expand on the key ideas by creating sub-branches, which branches out from the main branches further to expand on ideas and concepts. These sub-branches will also contain words that elaborate on the topic of the branch it stems from. These could be examples, models, theories, strengths, limitations, authors, important dates, or diagrams. Finally, you examine where you can add more sub-branches and information. Decide on where you can add lines to show connections between ideas.

Mind maps may not be useful for large amounts of text. The process helps you to keep visual track of lecture, and the relationships are easy to see. Remember that mind maps are very personal, and therefore, you may not understand the mind map of a fellow student or vice versa.

Charting Method

The charting method of taking notes is appropriate when you need to place-related information into different categories, arrange information in a certain order, or when you want to compare the relationships between different ideas.

In charting, columns in a table format are used with appropriate heading labels. Decide the categories, which may be covered in a lecture. Draw the columns with already determined categories as headings. Put the notes in appropriate columns. It is more applicable when the lecturer focuses on facts and relationships. It reduces the amount of writing needed and provides easy review mechanism. However, you need to learn how to develop effective categories while you are fully engaged in the lecture! You may find this charting method works best for you when you have a general knowledge of the content before the lecture begins.

Charting method of note taking is ideal for subjects that can be broken down into categories such as similarities and differences, events and impacts, and pros and cons. The method is ideal in situations where students are capable of identifying categories and draw tables with the categories before the lecture.

Outlining

In outlining method, notes are organized into hierarchies of points. Normally, text indentation is given to distinguish between main points, sub-points, and other details. The students begin with noting general study details at the left side of the paper, but more specific facts and clarifications are noted by indenting to the right. Listen critically, and then note down points in an organized pattern employing space indentation. Place major points farthest to the left and indent each more specific point to the right. Indicate the levels of importance by putting them away from major points. Simple indentation without any marking is the usual norm, or you can use roman numerals, English letters, or decimals.

Outline method is most suitable when the lecture is an organized and sequential one. There must be enough time during the lecture to think about and make organizational decisions. In outline method, normally, the need for editing does not arise, and it is easy to turn main points into questions. However, it requires accurate organization and you may lose some time for that while attending lectures. This method may become unwieldy, if the lecturer speaks fast. Outlining is useful to distinguish between key ideas and supporting details. Outlining may look disorganized, if you want to note too much information, and may be a waste of time, if you decide to rewrite the notes.

22.4 Listening and Note Taking

Just like successful teachers, students should also struggle to be successful learners. To be a successful student, you should be an active listener. If you consider attending a lecture is waste of time, it would be very difficult for you to concentrate. Students may also be taught how to be a good listener! If you decide confidently to hear and understand, you can improve your listening even with dull and boring subjects that may be difficult to comprehend. Students must understand that not all lecturers may be excellent in lecturing, but the information they are sharing is still valuable. Try to focus on what you are hearing, rather than how it is being presented. Distraction is the big enemy of active listening. Whatever be the distractions, be they domestic or personal worries, noises, or people, you take a deliberate decision to concentrate on the lecture. Understand that no one else can listen for you, and you alone would be the loser for procrastination!

When your mind strolls, try to refocus on the central issue. For example, if you feel confused or lose interest during a lecture, you can raise your hands for clarification or put a question mark in your notes. You can discuss such points later with your teacher or another student. In any case, do not stop your note taking efforts because of such distractions.

Ten Worst Listening Habits

To improve listening, you have to find out bad listening habits prevailing among students. Bad listening habits occur because of the listener and sometimes by the environment. Nichols and Stevens (1957) offer the following list as 10 worst listening habits. You will recognize that if you make a try, you can correct most of them.

1. ***Calling the subject uninteresting***

There is no point in calling a subject dull or uninteresting. Whether dull or not, you are destined to listen and study! The problem is that if you are convinced yourself that the topic is uninteresting, you turn to many other thoughts and concerns and most probably, you start daydreaming! A good listener, braving the temptation to dismiss it as uninteresting, may arrive at a different conclusion that it is necessary to listen and study. There would be something in it, which would be of future use.

2. ***Criticizing the speaker***

Many of us do this on a regular basis. We tend to criticize the speaker (at least mentally!) for not speaking distinctly, for talking too softly, for reading, for not looking at the audience in the eye, or something like that. We often do the same thing with the speaker's appearance. However, if we concentrate on what the teacher is saying, we may begin to get the message and interest may grow in us. Remember, the message is more important than the form or manner in which it is delivered.

3. ***Getting over-stimulated***

In lecture classes, this happens rarely only, if the lecturer is following the syllabus. Still, we may not agree with all that the teacher is saying. In such cases, our thoughts cause us to spend more time for developing counter arguments. If this happens, we may not listen to additional comments on the subject. In situations like this, our listening efficiency drops drastically because of over-stimulation. Try to hear the teacher before judging him/her.

4. ***Listening only for facts***

Many of us listen for facts, and while we may recall some isolated facts, we miss the primary comments or ideas the speaker is trying to make. Make sure that your concern for facts does not prevent you from hearing the speaker's primary points.

5. ***Sitting idle without taking notes***

Many lecturers are so unorganized that their comments really cannot be outlined in any logical manner. It is better to listen, in such a case, for the main points. As seen already, there are many systems of taking notes, and select the best one to fit a lecturer.

6. ***Faking attention***

This is probably one of the more common bad listening habits. While lecturing, if you notice that most of your audience is sitting with chin in hand staring at you, it is a good signal that attention is being faked. Although their eyes are on you but their minds are far away, a clear case of faking attention.

7. ***Tolerating or creating distractions***

People who whisper in an audience fall into this category. This kind of distractions can be corrected to improve the listening atmosphere.

8. ***Evading difficult materials***

Often, students tend to listen to things that are easy to comprehend and avoid things that are more difficult.

9. ***Allowing emotional words block the message***

Certain words or phrases used during lectures, especially, those related to religion, politics, or gender may have an emotional impact on the students. Do not allow emotional words hinder the way of hearing.

10. ***Wasting thought power (daydreaming)***

This is related to wasting the differential between thought speed and the speed at which most people speak. We are able to think much faster than we talk.

Most of us talk at the rate of about 125 words per minute. However, we have the mental capacity to understand someone speaking at 400 words per minute! This difference between speaking speed and thought speed means that when we listen to an average lecturer, we are using only 25 percent of our mental capacity. We still have 75 percent to do something else. Because of this difference in thought power, our minds wander. This means that we need to make earnest efforts to listen and concentrate more of our mental capacity on the act of listening.

As Nichols and Stevens (1957) stated, anticipating the lecturer's next point is a good remedy to overcome this problem of daydreaming. If you anticipate the points correctly, learning would be reinforced. If you anticipate incorrectly, you wonder why it is so, and this may help you to increase attention! Similarly, a good listener takes advantage of short pauses during lectures to summarize mentally what has been said already. These periodic summaries also strengthen learning immensely.

22.5 Note Making

Often, students make notes from textbooks and other reference sources to revise lessons before an examination, and to write an essay, assignment, or term paper. The purpose of note making is to put the material in such a form that it can be easily recalled and used for future purposes. Note making enables students to organize their thoughts better. It gives them an opportunity for active reading. They may not be reading passively, but condensing the points made by the author of the book and drawing their own conclusion about what is being presented. Note making is also necessary to plan a lecture, speech, discussion, or presentation.

In note making too, notes are not written in complete and conventionally correct sentences. Notes are typically much shorter than original texts. While you make notes, the information is condensed using certain devices like abbreviations, symbols, shorter words, and numbers. Certain types of words are often dropped. By adopting a systematic method of writing notes, majority of what you have read can be quickly put together. After going through a journal article or book, you may write down the main points in an abbreviated form. Later, when the need arises, you can prepare the draft in full sentences using the abbreviated notes, which would be substantially different from the original source. Preparing abbreviated notes like this manner helps you to paraphrase the material easily avoiding the blame of plagiarism.

Do not copy down chunks of text from a book. Make clear notes in your own words. You can also note your own ideas that have been stimulated by the text you have read. Write the main subject and important headings before you start, then fill in notes on these areas. Keep referring back to the title of the chapter or essay and make sure your notes are relevant. If more than one author makes similar statements, note the details of other authors in the margin next to your notes from the first author.

Methods of Recording Notes

Several methods of organizing notes are in vogue, but you should develop a note making system that suits your purpose. You should distinguish clearly between direct quotes and your own words. The recording of notes can be in the form of annotation, diagrams, paraphrasing, summary, quotation, and comments after evaluation.

Annotation: Annotation means adding comments to a text that explain or assess what you have read. These are written in the margins of textbooks and may accompany words you have highlighted to identify key information. Note that annotations should not be done on borrowed texts or library books. However, you can photocopy limited pages (without violating copyright laws), and on this, you can do the annotation.

Diagrams: Diagrams are a visual form of taking notes. They could be mindmaps, charts, tables, graphs, or perhaps a drawing to capture a process or cycle.

Paraphrasing: You write important details in your own words. This is ideal for details that support or oppose your viewpoints.

Summary: Sometimes, instead of paraphrasing the original information from an article or book, you may just give a gist of the information. If you state the contents of the article in a condensed form, it is called a summary. This is a common way of reporting, when information from several sources has to be evaluated and presented. Write the main points of the source in your own words.

Quotation: If you are reproducing the exact words of the author, it is called quotation. Sometimes, you may prefer to copy materials as such from sources to save time. In such a case, put quotation marks around everything that you have copied directly from the text to differentiate it from a paraphrased note. You can rewrite this text when paraphrasing later. Sometimes, you use the exact words from the source in your essay to buttress some of your arguments. For such uses of quotation in the text, it is important to copy the material exactly, it must be given in quotation marks and the source indicated.

Comment: Normally, you have to analyse and evaluate what you read and noted before using it for a review in a thesis or research article. You should shape your own ideas and opinions about different issues, and you are expected to support your arguments with literature in your field. If possible, add your own comments and responses to your notes from reading as you proceed. This will help you to recognize the difference between your own ideas and the ideas of other researchers.

It is essential that the information presented in support of an argument in an article is reliable and unbiased. If it is erroneous or biased, the readers easily recognize faults in your presentation. Therefore, take care to evaluate the source before using them for your works. You must develop the habit of interpreting the views of the writer, and noting down your own idea indicating agreement or disagreement. It is important to assess the credibility of each source as well.

Although both primary and secondary sources provide information, remember that primary sources are more reliable and convincing.

Note Down Bibliographic Details

If you are making notes for a literature review, enough bibliographic information must be noted. Include the following bibliographic details in the case of books and journals.

- Name of author/s, editor/s, or organization
- Title of article, book, or chapter (if it is an edited book)
- Volume and issue numbers (for journals)
- Edition and reprint (for books)
- Name of publisher (for books)
- Place of publication (for books)
- Year of publication
- Inclusive page numbers (for articles and chapter articles)
- Total pages (for books).

In the case of electronic publications, especially online publications, the details required are different. Note down the following details:

- Name of author/s, editor/s, or organization
- Title of the article
- Title of the site (i.e., homepage)
- Date of creation of the site
- Modified or updated date
- Date of access
- Full address of URL/DOI.

22.6 Note Making from Text Books

The way notes are made for literature review and everyday learning are different. The abbreviation, SQ3R, stands for a 5-step reading and study method—survey, question, read, recite, and review. It was originally developed by Francis Pleasant Robinson in his book *Effective Study* during 1940s (Robinson 1970). Robinson (1906–1983) was a professor of psychology at Ohio State University, and based on his research, devised the SQ3R method to help military personnel during World War II to learn specialized skills in limited time as possible.

The SQ3R strategy proved to be an efficient one for university students to learn and remember from textbooks for a long time. This is especially effective when the textbook is structured well and reader friendly with such features as boldface headings, italics, and study questions. If you are able to go through the five steps of SQ3R, you will be able to learn and remember the information.

For effective use of a textbook for learning, students should use the above activities from step 1 to 5 such as survey, question, read, recite, and review zealously.

1. **Survey: Survey the book to get an overview**
   A textbook is normally designed with chapters containing headed sections and subsections. During this first step, the survey, you must gather a firsthand broad understanding about the contents of the book. Scan through the book and note the layout, first and last chapters, preface, table of contents, index, the headings used, and familiarize yourself with the publications. Skim through the particular chapter you are about to read looking for its title, pictures, charts, graphs, and topic headings; and as you do this, you will become aware of a general outline the author has used in writing the chapter. This skimming also provides you with an idea of how long the material is, and approximately, how much time you require to read it. The survey gives you an overall impression about the book.
2. **Question: Ask one or more questions for each section in a chapter**
   As outlined by Robinson (1970), in step 2, you must try to turn each section heading given in the chapter into a question. Try to frame questions on the chapter and its contents, the manner in which the reading is structured, and think about the questions that you need to keep in mind while reading. Ask yourself what you already know about this topic. Framing questions like these helps you to reduce a reading to smaller, more meaningful ideas.
3. **Read: Read and mentally answer the questions**
   After framing questions, in step 3, you actively read the section to answer the questions. Remember that reading is a thinking process but comprehension is essential for learning. You engage in reading actively looking for the main points. When you approach a paragraph with heavy contents, slow down your pace for better comprehension. If you have read a section several times, and it still does not make sense to you, try reading it aloud. Similarly, make sure you understand each section before proceeding further.
4. **Recite: Recite and recall the answers to questions from your memory and write them down**
   After reading each section, you must recite the answer to the question from memory and note down this answer as a phrase or keyword on a sheet of paper. Underlining or highlighting also helps, but do it only after you finish the reading and understand what is written. Be selective; do not go for underlining everything making the page messy! Take care to mark only key words and phrases.
5. **Review: Review the chapter completely, by answering the chapter's questions from your memory**
   This last stage (stage 5) is for reviewing what you have done. Repeat the steps from 2 to 4 for each headed section of a given chapter. You can easily accomplish the review after completing the reading of a chapter by briefly looking over the notes to get the overall idea, and then reciting the main points of the chapter from your memory. Review the notes and try to elaborate the information they contain.

   Review should also include reflecting on what is learned, and then by making connection with previously acquired knowledge.

Many reports have shown that SQ3R method improves reading comprehension and learning. Nobody wants to read a textbook as you would read a literary work such as a novel. Literary works are for entertainment by arousing your passion. From a textbook, you must actively pick up information and digest it for understanding and learning. In the SQ3R system, the first step, survey, prepares you to read effectively. The second step, questioning, maintains your concentration on the subject, and the third step is actual reading. The fourth step, recitation of what you have read, allows you to check up on your learning. The last step, review, increases your understanding.

22.7 Assignments as a Part of Learning

Assignments have crucial role as an additional learning instrument. Often, they are compulsory part of the internal evaluation process, and points are set apart for them in both undergraduate and postgraduate studies. However, many students have a hostile attitude towards assignments as though these are thrust upon them from above without having any utility! If properly done, assignments would be of immense value as an additional learning instrument. However, assignments would be a big headache for the students, if they wait until the last minute to write the papers! Short essay assignments and long essays called term papers are the usual types of assignments given to the students. Many other types of assignments can also be given to the students, considering the need and leaning situations.

Students are entrusted with several types of assignments, each having its own structure and features. Examples include:

Essay: Probably, essay is the most ubiquitous among assignments especially in UG classes. Students are given specific topics, and the written essay must answer a question. The student is expected to present an argument based on facts. ***Term papers*** are also essays but these are lengthy essays on a subject submitted usually by the end of the term, the trimester or semester.

Literature review: A 'review of literature' is a critical and evaluative description of publications on a chosen topic and provides a narrative using the references in the bibliography. Appropriate for understanding current level of thinking and gaps in research in a particular area, this is ideal for PG classes.

Annotated bibliography: An annotated bibliography provides a brief account of the available literature on a given topic along with a brief summary of content and a short evaluation (usually about 150 words) in a paragraph, the ***annotation***. This helps to evaluate the usefulness of articles in relation to the topic by identifying key articles on a topic.

Critical review: This is similar to a peer review. Students are asked to critically review some sample research articles and submit a report as peer reviewers do. This is helpful to evaluate or appraise the data, research methods, and results of an article.

Case study: Appropriate to examine a situation by identifying positives and negatives and finally to make recommendations. If you want to get a detailed background view of a particular case or phenomenon, case study would be useful. The case may be a person, plant, group, process, disease, event, community, or any other similar unit.

Reflective journal: Reflective journals or reflective dairies are personal records of students' learning experiences. Students record learning-related incidents, for example, the people involved, the purpose of the event, their reflections on the event, their feelings, and so on. Reflective journals help to identify your understanding, reflect on your thinking, and realize how and what you have learned. Ideal for study tours and experiential learning-based courses.

Experiment write up: Sometimes, the assignment may be a small experiment. This is to explain what you did and to draw conclusions out of it.

Project report: Project reports include the details of activities undertaken as part of a specific work assigned to the student.

Herbarium: Herbarium preparation is a common assignment in plant taxonomy, weed science, plant diseases, agrostology, and forest studies where a large number of plants are involved.

Insect collection: Common in entomology courses, wherein students collect various insects and present them in specially designed insect boxes after identification.

In addition, in some disciplines, assignments would be to make *models, posters, popular articles, leaflets, processed products*, or others such as *visual aids.*

Depending on the learning situations and training needs, any of these assignments can be given. For PG students, the types like review of literature, annotated bibliography, critical review, case study, etc. would be more appropriate rather than simple essays. For UG teaching, essays, herbarium, insect collection, reflective journal, project reports, etc. are more common. Essay assignment is covered in detail here, as it is the most popular form of assignment.

22.8 Essay Assignments

Essay assignments have two major motives. One is related to the specific subject of the course; and the other one is based on your professional development. The first course-specific motive is to increase your expertise in some particular subject area. You will be benefited gainfully by doing literature search and by writing an essay on it. The effort will lead you in to the intricacies of a specific topic far beyond what is possible from lectures. Understand that the goal of essay writing is not to show off everything that you know about the topic, but to show that you understand and can think critically about your topic. The second motive behind essay assignment extends beyond the specific content of the course. The object is to refine your analytic and writing skills. As a professional, you should be in a position to find information, evaluate, and convey your conclusions and recommendations to others. You will learn to go through various source materials, synthesize them, criticize them, and come to your own conclusions.

A casual attitude to essay assignments is prevalent among students. Some students question the very usefulness of essays, going for blatant plagiarism by copying the entire essay of their friends! While writing the essay, students are supposed to follow good academic practices. Students must bestow special attention to avoid plagiarism. They should also learn the art of paraphrasing. Some ideas, which may help students to write good essays, are presented here.

The Process of Essay Writing

Academic essays are different from literary essays. You must be able to answer a question based on evidences available from different sources. Careful planning and note making will facilitate the writing process easier. This may be a daunting task for the beginners. By practice, anybody can master the art of essay writing in science. Essay writing involves, four phases—analysis and planning, literature collection, drafting, and editing.

1. **Analysis and planning**

Analysis of the topic and effective planning make the essay-writing job much easier for the students. Immediately after receiving your topic, start analyzing it and try to understand the problem. Decide your approach to the essay without postponing the preparation.

If the teacher has given you freedom to choose a topic, put some thought into it before settling for one. Make sure that any topic you select fulfils the objectives of the assignment. If possible, pick up a topic that attracts you. Similarly, ensure that selected topic fits into the length of essay that you intend to write and the reference resources are available to you. If you have trouble in choosing what to write about, start with a few ideas and choose the best one after several steps. You will get ideas by looking through the textbook, lecture notes, and journals and periodicals pertaining to the subject. Do not select very broad subjects, but select specific ones. Once you have a specific topic, you can write it as a question, which your essay should answer. Identifying key verbs in the title or question is essential before planning your answer. Essay titles may indicate key verbs such as analyse, describe, compare, discuss, evaluate, examine, explain, and justify, which inform you about how the question must be answered. Remember you can write essays as argumentative, informative, or analytical depending on the question you frame.

Examine the topic critically. Ask yourself what you already know about this topic. Decide on the time required to get the required information. If you require clarifications from the teacher, you can do it now. Maintain a positive attitude and work on the project. Understand that you may not be able to write an essay quickly without a plan. Provide enough time to think and organize thoughts before you start to write. Set a deadline for completion of background reading and literature search. When you have done the necessary planning and preparation, you are ready to involve thoroughly with actual literature collection. Decide how you proceed with literature collection, the sources, the types, etc., and allot sufficient time for literature collection. Sometimes, you need to revise your strategy after collecting some of the information sources. Making a framework for writing is also part of planning.

2. **Literature collection**

In essays, you must justify the statements you make. A major form of justification is through a reference to an authority who makes similar statements. Authoritative publications will always specify the date of publication and address of publishing houses. This allows you to decide whether the information they contain is current or out of date. Do not depend upon only one source to write an essay. There may be contradictory views on the topic assigned to you. Therefore, for searching the topic thoroughly, you must go through several books and articles.

Locating the sources of information, both online and print sources, is an important step in essay preparation. Two kinds of literature collection are necessary for an essay. For obtaining certain facts, you need to consult primary sources of information such as journal articles. You must also go through secondary sources such as textbooks, reviews, and monographs to find support for the argument presented in your essay. The main points you have identified in your analysis of the essay question should direct your literature hunt. Apply your critical reading skills during this phase of literature reading. Do not believe everything an author says, but evaluate them in the light of evidences collected. Sometimes, the chosen topic may involve contradictory views. You have to use your own judgement in such cases.

Books: You may find several kinds of books such as *textbook*, *handbook*, *monograph*, *treatise*, *manual*, *review*, and *yearbook* in the library, all of which are considered as secondary sources. A recent trend is mushrooming of 'predatory' publishing houses, which publish books of mediocre authors without any scrutiny. Check the credentials of both the authors and publishers before relying on any source for your assignment. Avoid 'predatory' publications lest they mislead you. The sources cited should be permanently available and dependable; therefore, do not cite transient Internet sources. If readers want, they must be able to check the accuracy of your writing, or find out more information on the topic. Many academic books and journals, which were first published in print form, would be available in electronic form too on a later date. These can be used with confidence, as they have the same status as print publications.

Journals and newspapers: Based on the nature of essay, you may refer scholarly journals as well, both print and online, which publish results of original works. Materials from the popular press such as newspapers and popular magazines may not always be reliable or authoritative as academic publications. They are not peer-reviewed and may have been written by persons who are not experts in the field. It is better to avoid such sources, or when used, exercise abundant caution. If the content is of academic importance, surely, you will be able to locate it from a reliable academic source such as a journal.

Internet sources: The diversity and variety of materials available on the Internet are incredible. You may find academically reliable materials on the Internet, but majority are not reliable. Before you use materials from the Internet for an essay, use good judgement about whether it is reliable and suitable for your purpose. You must ensure the possibility of the e-source continuing to be available, so that if some readers follow up your cited reference on a later date, they would be able to find it.

For Internet materials, besides usual bibliographic details, you must also note the web address and date of access.

You can also use the Online Public Access Catalogue (OPAC) of the library or databases such as CAB Direct, PubMed, or Google Scholar for searching. Use keywords for locating literature. When you have located relevant sources, conduct a survey first—check the contents pages, chapter titles, and the index to assess the relevance of the source to the topic. Get accustomed with the library catalogue and various facilities such as Consortium of e-Resources in Agriculture (CeRA) available in the library for faster literature search.

E-publications by institutions: Presently, most government departments maintain websites with useful information. Many publications in PDF format are available for free downloading. Sometimes, you may require the latest area, production, and productivity of crops. The website of Directorate of Economics & Statistics, Government of Kerala (https://www.ecostat.kerala.gov.in/) provides you the latest information on these details pertaining to Kerala. If you require the latest all India figures, refer the publications maintained by the Department of Agriculture, Cooperation and Farmers Welfare, Govt. of India (https://www.agricoop.nic.in/). For world figures, rely on FAO statistics. You can use these materials without any hesitation. Similarly, you can use confidently books downloadable from the sites of FAO, ICAR, and CGIAR sites such as IRRI.

When using the materials of private parties, exercise caution. Sometimes, it may be one sided, biased, or simply propaganda materials. Propaganda materials do not provide suitable academic support. Such materials are likely to change or vanish from their sites without any reason. These kinds of sites are frequently undated, and therefore, you cannot even determine whether the material is currently valid or not. It is better not to use such information at all.

Wikipedia: Wikipedia is a popular and free online encyclopaedia (https://www.wikipedia.org/). In Wikipedia, readers can act as editors and improve the contents. Anybody can write a piece for Wikipedia without knowing much about the subject. While many of the articles in Wikipedia may be correct and useful, lack of guarantee for genuineness is a problem. There is no way to distinguish between spurious materials and genuine materials provided, for example, by a reputed scientist. Wikipedia attempts to simplify entries, and this may lead to misrepresentation. For this reason, do not cite Wikipedia in essays. That does not mean that Wikipedia is of no use at all; it may be a valuable starting point for finding a suitable reference.

Making Notes from Information Sources

The objective of literature search is to locate materials that support the points or arguments you want to make in the essay. Your essay plan and the main essay points should guide your literature search. To assess the relevance of any reference item, for example, book, chapter, or article, apply your critical reading skills. Read the title, subheadings, and scan the pages for keywords. If a text is useful, the title or subheadings will be related to the essay topic. You need to be selective when searching. As you read a source material, marking and writing down the important point that you noticed will help you to remember them and understand them better.

You may note down background information about the topic, facts, examples, expert opinions, and arguments and explanations, which support or oppose your ideas.

Note down important points while reading journals and textbooks. If you want to quote sentences as such, be careful to copy accurately and put quotation marks at the beginning and end of the quote. See the section on note making described elsewhere.

If you find a source highly useful but do not have time to make notes, you can photocopy the material for later use. Do not attempt to 'cut and paste' from Internet documents. You can print them, and on these photocopies or printed Internet documents, you can highlight or underline important information. Put your remarks in the margin. This will help you when you are organizing and writing your essay later. Do not forget to put text copied from web pages in 'quotation' marks. If you do not do this, this may lead to unintentional plagiarism.

Use A4 size papers for writing down notes and keep them tagged together. Browse through the reference sources and select the most appropriate points before making notes. Selecting the most significant material is a skill to be nurtured. The title of the article, author's names, name of journal or book, volume number, page numbers, name of publishers, place of publication, etc., must be noted clearly for easy retrieval of the material and for preparing bibliography. If a reference book from the library has been used, the call number should also be noted to facilitate its location in future.

3. **Write the first draft**

Having completed the literature search and satisfied that you have collected enough background materials, you may write the draft copy. Examine the question carefully. This should include your first thoughts on the question, and think what you already know about this topic. By going through the literature, you can identify the gaps. Write the draft with a logical structure of information and ideas. Plan to write manageable, smaller chunks, and finally join them together logically. While you go on writing, make sure that you are aware of word limits of the essay. Further, ensure that you have a good idea of every paragraph or section, which you propose to include. Ensure that the essay is structured well with an introduction and a conclusion apart from the main points, which constitute the body. Write the draft in your own words avoiding the risk of plagiarism.

Parts of an Essay

An essay should have an *introduction*, *the body of the essay*, and a *conclusion*. However, you need not write the draft in that order. For example, you can write the introduction after completing all other sections. Write freely and attempt writing first those parts you find the easiest. At this stage, do not bother much about impeccable grammar and style. What is important here is to get your ideas down on the paper. The style and correct grammar can be addressed in the editing phase. The introduction and conclusion must be structured well so that after reading them a reader should be able to know what the essay is all about without having to read the main body of the text.

An '*introduction*' should introduce the central issue of the topic and provide background information to it. It must provide an outline of the main arguments. The 'introduction' is the key to help readers understand where they are directed and what they will accomplish. State in concise terms what the subject of the essay is, what you are going to discuss, and how you go about it. The introduction of an essay is usually 1–2 paragraphs long, but a bit longer for longer papers with additional background information. You can start the introduction in many ways; for example, using an anecdote that leads to your topic, with a surprising statement relating to the topic, or a quotation from a famous person or expert.

The '*main body*' of an essay should present the writer's argument through paragraphs, and sometimes as separate sections with headlines. In the body of essay, you examine, explain, and describe and provide supporting arguments citing sources. If the paper is long, it is often a good idea to divide the main body into sections designated by headings and subheadings. Write the essay assignment as though its readers would be reasonably intelligent and informed but not experts on the subject. Divide the body of the essay into sections and subsections, depending on the coverage.

A '*conclusion*' should summarize the main points presented in the body and provide a definitive statement of the writer's position. The conclusion should sum up what you have found and stress the evidence that supports your analysis. Based on what have already been discussed in the body of the essay, you can also suggest recommendations.

4. **Editing**

Editing is the last phase of the essay writing process before taking the final copy. This phase allows you to check that the essay is structured well, the argument is logical, and expression is grammatically correct. It is suggested that you start the editing process one or two days after writing the draft, so that you come to the editing task with a fresh mind. Editing stage is the time to make alterations and modifications.

Take advantage of the tools and resources within your reach. If the teacher permitted you to prepare the essays through computers, you are lucky. You can also ask for peer evaluations and ask your friends, teaching assistants, or professors to provide feedback. Confirm that you have satisfied all the essay requirements; the content is accurate, and the formatting and referencing are correct. Ensure that every work cited in your essay is listed in Bibliography and vice versa, and that the Bibliography entries are in alphabetical order following Name–Year system or any other system, which the teacher allows.

Take a print out, and do the final editing and proof reading on the printed copy. You will definitely see mistakes that you might not see on a computer screen. Do not mix up American and British spellings. In commonwealth countries, British spellings are preferred. Point your pen or pencil at each word, and think about the grammar, spelling, punctuation, capitalization, and meaning of every word. Although the spellcheck and grammar check facility available with computers are helpful, a manual check is inevitable, as computers cannot detect all the errors.

For taking the final copy, use A4 size paper with 1.5 lines spacing. Give a margin of 38 mm (1.5″) on the left side of the paper and 25 mm (1″) on all other sides. Number all the pages. Put the title of essay and your name, admission number, course title, etc., on a separate page.

22.9 Writing Style

Essay writing requires a formal style. Removing redundant words and modifying phrases to make them simple and short are widely accepted techniques in academic writing. Make sure that there is no ambiguity in meaning due to improper use of words. Before finalizing the manuscript, the language should be scrutinized for usage, flow, and clarity by checking for awkward phrasing, biased language, clichés, and jargons, which are inappropriate for an academic essay. For more information on style and grammar of academic writing, you may consult Chaps. 16–18.

Ensure accuracy and precision in what you present in the essay. Transitional words and phrases clearly state the relationship between two sentences. See common transitional words and phrases used in essays.

- *to show consequence/effect*—because, because of, due to, as a result of, so that, therefore, accordingly, consequently, thereby, under the circumstances, provided that, hence, as, so, unless
- *to introduce examples*—for example, for instance, such as, like, including, thus, especially, namely, particularly, notably
- *to add comparison*—similarly, alternatively, comparatively, likewise, in the same way, in similar fashion
- *to start with an idea*—first, first of all, to begin with
- *to add another idea*—in addition, furthermore, and, also, moreover, besides, as well as, likewise, similarly, secondly
- *to add a more important idea*—more importantly, what is worse, what is more
- *to add your last idea*—finally, most of all, most importantly
- *to contradict/contrast with the previous idea*—however, nevertheless, on the other hand, although, though, but, whereas, notwithstanding, in spite of, despite
- *to show the result of the previous idea*—therefore, thus, consequently, as a result
- *to emphasize an idea*—in fact, in particular
- *to show a time relationship between ideas*—first, second, then, next, finally
- *to generalize ideas*—often, usually, normally, generally, in general
- *to limit, restrict, and defining time*—as soon as, in the meantime, immediately, quickly, eventually, now, at present, presently, shortly, instantly, after, since, formerly, during
- *to qualify another point to the preceding statement*—perhaps, probably, nearly, possibly, always, almost.

Citation and Listing of References

Citation systems involve two aspects; the first is the way in which the items are cited or included in the text (*in-text citation* or *in-text reference*), and the second is the way in which the references are listed at the end of the text (*reference listing*). We may use ideas of others in a paraphrased form or include a direct quotation in our works; but in both cases, we have to acknowledge the sources appropriately in the text. Citation and listing of information sources must be as given for research papers.

As the Name-Year (N-Y) style of referencing is widely followed in agricultural literature, it is recommended for use in essay assignments too. For more information on references, refer to the Chap. 15.

Bibliography

Bligh, D. 2000. What's the Use of Lectures? Jossey Bass, San Francisco, 368p.

Boch, F. 2005. Note taking and learning: A summary of research. The WAC J. 16: 101–113.

Cottrell, S. 2013. The Study Skills Handbook (4th Ed.). Palgrave Macmillan, New York, 432p.

Davis, B.G. 2009. Tools for Teaching (2nd Ed). John Wiley & Sons, 608p.

Ebbinghaus, H. H. 1885. Memory: A Contribution to Experimental Psychology. Translated from German Ruger, H.A. and Bussenius, C. E. (translated from German). Teachers College, Columbia University, New York, 51p.

Ellis, D. B. 2013. Becoming a Master Student (14th Ed.). Wadsworth Cengage Learning, Boston, 402p.

Greetham, B. 2001. How to Write Better Essays. Palgrave Macmillan, New York, 285p.

Lynch, T. 2004. Study Listening – A Course in Listening to Lectures and Note taking (2nd Ed.). Cambridge University Press, Cambridge, 214p.

Murre, J.M.J. and Dros, J. 2015. Replication and analysis of Ebbinghaus' forgetting curve. PLOS ONE 10(7): 1–23. Available: https://journals.plos.org/plosone/article?id=10.1371/journal.pone. [Accessed 10 Feb. 2018].

Nichols, R. G. and Stevens, L.A. 1957. Are you listening? McGraw-Hill, New York, 235p.

Pauk, W. and Owens, R.J.Q. 2011. How to Study in College (10th Ed.). Wadsworth, Boston, USA, 406p.

Petty, G. 2009. Teaching Today: A Practical Guide (4th Ed.). Nelson Thornes, UK, 614p.

Race, P. 2007. The Lecturer's Toolkit: A Resource for Developing Assessment, Learning and Teaching (3rd Ed.). Routledge, Oxon, 264p.

Robinson, F. P. 1970. Effective Study (4th Ed.). Harper & Row, New York, 304p.

Rubens, P. 2004. Science and Technical Writing: A Manual of Style (2nd Ed.). Routledge, New York, 427p.

Chapter 23
Ethics and Related Issues in Research

Ethics is knowing the difference between what you have a right to do and what is right to do.
Potter Stewart (1915 –1985), US jurist

Professionals in any field, may they be medical practitioners, lawyers, engineers, agricultural officers, veterinarians, chartered accountants, or others—all are expected to follow certain ethics according to the nature of their profession. Scientists are no exception. *Ethics* are norms that distinguish between acceptable and unacceptable behaviour in society. *Research ethics* deal with the application of basic ethical principles to scientific research. Common ethical issues in science include the design and implementation of research involving human beings and animals; scientific misconduct such as fraud and plagiarism; improper practices; whistle blowing; and research management.

Scientists must conduct themselves according to the ethical norms of science in all of their scientific pursuits. As scientists, we presume that the results being reported by other scientists through research papers, conference proceedings, popular articles, and other means are true and original. The public also believes that the research findings reported are the results of sincere attempts by scientists. However, this faith in the system endures only if the scientific community devotes itself to characterize and spread ethical values associated with scientific conduct. Very often, the strongly held traditions of record keeping, openness, periodic reporting, and peer reviewing keep the majority of scientists within the limits of professional ethics. Nevertheless, some scientists withhold information or even falsify their findings because of their anxiety to get undue credit for the work. Sometimes, fraud occurs because of the ignorance and non-adoption of good research practices. In certain cases, the motive may be to secure coveted scientific posts or promotions. Some research students commit fraud wilfully to complete their thesis work. These kinds of violations of ethics, fraud, or wilful misconduct impede the progress of science, and therefore, the scientific community and the public must strongly condemn these practices. There are several issues to be considered as a part of research ethics.

C. G. Thomas, *Research Methodology and Scientific Writing*,
https://doi.org/10.1007/978-3-030-64865-7_23

23.1 Nuremberg Code and After

If you search the recent past history of research, you could find several issues of rights violations where ethics has been thrown to the winds. A well-known case was that of Edward Jenner who developed a vaccine for smallpox as reported by Wellington and Szczerbinski (2008). In 1796, Edward Jenner borrowed an eight-year old school boy named James Phipps and infected him with cowpox. Cowpox is relatively a harmless disease, but when the boy recovered from cowpox, he was infected with smallpox, a dreaded disease of that time. Fortunately, the boy recovered paving the way for accepting vaccination as the sure means for preventing smallpox. The vaccination drive was so effective that small pox was eradicated from the whole world. Although the research leading to vaccination was successful, Jenner employed an unethical method, which could not be accepted in the modern world. Another historic case is the horrendous and disgusting methods German scientists had used on captive human subjects during the World War II in appalling experiments. After the World War II, a consensus on the key ethical principles that should underlie research, especially those involving human subjects developed. This became known through the Nuremberg war crimes trials following World War II. Shocking cases like the above incidents forced scientists all over the world to develop a consensus on ethical standards involving human subjects. Everybody concerned with human experimentation felt the need of a clearly defined code of conduct governing the ethical aspects of human research. Although some countries had national policies, these were not enough. The *Nuremberg Code* formulated in 1947 is a set of ethical principles for human experimentation based on the subsequent Nuremberg trials (NIH 2008).

Opinions differ on the strict implementation of a legally bind code of conduct and its repercussions in research. Some scientists fear that excessively rigid controls may limit the effectiveness of research, possibly denying the society the answers to many important problems. However, if there are no restraints, chances are that some scientists may employ unethical experimental practices, which could infringe upon human rights. The *Declaration of Helsinki* was a right step, promoted by the World Medical Association (WMA), as a statement of ethical principles to provide guidance to medical community on research involving humans (Fathalla and Fathalla 2004). This vital document is considered as the first significant effort of the medical community to regulate research, and it formed the basis of most successive documents. The declaration was adopted in June 1964 in Helsinki, Finland, and it has since undergone seven revisions; the most recent was in October 2008.

23.2 Some Concepts in Ethics

Science ethics is an evolving field, and scientists must make ethical decisions with care. Some rules of scientific ethics are universal but some others are subjective.

Fraud, intellectual dishonesty, and theft of ideas are condemned by all. In the recent past, a number of concepts and terms have been developed to describe the system of ethics in science. Every researcher should have a basic idea about such concepts and terms. Therefore, a few common concepts related to ethical considerations in science are described below.

Voluntary Participation

Voluntary participation, a recently evolved ethical phrase, is applicable in the case of research involving people. It means that human subjects should not be forced into participating in research but must participate voluntarily. This is especially relevant now, as the society is more concerned with human rights. In the past, researchers had relied on captive subjects such as prisoners, slaves, war criminals, and refugees for doing experiments, especially related to medicine. These practices are against natural justice and should be condemned.

Informed Consent

In research involving human subjects, prospective participants must be fully informed about the procedures and risks involved, and their consent to participate must be obtained. This is called 'informed consent'. It is considered unethical to collect information without the knowledge of participants and their expressed willingness. This requirement is closely related to the notion of voluntary participation mentioned above. It is now established that research involving human subjects should be conducted only with the informed consent of the subjects. The people we wish to study should have full information about the research including why and how they have been selected to participate. Informed consent entails full disclosure of the risks and intended benefits of the research; and it also recognizes the right of subjects to refuse to participate. Researchers must not knowingly subject colleagues, co-workers, students, neighbours, or anybody to health risks without their knowledge and consent.

Risk of Harm

Research ethics also necessitate that researchers should not place volunteers or subjects in a condition where they might be at 'risk of harm' because of their involvement in the experiment. Harm can be both physical and psychological. When conducting research involving human subjects, adequate precautions should be taken to minimize harms and risks.

Confidentiality

The privacy and dignity of human subjects should also be protected. Two specified standards—*confidentiality* and *anonymity*—are applied to protect the privacy of those who participate in experimentation.

When human beings are made subjects in experimentation, the information gathered about respondents should be kept confidential. Sharing this information for purposes other than research is unethical. Most research workers guarantee confidentiality to the participants. Contravening this assurance, research workers should not reveal identifying details to anyone who is not directly involved in the study.

Anonymity

The principle of anonymity ensures that the participant will remain anonymous as long as the experiment is completed or even after the experiment, if required. It is unethical to identify a respondent with their opinions, problems, or defects. Although this standard is a strong guarantee of privacy, it is sometimes difficult to accomplish. For instance, in situations where participants are constantly exposed, especially when they are required to be measured at multiple time points as in a before and after study, the chances of revealing anonymity of a subject is greater.

In such cases, adequate care must be taken to keep anonymity intact.

Placebo and Placebo Effect

Placebo is an innocuous medication prescribed for the mental relief of a patient than for its actual effect, and *placebo effect* is the improvement in the condition of a sick person in response to placebo treatment. In many therapeutic cases involving patients, it has been established that the mere illusion of receiving treatment can play an important role in their recovery from an illness. Placebo effect is noticed in many other cases too, for example, those involving dietary studies. Therefore, experiments with human subjects should have some special safeguards to take care of the *placebo effect*. Often, the control treatment is either a placebo or no treatment.

Two ethical issues are important here. Firstly, giving only placebo to a seriously sick person as a part of experiment may invite problems. In drug trials, when you decide to give a placebo as control, you have to weigh the risk factors properly. In fact, you are denying the treatment to an individual! The second issue is how to administer a placebo. For example, when a new processed food product is made, it is common to evaluate its taste and acceptability. A psychological barrier can function here, if the ingredients or the method of preparation are revealed to the volunteers, who come forward to taste the products and give score values. A similar psychological effect with respect to the quality of several farm products is prevalent among lay people. For instance, most people believe that organic foods are of high quality compared to the farm products produced following integrated approaches. The mere knowledge that a particular product is 'organic' can induce a placebo effect, and because of this effect, when you conduct a test, it can suppress actual values. Therefore, to verify or falsify such claims with food items, you have to conduct single blind or double blind tests involving organoleptic tests (see Sect. 5.10).

Right to Service

Researchers should also address the ethical issue of a person's 'right to service'. As mentioned above, medical experiments often involve a control, requiring the use of a no-treatment control or placebo wherein a group of participant will not be given the proposed treatment under study. However, when the treatment under study may have positive effects, subjects assigned to the control group may feel that their rights to equal access to services are denied. You must assess the likely impact of your study and inform the participants accordingly.

Possible Harmful Effects of Results

Science ethics also include the possible harmful effects when the research results are applied to real situations. Although the long-term effects of the research results are difficult to predict, you can gather some idea by knowing who is interested in funding the research and from their expectations before sanctioning grants. For example, funding by corporate establishments and international advocacy groups may have some strings attached. Similarly, scientists working in military establishments and in R&D divisions of certain industrial concerns are expected to maintain strict secrecy. The secrecy maintained by the military or industrial houses may be acceptable to some scientists, but not to all. The willingness of a scientist to work on research topics of potential risk to humanity such as nuclear weapons or microbial warfare is mainly a case of personal ethics rather than professional ethics. However, as a rule, scientists should not agree to conduct research that is harmful to any sections of the society.

Environmental Impacts

Another important aspect of research, which the researchers must take into consideration even before the planning stage itself, is the environmental impact of research projects and the expected results. Research carried out in a more sustainable way allows you to generate more knowledge with the same resource. In the modern world, reducing waste of any sort of resource is an important reason for doing research. Try to design and conduct research projects in such a way that they may not have much negative impacts on the environment, especially in terms of carbon footprints. The same yardstick is important while utilizing the results of research for various developmental activities.

Care of Live Experimental Animals

Another concern of research ethics is related to possible harm or injury to live experimental animals that could occur from experiments. When using animals in research, show proper reverence and care for animals. Do not conduct unnecessary or poorly designed experiments with animals. The researcher should take ample care regarding the health, comfort, and well-being of animal subjects. This is especially important in veterinary research involving live animals. For example, damaging limbs or other vital organs of animals and conducting experiments to cure such instances are against ethics, which may also invite the wrath of animal lovers!

Intellectual Honesty

Scientists should also show intellectual honesty in perceiving ideas for research, conducting research, and reporting research results. Many people including scientists tend to ignore evidences that deviate from expected lines. 'Conditioned thinking' is a problem of established researchers. Some scientists always think in a particular direction, that is, they become conditioned, and may not notice new developments contrary to their way of thinking. They also want others to toe their conditioned line of thinking; sometimes, they even thwart colleagues and juniors to think differently. Often, this can prevent a person or establishment from adopting a new perspective that may be needed to solve a problem. This tendency can be overcome by continued awareness and evaluation of possible personal biases. Whenever we review research proposals, manuscripts, or research papers, we are supposed to show intellectual honesty objectively without any personal interest. All researchers should be open, well informed, and receptive to new ideas and impending results.

Fiscal Honesty
Usually, research projects are funded by government agencies, private donors, or international donors. The research funds should be used as agreed upon in the research proposal or protocol for meeting various expenses. The money must be used carefully, and any type of financial fraud and irregularities must be condemned and curbed. Expenditure has to be documented and subjected to audit by a competent agency. You cannot ensure scientific honesty unless you are sincere in fiscal matters too. You should also monitor your helpers who are dealing with money.

Non-discrimination
Non-discrimination is a vital human rights principle. Scientists and teachers should not discriminate their colleagues, juniors, or students based on gender, race, ethnicity, or other factors that are not related to their scientific competence and integrity. In a scientific establishment, there should not be any room for discrimination or rights violation of any hues.

23.3 Intellectual Property Rights

The phrase 'intellectual property rights' (IPR) is often used to refer to the rights given to a person or persons over the 'creations of their minds'. This 'rights' entail the creators an exclusive right over the use of their creation for a specific period. The creation can be in industrial, scientific, and artistic domains. In most counties, a system of intellectual property protection and enforcement is in existence. Intellectual property is often divided into two categories—***industrial property*** and ***copyright and related rights***. For a detailed account, consult Wadehra (2004) and IIFT (2010).

The intellectual property rights categorized as 'industrial property' includes the protection of patents, trademarks, geographical indications, plant varieties, industrial designs, and trade secrets. Protection of industrial property is envisaged mainly to stimulate innovation, design, and new technological breakthroughs.

The other category, 'copyright and related rights', includes creative works such as books including novels, poems, plays, films, musical works, computer software, and artistic works. These are the rights enjoyed by authors of literary and artistic works, and the rights of performers, singers, broadcasting agencies, etc. The main purpose of copyright laws is to promote and reward original and creative contributions.

Trade-Related Aspects of Intellectual Property Rights (TRIPS)
Laws related to intellectual property were enacted by many countries. Probably, the first intellectual property law was made in Venice in 1474 to protect inventors from unauthorized copying of their creations. Later, in England, the Statute of Monopolies was passed in 1624 conceding intellectual property rights to the inventor for a certain period. In spite of all these laws, the intellectual property system, as we know today, commenced only in 1883 with the Paris Convention for the Protection of Industrial Property. In this convention, patents, trademarks, and industrial designs were granted protection. The Paris Convention enabled individuals in one nation to obtain protection for their intellectual property globally. As a sequel to Paris Convention, the International Copyright Act was made in 1886, which was first accepted in Berne Convention (for the Protection of Literary and Artistic Works) held during the same year (1886).

The General Agreement on Tariffs and Trade (GATT) was formed in 1949 but was replaced by the World Trade Organization in 1995. The GATT came to stay as an outcome of the failure of negotiating governments to create the International Trade Organization (ITO). At present, the WTO is an important international organization for the development and protection of IPR.

The most comprehensive international agreement on intellectual property rights ever made is the Trade-Related Aspects of Intellectual Property Rights (TRIPS). TRIPS Agreement is an important agreement coming under the ambit of World Trade Organization (WTO). TRIPS came to its present form as part of the eighth round of multilateral trade negotiations under GATT generally known as the 'Uruguay Round' from 1986 to 1994. The final Act was signed at Marrakesh on 15 April 1994. The member countries of the WTO are bound to follow the agreement. The Agreement includes all types of intellectual property such as copyright, trademarks, patents, industrial designs, trade secrets, geographical indications, and exclusionary rights over new plant varieties.

The TRIPS Agreement took effect on 1 January 1995. As a first step towards compliance, for the developed countries, twelve months from the date of signing the agreement were granted to implement its various provisions. However, developing and transition countries were given more time, five years until 2000. After considering the difficulties of the least developed countries (LDCs), they were given 11 years until 2006. Because of popular pressure, for pharmaceutical patents in these LDCs, the term for compliance has been extended to 2016.

The Agreement under Trade-Related Aspects of Intellectual Property Rights (TRIPS) stipulates specific norms and standards in the following areas of intellectual property:

- Patents
- Trademarks
- Geographical indications

- Plant varieties
- Industrial designs
- Layout designs of integrated circuits
- Trade secrets (protection of undisclosed information)
- Copyright and related rights.

Brief details about the first four areas coming under industrial property—patents, trademarks, geographical indications, and plant varieties—and copyright and related rights are given below.

Patents

Patent is an intellectual property right relating to inventions. It is an exclusive right granted by law to the patentee for a specific period in exchange of full disclosure of his/her invention; at the same time, preventing all others from making, using, selling, or importing patented product or processes for business or other purposes. The objective of patenting is to encourage inventions by their promotion and utilization, especially for the development of industries. Patents ensure property rights for the invention.

Changes in the operation of intellectual property rights have made some changes in patenting regime in India too. In India, the Patent Act of 1856 was the first legislation relating to patents. However, the Act was repealed in 1857 as it had been enacted without the approval of the British Crown, but later re-enacted in 1859. In 1911, the Indian Patents and Designs Act was made repealing the earlier Patent Act. The government brought in the Indian Patent Act, 1970, which became effective from 20 April 1972, replacing older Acts. This new Act introduced several restraining changes related to patenting of inventions especially in the areas of chemicals, pharmaceuticals, agrochemicals, and food items.

Some of the most important and controversial provisions in the TRIPS agreement pertain to protection of patents. TRIPS parties are bound to make patents available for all inventions in all fields of technology without any discrimination. Earlier, only 'products' were eligible for patents; TRIPS have made it mandatory for 'processes' too, including those used in manufacturing products.

TRIPS also made it mandatory for a country that excludes plant varieties from patenting to provide an effective *sui generis* system of protection. Although the phrase *sui generis* (the Latin phrase *sui generis* means 'of its own kind') is not defined in the agreement, it is generally assumed that it enables member countries to design their own protection scheme for plants. If a country excludes plants, animals, and plant varieties from patenting, alternate protection must be ensured under an effective *sui generis* system as mandated by TRIPS. Plant Breeder's Rights system offered by UPOV Convention, plant patents, or licensing are the probable protection mechanisms (UPOV or the International Union for the Protection of New Varieties of Plants is an intergovernmental organization). It is possible to implement more than one form of protection in a given member country. India has enacted *sui generis* legislation granting rights to both breeders and farmers under the Protection of Plant Varieties and Farmers Rights Act, 2001.

In India, the patenting system works under the Controller General of Patents, Designs and Trademarks through the Patent Offices located at Kolkata, Mumbai, Delhi, and Chennai.

Trademark

A trademark is a sign or mark capable of distinguishing a product or service of one company from those of another company of any given business. The sign or mark accepted as trademark can be a unique word, letter, numeral, picture, drawing, shape, colour, sound, smell, logotype, or any combination of these. Often, a company uses a trademark extensively to reach out to customers by enabling them to identify and locate the product easily. Under the Indian Trademarks Act, 1999, a trademark is issued by a national office and is granted for a period of 10 years, which may be renewed from time to time.

Geographical Indication

Geographical indication (GI) is a name or sign used on certain goods that identifies a product as originating in a particular geographical location or a region, where a given quality, reputation, or other characteristic of the product is essentially attributable to its geographic origin. Such products can be in natural or manufactured form. Examples of products with GI status granted in India include *Darjeeling Tea, Alleppey Green Cardamom, Mysore Agarbathi, Mysore Silk, Aranmula Kannadi, Kullu Shawl, Mysore Sandal Soap, Chanderi Saree, Coorg.*

Orange, Monsoon Malabar Arabica Coffee, Malabar Pepper, Nilgiris Orthodox Tea, Paithani Saree, Kota Masuria, Kolhapuri Chappals, Bikaneri Bhujia, Agra Petha, and many others. The goods with a GI tag are supposed to possess distinct features related to their respective localities or regions. For example, Malabar pepper is grown exclusively in northern Malabar parts of Kerala and have certain specific features and quality compared to pepper from other regions. Similarly, Alleppey green cardamom is the traditional grade of cardamom from the high ranges much in demand internationally.

Before 2003, India did not have a system of protecting geographical indications of Indian origin. For complying with India's obligations in the TRIPS Agreement, India enacted the Geographical Indications of Goods (Registration and Protection) Act, 1999, which became effective from 15 September 2003. The present organization of geographical indications in India is run by the Geographical Indications of Goods (Registration and Protection) Act, 1999 and the Geographical Indication of Goods (Regulation and Protection) Rules, 2002.

Registration of geographical indication in India is not compulsory. However, it is always desirable to register the geographical indication as the certificate of registration is valid as evidence, and no further confirmation is required if some disputes arise. Registering a geographical indication in India bestows certain rights to the makers of the particular product. Contrary to other intellectual property rights owned mostly by individuals, a GI is owned by members of a community who make the products or goods.

The GI of a product may work as a guarantee that because of its geographical origin, the product enjoys certain reputation and has certain specific qualities. In most cases, it also ensures that the product has been made according to traditional methods. It can also prevent unauthorized use of the registered GI by others, for example, tea produced from Darjeeling hills can only be marketed a 'Darjeeling Tea', and if anybody market tea produced from any other parts as 'Darjeeling Tea', it a clear violation of GI rights. It is hoped that granting GI status to traditional products foster economic prosperity to producers of goods in a particular region.

In India, the GI Acts are managed by the Controller General of Patents, Designs and Trade Marks, who is also the Registrar of Geographical Indications. The Geographical Indications Registry is maintained at Chennai. Normally, a GI is registered for a period of 10 years, but is renewable after the elapse of this period.

Plant Varieties

New plant varieties also come under the purview of protection of intellectual property rights. The TRIPS Agreement tries to recognize breeders who evolve new plant varieties by giving them an exclusive right. To obtain such protection, the new varieties must satisfy certain specific criteria. In this context, please note that the *variety* is same as the *cultivar* (contraction for cultivated variety) as defined in the International Code of Nomenclature for Cultivated Plants (ICNCP). The cultivar or variety is a plant grouping within the lowest known rank, but it should be new, distinct, uniform, stable, and have a satisfactory denomination. The organization managing the protection of new plant varieties is known as UPOV (The International Union for the Protection of New Varieties of Plants). As discussed already, India has enacted *sui generis* legislation granting rights to both breeders and farmers under the Protection of Plant Varieties and Farmers Rights Act, 2001.

Copy Right and Related Rights

Copyright and related rights are to protect the rights of authors of literary and artistic works such as books, musical works, paintings, sculptures, computer programs, and films. In the same way, copyright is applicable for performers such as actors, singers, and musicians. Producers of sound recordings and broadcasting organizations are also eligible for copyright protection. The main purpose of copyright laws is to promote and reward original and creative contributions. Laws related to copyright give the owner the right to prevent others from copying, create derivative works, or use their works as such. Similar to patent rights, copyrights can also be divided, for example, by the right implicated, by specific geographic or market territories, or by some other specific criteria. Accordingly, each may be the subject of a separate license and royalty arrangements.

Sometimes, a book or article can be out of copyright, as it is old, by the nature of authorship, or because of some other reasons. In India, the duration of copyright for authors is life of the author plus 60 years after his/her death and for cinematograph films and sound recordings 60 years from the year of production. In the event of death of the owner, the rights pass on to his/her legal heirs. After this stipulated period, the work becomes public domain material.

India is a party to the Berne Convention on Copyrights and WTO. The relevant Act in India overseeing copyright and related issues is the Indian Copyright Act, 1957 as amended by Copyright (Amendment) Act, 1999. India is also a signatory to the Geneva Convention for the Protection of rights of Producers of Phonograms and to the Universal Copyright Convention. The copyright law is subject to amendments to keep pace with changing requirements.

While writing a book, an author must ensure that all material included in it is his/her own, or if some part is taken from another source, the author has permission to do so. Any material in your book that is borrowed from another source may require written permission. You should be able to distinguish between material for which permission is necessary and material that can be used without obtaining permission. In general, permission for reproduction is required for books, newspaper articles, magazine or journal articles, poems, plays, songs, websites, photographs, and other artwork such as cartoons, book or magazine covers and their design, logos, figures, graphs, charts, tables, software, screen shots from software or websites, out-of-print works, and advertisements.

In general, copyright laws do not protect the underlying 'ideas', and it does not protect 'facts' or 'common knowledge'. Although ideas cannot be copyrighted, original fictional characters, story lines, and settings can be copyrighted. Normally, copyright does not deter you from stating in your own words the ideas, facts, and common knowledge found in a book or journal you read. However, you should always give credit to the sources in which you found them, and understand that by not doing so, your action could amount to a violation of the Copyright Act. Moreover, what you have created maybe construed as a derivative work, unless you have obtained license or permission from the copyright holder to create the same. You may also be blamed for committing *plagiarism* (Chap. 24).

The relaxation for 'fair use' or 'fair deal' guarantees that copyright laws do not infringe upon your freedom of making a critical commentary on a work or using the work for teaching or research purposes. Examples of permitted use of texts under fair use are teaching, criticisms, comments, reporting, and research. It also allows limited quoting of copyrighted material. In most cases, it would allow creating a copy for personal use. However, you should particularly note that it does not grant anybody a right to do anything you want with the material and claims it is 'fair use'.

Royalty

The term 'royalty' originated from the practice of granting a right by a monarch to an individual or group to exploit specified natural resources. Presently, the term is used to denote a payment made to the legal owner for the use of an asset or property, especially patents, copyrighted works, or natural resources by those who wish to make use of it for the purpose of generating revenue or other desirable activities. If mutually agreed upon, payment of royalty at a specified rate is legally binding. Sometimes, the term 'running loyalty' is also used.

In the case of books, often, royalties are paid as a percentage of the list price or net price, for example, as a percentage of the gross price of a book in a shop or as a percentage of the publisher's net receipts from a book. Usually, royalties on books ranges from 10 to 15 percent. In this context, one should understand that 10 percent of what the retailer returns to the publisher would be less than 10 percent of gross receipts. The publisher determines an author's royalty rate; however, if the author is popular and a well-established figure in the field, he/she can bargain with the publisher for a higher royalty and advance. Sometimes, book authors can sell their copyright to the publisher and get a lump sum payment. Before taking up the publishing work, a publishing contract has to be signed incorporating various mutually agreed terms of publishing including royalty rate.

23.4 Fraud and Misconduct in Science

In scientific research, it is difficult to define a *fraud*. Generally, this includes passing off incorrect or unconfirmed data as factual data or passing off other's work as one's own. There are many known cases of fraud of these kinds in science. Fraud and misrepresentation of data are highly damaging to science. Even a single detected scientific fraud can severely erode the credibility and reputation of the scientist in question. This is also damaging to the reputation of his/her co-workers and the establishment to which that particular scientist belongs.

We must admit that scientists are not super-human beings, and all the emotions of simple human beings are inherent in them. Occasionally, we hear the cases of some scientists who are tempted to forge results or conduct experiments in a fraudulent manner for prestige or money. These kinds of wilful scientific fraud are unacceptable and must be booked with penal provisions. Nevertheless, because somebody is committing frauds, we cannot dismiss all the scientific findings as worthless. In many cases, frauds will eventually be detected by the repetition of results, which is routinely undertaken by scientists. In fact, undetected fraud in science is rare, and the existence of documented fraud is a good example of the self-correcting nature of science. We must also consider that most scientists are idealists, and a good majority of them perceive beauty in scientific truth. If this virtue had not been there, most of them would have gone into some other professions such as civil service or business management, which are definitely more lucrative in term of money and power! Nevertheless, problems arise when scientists are recruited on extraneous considerations such as nearness to some political leaders, when individuals join the profession as a means to gain respect, or simply when they could not get any other lucrative jobs! Some typical unacceptable practices that come under the purview of 'fraud' are mentioned below.

Research Papers

A major domain of scientific fraud is related to the publication of research papers. We know that a major activity of professional societies is to publish research journals. However, it is highly deplorable that sectarian activities in professional societies are rampant, and often, power groups seize the control of research journals. When dealing with professional societies, students and budding scientists should be vigilant about these unhealthy practices. Prospective authors should consider the impact factor or rating of journals (Sect. 9.6) and the past performance of research journal before sending their prestigious articles. The prevailing 'publish or perish' attitude also creates several unintended problems. Nevertheless, do not 'publish and perish' by adopting dubious practices. Authorship, duplication of papers, plagiarism, and refereeing are common problems. The case of plagiarism has already been discussed. Consider these issues when preparing research papers and choosing a journal for publication, and your efforts should be to 'publish and flourish'. For more information on these aspects on preparation of papers, refer to Chap. 11.

Authorship

In many cases, fraudulent authorship problems are serious. Names of persons who have not contributed anything to the paper are given as co-authors and sometimes as senior authors. There is a tendency to include the names of head of the department/institute as co-authors in a research paper prepared by subordinate scientists. Some heads of research establishments insist that all the subordinate scientists should include his/her name as co-author when they write papers. Sometimes, the names of friends and fellow workers are included just for gratification or for giving more credence to the article (a case of gift authorship!). Keep it a point that no one should be named as an author if s/he has not made substantial contribution to the work.

Duplication of Papers

Because of the 'publish or perish' syndrome, scientists and research students have a tendency to squeeze as many papers as possible out of the research work done. This is particularly serious when writing research papers out of thesis work. Writing different papers using the same data in different forms is unethical and can be put under the category of 'fraud'. This is not good for the advancement science; and now, having recognized this problem, many recruiting bodies insist on impact factors for journal papers. Similarly, writing two different versions of an article, one for a specialist audience and another for a general audience and sending them to two separate journals is also considered unethical.

Refereeing

Refereeing of research papers, popularly known as peer reviewing, is an issue of great concern, which affects the quality and genuineness of papers. Most journal establishments accept research papers for publication only after a thorough review for quality usually done by peers on honorary basis. If done properly, peer refereeing ensures some amount of quality control in journals (Sect. 11.6). It is an accepted norm that the reviewers should remain anonymous. It is deplorable that in the case of certain journals, this anonymity and confidentiality remain only in paper, and the reviewers often succumb to pressures from influential authors. Peer reviewing should be done with the seriousness it deserves, and in no case, substandard papers are accepted and published based on extraneous considerations.

Conflicts of Interest

Conflicts of interest in journal publishing can occur when individuals involved—authors, employers, reviewers, editors, funding agencies, journal establishment, etc.—hold conflicting interests that could bias editorial decisions. Academic, personal, financial, political, or religious considerations can affect editorial decisions and the way in which authors prepare the papers. For example, a conflict of interest can occur when author/authors or their employers have potential competing interests in the publication of their paper. In a journal establishment, there should be a policy on conflicts of interest for editors, reviewers, editorial board members, editorial staff, and authors, and these must be reviewed periodically.

When publishing a research article, most journals ask the authors to declare a statement regarding conflicts of interest, something like 'the researcher claims no conflicts of interest'. This is the author's affirmation that assures impartial and straightforward research. The idea in declaring 'no conflicts of research' is to protect the paper from skewed or corrupted reporting of results or their interpretation. If the research is funded by an outside agency, the author should issue a public statement that the research is free of bias. This is a way of adhering to the code of conduct needed for academic publishing, while acknowledging the agency's contribution. For example, in the medical research field, major conflicts of interest may arise between medical researchers and pharmaceutical corporations, and one of the most common points of conflict is the corporation influencing the researcher to exaggerate the effects of a drug or treatment. For avoiding these conflicts, researchers, grant agencies, supervisors, and publishers should jointly work and fulfil their own duties.

The authors can declare any interests affecting their paper, either in their cover letter, or by answering a question on conflict of interest in the submission form when a manuscript is sent for publication. The authors should also acknowledge properly all the sponsoring and funding agencies supporting the work.

Manipulation of Research

Fraud in research assumes notorious proportions in medical research, where drug manufacturers and other such companies sponsor motivated research and encourage suppressing or distorting data in order to support their own products. The claims of tobacco firms in some way or other asserting that smoking is harmless and the claims of drug manufactures on the safety and effectiveness of their products based on suppressed or bogus data are typical examples. In most cases, the claims on many health drinks, tonics, skincare products, cosmetics, and similar products are dubious. You may also notice such 'findings' from polluting enterprises. The perils of nuclear research are not adequately revealed to the society. Similarly, institutions are not forthright in revealing the intentions and methodologies of biotechnological research.

Mediocre Research and Political Appointments

This is a big problem in most third world countries. In scientific institutions and universities, often, merit takes a backseat. Institutions of higher learning and research should be freed of unjustifiable practices such as unfair appointments, undue promotions, unfair awards, and improper political interference. Often, directors and vice-chancellors are appointed based on extraneous considerations rather than merit. These aberrations, if prevalent in the establishment, would demoralize honest and brilliant scientists. Big claims are often made for results of average value. Awards or certificates of excellence are obtained from relatively lesser-known institutions abroad. If the scientists understand that rather than merit or research scholarship some other considerations are being counted, they will simply stop doing serious research. This may cause not only frustration among scientists but also rot in the scientific system. In scientific institutions and centres of higher learning, appointments and decisions should not be guided by caste, creed, colour, religion, gender, politics, or recommendations from influential persons from politics or administration.

Clamour for Administrative Posts

In most advanced countries, senior scientists and professors may not like to join any administrative posts as the administrative responsibilities curtail their research interests. However, it is essential that somebody must hold administrative posts. Therefore, most research establishments introduced rotation among senior scientists/professors for administrative posts such as heads of department and directors. In India also, rotation is being introduced for administrative posts in science establishments, but for wrong reasons! In India, because of the hierarchy in the science administration system, the clamour for administrative posts, which yields more power and less science, is so intense that if somebody becomes the head of the department or director, s/he would try to cling to that post by any means.

Educationists call for interdisciplinary convergence (cooperation between related disciplines) and intradisciplinary convergence (bringing together the multiple specialties within the same discipline). However, interdisciplinary convergence is impossible in the present set up of disciplinary organization of academic institutions in India. It is now a trend to form new watertight departments, faculties, and even universities based on false claims. It is often alleged that the lure of creation of more administrative posts is a major reason behind these tendencies and not science.

Improper Practices

In addition to wilful fraud, several improper practices can also accrue in research from many sources. It is very difficult to prove this kind of fraud. Most of these are unintentional, but that does not lessen the gravity of these unacceptable practices that happen mainly due to ignorance. Some of these are mentioned below. If you are thorough with the methodology and the art of technical writing, you can avoid most of these kinds of improprieties.

Errors and mistakes due to ignorance: These types of errors occur mainly because of ignorance about the methodology of experimentation and proper data collection techniques. Researchers should strive to have a basic understanding on the principles of experimentation and statistical techniques to avoid these kinds of problems (Chaps. 5 and 6).

Fallacies and grammatical errors: All scientists should master the art of good technical writing without committing grammatical errors (Chaps. 16–18). They should also take special attention not to commit fallacies in reasoning (Chap. 14).

False representation of data: This happens mainly due to ignorance about the various ways of representing data. The researchers should decide well in advance how to represent the data—text form, table, or through diagrams (Chap. 13).

Knowledge or belief on the part of the author that the representations are false: The author may be fully aware that the results and analysis are false. However, because of some extraneous considerations, s/he may conceal the truth.

An intention to induce others to rely on the false representation: If you are doubtful about the genuineness of the data, then do not induce others to rely on the observation or findings. You should overcome the temptation to publish the 'findings' with the doubtful data in the form of research papers or otherwise.

23.5 Unscientific Practices in Thesis Work

Thesis is a certificate that the student has been trained for doing independent research. In agriculture and allied sciences, to obtain Master's degree, the student has to conduct a research project and submit the thesis. A thesis is usually in the format of a research report with clear components and necessary certificates in a bound form. A thesis is mandatory to get 'Doctor of philosophy' in all scientific fields. A thesis or dissertation is also necessary for M.Phil., M.Tech, and MD degrees. A research student will be working under a *research guide*, sometimes also called *research supervisor* or *major advisor*. To assist the student, a multidisciplinary advisory committee will also be set up with the major advisor as the chairperson. The chairperson of advisory committee is primarily responsible for guiding the student. Scientists who act as major advisors or guides should have some previous experience in conducting independent research. In educational institutions, this is assessed by the number of peer-reviewed papers published by the scientists.

The student's primary interest in undertaking a research project is to prepare a thesis within the stipulated time and get a degree in a hassle-free environment. However, an ideal hassle-free environment is non-existent in most educational institutions, and research students have to face many constraints for completing their research work. When the students are exposed to constraints such as paucity of funds, lack of laboratory facilities, imperfect equipments, shortage of labour, and the like, s/he may lose confidence in research. These issues should be given due considerations from the stage of selection of the thesis topic itself. Select only those problems that can be completed under the existing constraints and facilities. Since the student is a novice in the field, s/he may not know much about the intricacies of research and the likely obstacles in the smooth execution of the research project. In such situations, it is the duty of the research guide to reassure the student and make available all the necessary facilities instead of promoting or conniving at unethical practices. Similarly, never give complicated research problems to students, which give them temptations for resorting to unfair means.

The pattern of guidance or counselling imparted to students is flawed in many cases. Many research supervisors do not take up guidance seriously. Some teachers are interested only in increasing the numbers of research students under them. Because of the power structure prevailing in the department or institution, the students may be afraid to complain and junior teachers may be helpless. All these affect the quality of student's work and the outcome may be simply spurious. However, such manipulative research guides are adept in finding convenient examiners for evaluating the thesis and conducting viva voce. Even if the examiner discovers that the thesis is worthless, s/he may not reject it because of several considerations ranging from closeness to the guide to the benevolence of the examiner. Some examiners who find the thesis worthless, instead of rejecting or suggesting amendments, recommend the award of degrees on 'compassionate grounds' or under pressure.

Chances of committing unscientific practices are greater in social sciences than natural sciences, as they generally use non-experimental methods such as cross-sectional studies involving surveys, wherein likelihood of errors is more. If not vigilant, wilful errors may become rampant, especially when conducting surveys and case studies. Follow only established and time honoured methods. While conducting surveys, do not write your opinion as the opinion of volunteers involved in the study. This is often a malady of public opinion surveys, pre-poll surveys, and attitude surveys. You may also come across 'sponsored surveys' to sway public opinion in a particular way. Conduct the surveys and case studies following scientific practices only. For example, if your research project specifies to interview 100 randomly selected farmers or patients, you must meet all the 100 individuals selected at random and record their responses. If you are manipulating the data by meeting only some individuals or not employing random selection procedures, you are committing a fraud. The research guide or supervisor has the responsibility to insist on good research practices. The supervisor must check the methods employed and the data generated by the students periodically.

Another area of concern is violation of statistical principles. Suppose you are expected to examine 400 cases. Instead of examining 400 cases, just examining 100 cases and then multiplying every observed value by 4 and using the statistical test based on 400 observations is a typical fraud of this kind. Similarly, some scientists and students conveniently sideline the principles of experimentation (Sect. 5.4), although statistical theory is used. For example, instead of having three replications as per design, including only two replications and using the average values as the third replication is a clear case of fraud. Similarly, violation of statistical principles is deliberately done in deciding the actual plot size and randomization of treatments, but these are concealed when reporting the methodology.

For a research student, conducting research and writing thesis are areas requiring greater attention and help. It is particularly important that the budding scientists learn only scientific and scrupulous methods of experimentation and reporting of research results. The research guide should bestow increased attention to these issues. The guide should not insist for agreement with some preconceived theory. Sometimes, manipulation of data is deliberately done, especially in cases where experiments cannot be easily duplicated. The students should be aware of the ethics involved and good principles of experimentation (Sect. 5.4). In any case, do not succumb to manipulation of results or 'cooking' data. Compared to laboratory or pot culture experiments, it may be difficult to ensure uniformity in field experiments. The practical considerations set out in Sect. 5.9 are helpful to avoid many difficulties one may face while conducting field experiments.

23.6 Ethics in Science Practicals

It is disturbing to note that science students are taught the 'art' of fraud and misconduct from the school days itself! Many unscientific practices are followed in the conduct of science practical classes. Often, sufficient number of instruments may not be available for all the students to perform the experiments prescribed by the syllabus. Even if available, they may not be working properly. Sometimes, the time allotted may not be sufficient to complete the exercise within the class hours. Nevertheless, all students are required to record the 'results' of these experiments in their practical books! In reality, only some students actually do the experiment while others just copy the results in their own record books. In many occasions, it also happens that the teacher demonstrates the experiment and the students write records as though they have actually conducted the experiment! Even in cases where students do the experiment, 'cooking' of results is freely encouraged. The teachers wrongly feel that getting observations near to those expected in theory is more important than to recording observations genuinely and sincerely. These kinds of mentalities among teachers and authorities are responsible for creating several wrong notions about science in the minds of young pupils. If students learn that fraud, forgery, and deceit can be done with the help of teachers, they may lose their faith in science and scientific methods.

Several reasons can be cited for this condemnable situation in schools and colleges. The administration is largely responsible for this state of affairs. The most common problem is inadequate funds for the purchase of instruments and chemicals. Other reasons such as apathy in maintenance of instruments, lack of facilities for repair of instruments, and inadequate supporting staff may also lead to unfair practices; and for those genuine teachers, conducting practicals will be a major headache. Some heads of educational institutions give less importance to practicals; they may think that it is a waste of time. In several cases, although touted as practical, the class may be simply repetition of theory. Students may not do any 'practical' at all. Another common malady is corruption in practical examinations with the connivance of teachers and authorities, which should be curbed by all possible means.

Another domain of fraud is 'projects'. In many institutions, projects have been made compulsory as a part of their academic curriculum, and the students have to submit a project report. However, just like practicals, the projects are not given due seriousness. In certain institutions, teachers simply dictate project details and students have to submit the project report using the dictated data. Sometimes, you can buy readymade project reports from 'ghost writers'. If the projects cannot be conducted properly in the institution, it is better that the authorities withdraw the requirement of a project. A fraudulent document is far more dangerous than the absence of it. It is highly deplorable to allow frauds of this nature to continue.

23.7 Whistle Blowing

The term 'whistle blowing' derives from the practice of police officers who would blow their whistles when they notice the occurrence of a crime. The whistle would alert other officers and the public of impending danger. You will find some whistle blowers in science establishments too. Whistle blowers may be *internal whistle blowers* or *external whistle blowers.* Those who report misconduct to a fellow employee or superior within their establishment are the internal whistle blowers. External whistle blowers, however, report misconduct to outside persons or authorities. In these cases, depending on the severity and nature of information, whistle blowers may report the misconduct to the media, law enforcement agencies, or other agencies. However, the scientist or employee who wants to blow the whistle must have reason to believe that his/her colleagues or superiors have violated some law, rule, regulation, or ethics.

Whistle blowers are often selfless individuals who suffer for public interest and organizational accountability. In fact, most whistle blowers are attempting to apply ethical principles braving several obstacles. However, many people do not bother to blow the whistle at all, because of fear of retaliation and the fear of losing their relationships at work and outside work. Although whistle blowers are doing a service to the society, harassment to whistle blowers is a serious issue of concern in many organizations. When exposed, they are prone to retaliation and harassment. Everybody wants to suppress reports of corruption, fraud, unethical treatment of patients, and sexual harassment at work. Although whistle blowing is a selfless service and require protection from persecution, there have been many cases where punishment for whistle blowing has occurred such as termination, suspension, downgrading, denying promotions, and harsh mistreatment from superiors and fellow employees. In India, the 'Right to Information Act' is serving yeomen service to the whistle blowers and the public.

Students can also be whistle blowers. They must blow the whistle whenever they come across improper practices or fraud in teaching, research, and evaluation. They may bring such unhealthy practices to the attention of the authorities, and if this fails, bring them out before the public.

23.8 Scientific Values: Needed a Code of Conduct

All researchers and students must be guided by some values and ethics. The society views scientists as idealists and selfless people who partake in scientific pursuits only. Researchers should not do anything against the expectations of the society. Having recognized the problem of ethical behaviour of scientists and students, the requirement of a code of conduct is often debated. It cannot be imposed but must be imbibed. Every researcher should abide by the research ethics and strive to uphold moral values in all their endeavours scrupulously. They must be curious and inquisitive about all natural phenomena, and try to find the truth about these phenomena by careful observation, experimentation, and logical thinking. In these pursuits, they should always be guided by scientific facts, logic, and reasoning and not fallacies. Scientists and students should not succumb to superstitious and prejudiced or preconceived notions prevalent in the society. Researchers should be courageous enough to question existing beliefs and practices. Try to inculcate the habit of recording observations honestly. Scientists and research students should not indulge in 'cooking' experimental results and copy or plagiarize observations of others. It is the duty of researchers to promote scientific temper, scientific culture, and scientific scholarship in society and promote quality in science. Students should desist from the lure of committing short cuts or unfair means when they plan and conduct research work for thesis purpose.

Scientists should not demean their work by indulging in unscientific activities. Their job is to unravel the unknown for the betterment of society and to solve the miseries of humankind. Research students always search for models in research around them. Their role models in research should be dedicated and devoted scientists who pursue higher scientific values and ethics and not those who may be in positions of power but do not follow ethical values in their work. Try to emulate the selfless traditions of great scientists of yesteryears who worked under hostile environments. Always remember that research is a noble profession, but science progresses only if you follow intellectual honesty and ethics.

Bibliography

Best, J.W. and Kahn, J.V. 1993. Research in Education (7th Ed., Indian Reprint, 2004). Prentice-Hall of India, New Delhi, 435p.

Bond, A. 2007. Your Masters Thesis: How to Plan, Draft, Write and Revise. Viva Books, New Delhi, 135p.

CSE [Council of Science Editors]. 2018. CSE's White Paper on Promoting Integrity in Scientific Journal Publications. Editorial Policy Committee, Council of Science Editors, Wheat Ridge, 81p. Available: https://www.councilscienceeditors.org/resource-library/editorial-policies/white-paper-onpublicationethics/ [Accessed 30 June 2018].

Davis, M. 2005. Scientific Papers and Presentations. Academic Press, Massachusetts, 356p.

Day, R.A. and Gastel, B. 2006. How to Write and Publish a Scientific Paper (6th Ed.). Cambridge University Press, Cambridge, 320p.

Fathalla, M.F. and Fathalla, M.M.F. 2004. A Practical Guide for Health Researchers. WHO Regional Publications Eastern Mediterranean Series 30. World Health Organization Regional Office for the Eastern Mediterranean, Cairo, 232p.

Gray, D.E. 2014. Doing Research in the Real World (3rd Ed.). Sage Publications, London, 729p.

Holmes, D., Moody, P., and Dine, D. 2010. Research Methods for the Bioscience (2nd Ed.). Oxford University Press, Oxford, 457p.

IIFT [Indian Institute of Foreign Trade]. 2010. Trade-Related Aspects of Intellectual Property Rights (TRIPS), FAQs. Centre for WTO Studies, Indian Institute of Foreign Trade, 28p. Available: https://wtocentre.iift.ac.in [Accessed 12 Jan. 2013].

Jarrard, R.D. 2001. Scientific Methods [E-book]. Available: https://www.emotionalcompetency.com/sci/sm_all.pdf. 236p. [Accessed 10 June 2018].

Lester, J.D. and Lester, J.D. Jr. 2009. Writing Research Papers: A Complete Guide (13th Ed.), Longman, 416p.

Matthews, J.R. and Matthews, R.W. 2007. Successful Scientific Writing: A Stepby-Step Guide for the Biological and Medical Sciences (3rd Ed.). Cambridge University Press, 256p.

NIH [National Institutes of Health, USA] 2008. Nuremberg Code [On line]. Available: https://ohsr.od.nih.gov/guidelines/index.html [Accessed 27 Aug. 2008].

Tiwari, P.N., Raman, T.S., and Gosh, S.K. (eds.) 1990. Scientific Values and Excellence in Science. Proceedings of a seminar held at the Indian National Science Academy, New Delhi (18 April 1989), Society for Scientific values, New Delhi, 52p.

Wadehra, B.L. 2004. Law Relating to Patents, Trade Marks, Copyright, Designs and Geographical Indications. Universal Law Publishing Company, Delhi, 674p.

Walliman, N. 2011. Your Research Project: Designing and Planning Your Work (3rd Ed.). Sage Publications, New Delhi, 378p.

Wellington, J. and Szczerbinski, M. 2008. Research Methods for the Social Sciences. Continuum International Publishing Group, London, 234p.

Chapter 24
Plagiarism: Prevention and Cure

Taking something from one man and making it worse is plagiarism.
George Augustus Moore (1852–1933), Irish novelist

Publication of research papers and other academic documents are the major accepted means to gain recognition for the author from fellow scientists and the public. Both quality and quantity of publications are important, but just to increase numbers, producing substandard works or those with plagiarized contents is unethical and improper. The sheer pressure to churn out publications has its own negative impacts too. Many times, scientists try to publish as many papers as possible by splitting major works. Proliferation of journals including the so-called predatory types and publication of spurious articles are other problems.

The phrase '*publish or perish*' is popular among academic circles to describe the pressure among scientists and academicians to publish research papers and similar publications continually to keep up or promote one's career. Along with this phrase, another phrase '*publish and perish*' has also come up to warn the scientific community of the perils involved in resorting to unhealthy practices in publication. The following are grave issues in research and publication circuits. Probably, you may 'perish', if you indulge in any of these practices!

- Plagiarism
- Manipulation and fabrication of data
- Duplicate manuscripts
- Redundant publications
- Fraudulent authorship
- Conflict of interest

All the above create 'publish and perish' kind of situations. Taking enough safeguards to avoid these issues and by publishing quality articles in reputed journals, you can '*publish and flourish*'! Most of these issues have been discussed in the previous chapter on ethics. In this chapter, issues related to plagiarism are considered in detail.

C. G. Thomas, *Research Methodology and Scientific Writing*,
https://doi.org/10.1007/978-3-030-64865-7_24

Broadly speaking, plagiarism is the action of an individual misappropriating another person's work and claiming that it is his/her own work. Most cases of plagiarism are reported from academic and research institutions, which produce documents such as essays, theses, research papers, textbooks, and reports. Plagiarism can also occur in other fields such as novels, poetry, cinema, short stories, features, art designs, and many others.

Earlier, there were no mechanisms to check whether the article or book contained copied materials. Similar were the cases with assignments, seminar papers, or theses submitted by the students. The students may have simply copied sentences or paragraphs from many sources and combined, often, with the overt or covert connivance of supervisors. After the arrival of computers and Internet, copying has become much professional and easier! For writing an assignment now, you may probably start searching for sources using a search engine such as Google, Google Scholar, PubMed, PubAg, or CAB Direct. The search engine will lead you to many sources, which contain anything you need. The student may copy the information bit by bit, and paste them into the body of the text. Ultimately, an assignment is completed in a hotchpotch manner. Modern copyright laws and ethics, however, brand this practice as simply stealing! You should understand that when you plagiarize, you are stealing from someone else and cheating the system thus putting others at a disadvantage. Apart from stealing, chances are that you may be caught with clear evidences! You should be aware that plagiarism checking software are now versatile, and if caught, its impact on your career would be disastrous. Therefore, never try to publish and perish, but instead publish and flourish!

24.1 What is Plagiarism?

When an individual lifts someone else's intellectual works such as ideas, opinions, art designs, or writings as such or in a modified form and passes them off as his/her own original work with or without permission, it is called *plagiarism*. This is nothing but some kind of kidnapping or stealing as the etymology of the word shows. Plagiarism committed by students, teachers, and scientists is a major problem in academic and research institutions. The failure on their part to acknowledge other people's ideas and statements must be viewed as an academic offence. The word plagiarism is frightening to the students as quite often they are caught and penalized for unintentional plagiarism! In most instances, they do so as they are not aware or not taught properly how to make plagiarism-free materials.

Plagiarism and *piracy* are related, but considered as two different aspects of copyright violation. Piracy is a clear case of copyright infringement, but in plagiarism, copyright violation may not always involve. For example, if you are reproducing a copyrighted material for large-scale distribution for financial gain or other benefits, it is piracy. However, if you are reproducing a book or essay, which is available in the public domain (non-copyrighted) for some benefits, it is legitimate and will not be called piracy. At the same time, if you use copyrighted or non-copyrighted material as such or in a partially changed form by putting your name on it as though it is your creation, it is plagiarism. Both piracy and plagiarism are unethical and illegal practices.

Language proficiency is a prerequisite to avoid unintentional plagiarism. For example, if you are not confident with English, you may find it difficult to write in your own words easily, and in such instances, chances of plagiarism are high. Procrastination, inadequate time for preparation, poor writing skills, pressure to produce publications, and ignorance about the implications of plagiarism are some other causes for increased instances of plagiarism. However, you cannot quote any of these reasons for committing plagiarism! Accidental plagiarism or unintentional plagiarism is equally disgraceful as that of intentional plagiarism. Students should be taught the ethics involved in writing and the perils of plagiarism during the undergraduate days itself. Teachers must ensure that the assignments and project reports prepared by the students are plagiarism-free.

In some cases, the boundaries of plagiarism are vague. For example, most scientists feel that it is not correct to start a research project inspired by communication from another scientist through a letter, conversation, or visit to his/her laboratory unless the other scientist has specifically asked you to do so. Similarly, it is not correct to allow your research plans be affected in any way by either a proposal or a manuscript that you have been asked to review. Nevertheless, bear in mind that researchers have no right to keep any field to themselves, and once it is published, anybody can pursue that research area, if they are interested.

Plagiarism: How Many Types?

In the academic world, plagiarism happens in many ways. For example, when students write an assignment, seminar paper, or thesis, they may copy whole paragraphs or sentences from books and articles and use them as such for their works. Such kinds of copying texts, ideas, concepts, research results, statistical tables, computer programs, designs, images, sounds, or any combination of these come under the purview of plagiarism. The extent of plagiarism can range from copying a single sentence or a part of it to a whole work as one's own. The following are explicit examples of plagiarism.

1. Copying sentences or paragraphs from books and articles without quotation marks
2. Copying from Internet sources through 'cut and paste'
3. Taking credit for a work created or done by somebody else
4. Making simple grammatical or word-order cosmetic changes to borrowed material and representing it as one's own work
5. Imperfect paraphrasing of the works of another person with minor changes, keeping the meaning, form, and progression of ideas of the original
6. Patching up works of others into a new one
7. Producing a written work as independent work when it has actually been produced with unacknowledged help of another person

8. Using any of your own previously published works such as texts, articles, and research results, make it look new (self-plagiarism).

The above types of plagiarism can be grouped into seven categories. AMA (2007) lists four types—*direct plagiarism, mosaic plagiarism, paraphrase plagiarism,* and *insufficient acknowledgement*. Along with these, three more common types have been added—*complete plagiarism, self-plagiarism,* and *accidental plagiarism.*

Direct plagiarism: In direct plagiarism, copying of whole paragraphs or longer passages from another source is done without quotation marks as though it is their creation. This type of plagiarism can be detected easily by using plagiarism checking software and comparing similarity index.

Paraphrase plagiarism: Probably, this is the most common type of plagiarism, wherein, the author paraphrases or restates a part of someone else's work, but without citing the source.

Mosaic plagiarism: This is a case of appropriating phrases from other sources or using synonyms to change some words or parts of a sentence; but general structure of the work remains the same. This kind of patch writing, whether intentional or not, is not a good academic practice.

Insufficient acknowledgement: In certain works, the authors may have mentioned the original source of only part of what has been taken or they may not have cited the source suitably. In both cases, the authors failed to acknowledge adequately, and the reader would not understand what is original and what has been borrowed.

Complete plagiarism: If you appropriate someone else's article as such and publish it under your name, it is a case of complete plagiarism. It is the most heinous form of plagiarism. Students sometimes commit complete plagiarism, for example, by submitting a copy of the assignment of another student as his/her own.

Self-plagiarism: This occurs when the author uses his/her own previous works such as texts, articles, and research results for a new work without mentioning that they have been published before. Most students publish papers based on their Masters and Doctoral thesis work, and these are permitted and encouraged. However, if the individual is producing a paper using the data of works already published, it is a clear case of self-plagiarism.

Accidental plagiarism: This is unintentional and includes the problems such as neglecting to cite the sources, misquoting the sources, or inadvertently using the source by similar words, groups of words, or sentence structure without attribution. Sometimes, plagiarism happens due to ignorance! Please note that accidental plagiarism is taken as seriously as any other plagiarism having the same range of concerns as other types of plagiarism!

To avoid inadvertent plagiarism, get familiar with good academic practices including the art of good technical writing (Chaps. 16–18). Use proper language and styles and learn to paraphrase properly. Learn to use citation in the correct format, both in-text citation and listing of sources (Chap. 15). Careful preparation of notes with full details of authors and publications is essential for proper citation

(Chap. 22). You can also use a suitable plagiarism checking software (free or priced) to detect and correct plagiarized materials, which inadvertently crept into your work.

Some popular techniques to avoid plagiarism are discussed below under the following heads. When you attempt an essay, research paper, or other works such as text books, you may have to use some or all of these techniques.

- Acknowledge sources appropriately
- Paraphrasing
- Direct and indirect quotations
- Summarizing
- Evaluation of text
- Use of plagiarism checking software.

24.2 Acknowledge Sources Appropriately

Reproduction of published material is a contentious issue when dealing with articles and books. There are certain rules and laws involved, especially related to copyright and related rights. Authors from India must make sure that their books or articles do not infringe upon any trademark, patent statutory rights, and proprietary rights of others and the provisions of the Indian Copyright Act of 1957 as amended by Copyright (Amendment) Act, 1999. Each country may have enacted similar laws applicable to their citizens.

When mentioning a previously published idea in a publication, due credit must be given to the originator unless the idea has become 'common knowledge'. You must indicate the name of the author or organization along with the year inside the text (citation) followed by full bibliographic information in the list of end references (see also Chap. 15).

Before using the works of others (e.g. a figure or table) as such or in modified form in your book or other publications with citation, you must confirm whether a written permission from the original author or publisher is required. Normally, permission is needed to use materials from books, articles, websites, poems, plays, songs, photographs, artworks such as diagrams and cartoons, newspaper articles, magazine or journal articles, book or magazine covers and their design, and copyright-protected logos, figures, graphs, charts, tables, and advertisements. Web materials shall be treated just like printed materials. Understand that online materials available through Internet and CD-ROMs are also copyrighted material, and simply reproducing them through 'cut and paste' is not acceptable. Take care to avoid 'cut and paste' and 'drag and drop' forms of verbatim copying from electronic sources.

Exceptions from Citations

For using certain kinds of materials for your work, limited reproduction of published materials without permission and exception from citations is permitted. For example, when you use common sense observations, folklore, legends, and historical events for your writing, they need no citations. Similarly, your own experiences, insights, musings, videos, presentations, music, and other such creations do not require citations. These are usually grouped as 'common knowledge' and 'materials in the public domain', and dealt differently. The scientific evidence you gathered after performing your own research experiments or surveys also need no citation because you are the owner of such information!

Dealing with Common Knowledge

In general, facts and information in common use, which can be verified from several places and are likely to be known by many people, are regarded as *common knowledge*. This usually includes facts available from various secondary sources such as textbooks, handbooks, manuals, dictionaries, directories, or encyclopaedias. When a common knowledge is paraphrased for use in a document, a source need not be cited. Although common knowledge or facts are not subject to copyright, the words used to express them may come under copyright restrictions, especially, if the wording is original or unique. Therefore, paraphrasing is a must for such sentences; however, if the information is the result of an investigative work or experimentation of some others that has not attained the status of common knowledge yet, it requires not only paraphrasing but also citation.

For strengthening and providing a firm base to an argument, we use facts as they cannot be disputed. Conversely, when a person makes an interpretation, it is often disputable. Some examples of interpretations are value judgments and explanations of causes and effects. Interpretations are generally used for strengthening the arguments. However, you must acknowledge the authors, if you are using their interpretations for buttressing your argument. In the discussion part of a thesis, you interpret your results, and to support your claims, you use interpretations of previous workers.

Principle of Fair Use

You should be able to distinguish between materials for which permission is necessary and those which can be used without permission. In general, consent is required for the reproduction of any material unless the *principle of fair use* or *public domain* is applied, thus allowing you to use the material without permission. The relaxation for 'fair use' or 'fair deal' guarantees that copyright laws do not infringe upon your freedom of making a critical commentary on a work or using the work for teaching or research purposes. Examples of permitted use of texts under fair use are teaching, learning, criticisms, comments, reporting, and research.

How to Integrate Information from Sources?

There are several methods for organizing and integrating information from various sources into your work. You can adopt one or more systems that suits your purpose, but you should be able to distinguish clearly between direct quotes (direct copying) and your own structure with apt words. You can write an essay, review, or discussion as (1) paraphrase, (2) quotation, (3) summary, or just as (4) comments after evaluating the source. Normally, for producing a plagiarism-free write up, an ingenuous mix of all these are employed.

24.3 Paraphrasing

If you copy paragraphs or sentences of another author without any change for your works, you are expected to mark them as 'quotations', meaning that they are not your own but borrowed. After the quotation, you are expected to cite the source and give the source details in the list of references. Quotes are, however, rarely used in science literature, and too many quotations from other sources are not feasible in a text book or research paper. Excessive use of quotations can make your writing look like a patchy work and may not reveal your understanding of the ideas or information being discussed. In scientific papers, you should express your own thoughts and interpretations, and not those of others. This problem can be solved easily, if you learn the art of paraphrasing and summarizing.

A *paraphrase* is a restatement of a piece of writing in your own words, syntax, and style, but preserving the tone of the original having almost the same length. You use the original author's idea but present it in your own language. Bear in mind that as you are borrowing someone's thoughts, you must document the source when using the paraphrase for a document. Proper and correct paraphrasing is important, otherwise, you would be blamed for committing plagiarism. Sometimes, you also try to paraphrase sentences of others to make it more comprehensible. A well written paraphrase is appreciably different from the original source in both wording and structure of sentences without substantially altering the intended meaning.

As discussed already, using 'common knowledge' or 'facts' for one's wok is handled differently. Because common knowledge is 'common' and known to many, it is not subject to copyright. However, if the wording is original or unique, you cannot use them with the original wording. For using it, you should paraphrase the information and write in your own words. For a native English speaker, this is easy, but for a non-native English speaker, s/he has to take extra effort for paraphrasing. Although paraphrasing is a part of English language teaching, it is not given the importance it deserves, and many research students find it difficult to do paraphrasing in a proper way. Researchers, teachers, and students should master the art of paraphrasing thoroughly for handling issues on plagiarism.

A common doubt expressed by many is the quantum of paraphrasing one needs to do to make sure that s/he is not blamed of plagiarizing. In fact, one cannot specify the number of words that should be added or changed to make a passage his/her own. What is important is to understand the passage and rewrite completely in one's own words. Often, occurrence of a maximum of five consecutive words identical to someone else's writing would not be regarded as plagiarism. For example, it is not fair to consider as plagiarism if you reproduce: '…sown on 16 Jan. 2018', or 'all students, teachers, and parents', although both cases have five consecutive words! In India, common knowledge or coincidental terms consisting of less than 14 consecutive words are excluded from plagiarism checks.

Please note that the sections such as references or bibliography, table of contents, preface, and acknowledgements of a thesis or research paper cannot be paraphrased.

These are usually exempted from similarity checks. So is the case with generic terms, laws, standard symbols, standards equations, phrases, jargon, and terms.

Tips for Paraphrasing

A reasonable command over English is essential for accurately paraphrasing the sentences of others. Paraphrasing can be done using the following techniques.

- Preparing abbreviated notes
- Using synonyms for the major words
- Using simpler words instead of difficult or archaic words
- Changing the class of words
- Changing the order of words
- Changing sentences in active voice to passive voice and vice versa
- Combining simple sentences
- Converting long and complex sentences into two or more simple sentences.

When you read a journal article or book, take down notes in an abbreviated form using devices such as abbreviations, symbols, shorter words, and numbers. Certain types of words like articles are often dropped. Later, you can prepare the draft in full sentences using the abbreviated notes. This would be substantially different from the original source (see also Chap. 22 on note making). Another technique is to use ingeniously a mix of other paraphrasing techniques listed above, for example, using synonyms, simpler words, changing the class or order of words, and changing the structure of sentences. For most words or expressions, you can rely on synonyms with similar meanings. A thesaurus is the best resource to look for synonyms. Presently, many online sources are also available to check out the most appropriate synonyms. However, keep vigil when using words that you are not familiar with as words can have multiple meanings depending on the context in which they are used.

You can also change the form of words, for example, from an adjective to a noun or from a noun to a verb. Changing the word form often necessitates changes in sentence structure and organization. You can also change sentences in active voice to passive voice or vice versa depending on the situation. Splitting long sentences into simple sentences or combining small sentences into complex sentences is also common. However, remember that technical writing demands simple, clear, and straightforward writing. Whatever be the method, strive to be original when you write an essay, review of literature, research paper, popular article, or a book based on the information gathered from several sources.

A few examples of paraphrasing are indicated below.

Example 1

Original script

'Organic agriculture generally implies alternate modes of agricultural production, which would denounce the use of synthetic fertilizers, pesticides, and herbicides; and it often includes philosophical overtones with regard to one's approach towards agriculture as a way of life'.

After paraphrasing

Broadly speaking, organic agriculture refers to alternate styles of agriculture that excludes the use of artificial fertilizers, pesticides, and herbicides. It also includes the philosophy behind one's attitude towards farming.

The above information is common knowledge, and therefore, the source need not be cited along with the paraphrased version.

Example 2

Original script

'During second crop, for higher yield in photosensitive high yielding varieties in Palakkad district, where assured irrigation is available, the crop commencement may be adjusted in such a way that it flowers only during the second fortnight of December, facilitating proper integration with the weather, better utilization of applied fertilizers and high filling percentage' (KAU 2011).

After paraphrasing

In irrigated areas of Palakkad district, adjust the time of planting of photosensitive high yielding varieties so that it flowers only during the second fortnight of December (KAU 2011). This helps to achieve proper integration with the weather, better use of applied fertilizers, and high filling percentage of grains.

In example 2, paraphrasing is attempted to make the writing easier to understand. Note that in the example, the source must be cited, as it is a specific recommendation.

Example 3

In the following example, paraphrasing is attempted for an entire paragraph containing common knowledge from a source. For easy understanding, the mode of paraphrasing of each sentence has been shown.

Original script

'Rice is the most important food crop of Kerala. It can be cultivated under a variety of climatic and soil conditions. Rice cultivation is conditioned by temperature parameters at the different phases of growth. The critical mean temperature for flowering and fertilization ranges from 16 to 20 °C, whereas, during ripening, the range is from 18 to 32 °C. Temperature beyond 35 °C affects grain filling. Rice comes up well in different soil types. For normal growth, a pH range of 5.0–8.0 is suitable. In general, rice can be grown as transplanted or direct sown crop during three seasons depending on the agro-climatic situations'.

Paraphrasing sentences

Rice is the most important food crop of Kerala (*In Kerala, rice is the main food crop*). It can be cultivated under a variety of climatic and soil conditions (*It is grown in a wide range of soil and climatic conditions*). Rice cultivation is conditioned by temperature parameters at the different phases of growth (*Prevailing temperature regimes during different growth phases affect the success of rice crop*). The critical mean temperature for flowering and fertilization ranges from 16 to 20 °C, whereas, during ripening, the range is from 18 to 32 °C (*For flowering and fertilization, the critical range of mean temperature is from* 16 *to* 20 °C; *however, during ripening, the preferred range is from* 18 *to* 32 °C). Temperature beyond 35 °C affects grain filling (*Temperature above* 35 °C *is not conducive as it affects grain filling*). Rice comes up well in different soil types. For normal growth, a pH range of 5.0–8.0 is suitable (*Rice can be grown in almost all soil types, and tolerates a range of pH from* 5.0 *to* 8.0). In general, rice can be grown as transplanted or direct sown crop during three seasons depending on the agro-climatic situations (*Based on the prevailing agro-climatic conditions, often, rice is grown either as transplanted or direct sown during the three seasons*).

Paraphrased version

In Kerala, rice is the main food crop. It is grown in a wide range of soil and climatic conditions. Prevailing temperature regimes during different growth phases affect the success of rice crop. For flowering and fertilization, the critical range of mean temperature is from 16 to 20 °C; however, during ripening, the preferred range is from 18 to 32 °C. Temperature above 35 °C is not conducive as it affects grain filling. Rice can be grown in almost all soil types and tolerates a wide range of pH from 5.0 to 8.0. Based on the prevailing agro-climatic conditions, often, rice is grown either as transplanted or direct sown during the three seasons.

Example 4

When you take information from a source, often, the assumption is that you have read the original source and not abstract or a secondary information. Rarely, you need to paraphrase some sentences from the review of literature of a different author. See the example and note how this can be handled.

Original script

'Biological invasions have been responsible for at least 3 of the 24 known extinctions of endangered species in the USA (Schmitz and Simberloff 1997)'.

After paraphrasing

Schmitz and Simberloff (1997) state that in the USA, minimum 3 of the 24 reported extinctions of endangered species is because of biological invasions.

The above information is part of a review by Holzmueller and Jose (2009). It is unethical to write as given above even after paraphrasing unless you have seen the original paper by Schmitz and Simberloff (1997). It is better you write as shown below:

Schmitz and Simberloff (1997) cited by Holzmueller and Jose (2009) state that in the USA, minimum 3 of the 24 reported extinctions of endangered species is because of biological invasions.

In the example, Holzmueller and Jose (2009) are responsible for what Schmitz and Simberloff (1997) reported. You shall then list Holzmueller and Jose (2009) instead of Schmitz and Simberloff (1997) in the list of references. This is a better practice instead of pretending that you have seen the original paper of Schmitz and Simberloff (1997).

24.4 Direct and Indirect Quotations

The reproduction of exact sentences or words of an author from documents such as thesis, research paper, book, or essay in your works is called *quotation*. Sometimes, instead of going for paraphrasing or evaluating the text and making comments, you may prefer to copy materials as such from sources as *direct quotation*. You can also do it as *indirect quotation,* which means that you are using the idea or information from an author, but in your own words as a paraphrase.

Direct quotation is exact copy of an original piece of text word for word. In writing, you may directly quote an author when:

- You feel that the original author's style is clear and appealing
- The views of the author exactly match or support your views
- You want your readers to know exactly what an author has said about a topic.
- The original phrasing reveals something about the author or creates a particular effect
- You want to avoid much explaining.

Although direct quotes are not common in research articles, when you write a paper related to history, current events, and similar fields, direct quotes or partial quotes may become necessary for a meaningful discussion of the subject. Still, it is not a good practice to quote large sections of text as such and rely too heavily on the original words of authors.

When you copy an author's words as such, use quotation marks around the exact words copied from the original source. The quotation marks, either single or double, must be consistent throughout the text. In India, the custom is to use double quotation marks for such borrowed material, and if you are including quotations within quotations, use single quotation marks. The ending commas or ending full stops should appear outside quotation marks unless the quotation forms a complete and independent sentence. Ensure that you have copied accurately, both words and punctuation marks as used by the author.

Note that if you are using direct quotations, it must be verbatim, and anything added or omitted by you must be indicated in square brackets. For example, if you are not reproducing the entire text but decided to drop a part of it, then you can use three dots or ellipses inside brackets [...] to show that some parts are missing. Sometimes, you may put square brackets enclosing the comment 'sic' as [sic] to inform the reader that the part is reproduced as such without any change, for example, with the same spelling or grammatical errors committed by the original author. Occasionally, square brackets as [he] are used to indicate that you have added some text to the original quote, for clarification. The last name or surname of the author, year, and page number on which the quotation is to be found must also be included.

When the quotations are short, less than 30 words or three lines, it is enough if you reproduce them exactly as they are printed in the original text enclosed within quotation marks. Introduce short quotations with a phrase in order to integrate them into your text naturally instead of appearing as abrupt additions. See two examples:

- Gardner et al. (1985:66) affirms that 'crop yield can be increased either by increasing the total dry matter produced in the field or by increasing the proportion of economic yield (the harvest index) or both'.

Note In the above example, the number given after the year is page number on which the quotation is to be found.

However, quotations of longer passages (quotes of more than 30 words) are displayed in block quotation, usually after indenting five spaces from the left. Leave an extra line of space at the beginning and at the end of the quote. Use a smaller size font, for example, if you are using 11 point size for the general text, use size 10 for the quote. For block quotations, quotation marks are not used. You must also include the last name of the author, year, and page number along with the quote. An example of the use of block quotation is shown below:

> Science and technology advanced to boundless levels during the last 500 years, bringing an ever-more comprehensive view of the world. This was possible because of some radical changes that took place in Europe by the end of Middle Ages putting an end to the long period of darkness. Thomas (2015: 15) attributes three major reasons for the upheaval. He writes:
>
> > In fact, three major inventions that happened during the fifteenth century were responsible for most of these changes. One of these was the invention of *printing*. Printing required the development of paper to replace parchment. Developed originally in China, the technique of papermaking reached Europe in 1200's, and by 1450, papermaking was common. The effective application of printing was achieved by Gutenberg in 1438. The printing press promoted literature and science. Because of the printing press, multiple copies of the Bible could be made available for the public. People began to read religious scriptures, and started questioning the teachings of religious authorities. Two other radical inventions that paved the way for revival were the *compass*, which helped increased trade, travel, and discovery of new continents and islands; and *gunpowder and firearms*, which increased the combative ability of humans.

Note that the writer introduces the quotation with a brief opening explanation to integrate it to the text. After the sentence preceding the quotation, a colon (:) shall be given.

When using a borrowed material, you should also take care whether you are infringing on copyright rules; in some cases, you should obtain permission for using the material as such. In general, consent is required for the reproduction of any material unless the principle of 'fair use' or 'public domain' is applied thus allowing the material to be used without permission.

Indirect Quotations Through Paraphrasing

Instead of quoting phrases or sentences exactly, the general trend in science literature, especially in review of literature, is to put the idea or comment of the author in a

Table 24.1 Selected announcing verbs for integrating and introducing quotations

Agree	Affirm	Argue	Assert	Believe
Claim	Comment	Conform	Consider	Contend
Contest	Contradict	Criticize	Debate	Defend
Describe	Disagree	Discuss	Dispute	Establish
Evaluate	Examine	Highlight	Imply	Illustrate
Insist	Instruct	Investigate	Justify	Maintain
Note	Observe	Offer	Propose	Point out
Predict	Question	Reason	Refute	Remark
Report	Show	State	Suggest	Uphold

paraphrased format. We generally follow two styles to incorporate a paraphrase in our review, *author-prominent* or *information-prominent.*

Prominence to Authors

You can write by giving prominence to authors, if they are well-recognized figures in their fields or their works are important. It is usually done by starting the sentences with the name of the authors immediately putting the year in parenthesis.

- Thomas (2008) stated that shade-tolerant fodder grasses could come up very well in coconut gardens.

Please note that instead of 'stated' in the example, you could use other suitable verbs such as 'remarked', 'reported', 'claimed', or 'confirmed' according to your liking. Similar is the case with 'highlighted'. These are called *announcing verbs* or *reporting verbs.* To avoid monotony, vary the verbs as frequently as possible. A list of selected announcing verbs, which can be used to introduce and integrate direct and indirect quotes in writing, is given in Table 24.1.

Author prominence can also be implied by starting the sentence with 'according to' or ending the sentence with the name of the author as a part of the sentence, as shown below:

- According to Thomas (2008), shade-tolerant fodder grasses come up very well in coconut gardens.
- Shade-tolerant fodder grasses come up very well in coconut gardens as reported by Thomas (2008).

Prominence to Information

If you want to highlight the information rather than the author, after writing the sentence containing the information, place the reference (name and year, a comma precedes the year) at the end of the sentence in parenthesis. Note that the author's name is not part of the sentence but added separately in parenthesis. Example:

- Shade-tolerant grasses can come up very well in coconut gardens (Thomas 2008).

24.5 Summarizing

Sometimes, instead of paraphrasing the original information from an article or book, you may just give a gist of the information. If you state the contents of the article in a condensed form in your words, it is called a *summary*. This is a common way of reporting, when information from several sources has to be evaluated and presented. Summarizing the source material prompts you to avoid the overuse of direct quotations and paraphrasing large chunks of the original text. You will be able to understand the meaning of what you have read and present the material using your own words. You should also add your own comments to show your analysis and interpretation of the work. The size of a summary is often debatable; it may vary from a paragraph to even pages depending upon the information it carries. Remember that even after summarizing, you still have to acknowledge the source of the information, if it is not common knowledge. See an example:

Original text: 'Modern lifestyle is responsible for the production of huge quantities of solid and liquid wastes. It is a major environmental issue in urban areas such as cities and towns. For waste disposal, the three principles of reduce, reuse, and recycle must be followed meticulously. In the case of biodegradable wastes, composting is an effective method to dispose of the wastes. It is perceived as a partial solution to solid waste crisis in both rural and urban areas. Several advantages can be listed for this strategy. Composting reduces the amount of garbage sent to the landfill, as it is recycled as valuable manure. In other words, it helps to convert all the organic wastes into good quality manure. Waste materials of various kinds like cereal straw, crop stubbles, banana wastes, cassava wastes, dry leaves, farm weeds, house refuse, kitchen wastes, food wastes, and litter from cattle sheds can be converted into quickly utilisable manure through composting.

Composting, which has sustained since the beginning of agriculture, is now entering into an era of renewed interest. Concerns on managing wastes and producing food in tune with the ways of nature have created much enthusiasm in developing appropriate composting systems, ranging from small-scale backyard composting to large-scale commercial and municipal composting systems. Escalating landfill costs and tighter restrictions on waste disposal have significantly increased the economic attractiveness of large-scale composting systems'.

Summary: Waste disposal is a major environmental issue in cities and towns. For waste disposal, emphasis has been placed on the three principles of reduce, reuse, and recycle. Concerns on managing wastes have led to renewed interest in composting systems, both backyard and commercial. As composting converts organic solid wastes into useful manure, it is perceived as a solution to the solid waste crisis. Many types of waste materials can be reused through composting. It is economically attractive due to high landfill costs and tighter restrictions on waste disposal.

24.6 Evaluation of Text

Before using a piece of information for a review in a thesis or research article, analysis and evaluation of what you read should be done. Do not be under the impression that all print and web sources are worth citing! Many of them would be biased, misleading, or simply wrong. You should shape your own ideas and opinions about different issues after reading a piece of information, and you are expected to support your arguments with literature in your field. If possible, add your own comments and responses to your notes from reading as you proceed. This will help you to recognize the differences between your own ideas and the ideas of other researchers. It is essential that the information presented in support of an argument in a thesis or article is reliable and unbiased. If it is erroneous or biased, readers easily recognize faults in your presentation.

You must develop the habit of interpreting the views of the writer, and noting down your own idea indicating agreement or disagreement. When you evaluate a piece of work, you need to consider whether the source is dealing with facts alone like the results section of a thesis, or simply interpretation like a critical essay or popular article on a finding, or a combination of both as in a research paper. It is important to assess the credibility of each source as well. Although both primary and secondary sources provide information, remember that primary sources are more reliable and convincing. Accordingly, you evaluate the sources and draw details from both types for your work.

Evaluate the contents of a textbook or article while reading them. It is common to see many pseudoscientific claims in the literature and arguments based on fallacies (see Chap. 9 for a detailed account on fallacies). Occasionally, you may also come across post-modern interpretations of science condemning what is called ‘reductionism’ (Sect. 2.9). It is essential that readers after going through such articles must record their own ideas agreeing or disagreeing with the authors. Some examples of evaluation are given below. See the results of evaluation in parenthesis.

- The productivity of rubber is the highest in India (fact, true).
- Green revolution was a failure. Organic farming is the only way out (not true, a biased interpretation).
- In the recent years, many farmers committed suicide. GM crops are responsible for this increased suicide rate among farmers. Farmers should revert back to traditional varieties (The first sentence is a fact and true. Although the second sentence is put as a fact, no evidences have been given. Therefore, it cannot be accepted. The third sentence is a call for action based on unsubstantiated claims).
- Biodynamic farming improves the spirituality of human beings (An opinion based on pseudoscientific hypothesis as the statement cannot be verified).

24.7 Plagiarism Checking

Detecting plagiarism within a document has become a great headache for the academics as well as publishers. The widespread use of computers and the advent of the Internet have made it easier to plagiarize the work of others. Plagiarism can be detected manually or by the use computer software. Manual detection is cumbersome as substantial human effort is needed, and in many cases, it is practically impossible where too many documents are to be compared. Computer-assisted detection permits checking for similarity comparing huge document collections within minimal time.

After the adoption of widespread plagiarism checking tools, unscrupulous elements found out some spurious ways to deceive the software. Remember that simply altering a published material by replacing synonyms for sufficient words to deceive plagiarism detection software is an unethical act, and it will not absolve you from the blame of plagiarism.

Regulations on Plagiarism

Fraud, misconduct, piracy, plagiarism, and similar issues in scientific publishing have been noted by big publishing groups, and they are taking several possible steps to prevent such malpractices. The following are the main international bodies, who issue publications and guidelines to govern misconduct in scientific publishing. For more information, please visit their homepages.

- The Committee of Publication Ethics (COPE)—https://publicationethics.org/
- The European Association of Science Editors (EASE)—https://www.ease.org.uk/
- The World Association of Medical Editors (WAME)—https://www.wame.org/
- The Council of Science Editors (CSE)—https://www.councilscienceeditors.org/.

The Council of Science Editors (CSE) has taken several steps to uphold ethics in research and research publishing. Its Editorial Policy Committee came out with a White Paper for promoting integrity in journal publications (CSE 2018). All who are involved in the scholarly publishing process must take responsibility for upholding integrity in scientific publishing. This White Paper gives you guidelines and an overview of measures and rules related to ethics and plagiarism prevailing in different countries.

Many countries have enacted laws with provisions for penalties to curb the menace of plagiarism. In India, the University Grants Commission (UGC) has issued directives to prevent plagiarism in academic institutions across the country (UGC 2018). To detect and manage cases of plagiarism in a department, a 'Departmental Academic Integrity Panel' (DAIP) shall be established at the department level. At the institutional or university level, an 'Institutional Academic Integrity Panel' (IAIP) shall be set up, to consider recommendations of the DAIP and take appropriate decisions about penalties to be imposed. It can investigate allegations of plagiarism at the institutional level;

If academic misconduct has been established by detecting cases of plagiarism, penalties have to be imposed on students pursuing Masters and Doctorate programmes and on researchers. In the case of theses and dissertations, considering the severity of plagiarism, the penalties would be as follows:

Level 0: Similarities upto 10%—Minor similarities, no penalty.
Level 1: Similarities above 10–40%—Such students need to submit a revised script within a stipulated period not exceeding 6 months.
Level 2: Similarities above 40–60%—Such students shall be debarred from submitting a revised script for one year.
Level 3: Similarities above 60%—The registration of students for that programme shall be cancelled.

Similar penalties have been prescribed for faculty members and researchers who indulge in the production of plagiarized research papers and other documents.

Plagiarism Detection Tools

TurnItIn, *iThenticate*, and *URKUND* are the most common anti-plagiarism software used by Indian universities. TurnItIn and iThenticate are provided by iParadigms, which share the same technology and database contents, but are designed for different categories of users. TurnItIn is intended for students in a classroom setting, but iThenticate is designed for teachers and researchers who want to organize and share their work in a folder-based system. URKUND is owned and developed by PrioInfo AB, Sweden. All the above software are effective but costly. Most often, only institutions are able to purchase such software.

Examples of some free or partially free plagiarism detection tools are *Dupli Checker, Copy Leaks, Paper Rater, Plagiarisma, Plagiarism Checker, Plagium, Plag Scan, Plag Tracker, Quetext*, and *Viper.* A major problem with free plagiarism detectors is that they have small databases, and may be less accurate than professional plagiarism checkers. In most cases, they would not allow you to enter more than 1000 words at a time. For small student's works, probably free plagiarism detection may be enough.

Please note that most plagiarism detection tools check for similarity and not for plagiarism. The plagiarism detection tools identify material in your work that matches text from documents in their extensive repositories. When you subject your article for a plagiarism check, the results give a percentage score, known as *Similarity Index*. The Similarity Index gives an indication of similar contents that has been found in other published materials or theses. If the author argues and proves that she/he has streamlined the arguments by citing and referencing the material, then that cannot be construed as plagiarism!

Materials Excluded from Similarity Check

As far as plagiarism is concerned, the ideal strategy must be zero tolerance. However, do not equate plagiarism with similarity. You cannot make someone guilty of plagiarism simply because the software identified a few consecutive words identical to someone else's writing. To safeguard such cases of inadvertent similarities, some limits are prescribed for exclusion. In India, the UGC prescribes exclusion of the following from similarity checks (UGC 2018):

- Information put in quotations with necessary permission or attribution
- References or bibliography, table of contents, preface, and acknowledgements
- Generic terms, laws, standard symbols, and standards equations
- Common knowledge or coincidental terms consisting of less than 14 consecutive words.

Usually, phrases, jargon, and terminology, which are commonly used in writing, may be matched by the software, but you can exclude them.

Exclusion of Small Matches
When writing an essay or any similar work, one can set the software to ignore small matches coming under common knowledge. If you do not enable this setting, you may observe many small text matches of very general information. Small matches commonly occur when you use specific terms with long names (e.g. 'person with physical abilities', 'all students, teachers, and parents', 'Departmental Academic Integrity Panel', 'College of Horticulture, Vellanikkara, Thrissur'). If you set the software to exclude small match cases below 14 words, every match the software finds to be less than 14 words would automatically be excluded from the report. This enables the users to focus more on relevant potential misconduct issues rather than struggling to remove trivial matches.

The plagiarism detection software generates an originality report, which is a statement of similarities between the work under scrutiny and a database of previously published work, websites, and other sources. If a match is found between a submitted work and an existing one, the software highlights the matching text. However, all cases of plagiarism explained in this chapter cannot be detected through a plagiarism detection tool. This is the main dilemma with such software. What is required is scientific honesty and ethical behaviour when dealing with the works of others!

Bibliography

AMA [American Medical Association]. 2007. AMA Manual of Style: A Guide for Authors and Editors (10th Ed.). Oxford University Press, New York, 1024p.

Barry, G. 2008. Plagiarism: Why it Happens, How to Prevent it. Heinemann, New Hampshire, 139p.

CSE [Council of Science Editors]. 2018. CSE's White Paper on Promoting Integrity in Scientific Journal Publications. Editorial Policy Committee, Council of Science Editors, Wheat Ridge, 81p. Available: https://www.councilscienceeditors.org/resource-library/editorial-policies/white-paperon- [Accessed 30 Nov. 2018].

Gardner, F.P., Pearce, R.B., and Mitchell, R.L. 1985. Physiology of Crop Plants. The Iowa University Press, 327p.

Harris, R. 2001. The Plagiarism Handbook: Strategies for Preventing, Detecting, and Dealing with Plagiarism. Pyrczak, Los Angeles, 208p.

Holzmueller, E.J. and Jose, S.2009. Invasive plant conundrum: what makes the aliens so successful? J. Tropic. Agric. 47 (1–2): 18-29.

KAU [Kerala Agricultural University]. 2011. Package of Practices Recommendations: Crop-2011. Kerala Agricultural University, Thrissur, 360p.

Lathrop, A. and Foss, K.E. 2005. Guiding Students from Cheating and Plagiarism to Honesty and Integrity. Libraries Unlimited, Westport, USA, 264p.
Marsh, B. 2007. Plagiarism: Alchemy and Remedy in Higher Education. State University of New York Press, 176p.
Mahuli, A.V, Mahuli, S.A, Patil, S, and Bhandi S. 2018. Plagiarism-related dilemmas in scientific writing. J. Contemp. Dent. Pract. 19(7):753-755.
Neville, C. 2016. The Complete Guide to Referencing and Avoiding Plagiarism (3rd Ed.). Open University Press, London, 229p.
Pecorari, D. 2013. Teaching to Avoid Plagiarism: How to Promote Good Source Use. Open University Press, London, 192p.
Randall, M. 2001. Pragmatic Plagiarism: Authorship, Profit, Power. University of Toronto Press, Toronto, 321p.
Schmitz, D.C. and Simberloff, D. 1997. Biological invasions: a growing threat. Issues Sci. Technol. 13: 33–40.
Thomas, C.G. 2008. Forage Crop Production in the Tropics (2nd Ed.). Kalyani Publishers, Ludhiana, 333p.
Thomas, C.G. 2015. Research Methodology and Scientific Writing. Ane Books, New Delhi, 531p.
Turabian, K.L. 2007. A Manual for Writers of Term Papers, Theses, and Dissertations (9th Ed.). The University of Chicago Press, 461p.
UGC [University Grants Commission] 2018. University Grants Commission (Promotion of academic integrity and prevention of plagiarism in higher educational institutions) Regulations, 2018. The Gazette of India notification No. F. 1–18/2010(CPP-II) dated 23rd July, 2018. Available: https://www.ugc.ac.in/pdfnews/7771545_academic-integrity-Regulation2018. [Accessed 23 Dec. 2018].
Zhang, Y. 2015. Against Plagiarism: A Guide for Editors and Authors. Springer, 162p.

Index

C. G. Thomas, *Research Methodology and Scientific Writing*,
https://doi.org/10.1007/978-3-030-64865-7

Ingram Content Group UK Ltd.
Milton Keynes UK
UKHW020134050723
424579UK00003B/252

9 783030 648640